LONDON MATHEMATICAL SOCIETY LECTURE NOTE SERIES

Managing Editor: Professor J.W.S. Cassels, Depart
and Mathematical Statistics, 16 Mill Lane, Cambridge CB2 1SB, England

The books in the series listed below are available from booksellers, or, in case of difficulty, from Cambridge University Press.

4 Algebraic topology, J.F.ADAMS
5 Commutative algebra, J.T.KNIGHT
8 Integration and harmonic analysis on compact groups, R.E.EDWARDS
11 New developments in topology, G.SEGAL (ed)
12 Symposium on complex analysis, J.CLUNIE & W.K.HAYMAN (eds)
13 Combinatorics, T.P.McDONOUGH & V.C.MAVRON (eds)
16 Topics in finite groups, T.M.GAGEN
17 Differential germs and catastrophes, Th.BROCKER & L.LANDER
18 A geometric approach to homology theory, S.BUONCRISTIANO, C.P.ROURKE & B.J.SANDERSON
20 Sheaf theory, B.R.TENNISON
21 Automatic continuity of linear operators, A.M.SINCLAIR
23 Parallelisms of complete designs, P.J.CAMERON
24 The topology of Stiefel manifolds, I.M.JAMES
25 Lie groups and compact groups, J.F.PRICE
26 Transformation groups, C.KOSNIOWSKI (ed)
27 Skew field constructions, P.M.COHN
29 Pontryagin duality and the structure of LCA groups, S.A.MORRIS
30 Interaction models, N.L.BIGGS
31 Continuous crossed products and type III von Neumann algebras,A.VAN DAELE
32 Uniform algebras and Jensen measures, T.W.GAMELIN
34 Representation theory of Lie groups, M.F. ATIYAH et al.
35 Trace ideals and their applications, B.SIMON
36 Homological group theory, C.T.C.WALL (ed)
37 Partially ordered rings and semi-algebraic geometry, G.W.BRUMFIEL
38 Surveys in combinatorics, B.BOLLOBAS (ed)
39 Affine sets and affine groups, D.G.NORTHCOTT
40 Introduction to Hp spaces, P.J.KOOSIS
41 Theory and applications of Hopf bifurcation, B.D.HASSARD, N.D.KAZARINOFF & Y-H.WAN
42 Topics in the theory of group presentations, D.L.JOHNSON
43 Graphs, codes and designs, P.J.CAMERON & J.H.VAN LINT
44 Z/2-homotopy theory, M.C.CRABB
45 Recursion theory: its generalisations and applications, F.R.DRAKE & S.S.WAINER (eds)
46 p-adic analysis: a short course on recent work, N.KOBLITZ
47 Coding the Universe, A.BELLER, R.JENSEN & P.WELCH
48 Low-dimensional topology, R.BROWN & T.L.THICKSTUN (eds)
49 Finite geometries and designs,P.CAMERON, J.W.P.HIRSCHFELD & D.R.HUGHES (eds)
50 Commutator calculus and groups of homotopy classes, H.J.BAUES
51 Synthetic differential geometry, A.KOCK
52 Combinatorics, H.N.V.TEMPERLEY (ed)
54 Markov process and related problems of analysis, E.B.DYNKIN
55 Ordered permutation groups, A.M.W.GLASS
56 Journêes arithmêtiques, J.V.ARMITAGE (ed)
57 Techniques of geometric topology, R.A.FENN
58 Singularities of smooth functions and maps, J.A.MARTINET
59 Applicable differential geometry, M.CRAMPIN & F.A.E.PIRANI
60 Integrable systems, S.P.NOVIKOV et al
61 The core model, A.DODD
62 Economics for mathematicians, J.W.S.CASSELS
63 Continuous semigroups in Banach algebras, A.M.SINCLAIR

64 Basic concepts of enriched category theory, G.M.KELLY
65 Several complex variables and complex manifolds I, M.J.FIELD
66 Several complex variables and complex manifolds II, M.J.FIELD
67 Classification problems in ergodic theory, W.PARRY & S.TUNCEL
68 Complex algebraic surfaces, A.BEAUVILLE
69 Representation theory, I.M.GELFAND *et al.*
70 Stochastic differential equations on manifolds, K.D.ELWORTHY
71 Groups - St Andrews 1981, C.M.CAMPBELL & E.F.ROBERTSON (eds)
72 Commutative algebra: Durham 1981, R.Y.SHARP (ed)
73 Riemann surfaces: a view towards several complex variables,A.T.HUCKLEBERRY
74 Symmetric designs: an algebraic approach, E.S.LANDER
75 New geometric splittings of classical knots, L.SIEBENMANN & F.BONAHON
76 Linear differential operators, H.O.CORDES
77 Isolated singular points on complete intersections, E.J.N.LOOIJENGA
78. A primer on Riemann surfaces, A.F.BEARDON
79 Probability, statistics and analysis, J.F.C.KINGMAN & G.E.H.REUTER (eds)
80 Introduction to the representation theory of compact and locally compact groups, A.ROBERT
81 Skew fields, P.K.DRAXL
82 Surveys in combinatorics, E.K.LLOYD (ed)
83 Homogeneous structures on Riemannian manifolds, F.TRICERRI & L.VANHECKE
84 Finite group algebras and their modules, P.LANDROCK
85 Solitons, P.G.DRAZIN
86 Topological topics, I.M.JAMES (ed)
87 Surveys in set theory, A.R.D.MATHIAS (ed)
88 FPF ring theory, C.FAITH & S.PAGE
89 An F-space sampler, N.J.KALTON, N.T.PECK & J.W.ROBERTS
90 Polytopes and symmetry, S.A.ROBERTSON
91 Classgroups of group rings, M.J.TAYLOR
92 Representation of rings over skew fields, A.H.SCHOFIELD
93 Aspects of topology, I.M.JAMES & E.H.KRONHEIMER (eds)
94 Representations of general linear groups, G.D.JAMES
95 Low-dimensional topology 1982, R.A.FENN (ed)
96 Diophantine equations over function fields, R.C.MASON
97 Varieties of constructive mathematics, D.S.BRIDGES & F.RICHMAN
98 Localization in Noetherian rings, A.V.JATEGAONKAR
99 Methods of differential geometry in algebraic topology, M.KAROUBI & C.LERUSTE
100 Stopping time techniques for analysts and probabilists, L.EGGHE
101 Groups and geometry, ROGER C.LYNDON
102 Topology of the automorphism group of a free group, S.M.GERSTEN
103 Surveys in combinatorics 1985, I.ANDERSEN (ed)
104 Elliptical structures on 3-manifolds, C.B.THOMAS
105 A local spectral theory for closed operators, I.ERDELYI & WANG SHENGWANG
106 Syzygies, E.G.EVANS & P.GRIFFITH
107 Compactification of Siegel moduli schemes, C-L.CHAI
108 Some topics in graph theory, H.P.YAP
109 Diophantine analysis, J.LOXTON & A.VAN DER POORTEN (eds)
110 An introduction to surreal numbers, H.GONSHOR
111 Analytical and geometrical aspects of hyperbolic space, D.B.A.EPSTEIN (ed)
112 Low dimensional topology and Kleinian groups, D.B.A.EPSTEIN (ed)
113 Lectures on the asymptotic theory of ideals, D.REES
114 Lectures on Bochner-Riesz means, P.A.TOMAS & Y-C.CHANG
115 An introduction to independence for analysts, H.G.DALES & W.H.WOODIN
116 Representations of algebras, P.J.WEBB (ed)
117 Homotopy theory, E.REES & J.D.S.JONES (eds)
118 Skew linear groups, M.SHIRVANI & B.WEHRFRITZ
119 Triangulated categories in representation theory of finite-dimensional algebras, D.HAPPEL
120 Lectures on Fermat varieties, T.SHIODA
121 Proceedings of Groups - St Andrews 1985, E.F.ROBERTSON & C.M.CAMPBELL (eds)

London Mathematical Society Lecture Note Series. 121

Proceedings of Groups – St Andrews 1985

Edited by E.F. ROBERTSON and C.M. CAMPBELL
The Mathematical Institute, University of St Andrews

CAMBRIDGE UNIVERSITY PRESS
Cambridge
London New York New Rochelle
Melbourne Sydney

CAMBRIDGE UNIVERSITY PRESS
Cambridge, New York, Melbourne, Madrid, Cape Town, Singapore,
São Paulo, Delhi, Dubai, Tokyo, Mexico City

Cambridge University Press
The Edinburgh Building, Cambridge CB2 8RU, UK

Published in the United States of America by
Cambridge University Press, New York

www.cambridge.org
Information on this title: www.cambridge.org/9780521338547

First published 1986

A catalogue record for this publication is available from the British Library

Library of Congress Cataloguing in Publication Data

ISBN 978-0-521-33854-7 Paperback

CONTENTS

INTRODUCTION

An international conference 'Groups - St. Andrews 1985' was held in the Mathematical Institute, University of St. Andrews, Scotland during the period 27 July to 10 August 1985. The initial planning for the conference began in August 1981 and in May 1983 invitations were given to Professor Seymour Bachmuth (Santa Barbara, California), Professor Gilbert Baumslag (City University, New York), Dr Peter Neumann (Queen's College, Oxford), Dr James Roseblade (Jesus College, Cambridge) and Professor Jacques Tits (Collège de France, Paris). They all accepted the invitation to give lecture courses consisting of four or five lectures each and these courses formed the main part of the first week of the conference. They have all contributed major articles to this volume based on these lecture courses. Automorphisms of soluble groups, the topic of Seymour Bachmuth's lectures, is written up in two articles, the second being contributed by his co-worker Horace Mochizuki.

In the second week of the conference there were twelve one-hour invited lectures. This volume contains articles based on eight of these talks. In addition there was a full programme of seminars, the remaining articles of this book arising from these lectures. A companion volume to this book, published as a part of the Proceedings of the Edinburgh Mathematical Society, contains articles based on two of the one-hour invited lectures and a number of articles based on some of the research seminars.

One of the major features of this book is the number of surveys, written by leading researchers, in the wide range of group theory covered. Jim Roseblade gives a group theorist's view of group rings in his survey describing work on group rings which has been motivated by problems in group theory. Jacques Tits provides a survey on buildings and group amalgamations considerably more major than his modest description of it as 'but a brief summary of five lectures...' would suggest.

References form an important component of any survey and since the majority of articles in this book are surveys there is a wealth of references. Gilbert Baumslag's article on one-relator groups wins the prize, containing over one hundred references. Only by taking the union of the references of two surveys on quite different aspects of the modular group does one get a more extensive bibliography.

The computing aspect of group theory is evident in a number of papers in this volume, in particular in the paper by Peter Neumann describing some algorithms for computing with finite permutation groups. At the conference this computing aspect was served by the availability of the CAYLEY group theory package and other group theory programs resident on the St. Andrews VAX. These facilities were well used and we would like to thank the Computing Laboratory of the University of St. Andrews for their assistance.

Groups - St. Andrews 1985 received financial support from the Edinburgh Mathematical Society, the London Mathematical Society and the British Council. We gratefully acknowledge this financial support and would also like to thank the Edinburgh Mathematical Society and the London Mathematical Society for their help with publishing. A total of 366 people from 43 different countries registered for the conference.

We would like to express our thanks to all our colleagues who helped in the running of the conference and in particular Tom Blyth, Peter Sumner and Trevor Walker. We would also like to thank our wives for their help and forbearance. We would like to thank Shiela Wilson who acted as conference secretary for the two weeks of the conference and who has undertaken the major task of transforming almost 800 pages of A4 typescript into the 350 camera-ready pages for this volume.

Our final thanks go to the authors who have contributed articles both to this volume and to the companion part in the Proceedings of the Edinburgh Mathematical Society. Because of space restrictions and our desire to include as many articles as possible, we have in some cases reduced the number of displays in our attempts to save space. We apologise to the authors and trust that this will not detract from the articles.

Edmund F. Robertson
Colin M. Campbell

Photograph on facing page. The photograph shows the five main speakers and the two conference organizers pictured at David Russell Hall, University of St. Andrews, during the conference. *Back row* (left to right): Dr James E. Roseblade, Dr Edmund F. Robertson, Professor Seymour Bachmuth, Dr Colin M. Campbell. *Front row* (left to right): Professor Jacques Tits, Professor Gilbert Baumslag, Dr Peter M. Neumann.

AUTOMORPHISMS OF SOLVABLE GROUPS, PART I

S. Bachmuth
University of California, Santa Barbara, CA 93106, USA

1.33 (A.I. Mal'tsev). *Describe the automorphism group of a free soluble group*

M.I. Karagapolov

- From the Kourovka Notebook 1965 [19]

0. *INTRODUCTION*

Prior to 1965 I know of no results concerned with automorphisms of free solvable groups as such. And until 1976 there was not much more than the following three theorems, two of which will play an important role for us later. These theorems look like very special and isolated results, but, in hindsight, it is precisely the restrictiveness of these results which made them harbingers of the developments to come and makes them rather interesting in their own right.

The first result by Smèlkin will not play any role in the sequel, but is particularly noteworthy in that it remains to this day the only result which does not require finite generation in the hypothesis. F denotes a non-cyclic free group, and $F = F^{(0)} \geqq F' \geqq \dots \geqq F^{(d)} \geqq \dots$ the derived series of F.

Theorem 1 (Smèlkin 1967 [30]). *Let* $G = F/F^{(d)}$ *be a free solvable group and* α *an automorphism of* G *which is the identity when restricted to* $F^{(d-1)}/F^{(d)}$. *Then,* α *is an inner automorphism of* G. *In fact,* α *is conjugation by an element of* $F^{(d-1)}/F^{(d)}$.

In 1976 Joan Dyer [16] gave an elegant proof of this theorem while at the same time widely generalizing it.

The second result restricts G to be of rank 3. We write $[x,y] = x^{-1}y^{-1}xy$ for the commutator of the group elements x and y, and F(n) denotes the free group of rank n.

Theorem 2 (Bachmuth 1965 [3]). *Let* $G = F(3)/F(3)^{(d)}$ *be a non-abelian free solvable group of rank* 3 *with free generators* a,b,c, *and let the automorphisms* α,β,γ, *of* G *be defined respectively by*

$$\alpha: \begin{array}{l} a \to a[b,c] \\ b \to b \\ c \to c \end{array} \qquad \beta: \begin{array}{l} a \to a \\ b \to b[a,c] \\ c \to c \end{array} \qquad \gamma: \begin{array}{l} a \to a \\ b \to b \\ c \to c[a,b] \end{array}$$

Then, α,β,γ *generate a free subgroup of rank* 3 *of* Aut(G), *the automorphism group of* G.

As a corollary, the corresponding three automorphisms for the case when G = F(3) also generate a free subgroup of rank 3, and it was to prove this corollary that the more general theorem was proved. In 1969, a different proof for the corollary was given by Chein [14]. At a casual glance, it is not clear to what extent these results are only rank 3 theorems, and, in fact, there still remains somewhat of a mystery about this situation, which we will take up in section 3. There we shall present evidence that there may not be an analogue to Theorem 2 when G is a free or a free solvable group of rank 4 or larger.

The third result, which also requires rank 3, is the very deep

Theorem 3 (Chein 1968 [13]). *There exists an automorphism of the free metabelian group* $F(3)/F(3)''$ *of rank* 3 *which is not induced by an automorphism of the free group* $F(3)$ *of rank* 3.

The automorphism of the free metabelian group $F(3)/F(3)'' = gp\langle a,b,c\rangle$ which Chein showed cannot be lifted to an automorphism of $F(3)$ is

$$\delta:\ \begin{array}{l} a \to a[b,c,a,a] \\ b \to b \\ c \to c \end{array}$$

As a result, Chein's Theorem certainly does not look like a rank 3 theorem, and there did not seem to be any reason why rank 3 could not be replaced by rank $\geqq 3$ in the statement of Chein's Theorem apart from the caution of not jumping to conclusions. Chein did not jump to conclusions and was careful to write near the bottom of page 605 in [13], "All the results of this paper apply in the case where F is a free group of rank 3; ..." But immediately following he added the innocent looking and believable statement "... most of them can be generalized without too much difficulty."

After these seemingly isolated results, a complete structure theorem for the automorphism groups of free solvable groups of rank 2 was proved in 1976 by E. Formanek, H. Mochizuki, and the writer.

Theorem 4 (BFM 1976 [6]). *If* $G = F(2)/F(2)^{(d)}$ *is a free solvable group of rank* 2, *then* $\mathrm{Out}(G) \cong GL_2(\mathbb{Z})$.

Out(G) is Aut(G) factored by the normal subgroup of inner automorphisms of G. Theorem 4 for the special case in which G is metabelian was already known in 1965 [3], and together with the first three theorems were virtually all that was known about automorphisms of free solvable groups prior to 1976, as mentioned earlier. Thus, in one stroke in 1976, our knowledge of automorphism groups of free solvable groups of any derived length was at least on a par with that of automorphism groups of polynomial rings over fields and superior to our knowledge of automorphism groups of most free objects. Malcev's problem received a great push, and one could look ahead to the prospect of studying the larger rank automorphism groups. But a foreboding fact about Theorem 4 was that it was more general than stated above. The actual result proved was for the groups $G = F(2)/R'$, where R is any normal subgroup of $F(2)$ subject only to the conditions that R be contained in $F(2)'$ and the integral group ring of $F(2)/R$ be a domain. (Theorem 4 is the case $R = F(2)^{(d-1)}$.) In other words, Theorem 4 was not so much a theorem about automorphisms of free solvable groups as much as one about two generator groups. Hence, there was no likelihood of penetrating insight or expectations for the automorphism groups of free solvable groups of larger rank following from Theorem 4.

What happened after 1976 was totally unexpected, close to unbelievable, and the subject of these talks. For motivation we will begin with a discussion of analogous phenomena in the general linear groups (GL_n) over polynomial rings and Laurent polynomial rings. Not only does this lead to a better understanding of the automorphism groups, but the results for GL_n have an interest and fascination of their own.

1. GL_n *AND* SL_n

Let R be a commutative ring with 1. $GL_n(R)$, $n \geqq 1$, denotes the multiplicative group of invertible $n \times n$ matrices over R, and $SL_n(R)$ the subgroup of $GL_n(R)$ of those matrices of determinant 1. An *elementary* $n \times n$ matrix is a matrix $E_{ij}(r)$, $i \neq j$, $r \in R$, whose diagonal entries are all 1, whose (i,j) entry is r and whose remaining entries are all zero. All elementary matrices are contained in $SL_n(R)$.

The *tame* elements of $GL_n(R)$ are those elements which lie in the subgroup $GE_n(R)$ generated by all elementary matrices and all invertible diagonal matrices. To avoid slight technicalities, it will be convenient to consider $SL_n(R)$ rather than $GL_n(R)$. In this case, the *tame* elements are the elements of $SL_n(R) \cap GE_n(R) = E_n(R)$, which is easily seen to be generated by the set of elementary matrices. Looking ahead to the next sections on the automorphism groups, we will see that the role of tame elements is played by those automorphisms which resemble automorphisms of the free group. Specifically, if $G \cong F(n)/N$ where N is a characteristic subgroup of F(n), then an automorphism is called *tame* if it is induced by an automorphism of F(n).

Intuitively, the tame elements are meant to be the obvious elements a system possesses. For example if we were to consider the automorphism group of the ring $K[x_1,\dots,x_k]$ of polynomials in k commuting indeterminates over a field K, the tame elements would be those in the subgroup generated by automotphisms which are linear or triangular. (An automorphism is called linear, respectively triangular, if it is defined by $x_i \to a_{i1}x_1 + \dots + a_{ik}x_k$, $1 \leqq i \leqq k$, where $(a_{ij}) \in SL_k(K)$, respectively $x_i \to x_i + f_i(x_1,\dots,x_{i-1})$, $1 \leqq i \leqq k$, where $f_i(x_1,\dots,x_{i-1}) \in K[x_1,\dots,x_{i-1}]$. See e.g. [26].) One of the outstanding problems in algebraic geometry is to determine if there are any non-tame automorphisms in the case $k \geqq 3$ - a problem completely and beautifully solved in the two variable case more than forty years ago with the conclusion that all automorphisms are tame. Similar comments could be made for other types of automorphism groups.

Thus, in whatever the setting, a concern of major importance is whether or not there are "non-tame" elements, and, if so, how one might distinguish between "tame" and "non-tame" elements, perhaps algorithmically or ...? A major triumph was Suslin's determination in 1977 [31] that for $n \geqq 3$, GL_n of various polynomial and Laurent polynomial rings over principal ideal domains (PID's) contain only tame elements, which is a considerable strengthening of earlier results by Bass [11] and Vaserstein [32]. We will return to these results after considering the case n = 2.

If K is a field and x an indeterminate over K, then K[x] is a

Euclidean domain, and hence trivially $SL_2(K[x]) = E_2(K[x])$. There is a much more definitive result due to Nagao [25]. Let $ST_2(R)$ be the 2×2 special triangular group over the commutative ring R, i.e. the subgroup of all upper triangular matrices in $SL_2(R)$. (Our notation and definition are at odds with that of some other mathematicians.)

Nagao's Theorem. $SL_2(K[x]) = SL_2(K) \underset{ST_2(k)}{*} ST_2(K[x])$. *In words,* $SL_2(K[x])$ *is the free product of* $SL_2(K)$ *and* $ST_2(K[x])$ *with amalgamation along their largest common subgroup* $ST_2(K)$.

Let $G = A *_U B$ be a decomposition of group G as an amalgamated free product. Then, the Reduced Form Theorem says that if $g \in G - U$ and if $g = g_1g_2 \ldots g_k = h_1h_2 \ldots h_\ell$ where the g_i, respectively the h_i, are alternately in A-U and B-U, then $k = \ell$ and g_i and h_i are both in A-U or in B-U, $1 \leqq i \leqq k$.

Let $\mathbb{Z}$ be the ring of rational integers and $\mathbb{Q}$ the field of rational numbers. We now apply Nagao's Theorem and the Reduced Form Theorem to $SL_2(\mathbb{Q}[x])$ to show

Theorem 5 (P.M. Cohn [15]). $SL_2(\mathbb{Z}[x]) \neq E_2(\mathbb{Z}[x])$.
Proof. We view the element of $SL_2(\mathbb{Z}[x])$

$$\begin{pmatrix} 1+2x & -x^2 \\ 4 & 1-2x \end{pmatrix} = \begin{pmatrix} 1 & 1/2x \\ 0 & 1 \end{pmatrix}\begin{pmatrix} 1 & 0 \\ 4 & 1 \end{pmatrix}\begin{pmatrix} 1 & -1/2x \\ 0 & 1 \end{pmatrix}$$

as being in $SL_2(\mathbb{Q}[x]) = SL_2(\mathbb{Q}) *_{ST_2(\mathbb{Q})} ST_2(\mathbb{Q}[x])$. If it were the case that $SL_2(\mathbb{Z}[x]) = E_2(\mathbb{Z}[x])$, then $SL_2(\mathbb{Z})$ and $ST_2(\mathbb{Z}[x])$ would generate $SL_2(\mathbb{Z}[x])$. Hence, by the Reduced Form Theorem, we could write

$$\begin{pmatrix} 1+2x & -x^2 \\ 4 & 1-2x \end{pmatrix} = \begin{pmatrix} 1 & f(x) \\ 0 & 1 \end{pmatrix}\begin{pmatrix} a & b \\ c & d \end{pmatrix}\begin{pmatrix} 1 & g(x) \\ 0 & 1 \end{pmatrix}$$

where $f(x), g(x) \in \mathbb{Z}[x]$ and $\begin{pmatrix} a & b \\ c & d \end{pmatrix} \in SL_2(\mathbb{Z})$. A straightforward calculation shows that such a factorization is impossible.

Before continuing, we observe that in contrast to $SL_2(\mathbb{Z}[x])$ it is not known if $SL_2(\mathbb{Z}[x,x^{-1}])$ has non-tame elements. We record this formally as:

Problem 1. *Does* $SL_2(\mathbb{Z}[x,x^{-1}]) = E_2(\mathbb{Z}[x,x^{-1}])$?

Even if $SL_2(\mathbb{Z}[x,x^{-1}]) \neq E_2(\mathbb{Z}[x,x^{-1}])$, one can still ask:

Problem 2. *Is* $SL_2(\mathbb{Z}[x,x^{-1}])$ *finitely generated?*

Notice that in the above theorem of Cohn that $\mathbb{Z}$ may be replaced by any integral domain P which is not a field. In the proof, all we need do is replace 2 by a non-unit element of P and replace $\mathbb{Q}$ by the quotient field of P in the use of Nagao's Theorem. Hence we have:

Theorem 5′. *Let* P *be an integral domain which is not a field. Then*

$SL_2(P[x]) \neq E_2(P[x])$.

A particularly interesting case of Theorem 5' is $P = K[y,y^{-1}]$, the group algebra of an infinite cyclic group over the field K, since it shows that $SL_2(K[x,y,y^{-1}])$ has non-tame elements; yet if we replace the Euclidean domain K[x] by the Euclidean domain $\mathbb{Z}$, as pointed out earlier, we are left in the dark.

If K is a field and x an indeterminate, then analogous to Nagao's Theorem, there is a factorization of $SL_2(K[x,x^{-1}])$ which is a consequence of Ihara's Theorem ([29], page 79, Corollary 1).

Theorem 6 ([29], page 80, example 2 of Corollary 1).

$$SL_2(K[x,x^{-1}]) = SL_2(K[x]) *_\Gamma SL_2(K[x])^T$$

where $T = \begin{pmatrix} x & 0 \\ 0 & 1 \end{pmatrix}$, *exponentiation by* T *means conjugation by* T *and* Γ *is the intersection of the two factors.*

This amalgamated free product decomposition remains valid if K is replaced by a PID (Bachmuth and Mochizuki [7], Proposition 3). Although this seems a modest enough extension of Ihara's Theorem, it is crucial for the applications that we have in mind.

Theorem 7. *Let* P *be a* PID, *and* x *an indeterminate over* P. *Then*

$$SL_2(P[x,x^{-1}]) = SL_2(P[x]) *_\Gamma SL_2(P[x])^T$$

where $T = \begin{pmatrix} x & 0 \\ 0 & 1 \end{pmatrix}$ *and* $\Gamma = SL_2(P[x]) \cap SL_2(P[x])^T$.

For what should be the most interesting case $P = \mathbb{Z}$, the significance of this theorem is hazy. It is of no help in deciding whether or not $SL_2(\mathbb{Z}[x,x^{-1}])$ has non-tame elements or is finitely generated, evidently because $\Gamma = SL_2(\mathbb{Z}[x]) \cap SL_2(\mathbb{Z}[x])^T$ is "too large" inside each factor, but it does show that $SL_2(\mathbb{Z}[x,x^{-1}])$ is generated by $SL_2(\mathbb{Z}[x])$ and $\begin{pmatrix} 1 & x^{-1} \\ 0 & 1 \end{pmatrix}$. Whether Theorem 7 and its consequences might be helpful in the ultimate determination of the generation of $SL_2(\mathbb{Z}[x,x^{-1}])$ is impossible to say at this time. Perhaps one should observe here that if P is a PID which is not a field and K is the field of quotients of P, then $E_2(P[x,x^{-1}]) = A *_U B$, where $A = E_2(P[x,x^{-1}]) \cap SL_2(K[x])$, $B = E_2(P[x,x^{-1}]) \cap SL_2(K[x])^T$, and $U = A \cap B = \Gamma \cap E_2(P[x,x^{-1}])$. But $A \supsetneq E_2(P[x])$. For example, if $P = \mathbb{Z}$, then it is not hard to see that $\begin{pmatrix} 1+2x & -x^2 \\ 4 & 1-2x \end{pmatrix} \in A$ but is not in $E_2(\mathbb{Z}[x])$ as we saw earlier.

However, in suitable situations, Theorem 7 enables one to use the Karrass-Solitar subgroup theorems for amalgamated free products (as a substitute for Nagao's Theorem and the Reduced Form Theorem in the previous considerations) to demonstrate that certain groups contain non-tame elements. For convenience we will state the form of the subgroup theorem needed.

Subgroup Theorem (Karrass-Solitar [18]). *Let* $G = A *_U B$ *where* A *and* B *are subgroups of* G *and* $U = A \cap B$, *and let* H *be a subgroup of* G. *Suppose one can find* $M_i \in A$ *and* $N_i \in (B-U) \cap H$, $i \geqq 1$, *such that*

(i) $\{M_i\}_{i \geqq 1}$ *is a subset of a set of double coset representatives of* $(H \cap A, U)$ *in* A, *and*

(ii) $\{M_i N_i M_i^{-1}\}_{i \geqq 1} \subseteq H$.

Then H *is not finitely generated.*

One proceeds by using the Subgroup Theorem to show that a given subgroup of $SL_2(P[x,x^{-1}])$ for a suitable PID P is not finitely generated, yet would have to be finitely generated if comprised entirely of tame elements. This method was first employed in [7] to prove that if D is an integral domain but not a field, then $SL_2(D[x,x^{-1},y,y^{-1}])$ is not finitely generated, and hence consists mainly of non-tame elements. (In Theorem 7, one puts $P = Q[y,y^{-1}]$ where Q is the quotient field of D.) Possibly the most important cases are $D = \mathbb{Z}$ and $D = K[z,z^{-1}]$, K a field, and it is extremely interesting that for these cases Suslin has shown that when $n \geqq 3$ SL_n of each of the two corresponding rings consists entirely of tame elements. We summarize these results in the following

Theorem 8 (Suslin [31], Bachmuth and Mochizuki [8]). (i) *If* $n = 1$ *or* $n \geqq 3$ *and if* $\ell \geqq 1$, *then* $SL_n(\mathbb{Z}[x_1,x_1^{-1},\ldots,x_\ell,x_\ell^{-1}])$ *consists entirely of tame elements and hence is finitely generated.*

(ii) *If* $j \geqq 2$, *then* $SL_2(\mathbb{Z}[x_1,x_1^{-1},\ldots,x_j,x_j^{-1}])$ *is not finitely generated, and hence any set of generators must contain infinitely many non-tame elements.*

The case $j = 1$ in Theorem 8 (ii) is unknown as we have already mentioned (Problems 1 and 2). Similarly, for K a field, it is an open question whether $SL_2(K[x_1,x_1^{-1},x_2,x_2^{-1}]) = E_2(K[x_1,x_1^{-1},x_2,x_2^{-1}])$. With this case also omitted, there is a companion to Theorem 8 for Laurent polynomial rings over fields.

Theorem 8' (Suslin [31], Bachmuth and Mochizuki [8]). *Let* K *be a field.*
(i) *If* $n = 1$ *or* $n \geqq 3$ *and if* $\ell \geqq 1$, *then* $SL_n(K[x_1,x_1^{-1},\ldots,x_\ell,x_\ell^{-1}])$ *consists entirely of tame elements and hence is finitely generated if* K *is finite.*

(ii) *If* $j \geqq 3$, *then* $SL_2(K[x_1,x_1^{-1},\ldots,x_j,x_j^{-1}])$ *is not finitely generated and hence any set of generators must contain infinitely many non-tame elements.*

There is an unsettling aspect to part (ii) of the above theorems and the general procedure derived from Theorem 7 of establishing the existence of non-tame elements. Namely, there is no algorithm known to us which decides whether or not a given matrix is tame. In fact, a set of generators for $SL_2(\mathbb{Z}[x_1,x_1^{-1},\ldots,x_j,x_j^{-1}]$, $j \geqq 2$, is not known, only that infinitely many are required. This discussion leads to the following

Scenario. For a given $\alpha \in SL_2(\mathbb{Z}[x_1,x_1^{-1},\ldots,x_j,x_j^{-1}]$, one is not able to decide, at present, whether or not α is a product of elementary matrices, although if $j \geqq 2$ and α is chosen randomly, it almost certainly cannot be. However

$$\beta = \begin{pmatrix} \alpha & 0 \\ 0 & 1 \end{pmatrix}$$

is a product of elementary matrices in $SL_3(\mathbb{Z}[x_1,x_1^{-1},\ldots,x_j,x_j^{-1}])$.

This scenario has been displayed and highlighted here for easy reference later, but mainly because it provides motivation for and helps in understanding the results on the automorphism groups that are the main concern of these lectures.

2. *THE AUTOMORPHISM GROUPS*

Before continuing with this article, the reader is advised to read the paper "Automorphisms of Solvable Groups - Part II" by H.Y. Mochizuki in this volume [24]. In [24], Mochizuki has kindly consented to write a thorough discussion of the results concerning automorphism groups of free solvable groups and related material. We will assume familiarity with the notation there. We begin by gathering together in one theorem some of the main results surveyed in [24]. Recall that $\phi(n)$ is the free metabelian group of rank n and the tame elements of Aut $\phi(n)$ are those induced by elements of Aut $F(n)$.

Theorem 9. (a) Aut $\phi(n)$ *consists entirely of tame elements if* $n = 2$ *or if* $n \geqq 4$.
(b) *Any generating set for* Aut $\phi(3)$ *contains infinitely many non-tame elements.*

Theorem 9 was of course formulated so that it reads the same as Theorem 8. For convenience we combine them in one statement.

Theorem 10. *Let* $\chi(n)$ *be either* Aut $\phi(n)$ *or* $GL_{n-1}(\mathbb{Z}[x_1,x_1^{-1},\ldots,x_k,x_k^{-1}])$, $k \geqq 2$.
(a) $\chi(n)$ *consists entirely of tame elements if* $n = 2$ *or if* $n \geqq 4$.
(b) *Any generating set for* $\chi(3)$ *contains infinitely many non-tame elements.*

In the future there are certain to be efforts toward extending Theorem 10 to other families $\chi(n)$. (In this regard see Section 7 of [24].) One might anticipate some degree of success in these endeavours if the family in question has characteristics such as commutativity properties, etc. similar to those of $\chi(n)$ in Theorem 10. The automorphism groups of polynomial rings over fields might be such an example. But it is too much to expect Theorem 10 to remain valid for inherently more complex families such as automorphism groups of free solvable groups of derived class larger than two, or even for GL_n of group rings of (non-abelian) free nilpotent groups. However, it may very well be the case that a more sophisticated version of Theorem 10 holds such as "$\chi(n)$ consists entirely of tame elements for almost all n". We formulate a much weaker version of this as:

Problem 3. *Let* $G(n)$ *be a free group of rank* n *in a locally solvable variety of groups. Which varieties have the property that* Aut $G(n)$ *is finitely generated for almost all* n?

Locally nilpotent varieties satisfy the condition of Problem 3 since the automorphism group of a finitely generated nilpotent group is always finitely generated (even finitely presented [2]). But for non-locally nilpotent, locally solvable varieties, the only other example known (another example has just been given by E.Stöhr. See section 7 of [24]) is the Jacobi variety $\mathfrak{J}$ which is defined by the law

$$[x,y,z][z,x,y][y,z,x] = 1.$$

Using known results about $\mathfrak{J}$ and Theorem 9, one establishes

Theorem 11. *Let* $J(n)$ *be the free group of rank* n *in* $\mathfrak{J}$.
(a) Aut $J(n)$ *is finitely generated for* $n = 2$ *or if* $n \geq 4$.
(b) Aut $J(3)$ *is not finitely generated.*

The easy proof of Theorem 11 appears in [9]. However, it is not known if all elements of Aut $J(n)$, $n \geq 4$ are tame. It is easy to show that all elements of Aut $J(n)$, $n \geq 4$ are tame if and only if the automorphism α of $J(4) = \text{gp}\langle x_1,x_2,x_3,x_4\rangle$ defined by

$$\alpha: \begin{array}{l} x_1 \to x_1[[x_1,x_2],\ [x_3,x_4]] \\ x_i \to x_i,\ i = 2,3,4 \end{array}$$

lifts to an automorphism of $F(4)$.

The final result in this section is an indication that some restrictions on the variety in Problem 3 are necessary if the automorphism groups of its free groups are to be finitely generated once a certain rank is reached. If k is an integer such that the ring $\mathbb{Z}/k\mathbb{Z} \cong \mathbb{Z}_k$ is not semi-simple and $G(n)$ is the free group of rank n in the variety of metabelian groups having commutator subgroup of exponent k, then it was shown by G. Baumslag, J. Dyer, H. Mochizuki and the writer in [5] that Aut $G(2)$ is not finitely generated. This group was but one example in a general procedure for producing non-finitely generated automorphism groups of various two generator metabelian groups. However, Mochizuki and the writer recently have extended this particular example to show that for all $n \geq 2$, Aut $G(n)$ is not finitely generated.

Theorem 12. *Let* k *be a positive integer which contains a square of a prime as a factor. Let* $G(n)$ *be the free group of rank* n *in the solvable variety defined by the two laws*

$$[[x,y],\ [z,w]] = 1 = [x,y]^k.$$

Then, Aut $G(n)$ *is not finitely generated for every integer* $n \geq 2$.

Thus, with regard to Problem 3, there does exist a locally solvable (even solvable) variety for which Aut $G(n)$ is not finitely generated for almost all n.

3. *SKIPPING OVER RANK 3*

The applications and generalizations thus far (June 1985) of Theorem 10 can be roughly compartmentalized into three categories. The first is the extension (as much as possible) of Theorem 10 to other families, an example of which is Theorem 11. This category is the easiest to visualize, particularly as concerns future applications. Results for automorphism groups of free algebraic systems, which either are consequences of Theorem 10 or use Theorem 10 only indirectly as a model, are predictable future endeavours. (Some examples are discussed in [24].)

The second type of application uses the rank 3 exception to deduce results that are "infinite" in nature. These applications are impossible or at least very difficult to predict in advance. For example, consider the statement "Aut $\phi(3)$ is infinitely generated implies that $GL_2(\mathbb{Z}[x,x^{-1}])$ is infinitely generated". Although at the moment there is no evidence for the validity of such a deduction, it is by no means out of the question. And there is no way to tell whether or not such a statement might ever be validated unless or until it actually is. Theorem 8 (i) is a concrete example of an application in this second category. Other examples can be found in Brunner and Ratcliff [12], McCullough [22], and McCullough and Miller [23], all of which used Theorem 8 (i) to obtain geometric theorems. ([24] contains a discussion of the McCullough-Miller Theorem.)

The third category of applications, a couple of which are discussed in this section, are those which skip over the troublesome rank 3 case in order to make use of the tameness of Aut $\phi(n)$. This phenomenon of skipping over low ranks because of an inability to deal with small dimensions has occurred in other contexts and perhaps is becoming a recurrent theme. For example in [17], M.H. Freedman remarks that during years of increased understanding of manifolds topologists "skipped over" the low dimensions 3 and 4. In [24], Mochizuki explains why researchers may be similarly forced to skip over low ranks in the study of automorphism groups such as $\mathrm{Aut}(F(n)/\gamma_k(F(n))')$. For matrix groups, e.g. K-theory, this is an old story.

Our first application concerns the automorphism groups of free groups. The aim here is only incidentally to illustrate a theorem which skips over rank 3, but primarily to highlight problems concerning these important groups. It is the opinion of this writer that the material in these lectures points to a dichotomy between the structures of Aut F(3) and Aut F(n) for $n > 3$. Recall the following easy consequence of Theorem 9(a), a proof of which is in section 2 of [24]. We use IA(G) to denote the kernel of the natural map $\mathrm{Aut}\ G \to \mathrm{Aut}\ (G/G')$ and $GL_{n-1}(\mathbb{Z}[x,x^{-1}],\ (x-1)\mathbb{Z}[x,x^{-1}])$ to denote the kernel of the natural map $GL_{n-1}(\mathbb{Z}[x,x^{-1}]) \to GL_{n-1}(\mathbb{Z})$ where x is mapped to 1.

Theorem 13. *For* $n > 3$, IA(F(n)) *maps homomorphically onto* $GL_{n-1}(\mathbb{Z}[x,x^{-1}],\ (x-1)\mathbb{Z}[x,x^{-1}])$.

The point of course is that we do not know whether or not $GL_2(\mathbb{Z}[x,x^{-1}]),\ (x-1)\mathbb{Z}[x,x^{-1}])$ is the image of IA(F(3)), and so we skip over this case. Nielson [27] showed that IA(F(2)) is just the group of inner automorphisms of F(2), but for $n \geq 3$, the structure of IA(F(n)) is very much a mystery. Theorem 13 is perhaps the first indication that the structure of IA(F(n)) and hence that of Aut F(n) may have significant differences when $n = 3$ and when $n > 3$. Thus, the spotlight is thrown onto Theorem 2 (or rather its corollary) of Section 0. Recall the Magnus-Nielsen Theorem concerning the generators for IA(F(n)) (Section 3 of [24]).

Theorem 14 (Nielsen [28], Magnus [21]). *Let* $F(n) = \mathrm{gp}\langle x_1,\dots,x_n\rangle$. *The following automorphisms of* F(n) *generate* IA(F(n)).

$$K_{ij} : \begin{array}{ll} x_i \to x_j x_i x_j^{-1}, & i \neq j \\ x_m \to x_m, & m \neq i \end{array}$$

and

$$K_{ijk} : \begin{array}{ll} x_i \to x_i[x_j,x_k], & i \neq j < k \neq i \\ x_m \to x_m, & m \neq i. \end{array}$$

It is clear that these are also a minimal set of generators for IA(F(n)). Before continuing, it is convenient to assign names to certain subgroups of IA(F(n)). We follow the notation used in Chein [14].

Definition. T(n) *is the subgroup of* IA(F(n)) *generated by the triply indexed automorphisms* K_{ijk}, *and* C(n) *is the subgroup generated by the doubly indexed automorphisms* K_{ij}.

Theorem 2 gives a complete description of T(3) but tells us very little about T(n) when n > 3. According to Theorem 14, T(n), $n \geqq 3$ requires $n\binom{n-1}{2}$ generators. Already T(4) requires 12 generating automorphisms, and all Theorem 2 tells us about them is that certain 3 element subsets of these 12 are free generators of a free group. There is no obvious generalization of Theorem 2 to the more general T(n) for n > 3. With hindsight it now seems that Theorem 2 reinforces the more recent Theorem 13 in the uneasy feeling that Aut F(3) and Aut F(n), n > 3 are fundamentally different.

Chein [14] showed that C(3) is the same group as the subgroup of Aut F(3) which takes any generator of F(3) into a conjugate of itself, and he reproved a theorem of Levinger giving a presentation for C(3). We state this as

Theorem 15 (Chein [14], Levinger [20]). C(3) *is a split extension of the group of inner automorphisms of* F(3) *by* gp $\langle K_{12},K_{13},K_{21}\rangle$. *This latter group is a free group of rank* 3. *In particular*

$$C(3) = \langle K_{12},K_{13},K_{21},K_{23},K_{31},K_{32} : K_{ij} \rightleftarrows K_{kj},\ K_{ij} \rightleftarrows K_{ik}K_{jk},\ i \neq j \neq k \neq i\rangle$$

The double arrows mean that the automorphisms commute. There is not much knowledge on how the subgroups T(3) and C(3) fit together. In Section 7 of [14], Chein made a couple of interesting conjectures and cited his thesis as containing some supporting evidence for them. One is that IA(F(3)) is not finitely presented, and the other is that the normal closure of C(3) in IA(F(3)) has trivial intersection with T(3). Hence a free group of rank 3 is a quotient of IA(F(3)).

Problem 4. *For* $n \geqq 3$, *decide whether* IA(F(n)) *is finitely related.*

Problem 5. *For* $n \geqq 3$, *determine the subgroups* $\overline{C(n)} \cap T(n)$, *where* $\overline{C(n)}$ *is the normal closure of* C(n) *in* IA(F(n)). *In particular, is this intersection trivial?*

Could it be that IA(F(3)) is not finitely presented yet IA(F(n)), n > 3 is finitely presented? Or vice versa? We are lacking much information about IA(F(n)) and hence about Aut(F(n)) for $n \geqq 3$.

We next turn our attention to the automorphism groups of free nilpotent groups and explain how Theorem 10 enables us to provide significant information about them for ranks larger than three. Recall that $\gamma_k(G)$ is the

k^{th} term of the lower central series of the group $G = \gamma_1(G)$. For $k \geqq 2$, let

$$K(G,k) = \text{kernel}\ \{\mathrm{Aut}(G/\gamma_k(G)) \to \mathrm{Aut}(G/\gamma_{k-1}(G)\}$$

and

$$K^*(G,k) = K(G,k) \cap \text{image}\ \{\mathrm{Aut}(G) \to \mathrm{Aut}(G/\gamma_k(G))\}.$$

If G is free of rank n, then $K(F(n),k)$ is free abelian and from Witt's formula [33] for the ranks of the quotient groups of the lower central series of $F(n)$, one has

$$\text{rank}\ K(F(n),k) = n \sum_{d/k-1} \left(\frac{1}{k-1}\right) n^d \mu\left(\frac{k-1}{d}\right) \qquad (1)$$

where the summation is over the divisors d of k-1 and μ is the Moebius function. It is well known that all automorphisms of $F(n)/\gamma_2(F(n))$ are induced by automorphisms of $F(n)$ (see [24] Section 3), and Andreadakis [1] showed that all automorphisms of $F(n)/\gamma_3(F(n))$ are also induced by automorphisms of $F(n)$. Hence $K^*(F(n),k) = K(F(n),k)$ when $k = 2$ or 3. For $k \geqq 4$, Andreadakis [1] showed that $K^*(F(n),k)$ is a proper subgroup of $K(F(n),k)$. He also computed the ranks of $K^*(F(3),k)$ when $k = 4$ and $k = 5$. Otherwise, nothing was known about the groups $K^*(F(n),k)$ when $n > 2$ and $k \geqq 4$. Theorem 10, however, enables us to determine the rank of $K^*(F(n),4)$ for any $n > 3$ precisely and to give nice estimates for the ranks of $K^*(F(n),k)$ for $n > 3$ and $k > 4$. This is done by first computing $K^*(\phi(n),k)$ when $\phi(n)$ is free metabelian of rank n and then invoking Theorem 10. We may restrict attention to $k \geqq 4$. It was computed in [4] that for $n \geqq 3$ and $k \geqq 4$,

$$\text{rank}\ K(\phi(n),k) = n(k-2)\binom{n+k-3}{n-2}. \qquad (2)$$

The faithful representation of $IA(\phi(n))$ as $n \times n$ matrices with entries in the ring $R = \mathbb{Z}[x_1, x_1^{-1}, \ldots, x_n, x_n^{-1}]$ as described in Section 2 of [24] may be modified to give a representation of $K(\phi(n),k)$. $K(\phi(n),k)$ is isomorphic to the group of matrices over R of the form $I + M$, where I is the $n \times n$ identity matrix, $I + M$ satisfies the same column conditions as in the representation of $IA(\phi(n))$, and the entries of M are in the k-2 power of the augmentation ideal of R. The fundamental difference is that no determinant condition has to be satisfied since the terms in M of degree larger than k-2 in the augmentation ideal of R are immaterial so that $I - M$ is the inverse of $I + M$. The main result in Section 5 of [4] is that for $k \geqq 4$, $(I + M) \in K(\phi(n),k)$ is in $K^*(\phi(n),k)$ if and only if the trace of M is zero. This has a consequence that for $n \geqq 3$ and $k \geqq 4$

$$\text{rank}\ K^*(\phi(n),k) = \text{rank}\ K(\phi(n),k) - \binom{n+k-3}{n-1}. \qquad (3)$$

Now there is the natural map (surjective)

$$K(F(n),k) \to K(\phi(n),k)$$

which induces a natural map

$$K^*(F(n),k) \to K^*(\phi(n),k)$$

which is surjective for $n \geqq 4$ by Theorem 10! Hence formulae (1) - (3) yield the promised bounds on the ranks of $K^*(F(n),k)$.

Theorem 16. *For* $n \geq 4$ *and* $k \geq 4$,

$$n \sum_{d/k-1} \left(\frac{1}{k-1}\right) n^d \mu\left(\frac{k-1}{d}\right) - \binom{n+k-3}{n-1} \geq \text{rank } K^*(F(n),k)$$

and

$$\text{rank } K^*(F(n),k) \geq n(k-2)\binom{n+k-3}{n-2} - \binom{n+k-3}{n-1}.$$

If $k = 4$, the upper and lower bounds in Theorem 16 coincide, and hence there is precise information for $K^*(F(n),4)$ for $n > 3$. But Andreadakis [1] showed that rank $K^*(F(3),4) = 18$, and thus we do not have to skip rank 3 for this value of k.

Corollary 17. *For* $n \geq 3$

$$\text{rank } K^*(F(n),4) = 2n\binom{n+1}{n-2} - \binom{n+1}{n-1}.$$

There are some obvious improvements possible in the lower bounds of Theorem 16 which are given in [10].

As an aid in extracting information about $K^*(F(3),k)$, Theorem 10 is useless - we have to skip over rank 3. However, about ten years before any part of Theorem 10 ever was established, Chein [13] proved a theorem which remains to this day as one of the most mysterious in the mathematical literature but which enables us to extend Theorem 16 to the rank 3 case.

Theorem 18 (Chein [13]). (a) *If* $k \neq 5$, *any element in* $K^*(\phi(3),k)$ *is induced by an automorphism of the free group* $F(3)$.
(b) *There is an element of* $K^*(\phi(3),5)$ *which is not induced by an automorphism of* $F(3)$.

Thus $k = 5$ is a genuine exception. This exception enabled Chein to deduce Theorem 3 (Section 0) as a corollary of Theorem 16. Thus the proof of Theorem 3 is very different from that of Theorem 9(b) as outlined in Section 4 of Mochizuki's survey [24]. Chein begins with a matrix in $K^*(\phi(3),5)$ and finds an exotic invariant which enables him to determine when this matrix cannot be lifted to an automorphism of $F(3)$. Recall that $\hat{\beta}$ in $K(\phi(3),5)$ can be represented as a 3×3 matrix over $\mathbb{Z}[s_1,s_1^{-1},s_2,s_2^{-1},s_3,s_3^{-1}]$ of the form $\beta = I + X$, where β satisfies the "column condition" and X is congruent to the zero matrix modulo the third power of the augmentation ideal, the ideal of $\mathbb{Z}[s_1,s_1^{-1},\ldots,s_3,s_3^{-1}]$ generated by $\sigma_i = (1 - s_i)$, $i = 1,2,3$. Chein defines the *critical sum* of the matrix β as the sum of the following four integers: the coefficient of $\sigma_1^2\sigma_3$ in the (2,1) entry of β, the coefficient of $\sigma_2^2\sigma_3$ in (1,2) entry of β, the coefficient of $\sigma_1\sigma_2\sigma_3$ in the (2,2) entry of β, and the coefficient of $\sigma_2\sigma_3^2$ in the (1,3) entry of β. Remember that β is in $K^*(\phi(3),5)$ if and only if the trace of M is zero. The key result of Chein is that if the critical sum of β is odd, then β does not lift to an automorphism of $F(3)$. The automorphism δ of $\phi(3)$ described in Section 0, if considered as an element in $K(\phi(3),5)$, is easily seen to be in $K^*(\phi(3),5)$ and to have odd critical sum. (The actual automorphism that Chein uses in [13] is a slight variation of δ.)

By combining Theorems 10 and 18, one arrives at a rather bizarre sounding corollary. It is perhaps fitting to close this survey with such a statement, which we call the

(3,5)-Theorem. *Any element of* $K(\phi(n),k)$ *which lifts to an automorphism of* $\phi(n)$ *also lifts to an automorphism of* $F(n)$ *except when* $n = 3$ *and* $k = 5$.

REFERENCES

1. S. Andreadakis, Automorphisms of free groups and free nilpotent groups, *Proc. London Math. Soc.* 15 (1965), 239-268.
2. L. Auslander & G. Baumslag, Automorphism groups of finitely generated nilpotent groups, *Bull. Amer. Math. Soc.* 73 (1967), 716-717.
3. S. Bachmuth, Automorphisms of free metabelian groups, *Trans. Amer. Math. Soc.* 118 (1965), 93-104.
4. S. Bachmuth, Induced automorphisms of free groups and free metabelian groups, *Trans. Amer. Math. Soc.* 122 (1966), 1-17.
5. S. Bachmuth, G. Baumslag, J. Dyer & H.Y. Mochizuki, Automorphism groups of 2-generator metabelian groups, to appear in *J. London Math. Soc.*
6. S. Bachmuth, E. Formanek & H.Y. Mochizuki, IA-automorphisms of certain two-generator torsion-free groups, *J. Alg.* 40 (1976), 19-30.
7. S. Bachmuth & H.Y. Mochizuki, IA-automorphisms of the free metabelian group of rank 3, *J. Alg.* 55 (1978), 106-115.
8. S. Bachmuth & H.Y. Mochizuki, $E_2 \neq SL_2$ for most Laurent polynomial rings, *Amer. J. Math.* 104 (1982), 1181-1189.
9. S. Bachmuth & H.Y. Mochizuki, The tame range of automorphism groups and GL_n, preprint.
10. S. Bachmuth & H.Y. Mochizuki, Lifting nilpotent automorphisms and the (3,5)-Theorem, to appear in *Arch. der Math.*
11. H. Bass, *Introduction to some methods of algebraic K-theory*, Regional conference series in mathematics No. 20, Amer. Math. Soc., Providence, R.I.
12. A.M. Brunner & J.G. Ratcliffe, Finite 2-complexes with infinitely-generated groups of self-homotopy equivalences, preprint.
13. O. Chein, IA-automorphisms of free and free metabelian groups, *Comm. Pure and Applied Math.* 21 (1968), 605-629.
14. O. Chein, Subgroups of IA automorphisms of a free group, *Acta Mathematica* 123 (1969), 1-12.
15. P.M. Cohn, On the structure of the GL_2 of a ring, *Inst. Hautes Etudes Sci. Publ. Math.* 30 (1966), 5-53.
16. J. Dyer, A criterion for automorphisms of certain groups to be inner, *J. Austral. Math. Soc. Ser* A 21 (1976), 179-184.
17. M.H. Freedman, There is no room to spare in four-dimensional space, *Notices of Amer. Math. Soc.* 31 (Jan 1984), 3-6.
18. A. Karrass & D. Solitar, The subgroups of a free product of two groups with an amalgamated subgroup, *Trans. Amer. Math. Soc.* 149 (1970), 227-255.
19. *The Kourovka Notebook*, American Mathematical Society Translations, Series 2, Volume 121.
20. B. Levinger, *A generalization of the braid group*, Ph.D. Thesis, New York University, 1968.
21. W. Magnus, Uber n-dimensionale Gittertransformationen, *Acta Math.* 64 (1934), 353-367.
22. D. McCullough, Compact 3-manifolds with infinitely-generated groups of self-homotopy-equivalences, preprint.
23. D. McCullough & A. Miller, The genus 2 Torelli group is not finitely generated, preprint.
24. H.Y. Mochizuki, Automorphisms of solvable groups, Part II, this volume.
25. H. Nagao, On GL(2,K[x]), *J. Poly. Osaka Univ.* 10 (1959), 117-121.
26. M. Nagata, *Polynomial rings and affine spaces*, Regional conference series in mathematics No. 37, Amer. Math. Soc., Providence, R.I.
27. J. Nielsen, Die Isomorphismen der allgemeinen unendlichen Gruppe mit zwei Erzeugenden, *Math. Ann.* 78 (1918), 385-397.
28. J. Nielsen, Die Gruppe der dreidimensionalen Gittertransformationen, *Kgl. Danske Videnskabernes Selskab., Math. Fys. Meddelelser* V. 12 (1924), 1-29.
29. J.P. Serre, *Trees*, translated from the French by J. Stillwell, Springer-Verlag, 1980.

30. A.L. Šmel'kin, Two notes on free solvable groups, (Russian), *Alg. i. Log.* 6 (1967), 95-109.
31. A.A. Suslin, On the structure of the special linear group over polynomial rings, *Isv. Akad. Nauk.* 11 (1977), 221-238.
32. L.N. Vaserstein, K_1-theory and the congruence problem, *Math. Zametki* 5 (1969); English translation in *Math. Notes* 5 (1969).
33. E. Witt, Treue Darstellung Lieschen Ringe, *J. Reine Angew. Math.* 177 (1937), 152-160.

AUTOMORPHISMS OF SOLVABLE GROUPS, PART II

H.Y. Mochizuki
University of California, Santa Barbara, CA 93106, USA

1. *INTRODUCTION*

Let F(n) denote the free group of rank n. If H is a group, the terms of its derived series are denoted by $H = H^0 \geqq H' \geqq H'' \geqq \dots H^{(k)} \geqq \dots$ and the terms of its lower central series by $H = \gamma_1(H) \geqq \gamma_2(H) \geqq \dots \geqq \gamma_k(H) \geqq \dots$.

An outstanding problem in group theory is the determination of the structure of the automorphism group Aut G of a group G if $G = F(n)/F(n)^{(\delta)}$ is free solvable of derived length δ and rank n or $G = F(n)/\gamma_{c+1}(F(n))'$ is free (abelian by nilpotent of class c) of rank n. As Problem 5.49 of the *Kourova Notebook* 1976, Remeslennikov asked if Aut G is finitely generated (f.g.). We begin by posing this and some other problems about Aut G.

Problem I. *Is* Aut G *f.g., respectively finitely presented (f.p.)? In particular, find a set of generators, respectively a presentation, for* Aut G.

The natural surjection $F(n) \to G$ induces a homomorphism Aut $F(n) \to$ Aut G, an automorphism of G being called *tame* if it is the image of an automorphism of F(n). We shall make Problem I more specific by asking

Problem I'. *Is each automorphism of* G *tame, i.e. is the map* Aut $F(n) \to$ Aut G *surjective?*

Problem II. *What can be said about the subgroups of* Aut G?

For example, does the Tits alternative hold for Aut G, are the periodic subgroups of Aut G finite or locally finite, are the nilpotent (resp. solvable) subgroups of Aut G of bounded nilpotency class (resp. derived length), etc.

Problem III. *Can* Aut G *be embedded in a well-understood group in a nice way?*

$\phi(n) = F(n)/F(n)''$ is called the free metabelian group of rank n. The major portion of this paper is devoted to the recent surprising developments in the study of Aut $\phi(n)$ in response to the preceding problems. The main results can be summarised as:

Theorem A. (i) (Bachmuth [1]) Aut $\phi(2)$ *is the extension of the group* Inn $\phi(2)$ *of inner automorphisms of* $\phi(2)$ *by* Aut $(\phi(2)/\phi(2)') \cong GL_2(\mathbb{Z})$.
(ii) (Chein [9]; Bachmuth and Mochizuki [5]) Aut $\phi(3)$ *is not f.g. Thus, the natural map* Aut $F(3) \to$ Aut $\phi(3)$ *is not surjective.*
(iii) (Bachmuth and Mochizuki [6]) *If* $n \geqq 4$, *then* Aut $F(n) \to$ Aut $\phi(n)$ *is surjective. In particular, one can give a finite generating set for* Aut $\phi(n)$.

The remarkable contrast between Aut $\phi(3)$ and Aut $\phi(n)$, $n \neq 3$, has parallels in several other setting such as the Torelli subgroup $\mathcal{I}(g)$ of the mapping class group of an orientable closed surface $S(g)$ of genus g, which we deal with in Section 6 of this paper, and linear groups over polynomial rings, which are discussed by Bachmuth in Section 1 of [2]. (Also, see [22] and [24].) Of course, Aut $\phi(3)$ provides a negative answer to Remeslennikov's question.

Theorem A(i) has a far reaching generalization, namely:

Theorem B. (Bachmuth, Formanek and Mochizuki [3]). *Let* R *be a normal subgroup of* F(2) *such that* $R \leqq F(2)'$. *If the integral group ring* $\mathbb{Z}(F(2)/R)$ *has no zero divisors, then* Aut $(F(2)/R')$ *is the extension of* Inn $(F(2)/R')$ *by* Aut $(F(2)/F(2)') \cong GL_2(\mathbb{Z})$. *Thus, all automorphisms of* $F(2)/R'$ *are tame.*

Note that Theorem B applies to the cases Aut $(F(2)/F(2)^{(\delta)})$, $\delta \geqq 3$, with $R = F(2)^{(\delta-1)}$ and Aut $(F(2)/\gamma_{c+1}(F(2)))$, $c \geqq 2$, with $R = \gamma_{c+1}(F(2))$.

Theorems A and B plus some heuristic evidence have led us to make the following:

Conjecture 1. *There is a non-negative integer-valued function* $f(\delta)$ *of the derived length* δ *such that all elements of* Aut $(F(n)/F(n)^{(\delta)})$ *are tame if* $n \geqq f(\delta)$.

Conjecture 2. *There is a non-negative integer-valued function* $g(c)$ *of the nilpotency class* c *such that all elements of* Aut $(F(n)/\gamma_{c+1}(F(n))')$ *are tame if* $n \geqq g(c)$.

Sections 2 through 5 contain background material and a discussion of the motivation for and the proof of Theorem A, parts (ii) and (iii). As mentioned earlier, Section 6 deals with the Torelli groups $\mathcal{I}(g)$ and contains an outline of proof of the McCullough-Miller theorem that $\mathcal{I}(2)$ is not f.g. Section 7 ends the paper by discussing some consequences of the above problems and verifying Conjectures 1 and 2.

If the reader has not done so already, he/she is advised to read the first two sections of Bachmuth's article [2] in this same volume before continuing with this paper. They contain some historical and motivational background besides the necessary material on linear groups of polynomial rings and Laurent polynomial rings.

2. *MAGNUS REPRESENTATIONS OF* F/R' *AND OF* Aut (F/R')

Let $F = F(n)$ be a free group of rank n with basis $\{x_1,\dots,x_n\}$, and let R be a fixed characteristic subgroup of F such that $R \leqq F'$. $s_1 = x_1R,\dots,s_n = x_nR$ and $a_1 = x_1R',\dots,a_n = x_nR'$ will serve as generators of F/R and F/R' respectively.

Let $\{t_1,\dots,t_n\}$ be a basis for a free right $\mathbb{Z}(F/R)$-module M of rank n where $\mathbb{Z}(F/R)$ is the integral group ring of F/R. Then

$$\begin{pmatrix} F/R & 0 \\ M & 1 \end{pmatrix} = \left\{ \begin{pmatrix} g & 0 \\ \sum_{i=1}^{n} t_i r_i & 1 \end{pmatrix} \;\middle|\; g \in F/R,\ r_1,\dots,r_n \in \mathbb{Z}(F/R) \right\}$$

forms a group under formal matrix multiplication.

Proposition 2.1. *The map*

$$a_j \longrightarrow \begin{pmatrix} s_j & 0 \\ t_j & 1 \end{pmatrix}, \quad 1 \leq j \leq n$$

extends to an injective homomorphism

$$\psi : F/R' \longrightarrow \begin{pmatrix} F/R & 0 \\ M & 1 \end{pmatrix}.$$

Moreover, an element $\begin{pmatrix} g & 0 \\ \sum_{i=1}^{n} t_i r_i & 1 \end{pmatrix}$ *of* $\begin{pmatrix} F/R & 0 \\ M & 1 \end{pmatrix}$ *is in* $\mathrm{Im}\psi$, *the image of* ψ, *if and only if* $\sum_{i=1}^{n} (s_i - 1) r_i = (g - 1)$.

The faithful representation ψ of F/R' in Proposition 2.1 is called the *Magnus representation* of F/R' after its discoverer W. Magnus [15]. The last statement of Proposition 2.1 was first proved by Bachmuth [1] for the special case $R = F'$ and was later generalized (evidently independently) by J. Birman [7, p. 111] and Remeslennikov and Sokolov [20].

We now turn our attention to the Magnus representation of Aut (F/R'). If $\alpha \in$ Aut (F/R), then α extends in a natural way to an automorphism of $\mathbb{Z}(F/R)$, which we also denote by α. The map

$$(r_{ij}) \to (\alpha r_{ij}), \ (r_{ij}) \in GL_n(\mathbb{Z}(F/R)),$$

defines an automorphism of $GL_n(\mathbb{Z}(F/R))$. Let $\mathfrak{G}$ denote the group consisting of all pairs $[(r_{ij}),\alpha]$ in $GL_n(\mathbb{Z}(F/R)) \times$ Aut (F/R) with multiplication defined by $[(s_{ij}),\beta][(r_{ij}),\alpha] = [(s_{ij})(\beta r_{ij}),\beta\alpha]$. $\mathfrak{G}$ will serve as the "home" of the Magnus representation of Aut (F/R').

Let $\phi \in$ Aut (F/R'). Then, ϕ canonically induces an automorphism $\bar{\phi}$ of F/R and thus an automorphism $\hat{\phi}$ of $\mathrm{Im}\,\psi$ (see Proposition 2.1) making the diagram

$$\begin{array}{ccc} F/R' & \xrightarrow{\phi} & F/R' \\ \psi\downarrow & & \psi\downarrow \\ \mathrm{Im}\,\psi & \xrightarrow{\hat{\phi}} & \mathrm{Im}\,\psi \end{array}$$

commute. Thus, $\hat{\phi}$ is defined by

$$\hat{\phi} : \begin{pmatrix} s_j & 0 \\ t_j & 1 \end{pmatrix} \longrightarrow \begin{pmatrix} \bar{\phi}s_j & 0 \\ \sum_{i=1}^{n} t_i r_{ij} & 1 \end{pmatrix}, \quad 1 \leq j \leq n,$$

where $\sum_{i=1}^{n} (s_i - 1) r_{ij} = (\bar{\phi}s_j - 1), \quad 1 \leq j \leq n.$

Proposition 2.2. *The map* $\Theta : \mathrm{Aut}\,(F/R') \to \mathfrak{G}$ *defined by*

$$\Theta : \phi \to [(r_{ij}), \bar{\phi}]$$

for all $\phi \in \mathrm{Aut}\,(F/R')$, *where* $\bar{\phi} \in \mathrm{Aut}\,(F/R)$ *is induced by* ϕ *and* $\sum_{i=1}^{n} (s_i - 1) r_{ij} = (\bar{\phi} s_j - 1)$, $1 \leqq j \leqq n$, *is an injective homomorphism. Moreover, an element* $[(a_{ij}), \alpha]$ *of* $\mathfrak{G}$ *is in* $\mathrm{Im}\,\Theta$, *the image of* Θ, *if and only if there is* ϕ *in* $\mathrm{Aut}\,(F/R')$ *such that* $\bar{\phi} = \alpha$ *and* $\sum_{i=1}^{n} (s_i - 1) a_{ij} = (\bar{\phi} s_j - 1)$, $1 \leqq j \leqq n$.

Let C be a characteristic subgroup of F such that $R \leqq C \leqq F'$. Then, we have a natural homomorphism $\mathrm{Aut}\,(F/R') \to \mathrm{Aut}\,(F/C)$, and we let $\mathrm{Aut}\,(F/R'; F/C)$ denote the kernel of this map. Under restriction, Θ induces a faithful representation $\Theta_C : \mathrm{Aut}\,(F/R'; F/C) \to \mathfrak{G}$.

We shall call the faithful representation Θ of $\mathrm{Aut}\,(F/R')$, respectively Θ_C of $\mathrm{Aut}\,(F/R'; F/C)$, the Magnus representation of $\mathrm{Aut}\,(F/R')$, respectively of $\mathrm{Aut}\,(F/R'; F/C)$.

Let $\sigma_j = s_j - 1$. For the special case $C = R$ and using the notation of Proposition 2.2, we have:

Proposition 2.3. *The map*

$$\Theta_R : \mathrm{Aut}\,(F/R'; F/R) \to GL_n(\mathbb{Z}\,(F/R))$$

defined by $\Theta_R : \phi \to (r_{ij})$ *for all* $\phi \in \mathrm{Aut}\,(F/R'; F/R)$ *is a faithful representation of* $\mathrm{Aut}\,(F/R'; F/R)$. *Moreover, an element* (r_{ij}) *of* $GL_n(\mathbb{Z}\,(F/R))$ *lies in* $\mathrm{Im}\,\Theta_R$ *if and only if* $\sum_{i=1}^{n} \sigma_i r_{ij} = \sigma_j$, $1 \leqq j \leqq n$.

For this paper, we are interested in the even more special case $R = F'$. We use the original notation $IA(G)$ instead of $\mathrm{Aut}\,(G; G/G')$ for the kernel of the natural map $\mathrm{Aut}\, G \to \mathrm{Aut}\,(G/G')$.

Let E_j denote the jth column of the identity $n \times n$ matrix I_n. As a consequence of Proposition 2.3, we have:

Corollary (cf. Theorem 1 of [1]). *An* $n \times n$ *matrix* (r_{ij}) *over* $\mathbb{Z}\,(F/F')$ *is in the image of* $\Theta_{F'} : IA(F/F') \to GL_n(\mathbb{Z}\,(F/F'))$ *if and only if*

(a) *for each* j, *the* jth *column of* (r_{ij}) *has form*

$$E_j + \sum_{1 \leqq i < k \leqq n} (E_k \sigma_i - E_i \sigma_k) s_{ik},$$

where $s_{ik} \in \mathbb{Z}(F/F')$, *and*

(b) $\det(r_{ij}) = s_1^{k_1} s_2^{k_2} \ldots s_n^{k_n} \in F/F'$ *where the* k's *are integers.*

We now have the necessary machinery to prove

Proposition 2.4. $GL_{n-1}(\mathbb{Z}\,[s_1, s_1^{-1}],\ \sigma_1 \mathbb{Z}\,[s_1, s_1^{-1}])$ *is a homomorphic image of* $IA(\phi(n))$ *for* $n \geqq 2$.

Proof. According to the Corollary to Proposition 2.3, under the homomorphism $GL_n(\mathbb{Z}\,(F/F')) \to GL_n(\mathbb{Z}\,[s_1, s_1^{-1}])$ obtained by putting s_i equal to 1 for $2 \leqq i \leqq n$, each element (r_{ij}) of $\mathrm{Im}\,\Theta_{F'}$ is mapped into a matrix in $GL_n(\mathbb{Z}\,[s_1, s_1^{-1}],\ \sigma_1 \mathbb{Z}\,[s_1, s_1^{-1}])$ of the form

$$\begin{pmatrix} 1 & 0 & \cdots & 0 \\ * & 1+\sigma_1 b_{22} & \cdots & \sigma_1 b_{2n} \\ \vdots & \vdots & & \vdots \\ * & \sigma_1 b_{n2} & \cdots & 1+\sigma_1 b_{nn} \end{pmatrix}.$$

Thus, there is a homomorphism $\Lambda : \mathrm{Im}\,\Theta_{F'} \to GL_{n-1}(\mathbb{Z}[s_1,s_1^{-1}],\ \sigma_1\mathbb{Z}[s_1,s_1^{-1}])$ such that $\Lambda(\bar{r}_{ij}) = \begin{pmatrix} 1+\sigma_1 b_{22} & \cdots & \sigma_1 b_{2n} \\ \vdots & & \vdots \\ \sigma_1 b_{n2} & \cdots & 1+\sigma_1 b_{nn} \end{pmatrix}$. Each matrix in $GL_n(\mathbb{Z}(F/F')$ of the form

$$\begin{pmatrix} 1 & -\sum_{i=2}^{n} \sigma_i b_{i2} & \cdots\cdots & -\sum_{i=2}^{n} \sigma_i b_{in} \\ 0 & 1+\sigma_1 b_{22} & \cdots\cdots & \sigma_1 b_{2n} \\ \vdots & \vdots & & \vdots \\ 0 & \sigma_1 b_{n2} & \cdots\cdots & 1+\sigma_1 b_{nn} \end{pmatrix}$$

where the b_{ij} are in $\mathbb{Z}[s_1,s_1^{-1}]$ satisfies conditions (a) and (b) of the Corollary to Proposition 2.3. Moreover, it is clear that the set of these matrices forms a subgroup of $\mathrm{Im}\,\Theta_{F'}$, which is mapped isomorphically onto $GL_{n-1}(\mathbb{Z}[s_1,s_1^{-1}],\ \sigma_1\mathbb{Z}[s_1,s_1^{-1}])$ by the homomorphism Λ. The proposition is now established.

The proof of Proposition 2.3 contains more information than in the statement of the proposition, namely, that the homomorphism

$$\Lambda\cdot\Theta_{F'} : IA(\phi(n)) \to GL_{n-1}(\mathbb{Z}[s_1,s_1^{-1}],\ \sigma_1\mathbb{Z}[s_1,s_1^{-1}])$$

splits.

3. Aut F *AND* IA(F)

We let $F = F(n)$ denote the free group of rank n with basis $\{x_1,\ldots,x_n\}$ and recall that for a group G, IA(G) denotes the kernel of the natural map Aut G $\to$ Aut (G/G'). J. Nielsen [17,18] gave a finite presentation for Aut F. Furthermore, he proved that the natural map Aut F $\to$ Aut (F/F') is surjective, more specifically, that a finite presentation for Aut (F/F') is obtained from his presentation by adding one more relation. In other words, IA(F) is the normal closure in Aut F of one element.

Following Nielsen, the set of automorphisms below is sufficient to generate Aut F:

For all $1 \leq i, j \leq n$, $i \neq j$,

$$\rho_{ij}: \begin{cases} x_i \to x_i x_j \\ x_k \to x_k, \quad k \neq i \end{cases}$$

$$\lambda_{ij}: \begin{cases} x_i \to x_j x_i \\ x_k \to x_k, \quad k \neq i \end{cases}$$

$$\delta_i: \begin{cases} x_i \to x_i^{-1} \\ x_k \to x_k, \quad k \neq i \end{cases}$$

F/F' is free abelian with basis $s_1 = x_1F', \ldots, s_n = x_nF'$, and so $\mathrm{Aut}\,(F/F') \cong GL_n(\mathbb{Z})$, this isomorphism depending on the choice of basis for F/F'. Relative to basis $\{s_1, \ldots, s_n\}$, the map $\mathrm{Aut}\,F \to \mathrm{Aut}\,(F/F')$ induces a surjective homomorphism $\mathrm{Aut}\,F \to GL_n(\mathbb{Z})$. The image of ρ_{ij} and of λ_{ij} is the elementary matrix E_{ij}, which equals the identity $n \times n$ matrix except for 1 in the (i,j) position, and the image of δ_i is the diagonal matrix D_i, which equals the identity $n \times n$ matrix except for -1 in the (i,i) position. It is well-known that the set consisting of all the E_{ij}, $i \neq j$, and all the D_i generates $GL_n(\mathbb{Z})$.

The structure of Aut F(2) is fairly transparent. J. Nielsen [17] showed that IA(F(2)) consists entirely of inner automorphisms of F(2). We should also mention that Theorem B of Section 1 is a generalization of this result. By contrast, Aut F(n), $n > 2$, is still a relatively unknown wilderness. (See Chapter I of [14] for information about Aut F(n).) It is known that IA(F(n)) is torsion-free and a finite set of generators for IA(F(n)) consists of the following automorphisms:

For all $1 \leq i, j \leq n$, $i \neq j$,

$$K_{ij}: \begin{cases} x_i \to x_j x_i x_j^{-1} \\ x_\ell \to x_\ell, \quad \ell \neq i \end{cases}$$

For all $1 \leq i, j, k \leq n$ with i, j, k distinct,

$$K_{ijk}: \begin{cases} x_i \to x_i [x_j, x_k] \\ x_\ell \to x_\ell, \quad \ell \neq i \end{cases}$$

A fuller discussion of IA(F(n)) is given in Section 3 of the paper by Bachmuth in this volume [2].

4. Aut $\phi(3)$ *IS NOT* f.g.

In this section we purport to discuss the proof of Theorem A (ii) of the Introduction, which says that Aut $\phi(3)$ is not f.g. For the rest of this section, let $\phi = \phi(3) = F(3)/F(3)''$, and consider the exact sequence $1 \to IA(\phi) \to \mathrm{Aut}\,\phi \to \mathrm{Aut}(\phi/\phi') \to 1$. A result of P. Hall tells us that since $\mathrm{Aut}(\phi/\phi') \cong GL_3(\mathbb{Z})$ is finitely presented, if Aut ϕ is f.g., then IA(ϕ) is f.g. as a *normal* subgroup of Aut ϕ. Thus, it is sufficient to prove that IA(ϕ) is *not* f.g. as a normal subgroup of Aut ϕ.

The Magnus representation of Aut ϕ (Proposition 2.2) and subgroup theorem for amalgamated free products due to Karrass and Solitar (see Section 1 of [2]) figure prominently in the proof. Unfortunately, the details of the proof are too technical to outline effectively. Since the preceding two ideas, or at least variants of them, also play a major role in showing that $IA(\phi)$ is not f.g., we content ourselves with a sketch of this latter fact.

$\phi/\phi' \cong F/F'$ in a natural way, and we recall the notation $\mathbb{Z}(F/F') = \mathbb{Z}[s_1,s_1^{-1},s_2,s_2^{-1},s_3,s_3^{-1}]$ where $s_i = x_iF'$, $1 \leqq i \leqq 3$. Let $\Theta = \Theta_{F'}$: $IA(\phi) \to GL_3(\mathbb{Z}(F/F'))$ be the Magnus representation described in the corollary to Proposition 2.3. Thus, if $\alpha \in \mathrm{Aut}\,\phi$, then

$$\Theta\alpha = (r_{ij}) \in GL_3(R), \quad \sum_{i=1}^{n} \sigma_i r_{ij} = \sigma_j, \quad 1 \leqq j \leqq 3,$$

where $\sigma_i = s_i - 1$. By the Corollary, (r_{ij}) has the form

$$\begin{pmatrix} 1+\sigma_2u_1+\sigma_3u_2 & -\sigma_2v_3-\sigma_3v_1 & -\sigma_3w_3+\sigma_2w_2 \\ -\sigma_1u_1+\sigma_3u_3 & 1+\sigma_3v_2+\sigma_1v_3 & -\sigma_3w_1-\sigma_1w_2 \\ -\sigma_1u_2-\sigma_2u_3 & -\sigma_2v_2+\sigma_1v_1 & 1+\sigma_1w_3+\sigma_2w_1 \end{pmatrix} \tag{1}$$

where the u's, the v's and the w's are in $\mathbb{Z}(F/F')$. In fact, if $y_j = x_jF''$, $1 \leqq j \leqq 3$, α is defined by

$$\alpha : \begin{cases} y_1 \to y_1[y_1,y_2]^{u_1}\,[y_1,y_3]^{u_2}\,[y_2,y_3]^{u_3} \\ y_2 \to y_2[y_1,y_2]^{-v_3}\,[y_1,y_3]^{-v_1}\,[y_2,y_3]^{v_2} \\ y_3 \to y_3[y_1,y_2]^{w_2}\,[y_1,y_3]^{-w_3}\,[y_2,y_3]^{-w_1} \end{cases} \tag{2}$$

Let $Q = Q(\mathbb{Z}(F/F'))$ denote the field of quotients of $\mathbb{Z}(F/F')$. In $GL_2(\mathbb{Z}(F/F\ )[\sigma_1^{-1}]) \leqq GL_2(Q)$, we conjugate the matrix (1) by the matrix

$$E^{-1} = \begin{pmatrix} \sigma_1^{-1} & -\sigma_1^{-1}\sigma_2 & -\sigma_1^{-1}\sigma_3 \\ 0 & 1 & 0 \\ 0 & 0 & 1 \end{pmatrix}$$

to obtain

$$\begin{pmatrix} 1 & 0 & 0 \\ -u_1+\sigma_1^{-1}\sigma_3u_3 & 1+\sigma_3v_2+\sigma_1v_3 & -\sigma_3w_1-\sigma_1w_2 \\ & +\sigma_2u_1-\sigma_1^{-1}\sigma_2\sigma_3u_3 & +\sigma_3u_1-\sigma_1^{-1}\sigma_3^2u_3 \\ -u_2+\sigma_1^{-1}\sigma_2u_3 & -\sigma_2v_2+\sigma_1v_1 & 1+\sigma_1w_3+\sigma_2w_1 \\ & +\sigma_2u_2+\sigma_1^{-1}\sigma_2^2u_3 & +\sigma_3u_2+\sigma_1^{-1}\sigma_2\sigma_3u_3 \end{pmatrix} \tag{3}$$

We focus our attention on the lower right 2×2 submatrix of matrix (3), namely

$$\begin{pmatrix} 1+\sigma_3 v_2+\sigma_1 v_3 & -\sigma_3 w_1-\sigma_1 w_2 \\ +\sigma_2 u_2-\sigma_1^{-1}\sigma_2\sigma_3 u_3 & +\sigma_3 u_1-\sigma_1^{-1}\sigma_3^2 u_3 \\ -\sigma_2 v_2+\sigma_1 v_1 & 1+\sigma_1 w_3+\sigma_2 w_1 \\ +\sigma_2 u_2+\sigma_1^{-1}\sigma_2^2 u_3 & +\sigma_3 u_2+\sigma_1^{-1}\sigma_2\sigma_3 u_3 \end{pmatrix} \tag{4}$$

The homomorphism (denoted by upper bars) of $\mathbb{Z}(F/F')[\sigma_1^{-1}]$ onto $\mathbb{Z}[s_1,s_1^{-1},s_2,s_2^{-1}][\sigma_1^{-1}]$ that maps s_3 into 1 and $\mathcal{S} = \mathbb{Z}[s_1,s_1^{-1},s_2,s_2^{-1}]$ identically onto itself induces a homomorphism of $GL_2(\mathbb{Z}(F/F')[\sigma_1^{-1}])$ onto $GL_2(\mathcal{S}[\sigma_1^{-1}])$ which when followed by conjugation by $\begin{pmatrix}\sigma_1 & 0\\ 0 & 1\end{pmatrix}$ maps the matrix (4) into

$$\begin{pmatrix} 1+\sigma_1\bar{v}_3+\sigma_2\bar{u}_2 & -\bar{w}_2 \\ -\sigma_1\sigma_2\bar{v}_2+\sigma_1^2\bar{v}_1 & 1+\sigma_1\bar{w}_3+\sigma_2\bar{w}_1 \\ +\sigma_1\sigma_2\bar{u}_2+\sigma_2^2\bar{u}_3 & \end{pmatrix} \tag{5}$$

an element of $GL_2(\mathcal{S})$. Hence, we have derived a representation of $IA(\phi)$ into $GL_2(\mathcal{S})$ which maps the automorphism (2) into matrix (5). We denote the image of this representation by $\mathfrak{a}$. To show that $IA(\phi)$ is not f.g., it is sufficient to show that $\mathfrak{a}$ is not f.g.

The group of units of $\mathcal{S}$ is $\{\pm s_1^i s_2^j : i,j \in \mathbb{Z}\}$, and each element $\mathfrak{a}$ has determinant of form $s_1^i s_2^j$. Let $\mathfrak{B} = \mathfrak{a} \cap SL_2(\mathcal{S})$ be the kernel of the determinant map on $\mathfrak{a}$.

The matrices

$$S_1 = \begin{pmatrix} s_1 & 0\\ 0 & 1\end{pmatrix},\ S_2 = \begin{pmatrix} s_2 & 0\\ 0 & 1\end{pmatrix},\ T_1 = \begin{pmatrix} s_1 & 0\\ 0 & s_1^{-1}\end{pmatrix},\ T_2 = \begin{pmatrix} s_2 & 0\\ 0 & s_2^{-1}\end{pmatrix}$$

represent the automorphism whose Magnus representations in $GL_3(\mathbb{Z}(F/F'))$ are

$$\begin{pmatrix} 1 & -\sigma_2 & 0\\ 0 & 1+\sigma_1 & 0\\ 0 & 0 & 1\end{pmatrix},\quad \begin{pmatrix} 1+\sigma_2 & 0 & 0\\ -\sigma_1 & 1 & 0\\ 0 & 0 & 1\end{pmatrix},$$

$$\begin{pmatrix} 1 & -\sigma_2 & s_1^{-1}\sigma_3\\ 0 & 1+\sigma_1 & 0\\ 0 & 0 & 1-s_1^{-1}\sigma_1\end{pmatrix},\quad \begin{pmatrix} 1+\sigma_2 & 0 & 0\\ -\sigma_1 & 1 & s_2^{-1}\sigma_3\\ 0 & 0 & 1-s_2^{-1}\sigma_2\end{pmatrix},$$

respectively. Thus, $S_1,S_2 \in \mathfrak{a}$ and $T_1,T_2 \in \mathfrak{B}$, and, moreover, it is easy to see that $\mathfrak{a}$ is the split extension of $\mathfrak{B}$ by the subgroup generated by S_1 and S_2.

Suppose that $\mathfrak{a}$ is f.g. Then $\mathfrak{a}$ is generated by S_1,S_2 and a finite set of matrices M_i, $1 \leq i \leq n$, of $\mathfrak{B}$. Evidently, the set

$$\left\{M_i^{S_1^j S_2^k} : 1 \leq i \leq n \text{ and } j,k \in \mathbb{Z}\right\}$$

generates $\mathfrak{B}$. But, $M^{T_j^{\pm 1}} = M^{S_j^{\pm 2}}$ for any $M \in \mathfrak{B}$. Therefore, $\mathfrak{B}$ is generated by the finite set of matrices

$$T_1, T_2, M_i, M_i^{S_1}, M_i^{S_2}, M_i^{S_1 S_2}, \ 1 \leq i \leq n.$$

To complete the proof, we first apply Theorem 7, Section 1 of [2] to embed $\mathfrak{B}$ in the following amalgamated free product,

$$\mathfrak{B} \subseteq SL_2(\mathbb{Q}[s_1, s_1^{-1}, s_2, s_2^{-1}]) = A *_U B$$

where $A = SL_2(\mathbb{Q}[s_1, s_1^{-1}, s_2])$, $B = SL_2(\mathbb{Q}[s_1, s_1^{-1}, s_2])^{S_2}$ and $U = A \cap B$. Then, we use the Karrass-Solitar subgroup theorem as stated in Section 1 of the same paper to obtain a contradiction, as we now do.

The set of matrices

$$\left\{ M(\pi) = \begin{pmatrix} 1 & 0 \\ \sigma_1/\pi & 1 \end{pmatrix} : \pi \text{ is a prime integer} \right\}$$

can be chosen as a part of a set of double coset representatives of $(\mathfrak{B} \cap A, U)$ in A (in fact, of $(SL_2(\mathbb{Z}[s_1, s_1^{-1}, s_2]), U)$ in A). Each matrix

$$N(\pi) = \begin{pmatrix} 1 & \pi^2 \sigma_1^2 s_2^{-1} \\ 0 & 1 \end{pmatrix}$$

is in $\mathfrak{B} \cap B$ since it represents the automorphism whose Magnus representation is

$$\begin{pmatrix} 1 & 0 & -\pi^2 \sigma_2^2 s_2^{-1} \\ 0 & 1 & \pi^2 \sigma_1^2 s_2^{-1} \\ 0 & 0 & 1 \end{pmatrix}$$

$$M(\pi) N(\pi) M(\pi)^{-1} = \begin{pmatrix} 1 - \pi \sigma_1^3 s_2^{-1} & \pi^2 \sigma_1^2 s_2^{-1} \\ -\sigma_1^4 s_2^{-1} & 1 + \pi \sigma_1^3 s_2^{-1} \end{pmatrix}$$

is not in $\mathfrak{B} \cap A$, yet represents the automorphism whose Magnus representation is

$$\begin{pmatrix} 1 & \pi \sigma_2 \sigma_1^2 s_2^{-1} + \sigma_3 \sigma_1^2 s_2^{-1} & -\pi^2 \sigma_2 \sigma_1^2 s_2^{-1} - \pi \sigma_3 \sigma_1^2 s_2 \\ 0 & 1 - \pi \sigma_1^3 s_2^{-1} & \pi^2 \sigma_1^3 s_2^{-1} \\ 0 & -\sigma_1^3 s_2^{-1} & 1 + \pi \sigma_1^3 s_2^{-1} \end{pmatrix}.$$

Invoking the Karrass-Solitar subgroup theorem, we conclude that $\mathfrak{B}$ is not f.g., a contradiction.

Remarks 1. We note that the general procedure used here for showing that a group G is not f.g. is to construct a representation of G in a free product of groups with amalgamation and then to apply the Karrass-Solitar subgroup theorem (Section 1 of [2]). This procedure has been used successfully in several context, in particular, by McCullough and Miller to prove that the Torelli group T(2) of genus 2 is not f.g. (See Section 6 below).

Remarks 2. Since $\mathrm{Aut}\phi(3)$ is not f.g., any generating set for $\mathrm{Aut}\phi(3)$ must contain infinitely many non-tame elements, and there exist generating sets in which all but finitely many elements are non-tame. However, no one has constructed a generating set for $\mathrm{Aut}\phi(3)$.

As with matrices in $SL_2(\mathbb{Z}[s_1,s_1^{-1},\ldots,s_j,s_j^{-1}]$, $j \geqq 2$, there is no algorithm known to us which decides whether or not a given automorphism α of $\phi(3)$ is tame, although almost certainly it cannot be. However, if one extends α to an automorphism β of $\phi(4)$ in any way, β is tame in $\mathrm{Aut}\phi(4)$. *Note that this is exactly the "Scenario" at the end of Section 1 of Bachmuth's article in this volume as translated to automorphism groups of metabelian groups.*

5. $\mathrm{Aut}F(n) \to \mathrm{Aut}\phi(n)$ *IS SURJECTIVE*

As indicated in the Introduction, we shall discuss the motivation and highlights of the proof of Theorem A (iii), which is stated as the title of this section. For simplicity, we shall restrict ourselves to the case $n = 4$. We set $F = F(4)$ and $\phi = \phi(4)$ for the remainder of this section.

Examination of the diagram with exact rows

$$\begin{array}{ccccccccc} 1 & \to & IA(F) & \to & \mathrm{Aut}\,F & \to & \mathrm{Aut}(F/F') & \to & I \\ & & \downarrow & & \downarrow & & \downarrow \cong & & \\ 1 & \to & IA(\phi) & \to & \mathrm{Aut}\,\phi & \to & \mathrm{Aut}(\phi/\phi') & \to & 1 \end{array}$$

makes it clear that to prove that $\mathrm{Aut}\,F \to \mathrm{Aut}\,\phi$ is surjective, it is (necessary and) sufficient to prove:

Theorem 5.1. $IA(F(n)) \to IA(\phi(n))$ *is surjective for each* $n \geqq 4$, i.e. *all elements of* $IA(\phi(n))$ *are tame.*

Digressing for a moment, we can state:

Corollary. *For each* $n \geqq 4$, $GL_n(\mathbb{Z}[s_1,s_1^{-1}], (s_1 - 1)\mathbb{Z}[s_1,s_1^{-1}])$ *is a homomorphic image of* $IA(F(n))$.

Proof. Immediate from Proposition 2.4 and Theorem 5.1.

Sketch of proof of Theorem 5. *We identify* $IA(\phi)$ *with its image* $\mathfrak{a} = \mathrm{Im}\theta_{F'}$ *under the Magnus representation* $\theta_{F'}$ (see Section 2). Thus, by Proposition 2.3 and its Corollary, $\mathfrak{a}$ consists of all matrices $I_4 + (r_{ij})$ in $GL_4(\mathbb{Z}[s_1,s_1^{-1},\ldots,s_4,s_4^{-1}])$ such that $\sum_{i=1}^{4} \sigma_i r_{ij} = 0$, where $\sigma_i = s_i - 1$. Let $\mathfrak{d}$ denote the subgroup of $\mathfrak{a}$ of all tame elements, i.e. "automorphisms" of $\mathfrak{a}$ which are induced by automorphisms of F.

We use the following notation:

M_{ij} $(i \neq j)$ denotes the matrix which equals the identity matrix except for $s_j = 1 + \sigma_j$ in the (i,i) position and $-\sigma_i$ in the (j,i) position.

$M_{i,j,k}(r)$ (i,j,k distinct and $r \in \mathbb{Z}[s_1,s_1^{-1},\ldots,s_n,s_n^{-1}]$) denotes the matrix which equals the identity matrix except for $\sigma_k r$ in the (j,i) position and $-\sigma_j r$ in the (k,i) position.

$N_{i,j}(r)$ ($i \neq j$ and $r \in \mathbb{Z}[s_1,s_1^{-1},\ldots,s_4,s_4^{-1}]$ denotes the matrix which equals the identity matrix except for $1+\sigma_i\sigma_j r$, $\sigma_j^2 r$, $-\sigma_i^2 r$ and $1-\sigma_i\sigma_j r$ in the (i,i), (i,j), (j,i) and (j,j) positions, respectively.

For example,

$$M_{23} = \begin{pmatrix} 1 & 0 & 0 & 0 \\ 0 & 1+\sigma_3 & 0 & 0 \\ 0 & -\sigma_2 & 1 & 0 \\ 0 & 0 & 0 & 1 \end{pmatrix},$$

$$M_{134}(r) = \begin{pmatrix} 1 & 0 & 0 & 0 \\ 0 & 1 & 0 & 0 \\ \sigma_4 r & 0 & 1 & 0 \\ -\sigma_3 r & 0 & 0 & 1 \end{pmatrix}, \text{ and}$$

$$N_{24}(r) = \begin{pmatrix} 1 & 0 & 0 & 0 \\ 0 & 1+\sigma_2\sigma_4 r & 0 & \sigma_4^2 r \\ 0 & 0 & 1 & 0 \\ 0 & -\sigma_2^2 r & 0 & 1-\sigma_2\sigma_4 r \end{pmatrix}.$$

Each of the matrices M_{ij}, $M_{ijk}(r)$ and $N_{ij}(r)$ is contained in $\mathbf{a}$, and $M_{ijk}(r)^{-1} = M_{ikj}(r)$, $N_{ij}(r)^{-1} = N_{ji}(r)$. Moreover, with the notation at the end of Section 3, M_{ij}, respectively $M_{ijk}(1)$, is the image of K_{ij}, respectively K_{ijk}, under the map $IA(F) \to \mathbf{a}$. Since the set of all K_{ij} and all K_{ijk} generates IA(F), the set of all M_{ij} and all $M_{ijk}(1)$ generates $\mathbf{B}$.

The first step in verifying that $\mathbf{a} = \mathbf{B}$ was to prove that each $M_{ijk}(r)$ is tame. We illustrate the type of computations required by considering a specific example.

Chein [9] showed that the matrix $\begin{pmatrix} 1 & 0 & 0 \\ \sigma_3\sigma_1^2 & 1 & 0 \\ -\sigma_3\sigma_1^2 & 0 & 1 \end{pmatrix}$ represents a non-tame automorphism in IA(ϕ(3)). This automorphism is denoted by δ in Bachmuth's paper, this volume. If we extend δ to an automorphism δ of ϕ by mapping a_4 into itself, the resulting matrix in $\mathbf{a}$ is

$$\begin{pmatrix} 1 & 0 & 0 & 0 \\ \sigma_3\sigma_1^2 & 1 & 0 & 0 \\ -\sigma_2\sigma_1^2 & 0 & 1 & 0 \\ 0 & 0 & 0 & 1 \end{pmatrix} = \begin{pmatrix} 1 & 0 & 0 & 0 \\ \sigma_3 s_1^2 & 1 & 0 & 0 \\ -\sigma_2 s_1^2 & 0 & 1 & 0 \\ 0 & 0 & 0 & 1 \end{pmatrix} \begin{pmatrix} 1 & 0 & 0 & 0 \\ -\sigma_3 s_1 & 1 & 0 & 0 \\ \sigma_2 s_1 & 0 & 1 & 0 \\ 0 & 0 & 0 & 1 \end{pmatrix}^2 \begin{pmatrix} 1 & 0 & 0 & 0 \\ \sigma_3 & 1 & 0 & 0 \\ -\sigma_2 & 0 & 1 & 0 \\ 0 & 0 & 0 & 1 \end{pmatrix}$$

$$= M_{123}(s\ )M_{123}(s\ )^{-2}M_{123}(1).$$

Therefore, to show δ is tame, it is sufficient to prove that $M_{123}(g)$ is tame for all invertible elements g of $\mathbb{Z}[s_1,s_1^{-1},\ldots,s_4,s_4^{-1}]$. Indeed, since $M_{123}(f)$ is additive in the variable f, it would show that $M_{123}(r)$ is tame for all $r \in \mathbb{Z}[s_1,s_1^{-1},\ldots,s_4,s_4^{-1}]$.

We show that if, for fixed a, all $M_{ijk}(a)$ are tame, then $M_{123}(as_i^{\pm 1})$ is tame. By a similar argument, one can also show all $M_{ijk}(as_i^{\pm 1})$ are tame, whence an induction argument proves that $M_{123}(g)$, indeed $M_{ijk}(g)$ is tame for all units g. If $i \neq 1$, then

$$M_{1i}^{\mp 1}M_{123}(a)M_{i1}^{\pm 1} = M_{123}(as_i^{\pm 1})$$

is tame. Also,

$$M_{14}^{-1}M_{423}(a)^{-1}M_{14}M_{423}(a)M_{123}(a) = M_{123}(as_1)$$

and

$$M_{423}(as_1^{-1}) = M_{41}M_{423}(a)M_{41}^{-1}$$

are tame. Thus

$$M_{14}^{-1}M_{423}(as_1^{-1})M_{14}M_{423}(as_1^{-1})^{-1}M_{123}(a) = M_{123}(as_1^{-1})$$

is tame, whence $M_{123}(as_i^{\pm 1})$ is tame for $1 \leqq i \leqq 4$.

The next step was to show that each matrix in $\mathfrak{a}$ with exactly one nontrivial column is tame, the proof of which we shall skip. The third step was to verify that all $N_{ij}(r)$ are tame. For example,

$$N_{12}(r) = M_{412}(-r\sigma_4)M_{14}M_{24}M_{412}(r)M_{14}^{-1}M_{24}^{-1}M_{412}(-r)$$

is tame.

The crucial and by far most difficult part of the proof that $\mathfrak{a} = \mathfrak{B}$ is what we call in [6]:

Main Lemma. *Each element of* $\mathfrak{a}$ *of form*

$$\begin{pmatrix} 1 & -\sum_{i=1}^{3}\sigma_i b_{i2} & -\sum_{i=1}^{3}\sigma_i b_{i3} & -\sum_{i=1}^{3}\sigma_i b_{i4} \\ 0 & 1+\sigma_1 b_{22} & \sigma_1 b_{23} & \sigma_1 b_{24} \\ 0 & \sigma_1 b_{32} & 1+\sigma_1 b_{33} & \sigma_1 b_{34} \\ 0 & \sigma_1 b_{42} & \sigma_1 b_{43} & 1+\sigma_1 b_{44} \end{pmatrix}$$

is tame.

The rest of the proof of Theorem 5.1 boils down to factoring each element of $\mathfrak{a}$ as a matrix in $\mathfrak{B}$ times a matrix of the form in the Main Lemma.

Many of the ideas and methods used in the proof of Theorem 5.1 are heavily influenced by the work of Suslin [22] and Vaserstein [24] and their results about matrices and matrix groups over integral Laurent polynomial rings. Indeed, much of the matrix factorizations we employ are analogues of matrix calculations of Suslin and Vaserstein.

6. *TORELLI GROUPS*

In this section we discuss a situation which is analogous to that of $IA(\phi(n))$ and of GL_{n-1} of a Laurent polynomial ring and, as mentioned earlier, is a nice application of some theorems in Section 1 of Bachmuth's article [2].

Let $M(g)$, $g \geq 1$, be the mapping class group of a closed orientable surface $S(g)$ of genus g. Thus, $M(g)$ is the group of orientation preserving homeomorphisms of $S(G)$ modulo the normal subgroup of free isotopies of $S(g)$. Dehn [10] established that $M(g)$ is generated by so called Dehn twists of $S(g)$, and Lickorish [13] later showed that a certain set of $(3g-1)$ twists suffices to generate $M(g)$. We are interested in the subgroup $\mathcal{I}(g)$ of $M(g)$ of all mapping classes which induce the identity map on the homology of $S(g)$ i.e. on the abelianized fundamental group of $S(g)$. $\mathcal{I}(g)$ is called the *Torelli group of genus* g.

Of course, $M(1) \cong GL_2(\mathbb{Z})$ and $\mathcal{I}(1) = 1$. Powell [19] found an infinite set of generators for $\mathcal{I}(g)$, $g \geq 2$, consisting of BSCC maps, i.e. twists on bounding simple closed curves (BSCC's) and BP maps, which are opposite twists on bounding pairs of disjoint homologous simple closed curves, each being nonbounding. D. Johnson [11] proved that for $g \geq 3$, $\mathcal{I}(g)$ is generated by a particular finite set of BP maps.

Thus, the problem of whether or not $\mathcal{I}(2)$ is f.g. remained. We note that all BP maps in $\mathcal{I}(2)$ are trivial modulo isotopies, and hence by Powell's result, $\mathcal{I}(2)$ is generated by BSCC maps. In 1984, a startling theorem of McCullough and Miller [16] appeared, which said that $\mathcal{I}(2)$ is *not* f.g. We gather the preceding results on $\mathcal{I}(g)$ in:

Theorem 6.1 ([11], [16]). (i) $\mathcal{I}(1)$ *is trivial.*
(ii) $\mathcal{I}(2)$ *is not finitely generated.*
(iii) *For* $g \geq 3$, $\mathcal{I}(g)$ *is finitely generated.*

We briefly discuss the proof of part (ii). Let $(S,*)$ be an orientable closed surface of genus 2 with base point $*$, and let a_1, b_1, a_2, b_2 be canonical generators of the fundamental group $\pi_1 = \pi_1(S,*)$ of $(S,*)$ where a_1, a_2 correspond to meridians and b_1, b_2 correspond to latitudes. If N denotes the normal subgroup of π_1 generated by a_1, b_1, then π_1/N is a free abelian group of rank 2, generated, say, by $s = a_2N$ and $t = b_2N$. π_1/N acts on N/N' by conjugation, and N/N' is, in fact, a free $\mathbb{Z}[s,s^{-1},t,t^{-1}]$-module of rank 2 with basis $\bar{a}_1 = a_1N'$, $\bar{b}_1 = b_1N'$.

Let $\mathcal{I}(*)$ denote the group of all base point preserving mapping classes of $(S,*)$ which induce the identity on π_1/π_1'. Thus, the elements of $\mathcal{I}(*)$ are basepoint preserving homeomorphisms of $(S,*)$ modulo basepoint preserving isotopies of $(S,*)$. If $T(*)$ is the subgroup of $\mathcal{I}(*)$ generated by BSCC maps where the curves do not pass through $*$, then the forgetful homomorphism $T(*) \to \mathcal{I} = \mathcal{I}(2)$ which maps a basepoint preserving isotopy class to its respective free isotopy class is surjective by Powell's result that BSCC maps generate .

Let $(\tilde{S},\tilde{*})$ be the covering of $(S,*)$ corresponding to N. Then, $\pi_1(\tilde{S},\tilde{*}) = \tilde{\pi}_1$ can be identified with N and $\tilde{\pi}_1/\tilde{\pi}_1'$ with N/N'. If t_D is a twist about a BSCC D not passing through $*$, then McCullough and Miller showed that the lift $\tilde{t}_D$ of t_D to $(\tilde{S},\tilde{*})$ induces an R-module automorphism of $\tilde{\pi}_1/\tilde{\pi}_1'$. Since the group of R-module automorphisms of $\tilde{\pi}_1/\tilde{\pi}_1'$ is isomorphic to $GL_2(\mathbb{Z}[s,s^{-1},t,t^{-1}])$, one has a representation $T(*) \to GL_2(\mathbb{Z}[s,s^{-1},t,t^{-1}])$, which they showed was into $SL_2(\mathbb{Z}[s,s^{-1},t,t^{-1}])$ and induces a representation $\mathcal{I} \to SL_2(\mathbb{Z}[s,s^{-1},t,t^{-1}])$ with image denoted by K.

As noted in Section 1 of [2]

$$K \leqq SL_2(\mathbb{Q}[s,s^{-1},t,t^{-1}]) = SL_2(\mathbb{Q}[s,s^{-1},t]) *_U SL_2(\mathbb{Q}[s,s^{-1},t])^{\begin{pmatrix} t & 0 \\ 0 & 1 \end{pmatrix}}$$

where U is the intersection of the factors. McCullough and Miller proceed to construct BSCC maps S_i and T_i, $i \geqq 1$, of S whose images M_i and N_i, respectively, in K under the representation satisfy the hypotheses of tne Karrass-Solitar subgroup theorem of Section 1 of [2] with the proper identification of G,A,B,U and H. Thus, K is not f.g., whence $\mathcal{I}$ is not f.g.

7. *FURTHER COMMENTS*

Even though the generation part of Problem I for $Aut(\phi(n))$ has been answered, whether or not $Aut(\phi(n))$, $n > 3$, is finitely presented is a completely open problem. In this connection, whether or not $IA(F(n))$, $n > 3$, respectively $IA(\phi(n))$, $n > 3$, is finitely presented is also unknown.

It has long been known that $IA(F(n))$ is torsion free. The same result holds for $IA(\phi(n))$ and follows from the faithful Magnus representation of $IA(\phi(n))$ in $GL_n(\mathbb{Z}[F(n)/F(n)'), \sigma_1\mathbb{Z}(F(n)/F(n)') + \ldots + \sigma_n\mathbb{Z}(F(n)/F(n)'))$. Thus, questions about periodic subgroups of $Aut\phi(n)$ reduce to questions of periodic subgroups of $GL_n(\mathbb{Z}) \cong Aut(\phi(n))/IA(\phi(n))$. Since $Aut\phi(n)$ is linear-by-linear, certain properties of nilpotent, respectively solvable, subgroups of $Aut(\phi(n))$ follow from the well-known theory of linear groups over fields (or finitely generated integral domains). Also, the Tits alternative obviously holds for $Aut\phi(n)$, i.e. any f.g. subgroup of $Aut\phi(n)$ contains either a noncyclic free subgroup or a solvable subgroup of finite index. It is unknown if $AutF(n)$ satisfies the Tits alternative. Of course, an important question is whether or not $Aut\phi(n)$, respectively $AutF(n)$, is linear.

Bachmuth [2] discusses some other applications, e.g. see Section 3 and Theorem 11 of Section 2 of his paper in this volume. Very recently, a preprint was sent to us in which Elena Stöhr proved the following.

Theorem 7.1 [21]. *Let* $G(n) = F(n)/[F(n)'',F(n)]$ *be the free center-by-metabelian group of rank* n. *Then* (i) $AutG(2)$ *and* $AutG(3)$ *are each not f.g.* (ii) *For* $n \geqq 4$, $AutG(n)$ *is f.g.*

Thus, the center-by-metabelian variety provides another example which answers Problem 3, Section 2 of [2]. It is worth noting that as with $AutJ(n)$, it is unknown if all the elements of $AutG(n)$ are tame for $n \geqq 4$. Also, the contrast of the non-finite generation of $AutG(2)$ with Theorem B in Section 1 is intriguing.

It is a natural step that one should next study automorphism groups of f.g. free (abelian-by-nilpotent) groups $F(n)/\gamma_{c+1}(F(n))'$, where $\gamma_{c+1}(F(n))$ denotes the (c+1)th term of the lower central series of F(n). K. Brown, Lenigan and Stafford [8] have shown that $GL_r(F(n)/\gamma_{c+1}(F(n))')$ is generated by elementary and diagonal matrices for all sufficiently large r, depending on the size of both n and c. If the method for study of $Aut\phi(n)$ is to generalize, then it seems clear that one needs to show that $GL_r(F(n)/\gamma_{c+1}(F(n))')$ is generated by elementary and diagonal matrices for all sufficiently large r, depending on the size of c alone. Then, one can hope to show that Conjecture 2 of Section 1 is true.

REFERENCES

1. S. Bachmuth, Automorphisms of free metabelian groups, *Trans. Amer. Math. Soc.* 118 (1965), 93-104.
2. S. Bachmuth, Automorphisms of solvable groups: Part I, this volume.
3. S. Bachmuth, E. Formanek & H.Y. Mochizuki, IA-Automorphisms of certain two-generator torsion free groups, *J. Alg.* 40 (1976), 19-30.
4. S. Bachmuth & H.Y. Mochizuki, IA-Automorphisms of the free metabelian group of rank 3, *J. Alg.* 55 (1978), 106-115.
5. S. Bachmuth & H.Y. Mochizuki, The non-finite generation of Aut(G), G free metabelian of rank 3, *Trans. Amer. Math. Soc.* 270 (1982), 693-700.
6. S. Bachmuth & H.Y. Mochizuki, $Aut(F) \to Aut(F/F'')$ is surjective for F free of rank ≥ 4, to appear in *Trans. Amer. Math. Soc.*
7. J. Birman, *Braids, links and mapping class groups*, Ann. Math. Studies 82, Princeton University Press, Princeton, N.Y. (1974).
8. K.A. Brown, T.H. Lenigan & J.T. Stafford, K-theory and stable structure of some Noetherian group rings, *Proc. London Math. Soc.* 42 (1981), 193-230.
9. O. Chein, IA automorphisms of free and free metabelian groups, *Comm. Pure Appl. Math.* 21 (1968), 605-629.
10. M. Dehn, Die Gruppe der Abbildungsklassen, *Acta Math.* 69 (1938), 135-206.
11. D. Johnson, The structure of the Torelli group-I: A finite set of generators for $\mathcal{I}$, *Ann. Math.* 118 (1983), 423-442.
12. W.B.R. Lickorish, A representation of orientable, combinatorial 3-manifolds, *Ann. Math.* 76 (1962), 531-540.
13. W.B.R. Lickorish, A finite set of generators for the homeotopy group of a 2-manifold, *Proc. Camb. Phil. Soc.* 60 (1964), 769-779. Also, Corrigendum, 62 (1966), 679-681.
14. R.C. Lyndon & P.E. Schupp, *Combinatorial Group Theory*, Springer-Verlag, Berlin-Heidelberg-New York (1977).
15. W. Magnus, On a theorem of Marshall Hall, *Ann. Math.* (2) 40 (1939), 764-768.
16. D. McCullough & A. Miller, The genus 2 Torelli group is not finitely generated, preprint.
17. J. Nielsen, Die Isomorphismen der Allgemeinen undendlichen Gruppe mit zwei Erzeugenden, *Math. Ann.* 78 (1918), 385-397.
18. J. Nielsen, Die Isomorphismengruppe der freien Gruppen, *Math. Ann.* 91 (1924), 169-209.
19. J. Powell, Two theorems on the mapping class group of surfaces, *Proc. Amer. Math. Soc.* 68 (1978), 347-350.
20. V.N. Remeslennikov & V.G. Sokolov, Some properties of a Magnus embedding, *Algebra i Logika* 9 (1970), 566-578; English transl. in *Algebra and Logic* 9 (1970), 342-349.
21. E. Stöhr, On automorphisms of free centre-by-metabelian groups, preprint.
22. A.A. Suslin, On the structure of the special linear group over polynomial rings, *Math. USSR Izvestia* 11 (1977), 221-238.
23. L.N. Vaserstein, K_1-theory and the congruence problem, *Math. Notes* 5 (1969), 141-148.
24. L.N. Vaserstein, On the stabilization of the general linear group over a ring, *Math. USSR Sb.* 8 (1969), 383-400.

A SURVEY OF GROUPS WITH A SINGLE DEFINING RELATION

Gilbert Baumslag
City College, City University of New York, New York, NY 10031, U.S.A.

A. *INTRODUCTION*

A.1 *A little history*

A group G is termed a one-relator group or a group with a single defining relation if it can be presented in the form

$$G = \langle x_1, x_2, \ldots ; r = 1 \rangle.$$

Notice that the possibility that there are infinitely many generators is not excluded here.

The study of groups with one defining relation goes back at least to the latter part of the 19th century to the theory of functions of a complex variable. Let me remind you of the details. First of all recall that a group G of homeomorphisms of a topological space X is said to act properly discontinuously on X if each point $x \in X$ has a neighbourhood N such that

$$Ng \cap N = \emptyset \text{ for all } g \in G,\ g \neq 1.$$

Now suppose that X is the upper half of the complex plane and that G is a group of real linear fractional transformations acting properly discontinuously on X. Thus the elements of G are transformations of the form

$$z \longmapsto (az + b)/(cz + d) \quad (a,b,c,d \text{ real},\ ad - bc = 1)$$

i.e., G is a subgroup of $PSL_2(\mathbb{R})$. Such a group G is termed a fuchsian group. Fricke and Klein [37] proved in 1897 that if G is finitely generated then it is finitely presented. Indeed they proved the following.

Theorem A.1. *A finitely generated fuchsian group* G *can be presented as*
$$G = \langle c_1, \ldots, c_n, a_1, b_1, \ldots, a_k, b_k;\ c_i^{e_i} = 1\ (1 \leq i \leq n), c_1 \ldots c_n [a_1, b_1] \ldots [a_k, b_k] = 1 \rangle.$$
Here $[x,y] = x^{-1}y^{-1}xy$, the exponents e_i are either integers at least 2 or 0 and

$$2g - 2 + \sum (1 - e_i^{-1}) > 0.$$

Now every finitely generated subgroup of a fuchsian group is, by the very definition itself, a finitely generated fuchsian group and so by Theorem A.1 finitely presented. Conversely every abstract group with a finite presentation of the above form which satisfies the requisite conditions is again a finitely generated fuchsian group. So, in particular, the fuchsian groups include the one-relator groups

$$G_k = \langle a_1, b_1, \ldots, a_k, b_k; [a_1,b_1]\ldots[a_k,b_k] = 1\rangle.$$

Thus a finitely generated subgroup of a G_k is again finitely presented - indeed is either free or has a presentation of exactly the same kind. Thus this work of Fricke and Klein yields a remarkable theorem about the subgroups of a special class of one-relator groups. I will have more to say about this later on.

Now $PSL_2(\mathbb{C})$ maps the complex upper half plane $y > 0$ into itself. Hence it can be viewed as acting properly discontinuously on the hyperbolic plane. In particular, this is true also of the fundamental group G_k of a two dimensional orientable surface of genus k. It turns out that G_k has a fundamental region which is a 4k-gon. This was the starting point of Dehn's paper [31] in 1910 where he obtained a purely geometric solution of the word problem for these groups. In a subsequent paper [32] in 1912 Dehn proved, again by geometric means, that all the relators in the G_k $(k > 1)$ share a purely algebraic property which leads to combinatorial solutions of the word and conjugacy problems. The ideas involved have now grown into an important area of combinatorial group theory called small cancellation theory. This theory applies, in particular, to certain one-relator groups and I will briefly discuss this aspect of the theory in due course. The point that I want to make now is that Gromov [43] has very recently developed a theory of what he calls hyperbolic groups which encompasses much of small cancellation theory. All of these hyperbolic groups, like the G_k, have solvable word and conjugacy problem, again because of the underlying geometry! How this affects the study of one-relator groups is unclear. But the theory seems to me to be so far-reaching that I felt it worthwhile to mention it here.

The history then of this subject is rooted in complex analysis and in hyperbolic geometry. In fact one-relator groups occur, as we have already seen, as fuchsian groups in analysis, as fundamental groups in topology and also as dense subgroups of Galois groups in number theory. In addition they play an important part in the study of finitely presented groups.

In this series of 4 lectures I will not have the time to survey the entire field of modern day one-relator group theory. Instead I have chosen to focus on a few of the facets of the theory that either arise from other areas of mathematics or seem to have wider application than to one-relator groups alone. Much of what I say will involve the prevalence of free subgroups and the way in which properties of free groups persist in one-relator groups. It is clear to me that I will have left many important areas untouched and that often too little time will be devoted to work that deserves more. Indeed there are many important papers that will not even be referred to here and the bibliography is by no means a comprehensive one.

The first theorem concerned with one-relator groups as a whole was the so-called Freiheitssatz, proved by Wilhelm Magnus in his paper [68] in 1930. In a subsequent paper [70] in 1932 Magnus simplified some of his earlier work. Indeed in this second paper Magnus created a method, which I will refer to as Magnus' method, of breaking a one-relator group down into simpler one-relator groups. It is with this method, which has turned out to be an almost indispensable tool in the study of these groups, that I would like to begin the first part of my survey.

Before doing so, however, I would like to recall, for completeness, two well-known definitions.

A.2 *Generalised free products and HNN-extensions*

Suppose that the groups A and B are presented as follows

$$A = \langle X;R \rangle, \quad B = \langle Y;S \rangle.$$

Suppose further that H is a subgroup of A generated by the X-words $v_1, v_2, \ldots$ and that K is a subgroup of B generated by the equally indexed set of Y-words $w_1, w_2, \ldots$. Finally suppose that the mapping $v_i \longmapsto w_i$ $(i = 1,2,\ldots)$ defines an isomorphism α from H to K. Then the group

$$G = \langle X \cup Y;\ R \cup S \cup \{v_i w_i^{-1} \mid i = 1,2,\ldots\} \rangle$$

is called the generalised free product of A and B amalgamating H with K or sometimes simply the generalised free product of A and B amalgamating H (with K). G is then denoted either by

$$\{A * B;\ H = K\} \text{ or by } \{A * B;\ H\}.$$

The various properties of such generalised free products are detailed in Lyndon and Schupp [66] (see also Magnus, Karrass and Solitar [77]).

Next suppose that B is a given group, called the base group, that I is a non-empty index set and that for each $i \in I$, α_i is an isomorphism from the subgroup L_i of B to another one of its subgroups R_i. These subgroups L_i and R_i are called associated subgroups. We now define

$$G = \langle Y \cup \{t_i \mid i \in I\};\ S \cup \{t_i^{-1} a t_i = a\alpha_i \mid a \in L_i,\ i \in I\} \rangle,$$

where here $B = \langle Y;S \rangle$ is a presentation of B. G is called an HNN extension with stable letters t_i $(i \in I)$. Again the properties of such HNN extensions are discussed in [66].

The relevance of these constructions is, as we shall shortly see, that groups with one defining relation can be built up from cyclic groups using one or other of these constructions, by Magnus' method.

B. *REFLECTIONS ON THE FREIHEITSSATZ*

B.1 *The Freiheitssatz*

In 1930 Magnus proved the following

Theorem B.1 (The Freiheitssatz). *Let* $G = \langle a_1, a_2, \ldots ;\ r = 1 \rangle$ *be a group with a single defining relator* r. *If* r *is cyclically reduced,* i.e. *the first and last letters of* r *are not inverses, then any proper subset of the generators that appear in* r *freely generates a free group.*

According to Magnus and Chandler ([76], p.114) Dehn suggested the possibility of such a theorem as well as the possibility that groups with a single defining relation could be built up from free groups. Magnus carried out these suggestions in his thesis.

B.2 Magnus' method

Our objective now is to describe a variation of Magnus' method for breaking down a one-relator group into simpler pieces, due to Moldavanskii [86]. To this end, let $G = \langle t,a,\ldots,c\ ;\ r=1\rangle$ be a group with a single defining relator r. Suppose that r is cyclically reduced and involves all of the generators $t,a,\ldots,c$. In addition suppose that t occurs with exponent sum zero in r. This means that if we put

$$a_i = t^{-i}at^i,\ldots,c_i = t^{-i}ct^i\quad (i \in \mathbb{Z}),$$

then r can be re-expressed as a new word, say s, in these new generators. Let m be the minimum of the a-subscripts occurring in s, M the maximum a-subscript,..., n the minimum c-subscript and N the maximum c-subscript. It is not hard to see that s is a cyclically reduced word in the generators $a_m,\ldots,a_M,\ldots,c_n,\ldots,c_N$. The thing to notice here is that some of these generators may not appear in s and, more importantly, that s is of smaller length than r. Let now H be the one-relator group

$$H = \langle a_m,\ldots,a_M,\ldots,c_n,\ldots,c_N\ ;\ s=1\rangle.$$

Inductively both $L = gp(a_m,\ldots,a_{M-1},\ldots,c_n,\ldots,c_{N-1})$ and $R = gp(a_{m+1},\ldots,a_M,\ldots,c_{n+1},\ldots,c_N)$ are free. So we can form the HNN extension

$$G^* = \langle t,H;t^{-1}a_mt = a_{m+1},\ldots,t^{-1}a_{M-1} = a_M,\ldots,t^{-1}c_n^{\,t} = c_{n+1},\ldots,t^{-1}c_{N-1}t = c_N\rangle.$$

It is not hard now to verify that $G \cong G^*$ and thence that the subgroup of G generated by $a,\ldots,c$ is free on these generators. This argument, due to McCool and Schupp [84], can be expanded to give a complete proof of the Freiheitssatz and also the following theorem of Moldavanskii [86].

Theorem B.2. *Every group with one cyclically reduced defining relator involving at least two generators, is a subgroup of an HNN extension with a single stable letter. The base of this HNN extension is a one-relator group, whose defining relator is of smaller length than the original relator, and whose associated subgroups are free.*

This theorem, although not stated in this way, is implicit in Magnus' original paper in 1932 and is, in essence, what I have referred to as Magnus' method. It has been the basis for the proof of a number of versions of the Freiheitssatz, of which I want only to mention one here, due to Newman [87] (see also [88]).

Theorem B.3. *Let*

$$G = \langle t,a,b,\ldots,c\ ;\ r=1\rangle$$

be a group with a single defining relation. Suppose that r *is cyclically reduced and involves at least three generators* t,a,b *and that* t *occurs with exponent sum zero in* r. *Then one of these generators, say* a, *has the property that there exists an integer* k *such that for every* $m > k$ *the elements*

$$t,\ a^m,\ b, \ldots, c$$

freely generate a free group.

I would like now to turn to some other variations on the Freiheitssatz which are of independent interest, and which offer further applications to one-relator groups.

B.2 Brodskii's variation

Recall to begin with that a group G is called locally indicable if every finitely generated non-identity subgroup of G has a homomorphism onto the infinite cyclic group. Locally indicable groups were introduced first by Higman in his paper [47] on zero divisors in group rings. They include, among others, free groups and torsion-free abelian groups. By, in part, mimicking Magnus' original argument for one-relator groups, Brodskii [21] was able to prove the following.

Theorem B.4. *Let* A *and* B *be locally indicable groups and let* $F = A * B$ *be their free product. Furthermore, let* r *be a cyclically reduced element of* F *of (free product) length at least two - so* r *begins and ends with elements out of different factors. Then*

$$A \cap gp_F(r) = 1 = B \cap gp_F(r)$$

(where $gp_F(r)$ *here denotes the normal closure of* r *in* F*).*

Quotients of the form $F/gp_F(r)$, where we have adopted here the notation above, have been termed one-relator products by Howie [50]. Thus Theorem B.4 may be viewed as a theorem about one-relator products, which have been investigated by Brodskii [21], by B. Baumslag [5] and most recently by Howie [51]. The upshot of this is that many theorems about one-relator groups have been carried over to one-relator products of locally indicable groups.

Brodskii's Theorem B.4 contains the Freiheitssatz as a special case. To see this, let

$$G = \langle X \ ;\ r = 1 \rangle$$

be a one-relator group, where r is cyclically reduced of length at least two. Let Y be a proper subset of the generators appearing in r and let Z be its complement in X. Now let A be the free group on Y, B the free group on Z and F the free product of A and B. Then r, replaced by one of its conjugates if necessary, can be thought of as a cyclically reduced element of F, of length at least two and G can be identified with $F/gp_F(r)$. It follows then from Brodskii's theorem that A is embedded in $F/gp_F(r)$, i.e. Y freely generates a free subgroup of G.

There is another way of viewing this application of Brodskii's theorem to one-relator groups. Let x be any element of X that appears in r, the relator in the discussion above, and let Y now be the set consisting of the remaining elements of X. The relation $r = 1$ then can be viewed as an equation over the free group on Y in the single group variable x. The point then of both Brodskii's theorem and the Freiheitssatz is that every such equation over a free group in a single variable has a solution in some supergroup. It is still an unsolved problem as to whether every equation in a single variable over an arbitrary torsion-free group also has a solution.

There are two related results concerned with the solution of equations that are relevant here. The first of these, which is due to Levin [61], involves equations of the form

$$xg_1xg_2\dots xg_n = 1.$$

Here $g_1, g_2, \dots, g_n$ are elements belonging to an arbitrary group G. Notice that the variable x occurs only with positive exponents and so we shall refer to such equations as positive equations. Levin's theorem then can be couched as follows.

Theorem B.5. *Every positive equation over an arbitrary group* G *can be solved in the wreath product*

$$W = G \text{ wr } C_n$$

of G *with a cyclic group* C_n *of order* n, *where* n *is the exponent sum of the variable in the given equation.*

Thus positive equations over arbitrary groups can be solved in such a way that the solutions lie in very simple supergroups. Besides the fact that Levin's theorem yields a proof of the Freiheitssatz in the special case where one of the generators occurs only with positive exponents, it can be used also to prove that certain one-relator groups are residually solvable. I shall have more to say about this subject later on in my talks. Levin's argument then, unlike Brodskii's, is completely independent of Magnus' ideas.

I want to discuss two other approaches to the Freiheitssatz which are radically different from that adopted by Magnus in 1930.

The first of these is due to Gerstenhaber and Rothaus [40] and can be reformulated so as to fit our needs as follows.

Theorem B.6. *Let* F *be a finite group and let* $r = 1$ *be an equation over* F *in the single variable* x. *If* x *occurs with non-zero exponent sum in* r, *then the equation* $r = 1$ *can be solved in a finite supergroup* G *of* F.

Let me give a brief sketch of the proof. First we embed F in an appropriately chosen unitary group $U_n(\mathbb{C})$ over the field $\mathbb{C}$ of complex numbers. Recall that $U_n(\mathbb{C})$ consists of all non-singular $n \times n$ matrices M over $\mathbb{C}$ satisfying the additional condition $M^{-1} = M'$. The relator r then can be viewed as a mapping ρ of $U_n(\mathbb{C})$ into itself. Now $U_n(\mathbb{C})$ is a connected topological space and so for each group element occurring in r, there is a homotopy taking that element to 1. So if k is the exponent sum of x in r, then ρ is homotopic to the k-th power map κ of $U_n(\mathbb{C})$ into itself. Now $U_n(\mathbb{C})$ is a finite dimensional compact manifold of dimension d $(= 1 + 3 + \dots + 2n-1)$. So $H_d(M, \mathbb{Z}) = \mathbb{Z}$ and by a theorem of Hopf [49], κ has degree k^n. Since ρ is homotopic to κ it too has degree k^n. Now the image of any continuous map of

$U_n(\mathbb{C})$ into itself is compact and so closed. Any proper closed subspace of a manifold of dimension d has zero homology in that dimension. It follows that ρ is onto; otherwise on the d-th homology group of $U_n(\mathbb{C})$ ρ factors through the zero map contradicting the fact that it is of non-zero degree. It follows that 1 is in the image of ρ. By a theorem of Malcev [78] every finitely generated group of matrices over a commutative field is residually finite. The theorem of Gerstenhaber and Rothaus follows from this remark.

B.3 Affine sets of characters

The work of Gerstenhaber and Rothaus and indeed the history of the subject itself suggests that linear methods might well be of value in the study of one-relator groups. In 1968 Ree and Mendelsohn [93] used linear methods to prove the following theorem.

Theorem B.7. *Let*

$$G = \langle a,b \ ; \ w^m = 1 \rangle$$

be a one-relator group, where w *involves both* a *and* b, *is cyclically reduced of length at least two and* $m > 1$. *Then if* n *is sufficiently large,* a *and* b^n *freely generate a free group of rank two.*

The proof depends on the fact that in $SL_2(\mathbb{C})$ a matrix M has order $m > 2$ if and only if

$$\text{trace}(M) = \mu + \mu^{-1} \ (= \tau),$$

where μ is a primitive mth root of unity. Thus we suppose $m > 2$ (the case $m = 2$ can be handled by taking $m = 4$ and then going over to $PSL_2(\mathbb{C})$). Now let

$$A = \begin{pmatrix} 1 & x \\ 0 & 1 \end{pmatrix} \quad \text{and} \quad B = \begin{pmatrix} 1 & 0 \\ x & 1 \end{pmatrix}$$

with x still to be determined. On computing s(A,B) we find that

$$\text{trace}(s(A,B)) = f(x)$$

is a polynomial in x of positive degree with constant term 2. Since $\tau \neq 2$ there is a non-zero solution $\sigma \in C$ of the equation $f(x) = \tau$. So if we put $x = \sigma$ in both A and B then it follows that $s(A,B)^m = 1$.

The next thing to notice is that if

$$C = \begin{pmatrix} 1 & \alpha \\ 0 & 1 \end{pmatrix} \quad \text{and} \quad D = \begin{pmatrix} 1 & 0 \\ \beta & 1 \end{pmatrix}$$

then C and D generate a free group of rank two whenever $|\alpha\beta| > 4$. So if n is chosen sufficiently large A and B^n freely generate a free group. The theorem of Ree and Mendelsohn follows immediately from these remarks.

This theorem was later taken up by a number of people, for example by Newman [87] (see, in particular his Theorem B.3), who used Magnus' original method and then by Magnus himself in [75], who used the Ree-Mendelsohn approach. More recently, Baumslag, Morgan and Shalen [15] have used a similar approach to prove that certain so-called generalised triangle groups are non-trivial.

The scope of this representation-theoretic technique can be more easily gauged by examining all the representations of a given one-relator group in some special linear group. This approach has been used with great effect by Culler and Shalen [29] in their work on three manifolds. The ideas themselves are pretty old, and go back to Poincare, to Fricke and Klein and to Vogt (see e.g. Magnus [74]).

I want to remind you of the details. Let K be a fixed algebraically closed field,

$$A = K[x_1,\ldots,x_q]$$

the polynomial algebra in q variables over K and

$$K^q = \{(a_1,\ldots,a_q) \mid a_i \in K\}.$$

K^q is sometimes referred to as affine q-space over K. We term

$$P = (a_1,\ldots,a_q)$$

a point in K^q; such a point P is termed a zero of $f = f(x_1,\ldots,x_q) \in A$ if $f(P) = f(a_1,\ldots,a_q) = 0$. If S is any subset of A, then the zero set Z(S) of S consists of all points in K^q which are zeroes of all the polynomials in S. A subset X of K^q is called an affine algebraic set if X = Z(S) for some subset S of A. If one takes these affine algebraic sets as the closed sets of K^q this defines a topology on K^q, the so-called Zariski topology. Next, if Y is any subset of K^q one can form I(Y), the set of all polynomials in A which vanish on Y. The set A[X] of all polynomial functions from an affine algebraic set X to K is a K-algebra, called the coordinate algebra of X. It is not hard to identify A[X]:

$$A[X] \cong A/I[X].$$

A[X] is therefore a finitely generated K-algebra which turns out to have radical zero, i.e. A[X] has no nilpotent elements, and so A[X] is what is now commonly termed an affine K-algebra. It turns out that there is a categorical equivalence between affine algebraic sets and such affine K-algebras. The dimension of X as a topological space can then be identified with that of A[X], i.e. with the length of the longest chain of prime ideals in A[X]. If A[X] is an integral domain, then X is termed an affine algebraic variety and the dimension of X (or of A[X]) is simply the transcendence degree of the quotient field of A[X] over K. If now $SL_n(K)$ acts on an affine algebraic set in an affine algebraic way, then $SL_n(K)$ also acts as a group of K-automorphisms of the K-algebra A[X]. It turns out that the set of fixed points, or so-called invariants, under this action is again an affine K-algebra (Haboush [45]). The affine algebraic set corresponding to this affine K-algebra under the categorical equivalence mentioned above is then referred to as the categorical quotient of X under this action of $SL_n(K)$.

Now suppose that G is a group finitely generated by the n elements a,...,c and suppose that $d > 1$ is a given integer. The set $R_d(G)$ of all representations ρ of G in $SL_d(K)$ carries with it the structure of an affine algebraic set in affine q-space over K, where $q = nd^2$. Indeed let us associate to ρ the point

$$(a_{11},\ldots,a_{1d},\ldots,a_{d1},\ldots,a_{dd},\ldots,c_{11},\ldots,c_{1d},\ldots,c_{d1},\ldots,c_{dd})$$

in K^q, where

$$\rho(a) = (a_{ij}), \ \ldots \ , \rho(c) = (c_{ij}).$$

Thus $R_d(G)$ has been parametrised by a set of points in affine q-space over K, indeed as it turns out, by an affine algebraic set. The polynomials that define this affine algebraic set arise from the possibly infinitely many defining relations of G. Henceforth we identify $R_d(G)$ with this affine algebraic set.

The next thing to notice is that $SL_d(K)$ acts on $R_d(G)$ by conjugation in the natural way. So one can form the categorical quotient $X_d(G)$ of $R_d(G)$ under this action. $X_d(G)$ is termed the affine algebraic set of characters of G. It can be thought of as a parametrisation of the inequivalent semi-simple representations of G in $SL_d(K)$. If G can be defined by h relators in the n given generators of G, then it is not hard to estimate the dimension of $X_d(G)$, provided only that there is at least one irreducible representation of G in $SL_d(K)$:

$$\dim X_d(G) \geqslant (n-1)(d^2-1) - hd^2.$$

In our discussion of the theorem of Ree and Mendelsohn $K = \mathbb{C}$, $n = 2$, $h = 1$ and the relator is an m-th power, with $m > 2$. It is possible in this case to improve the estimate on the dimension of the affine algebraic set of characters of G. In fact one can show that

$$\dim X_2(G) \geqslant 3 - 1 > 0$$

which means that G has a plethora of inequivalent representations in $SL_2(\mathbb{C})$, which helps to explain why the argument of Ree and Mendelsohn works so well. It also suggests that a more careful investigation of $X_2(G)$ will turn out to be fruitful. The interested reader should consult the book by Hartshorne [46] for more details concerning the algebraic geometry involved in the discussion above, the recent account by Lubotsky and Magid [64] and the paper by Culler and Shalen [29] cited earlier, where these techniques are used to produce graph-product decompositions of three-manifold groups.

I want now to turn to the second topic in this survey, the subgroup structure of one-relator groups. Before doing so I would like to mention the geometric methods of Schupp [73], which not only yield another proof of the Freiheitssatz but also suggest proofs of possible generalisations. Indeed the work of Howie [51] is witness to the fruitfulness of Schupp's ideas.

C. *SUBGROUPS OF ONE-RELATOR GROUPS*

C.1 *The theorems of Magnus and Brodskii*

In 1932 Magnus deduced the following theorem from his Freiheitssatz [70].

Theorem C.1. *Every group with a single defining relation has a solvable word problem.*

Somewhat surprisingly the conjugacy problem for one-relator groups is still unsolved. However a good deal of progress has been made, part of this through the aid of small cancellation theory, a theory that I alluded to earlier. It seems appropriate to describe one part of this theory here, without going into any of the applications to the study of one-relator groups. To this end suppose that the group G has a finite presentation

$$G = \langle X \; ; \; R \rangle,$$

where each $r \in R$ is a reduced word. The given presentation is termed symmetrised if it is closed under inversion and cyclic permutations. It is called a one-sixth presentation if it is symmetrised and satisfies the following condition:

> *if* $r, s \in R$ *and if either more than one-sixth the length of* s *cancels on computing the reduced word representing* rs, *then* $r = s^{-1}$.

The following theorem of Greendlinger [41] (see also Greendlinger [42]) then holds.

Theorem C.3. *Suppose that the group* G *has the one-sixth presentation above. Furthermore, suppose that* w *is a reduced word in the generators* X. *If* $w = 1$ *in* G, *then there exists a relator* $r = a_1 \ldots a_p \in R$ *such that*

$$w = b_1 \ldots b_m a_1 \ldots a_n c_1 \ldots c_s$$

where $n > p/2$, i.e. w *contains more than half a (subword of a) relator (here the* $a_i, b_j, c_k \in X \cup X^{-1}$).

Theorem C.2 provides an algorithm for solving the word problem in G. For if w is a reduced word in the generators X, by inspecting the finitely many elements of R, we can determine whether more than the first half of one of them is a subword of w. If this is not the case, $w \neq 1$ in G. If this is the case, $w = tuv$ where u is more than the first half of $r = us \in R$. So $w = ts^{-1}v$ in G and the reduced form w_1 of $ts^{-1}v$ is of smaller length than that of w. We can therefore repeat the above process with w_1 in place of w. This then is an algorithm which solves the word problem for G. A similar algorithm applies also to the conjugacy problem.

Algorithms like these were first used by Dehn [32] in his solution of the word and conjugacy problems for the fundamental groups of surfaces. Variations of the one-sixth condition and also of these algorithms have been used in studying the conjugacy problem in a variety of one-relator groups. I do not have time to go into this here but would like to refer the reader to the papers by Juhasz [52], [53], by Pride [92], by Hill, Pride and Vella [48] as well as others. The interested reader should consult the book by Lyndon and Schupp [66] for a detailed discussion of small cancellation theory.

Now a finitely generated subgroup of a free group is again free and so is finitely presented. Whether this is true more generally for one-relator groups is still an open problem.

Problem 1. *Is every finitely generated subgroup of a one-relator group finitely presented?*

As I pointed out before, the finitely generated subgroups of the fundamental groups of two dimensional orientable surfaces are finitely presented. Before I discuss this facet of the subgroup problem further I would like to point out that the countable subgroup structure of one-relator groups is, not surprisingly, extremely complicated. Indeed most one-relator groups contain continuously many non-isomorphic subgroups. This is true, e.g. of $G = \langle a,b;a^{-1}b^2a = b^4 \rangle$ - this observation is due to Miller and myself [14]. In fact it seems likely that a finitely generated one-relator group G has only a countably infinite number of non-isomorphic subgroups if and only if it is either solvable or if every subgroup is either of finite index or free or if it has a non-trivial center.

On the other hand the controlled nature of the subgroup structure of one-relator groups is underlined by the following theorem of Brodskii [21].

Theorem C.4. *One-relator groups are locally indicable.*

This important theorem has subsequently been reproved by rather different means by Howie [50].

Despite Brodskii's theorem, not enough is known of the general subgroup structure of one-relator groups. For example the following problem is still open.

Problem 2. *Can a one-relator group contain a non-abelian simple subgroup?*

On the other hand a great deal is known about the presence of free subgroups in one-relator groups. Our reflections on the Freheitssatz may be viewed as a contribution to this topic to which I want now to turn more generally.

C.2 Free subgroups

It follows from the Freiheitssatz that non-abelian free subgroups abound in one-relator groups. Indeed even more is true by virtue of the following theorem of Karrass and Solitar [57], 1971.

Theorem C.5. *A subgroup of a one-relator group is either solvable or contains a free subgroup of rank two.*

Thus one-relator groups satisfy what has come to be known as the Tits alternative because of the following theorem of Tits [105], 1972.

Theorem C.6. *Let* G *be a finitely generated group of matrices over a commutative field. Then* G *is either virtually solvable,* i.e. *contains a solvable subgroup of finite index, or else contains a free subgroup of rank two.*

Notice that G here need not be finitely presented. It is by no means the case that finitely presented and, less restrictively, finitely generated groups satisfy the Tits alternative. Finitely generated counter-examples are plentiful and although less common, finitely presented ones exist as well.

There is however a version of the Tits alternative for finitely presented groups, due to Bieri and Strebel [20], 1978.

Theorem C.7. *Let* G *be a finitely presented group and suppose that* G_{ab}, *the abelianisation of* G, *is infinite. Then either* G *contains a free subgroup of rank two or else* G *is a descending HNN extension with a finitely generated base* B, i.e.,

$$G = \langle B, t \ ; \ tBt^{-1} < B \rangle.$$

Notice that even though G is finitely presented B need not be.

Now it can be proved that a one-relator group with at least three generators is not a descending HNN extension with a finitely generated base. So the Bieri-Strebel theorem yields a host of free subgroups once again.

So too does the following theorem of B. Baumslag and Pride [7], 1978.

Theorem C.8. *Suppose* G *is a finitely presented group with at least two more generators than relations. Then* G *contains a normal subgroup with finite cyclic factor group which maps onto the free group of rank two.*

The last two theorems indicate how one-relator groups fit into the more general context of finitely presented groups. They take on a somewhat sharper form, however, for one-relator groups as further work of B. Baumslag and Pride [8], Stohr [103] and Edjvet [36] shows.

There is a more direct link between free subgroups and one-relator groups that comes out of Tits' theorem and a computation of the dimension of the affine algebraic set $X_2(G)$ of characters of G. In fact the character set approach yields the existence of a large number of representations in $SL_2(\mathbb{C})$ which are not virtually solvable, of every one-relator group with at least four generators. Tits' alternative can then be invoked.

C.3 The general subgroup theorems

According to Moldavanskii's Theorem B.2 every finitely generated one-relator group G, where at least two generators appear in the cyclically reduced defining relator r of G, can be embedded in another one-relator group H with the following properties: H is an HNN extension with one stable letter, with base a one-relator group whose single defining relator is of smaller length than that of r and whose associated subgroups are finitely generated free groups.

From a topological standpoint HNN extensions and free products of two groups with one amalgamation are much the same. So it comes as no surprise to find that many one-relator groups can be decomposed into generalized free products. This is a consequence of the following theorem of Shalen and myself [16].

Theorem C.9. *Let* G *be a finitely presented group of deficiency* d. *If*

$$3d - 3 > \dim(H_1(G, \mathbb{Z}/2\mathbb{Z})),$$

then G *is a free product with amalgamation*

$$G = \{A * B \,;\, C\},$$

where C *is a proper subgroup of* A *and is of infinite index in* B.

As usual the deficiency d of a finitely presented group G is the maximum difference between the number of generators and the number of defining relations taken over all finite presentations of G. And $H_1(G,\mathbb{Z}/2\mathbb{Z})$ is the first homology group of G with coefficients in the field of two elements, with $\dim(H_1(G,\mathbb{Z}/2\mathbb{Z}))$ its dimension as a vector space over $\mathbb{Z}/2\mathbb{Z}$ So if G is a one-relator group with at least 4 generators the hypothesis of the theorem is in effect. Thus we have the following consequence of Theorem C.9.

Corollary C.9. *Every one-relator group with at least* 4 *generators can be decomposed into a generalised free product of two groups where the amalgamated subgroup is proper in one factor and of infinite index in the other.*

In fact if one asks only that the amalgamated subgroup be proper in both factors it is not hard to show that every one-relator group with at least 3 generators can be decomposed into a generalised free product. This then together with the corollary above partially answers a question raised by Rosenberger in [94](Problem F2, page 386).

In 1970 Karrass and Solitar determined the subgroups of the generalised free product of two groups with one amalgamated subgroup. They subsequently exploited this theorem to obtain a precise description of the subgroups of an HNN extension and then applied their subgroup theorem to one-relator groups. Then, in 1974, Karrass, Pietrowski and Solitar simplified and improved this theorem on the subgroups of HNN extensions [55]. Later the same year Cohen [24] pointed out that Serre's theory of groups acting on trees yielded some similar results about the subgroups of HNN extensions as well as the subgroups of the free product of two groups with one amalgamated subgroup.

Serre's theory of groups acting on trees was developed in a course given at the College de France in 1968/69. His work overlaps and, to some extent, anticipates much of the work that I have been discussing. Although he did not work out all the details of his theory of groups acting on trees, he suggested them in the course. Subsequently Bass created the appropriate framework and completed this work that Serre had begun (cf. [98]).

Thus at this time there are really two approaches to the subgroup structure of both HNN extensions and generalised free products of two groups the generator and relation approach of Karrass and Solitar and the global, less detailed but perhaps more revealing approach of Serre and Bass embodied in groups acting on trees. I will, before discussing the applications of these structure theorems to groups with one defining relation, describe some of the details of the Bass-Serre theory. Much of the Karrass-Solitar theory can be deduced on making use of their carefully selected representative systems.

Suppose then that X is an oriented graph with the inverse of an oriented edge e denoted e^{-1}. A group G is said to act without inversion on a graph X if it comes equipped with a homomorphism into Aut X, the group of graph automorphisms of X, such that $ge \neq e^{-1}$ for every $g \in G$ and every edge e of X. If a group G acts on a graph X without inversion, then one can form the quotient graph $G\backslash X$. One also has the usual notions of connectedness etc. Then an oriented graph X is termed a tree if it is connected and has no non-trivial circuits.

Every group G that acts without inversion on a graph X can be built up from the stabilisers of the vertices and edges of X under the action of G by repeatedly forming generalised free products of two groups and HNN extensions. The manner in which this is done can be codified in terms of a so-called graph of groups $(\mathfrak{G},Y)$. The definition is as follows. Y is an oriented graph and $\mathfrak{G}$ is a map from the set of vertices and edges of Y into a set of groups. Thus for each edge e with origin P and terminus Q we have the groups $\mathfrak{G}(e)$, $\mathfrak{G}(P)$ and $\mathfrak{G}(Q)$. In addition we have two monomorphisms

$$\alpha(e) : \mathfrak{G}(e) \to \mathfrak{G}(P), \quad \beta(e) : \mathfrak{G}(e) \to \mathfrak{G}(Q).$$

Let T be a maximal tree in Y. Then the fundamental group

$$\pi(\mathfrak{G},Y,T) \text{ of } (\mathfrak{G},Y) \text{ at } T$$

which depends, up to isomorphism, only on the graph of groups $(\mathfrak{G},Y)$, is defined in two stages as follows. Since Y is oriented we can partition the edges E of Y as $E_+ \cup E_-$, where $E_+^{-1} = E_-$. We then form the generalised free product G_T of all of the groups $\mathfrak{G}(P)$, $\mathfrak{G}(Q)$ amalgamating $\mathfrak{G}(e)\alpha(e)$ with $\mathfrak{G}(e)\beta(e)$ according to the isomorphism $\alpha(e)^{-1}\beta(e)$, where here e ranges over all the edges of E_+ in T. Now choose a stable letter t_e for each edge in E^+ not in T and form the HNN extension

$$\pi(\mathfrak{G},Y,T) = \langle G_T, t_e \ (e \notin T, e \in E_+) \mid t_e \mathfrak{G}(e)\alpha(e) t_e^{-1} = \mathfrak{G}(e)\beta(e) \rangle.$$

The thing to observe is that if $G = \pi(\mathfrak{G},Y,T)$, then G acts on a tree X. And, conversely if G acts on a tree X, then $G = \pi(\mathfrak{G},Y,T)$, for a suitable choice of the graph of groups $(\mathfrak{G},Y)$. In fact Y is simply the quotient graph $G\backslash X$. The vertex and edge groups of Y are the stabilisers in X of a suitably selected set of vertices and edges of X.

There are two special cases which are of some interest here. The case where

$$Y : \quad \underset{\mathfrak{G}(P) = A}{\bullet} \xrightarrow[e]{\mathfrak{G}(e) = C} \underset{\mathfrak{G}(Q) = B}{\bullet}$$

is a segment, and the case where

is a loop. In the first case $T = Y$ and $\pi(\mathfrak{G},Y,T) = \{A * B ; C\}$, while in the second $T = \{P\}$ and $\pi(\mathfrak{G},Y,T) = \langle B,t;\ tCt^{-1} = C\beta \rangle$ is an HNN extension, where C is a subgroup of B and β is an injection of C into B. So both the generalised free product of two groups with a subgroup amalgamated and an HNN extension with a single stable letter act without inversion on trees. Hence every subgroup of such groups is isomorphic to the fundamental group of a graph of groups. This remark should be compared with the one made about the finitely generated subgroups of fuchsian groups.

There are two consequences of these subgroup theorems for one-relator groups which can be extracted from the discussion above. They are both due to Karrass and Solitar ([56], [57]) and have been joined together in the following theorem.

Theorem C.10. *Let* G *be either a free product of two free groups with a cyclic subgroup amalgamated or an HNN extension with a free base and cyclic associated subgroups. Then the finitely generated subgroups of* G *are finitely presented.*

Here is an explicit description of the subgroups of an HNN extension contained in the paper by Karrass, Pietrowski and Solitar [55]. It is an improved version of the subgroup theorem obtained previously by Karrass and Solitar [57]. A similar theorem holds also for the subgroups of a generalised free product of two groups (Karrass and Solitar [56]; see also Cohen [24]).

Theorem C.11. *Let* G *be an HNN extension with base* B*, stable letter* t *and associated subgroups* L *and* R. *Then every subgroup* H *of* G *is also an HNN extension with possibly infinitely many stable letters and base* C. *Indeed there exists a set* S *of representatives of the right cosets* Hg *of* H *in* G *such that* ($\overline{g}$ *denoting throughout the representative of* Hg *in* S):

(i) *the base group* C *of* H *is generated by a select set of intersections*

$$H_d = dBd^{-1} \cap H,$$

where d *ranges over a subset* D *of* S;

(ii) *if* $d,\delta \in D$ *and if* $\delta = \overline{det}$, *where* e *is a word in the generators of* B, *then* H_d *and* H_δ *generate their generalised free product amalgamating*

$$deR(de)^{-1} \cap H \text{ with } \delta L\delta^{-1} \cap H;$$

(iii) *the stable letters in this HNN extension consist of the elements*

$$det\,\overline{det}^{-1}\ (\overline{det} \neq det)$$

where de *ranges over a subset of* S; *the associated subgroups corresponding to such a* $det\,\overline{det}^{-1}$ *are*

$de\,R(de)^{-1} \cap H$ *and* $\overline{de t}\, L\, \overline{de t}^{\,-1} \cap H$.

Notice that we allow here the degenerate case where there are no stable letters. Then $H = C$ and has the structure described in (i) and (ii).

D. *CLASSIFICATION AND PROPERTIES OF VARIOUS ONE-RELATOR GROUPS*

D.1 Free groups

Free groups play a dual role in the study of groups with a single defining relation. On the one hand much of the theory of one-relator groups depends on a deeper knowledge of free groups. And on the other hand many theorems about free groups either have counterparts for one-relator groups, or suggest fruitful avenues for further investigation. Much of the following discussion should be viewed in this light.

In 1935 Magnus [72] proved that free groups are residually nilpotent by finding faithful representations in appropriate power series rings.

Theorem D.1. *Let* F *be the free group on* $a_1,\ldots,a_n$ *and let* R *be the ring of power series in the non-commuting variables* $x_1,\ldots,x_n$. *Then the mapping*

$$a_i \longmapsto 1 + x_i \quad (i = 1,\ldots,n)$$

defines a faithful representation of F *in the group of units of* R.

One of the consequences of this theorem, also due to Magnus [71], is the following theorem.

Theorem D.2. *Suppose that the group* G *is given by the finite presentation*

$$G = \langle x_1,\ldots,x_n;\ r_1,\ldots,r_m \rangle.$$

If G *can be generated by* $n-m$ *elements, then* G *is free.*

The relevance of this theorem to one-relator groups then is that if a group G is presented on n generators and 1 defining relation, then either G cannot be generated by $n-1$ elements or else is free.

There is another theorem, due to Stallings [101], which is closely connected to the residual nilpotence of free groups.

Theorem D.3. *Let* G *be a group and suppose that* $H_1(G,\mathbb{Z})$ *is free abelian and that* $H_2(G,\mathbb{Z}) = 0$. *Then any set of elements of* G *whose homology classes are independent in* $H_1(G,\mathbb{Z})$ *freely generate a free group.*

Now let $G = \langle x_1,\ldots,x_n;\ r = 1 \rangle$ be a one-relator group and suppose that r, viewed as an element in the free group F on $x_1,\ldots,x_n$, is not a proper power modulo the derived group of F. Then Stallings' theorem applies to G. So we have here another variant of the Freiheitssatz. In fact it turns out that such a one-relator group G has the same lower central factor groups as a free group (of rank $n-1$). A residually nilpotent group with this property is termed a parafree group. I will say a little more about such groups later. The point here is that this is an illustration of one way how homology intertwines with group theory. In fact Stallings [100] has obtained a most extraordinary cohomological characterisation of finitely generated free groups.

Theorem D.4. *A finitely generated group* G *is free if and only if it is of cohomological dimension at most* 1, i.e. $H_2(G,A) = 0$ *for all coefficient groups* A.

The analogous characterisation of all free groups is due to Swan [104]. On combining the Stallings-Swan theorem with one of Serre [99] it can be shown that:

Theorem D.5. *A torsion-free group with a free subgroup of finite index is free.*

Theorem D.5 applies also to one-relator groups and can therefore be viewed as a contribution to the isomorphism problem for one-relator groups.

More recently in their work on Poincare duality groups of dimension two, Eckmann and Muller [35] have obtained a counterpart to Theorem D.5.

Theorem D.6. *Let* G *be a torsion-free group which contains a surface group as a subgroup of finite index. Then* G *is itself a surface group.*

Here a group is termed a surface group if it is the fundamental group of a closed two dimensional manifold (see e.g., Massey [79]).

Whether analogues of these finite index theorems exist for other classes of one-relator groups is as yet unknown.

I want to turn next to automorphism groups of free groups. Gersten has made significant progress in the study of these groups in his recent work [38], [39] on the fixed points of automorphisms and Whitehead's algorithm. The first of his theorems may be viewed as a substantial generalisation of important earlier work of McCool [80], [81], [82].

Theorem D.7. *Let* A *be the automorphism group of a finitely generated free group* F, *let* S *be the set of conjugacy classes of finitely generated subgroups of* F *with natural* A*-action and let* S^n *be the cartesian product of* n *copies of* S *with diagonal* A*-action. Then*

(i) *there exists an effective procedure for determining when two elements of* S^n *are in the same* A*-orbit;*

(ii) *the stabiliser in* A *of an element in* S^n *is finitely presented and a finite presentation of this stabiliser can be effectively determined.*

It follows then in particular that the automorphism group of a finitely generated free group is finitely presented, a result which was proved first by Nielsen [89]. This suggests the following.

Problem 3. *When is the automorphism group of a finitely generated one-relator group finitely presented?*

This seems to be a hard problem (see Collins and Levin [26]).

I will talk, in some detail, about another of Gersten's theorems in a few minutes. However I would like first to record a theorem of a rather different kind, that was proved by Culler and Vogtmann [30].

Theorem D.8. *Let F be a finitely generated free group. Then the automorphism group A of F is virtually of finite cohomological dimension,* i.e. A *contains a subgroup of finite index which is of finite cohomological dimension.*

This theorem should be compared with the corresponding result by Alperin and Shalen [4] for linear groups over integral domains of characteristic zero.

Now a subgroup of a group of cohomological dimension n is of cohomological dimension at most n. This puts considerable restrictions on the subgroup structure of such groups. It follows, for example, that every finitely generated abelian subgroup of the automorphism group of a finitely generated free group has a boundedly finite number of generators. I do not know how to prove this statement directly.

I want to turn now to the other theorem of Gersten that I mentioned a couple of minutes ago, proved in [38].

Theorem D.9. *Let F be a finitely generated free group, α an automorphism of* F. *Then the fixed subgroup* F_α *of α, i.e. the set of those elements of* F *fixed by α, is finitely generated.*

Gersten's proof is geometric, using graph-theoretic ideas. I would like to give here a sketch of a rather different proof of Gersten's theorem which is due to Cooper [27] and which might carry over also to some one-relator groups.

Thus suppose that F is free on $X = \{x_1, \ldots, x_n\}$, and that α is an automorphism of F. Every element $f \in F$ can be expressed as a reduced X-product of these generators. Let Y be the set of all reduced, possibly infinite, X-products:

$$c_1c_2\ldots \quad (c_i \in X \cup X^{-1},\ c_ic_{i+1} \neq 1 \ (i = 1,2,\ldots)).$$

Then Y can be turned into a complete metric space by defining

$$d(y,z) = 1/(1+\delta(y,z)) \quad (y,z \in Y,\ y \neq z),$$

where $\delta(y,z)$ is the length of the biggest common initial segment of y and z. In fact Y is a compact metric space and α extends to a homeomorphism of Y to itself, again denoted α. Thus, in particular, α is a continuous map of Y onto Y and so is also uniformly continuous. The uniform continuity of α allows one to prove that if y and z are elements of F and if there is no cancellation on forming the reduced X-product of y and z, then there is a bounded amount of cancellation in the computation of the reduced X-product of $y\alpha$ and $z\alpha$, independent of y and z.

Here is a very brief sketch of the rest of Cooper's proof. Suppose, if possible, that F_α is not finitely generated. Then it is free on say

$$a_1,\ a_2,\ \ldots\ .$$

We now choose a new set $b_1, b_2, \ldots$ of free generators for F_α as follows. First we set $b_1 = a_1$ and inductively choose b_i to be an element of minimal length in the coset $gp(a_1, \ldots, a_{i-1})a_i$. Since Y is compact there is a convergent subsequence $b_{m(1)}, b_{m(2)}, \ldots$ converging to some element b, say, of Y. Since the

$b_k \in F_\alpha$ it follows from the continuity of α that b is fixed by α. Now we may assume also that

$$\delta(b_{m(i)},b) < \delta(b_{m(j)},b) \text{ if } i < j,$$

and define c_j to be the biggest common initial segment of $b_{m(j)}$ and b. Notice that c is an initial segment of c_j and that $\ell(c_i) < \ell(c_j)$ whenever $i < j$, where here $\ell(f)$ is the length of $f \in F$ in terms of the given basis X of F. Now very little cancellation takes place in the computation of the reduced X-product of $y\alpha$ and $z\alpha$ (see the remark above concerning the uniform continuity of α). So the fact that the $b_{m(i)}$ and b are left fixed by α, implies that

$$c_i\alpha = c_i d_i$$

where the d_i are of bounded length, independent of i. Since there are only a finite number of elements of length less than or equal to any given integer it follows that there is a pair of positive integers h and k, $h < k$, such that $d_h = d_k$. Hence $c_h c_k^{-1} = w$ is left fixed by α, i.e. $w \in F_\alpha$. We now express w as a word in the original generators $a_1, a_2, \ldots$ of F_α. Choose $m(q)$ to be bigger than all the subscripts involved in this expression for w and also bigger than $m(k)$. Clearly $wb_{m(q)} \in F_\alpha$ and

$$w \in gp(a_1, \ldots, a_{m(q-1)}).$$

Thus by the very definition of $b_{m(q)}$,

$$\ell(wb_{m(q)}) \geq \ell(b_{m(q)}).$$

On the other hand, noting that c_k is an initial segment of $b_{m(q)}$ and that $\ell(c_h) < \ell(c_k)$, we find that

$$\ell(wb_{m(q)}) \leq \ell(b_{m(q)}) - \ell(c_k) + \ell(c_h) < \ell(b_{m(q)}),$$

a contradiction.

Other proofs of Gersten's theorem have been obtained by Culler [28] and by Krstic [58].

D.2 Cohomology of one-relator groups

The cohomological characterisation of finitely generated free groups by Stallings leads one to suspect that cohomology might well play an important role in the study of one-relator groups.

The following theorem of Lyndon [65] shows that the calculation of these cohomology groups can be restricted to dimension one.

Theorem D.10. *Let* $G = \langle X ; r \rangle$ *be a one-relator group and suppose that* $r = s^n$ *where* s *is not a proper power in the free group on* X. *Let* $S = gp(s)$. *Then for every coefficient module* A

$$H^i(G,A) = H^i(S,A) \text{ and } H_i(G,A) = H_i(S,A) \quad (i > 1).$$

Thus Lyndon's theorem reduces the cohomology and homology of one-relator groups in dimensions greater than one to that of finite cyclic groups, which is well known (cf. e.g., MacLane [67]). In particular then torsion-free one-relator groups are of cohomological dimension at most two and hence so too are their subgroups. As I have already remarked in my discussion about the automorphism groups of free groups, this cohomological condition imposes a severe restriction on the subgroups of one-relator groups. Thus it follows from Lyndon's theorem that the one-relator group G is torsion-free if and only if r (above) is not a proper power, that the solvable subgroups are all of derived length at most two and that the abelian subgroups are of rank at most two (see, e.g. Gruenberg [44]).

The abelian subgroups of one-relator groups are now completely understood because of the following theorem of Newman [88], 1968.

Theorem D.11. *The abelian subgroups of a one-relator group are of three types:*

(i) *cyclic;* (ii) *free abelian of rank two;* (iii) m-*adic for some positive integer* m, i.e., *isomorphic to the subgroup of the additive group of rational numbers comprised of rationals of the form* j/m^k, m *fixed,* j *and* k *arbitrary integers. Moreover in the event that there is torsion, the abelian subgroups are all cyclic.*

As I have already pointed out, Lyndon's Theorem D.10 reveals that a one-relator group has non-trivial torsion if and only if its defining relator, assumed non-trivial, is a proper power (in the underlying free group). In fact the torsion in one-relator groups has been worked out by Karrass, Magnus and Solitar [54].

Theorem D.12. *Let* $G = \langle X ; r \rangle$ *be a one-relator group. Suppose that* r *is a non-trivial word in the free group* F *on* X *and that*

$$r = s^n \quad (n \geq 1)$$

where s *is not a proper power in* F. *Then* G *has non-trivial torsion if and only if* $n > 1$ *and in this case* s *is of order* n *and every element of finite order in* G *is conjugate to an appropriate power of* s.

There is another application of cohomology to the subgroup structure of one-relator groups which comes out of the following theorem of R. Bieri [19].

Theorem D.13. *A finitely presented normal subgroup of a finitely generated group of cohomological dimension at most two is either free or of finite index.*

It is easy then to deduce the:

Corollary D.13. *A finitely presented normal subgroup of a one-relator group is either free or of finite index.*

D.3 Classification

Suppose that F is free on a finite set X and that the one-relator group G is presented as a one-relator group on X in two ways:

$$F/gp_F(r) \cong G \cong F/gp(s).$$

At one time Magnus suggested that there might be an automorphism of F which carries r either to s or to s^{-1}. This suggestion was prompted by the following theorem which he proved in 1931 [69].

Theorem D.14. *Let* F *be a free group and suppose that the elements* r *and* s *in* F *have the same normal closures. Then* r *is conjugate to either* s *or* s^{-1}.

In fact there are many presentations of one-relator groups which do not fit in with Magnus's suggestion. The first counter-examples were discovered by Zieschang [108] and McCool and Pietrowski [83]. However by examining the effect of automorphisms on one-relator presentations Pride [91] was able to prove the following important theorem about the isomorphism problem.

Theorem D.15. *The isomorphism problem for two-generator one-relator groups with torsion is solvable.*

No such theorem exists for two-generator one-relator groups in general.

Now the class of one-relator groups with torsion is a recursive subclass of the class of all one-relator groups; this follows immediately from Theorem D.12. The same remarks holds also for one-relator groups with non-trivial centre (G. Baumslag and Taylor [18]) and for such groups the isomorphism problem is also solvable. This was proved by Pietrowski [90] - surprisingly the proof is quite involved.

Theorem D.16. *The isomorphism problem for one-relator groups with non-trivial centre is solvable.*

There is another class of groups, the parafree groups, which are at the other extreme to one-relator groups with center. Recall that a group G is parafree if G is residually nilpotent and if $G/\gamma_n G \cong F/\gamma_n F$ for every n, where F is free and $\gamma_n G$ denotes the nth term of the lower central series of G. Non-free parafree groups exist in abundance, but are very hard to distinguish one from the other (cf. G. Baumslag [10], [11]). In this connection I want to mention the following theorem of Meskin [85], which is based in part on work of Rosenberger [95].

Theorem D.17. *Let*

$$G = \langle t,a,\ldots,b;\ t^k a^m \ldots b^n = 1\rangle$$

where all of the exponents involved are greater than one. Then any other one-relator group with a relator of the same form is isomorphic to G *if and only if the sequence of its exponents is a permutation of* k,m,...,n. *The relevance of this to parafree groups is that each of the above groups is parafree whenever the greatest common divisor of the exponents involved is one* (G. Baumslag [10]).

In 1957 Lazard [60] defined for each group G an associated lie ring ΓG as follows. We recall the definition. For each $n > 0$ we denote $\gamma_n G/\gamma_{n+1} G$ by $\Gamma_n G$. Then additively ΓG is the direct sum of all the $\Gamma_n G$:

$$\Gamma(G) = \bigoplus_{n=1}^{\infty} \Gamma_n G.$$

If $\alpha = a\gamma_{m+1}G \in \Gamma_m G$ and $\beta = b\gamma_{n+1}G \in \Gamma_n G$, then we define

$$(\alpha,\beta) = [a,b]\gamma_{m+n+1}G,$$

and extend this multiplication to all of ΓG by distributivity. It turns out that ΓG is indeed a lie ring, reflecting the structure of the lower central series of G and thereby some of the properties of G. For example if G is finitely generated, so is ΓG and every automorphism of G defines an automorphism of ΓG. However if G is finitely related, ΓG need not be - the infinite dihedral group provides a simple counter-example. In the case of free groups the connection between the group and its associated lie ring is satisfying close, as the following theorem of Magnus [73] and Witt [107] shows.

Theorem D.18. *If F is a free group, then ΓF is a free lie ring.*

The converse of Theorem D.18 is false. And this is where parafree groups come into the picture. Indeed it follows immediately that a residually nilpotent group is parafree if and only if its lie ring is free. It seems natural then to put forward the following.

Problem 4. *Is the isomorphism problem for parafree one-relator groups solvable?*

Whether every finitely generated parafree group is finitely presented, and whether the two dimensional integral homology groups of parafree groups are always zero are two other unresolved problems about parafree groups. This lack of knowledge about parafree groups helps to explain why I have formulated Problem 4 in such restrictive terms. Indeed I do not know whether the class of all finitely generated one-relator parafree groups is a recursive subclass of the class of all finitely generated one-relator groups.

There is an important generalization of Theorem D.18 to one-relator groups by Labute [59].

Theorem D.19. *Let F be the free group on* $a,b,\ldots,c$ *and let r be a reduced word in* $a,b,\ldots,c$. *Suppose that* $r \in \gamma_n F$, $r \notin \gamma_{n+1}F$ *and that* $r\gamma_{n+1}F$ *is not a proper power in* $F/\gamma_{n+1}F$. *Then the lie ring of the one-relator group*

$$G = \langle a,b,\ldots,c;r \rangle$$

is a lie ring that can be defined by a single defining relation.

Surprisingly, Theorem D.19 has as yet found little application.

There is another theorem that is somehow linked to the classification problem, due to E. Dyer and Vasquez [33]. Let me recall some of the details.

As usual a pathwise connected topological space X is termed aspherical if its higher homotopy groups $\pi_n(X,*) = 0$ for $n > 1$. Suppose next that X is a CW-complex with fundamental group G. Then X is termed a K(G,1) (Eilenberg-MacLane space) if it is aspherical. If Y is another K(G,1) then X and Y have the same homotopy type. Thus a K(G,1) may be viewed as an invariant of the group G. If there is a K(G,1) which has finite topological dimension and finitely many cells in each dimension, then G is said to be geometrically finite.

Now suppose that

$$G = \langle a,b,\dots,c;\ s^m \rangle \quad (m \geq 1)$$

is a one-relator group where s is a non-trivial reduced word in a,b,...,c which is not a proper power in the free group on a,b,...,c. Let W be the wedge of n 1-spheres $S_1,\dots,S_n$, where n is the number of generators a,b,...,c of G. Notice that if * is the point common to each of these circles, then

$$\pi_1(W,*) = \langle a,b,\dots,c \rangle$$

is a free group of rank n. Next let C_m be the cyclic group of order m. Then there exists a $K(C_m,1)$ (for example one of the infinite dimensional lens spaces, see e.g. Steenrod [102], p.67) with the 1-sphere S^1 as the 1-skeleton of this $K(C_m,1)$. Let

$$f : S^1 \to W$$

be chosen so that a generator x of $\pi_1(S^1,*)$ is mapped by f_* onto s. Form the quotient D(G) of the disjoint union of W and $K(C_m,1)$ obtained by identifying the elements of S^1 with its images under f:

$$D(G) = (K(C_m,1) \vee W)/f.$$

It follows then from the Seifert-van Kampen theorem (see e.g. Massey [79]) that

$$\pi_1(D(G)) \cong G.$$

Moreover D(G) turns out to be aspherical. These remarks constitute part of the following theorem of E. Dyer and Vasquez cited above.

Theorem D.20. *Let* $G = \langle a,b,\dots,c;\ s^m \rangle$ *be a one-relator group where* s *is not a proper power. Then* D(G) *is a* K(G,1). *In particular if* m = 1, i.e. *if* G *is torsion-free, then* D(G) *can be chosen to be a finite CW-complex of dimension two.*

D.4 One-relator groups with torsion

One-relator groups with torsion seem to be much less complicated than torsion-free one-relator groups. For example, Newman [87] has proved two important theorems about these groups, that bear out this remark. The first is his so-called "Spelling Theorem", which provides a sharpened solution of the word problem for one-relator groups with torsion that is simpler than the solution of the word problem for one-relator groups in general.

Theorem D.21. *Let* $G = \langle t,a,\ldots,c;\ s^m = 1 \rangle$ *be a one-relator group where* s *is cyclically reduced and* $m > 1$. *Suppose that* $w = v$ *in* G *where* w *is a freely reduced word in the given generators and* v *does not contain one of the generators occurring in* w. *Then* w *contains a subword* u *such that* u *is also a subword of either* s^m *or* s^{-m} *and the length of* u *is greater than* $(n-1)/n$ *times the length of* s^m.

Furthermore although the conjugacy problem for one-relator groups as a whole remains intractable, Newman [87] has proved that it is solvable for one-relator groups with torsion.

Theorem D.22. *One-relator groups with torsion have solvable conjugacy problem.*

There is another point worth making here. It is now a familiar fact that the one-relator group

$$G = \langle a,b;\ a^{-1}b^2a = b^3 \rangle$$

is non-hopfian and hence not even residually finite (G. Baumslag and Solitar [17]). However if one replaces the defining relator, say r, of this group G by a proper power of itself, then the resultant one-relator group turns out to be residually finite (B. Baumslag and Levin [6] and also Allenby, Moser and Tang [1]). This suggests the following problem, which is still unresolved.

Problem 5. *Is every one-relator group with torsion residually finite?*

The evidence in support of a positive solution to this problem is still rather slim. Besides the work of Allenby and Tang [2], [3] the most persuasive work here is due to Pride [91].

Theorem D.23. *Every two-generator one-relator group with torsion is hopfian.*

It may in fact turn out that one-relator groups with torsion are really very simple indeed. This is the suggestion implicit in the following problem.

Problem 6. *Is every one-relator group with torsion virtually free-by-cyclic?*

This would then imply residual finiteness by invoking G. Baumslag [13]. It seems to me also that there is a certain rigidity about one-relator groups with torsion. Perhaps this will be reflected in the fact that most one-relator groups with torsion have relatively few outer automorphisms.

D.5 Some special classes of one-relator groups

The isomorphism problem is unresolved even for very special classes of one-relator groups. I want to talk a little about some of these classes and their properties in this, the last section of my talk.

Let us term

$G = \{A * B;\ a = b\}$,

the free product of two free groups A and B amalgamating a non-trivial cyclic subgroup gp(a) of A with a non-trivial cyclic subgroup gp(b) of B, a cyclicly pinched group. Cyclicly pinched groups may be viewed as generalisations of the fundamental groups of two dimensional surfaces. And like these groups they turn out to be residually finite (G. Baumslag [9]). In fact they enjoy a very strong form of residual finiteness as the following theorem of Brunner, Burns and Solitar [22] clearly shows.

Theorem D.24. *Let* G *be a cyclicly pinched one-relator group. If* H *is any finitely generated subgroup of* G *and* g *is any element of* G *not in* H, *then there is a homomorphism of* G *onto a finite group such that the image of* g *lies outside the image of* H.

Cyclicly pinched one-relator groups share another property with the fundamental groups of surfaces provided the elements a and b are not proper powers as Wehrfritz [106] has shown.

Theorem D.25. *If* G *is the generalised free product of two free groups with a maximal cyclic subgroup of each factor amalgamated, then* G *has a faithful linear representation over a commutative field.*

There are a large number of residual properties of various classes of cyclicly pinched groups that have been established over the years but I do not have time to go into the matter here.

In view of the fact that Thurston has apparently proved that most of the fundamental groups of three dimensional closed manifolds have the property that any pair of finitely generated subgroups have finitely generated intersection, the following theorem of Burns [23] has some added interest.

Theorem D.26. *The intersection of two finitely generated subgroups of a cyclicly pinched group is again finitely generated provided one of the amalgamated subgroups is maximal cyclic.*

Cyclicly pinched one-relator groups have solvable conjugacy problem (Lipschutz [63]). More generally, J. Dyer [34] has proved the:

Theorem D.27. *Let* G *be cyclicly pinched. Then two elements of* G *are conjugate if and only if they are conjugate in every finite factor group of* G.

Despite the many results about these cyclicly pinched one-relator groups the isomorphism problem for them is still unsolved. Again I do not know whether the class of cyclicly pinched one-relator groups is a recursive subset of the class of all one-relator groups.

There is another class of one-relator groups that seems to be worth studying, the so-called positive one-relator groups, i.e. the groups which can be presented in the form $G = \langle X; r \rangle$, where r is a positive word, i.e., one in which no negative exponents occur. Positive one-relator groups share a residual property with free groups, as the following theorem of G. Baumslag [12] shows.

Theorem D.28. *Positive one-relator groups are residually solvable.*

The proof of this theorem depends on Levin's theorem on the solution of positive equations over groups. Similarly one might be able to prove that some suitably chosen one-relator groups are residually finite by using the theorem of Gerstenhaver and Rothaus.

I want to end this survey by mentioning one more theorem that did not quite fit into the scheme of these lectures, but which is a remarkable application of the techniques developed by Cohn and Bergman (see Cohn [25]).

Theorem D.29. *Let* G *be a torsion-free one-relator group. Then* $\mathbb{Z}[G]$, *the integral group ring of* G, *can be embedded in a division ring.*

This theorem is the result of joint work of Lewin and Lewin (nee Taylor) [62] and makes heavy use of Magnus' basic method of dealing with one-relator groups. So my survey ends fittingly, as it began, with the work of Wilhelm Magnus.

Acknowledgement. The author gratefully acknowledges the hospitality of the University of St Andrews, their support and the support of the National Science Foundation.

REFERENCES

1. R.B.J.T. Allenby, L.E. Moser & C.Y. Tang, The residual finiteness of certain one-relator groups, *Proc. Amer. Math. Soc.* 78 (1980), 8-10.
2. R.B.J.T. Allenby & C.Y. Tang, The residual finiteness of some one-relator groups with torsion, *J. Algebra* 71 (1981), 132-140.
3. R.B.J.T. Allenby & C.Y. Tang, Residually finite one-relator groups with torsion, *Arch. Math. (Basel)* 37 (1981), 97-105.
4. R.C. Alperin & P.B. Shalen, Linear groups of finite cohomological dimension, *Invent. Math.* 66 (1982), 89-98.
5. B. Baumslag & F. Levin, A class of residually finite one-relator groups with torsion, (1980), preprint.
6. B. Baumslag & S.J. Pride, Groups with two more generators than relators, *J. London Math. Soc.* (2) 17 (1978), 425-426.
7. B. Baumslag & S.J. Pride, Groups with one more generator than relators, *Math. Z.* 167 (1979), 279-281.
8. B. Baumslag & S.J. Pride, A generalisation of the Freheitssatz, *Proc. Cambridge Philos. Soc.* 89 (1981), 35-41.
9. G. Baumslag, On the residual finiteness of generalised free products of nilpotent groups, *Trans. Amer. Math. Soc.* 106 (1963), 193-209.
10. G. Baumslag, Groups with the same lower central sequence as a relatively free group. I. The groups, *Trans. Amer. Math. Soc.* 129 (1967), 308-321.
11. G. Baumslag, Groups with the same lower central sequence as a relatively free group. II. Properties, *Trans. Amer. Math. Soc.* 142 (1969), 507-538.
12. G. Baumslag, Positive one-relator groups, *Trans. Amer. Math. Soc.* 156 (1971), 165-183.
13. G. Baumslag, Finitely generated cyclic extensions of free groups are residually finite, *Bull. Austral. Math. Soc.* 5 (1971), 87-94.
14. G. Baumslag & C.F.M. Miller, A remark on the subgroups of one-relator groups, *Illinois J. Math.*, to appear.
15. G. Baumslag, J. Morgan & P.B. Shalen, Generalised triangle groups, to be submitted for publication.
16. G. Baumslag & P.B. Shalen, Group deficiencies and generalised free product decompositions, to be submitted for publication.

17. G. Baumslag & D. Solitar, Some two-generator one-relator non-hopfian groups, *Bull. Amer. Math. Soc.* 68 (1962), 199-201.
18. G. Baumslag & T. Taylor (=Lewin), The centre of groups with one defining relator, *Math. Ann.* 175 (1968), 315-319.
19. R. Bieri, Normal subgroups in duality groups and in groups of cohomological dimension 2, *J. Pure Appl. Algebra* 7 (1976), 35-51.
20. R. Bieri & R. Strebel, Almost finitely presented solvable groups, *Math. Helv.* 53 (1978), 258-278.
21. S.D. Brodskii, Equations over groups and groups with a single defining relation, *Uspekhi Math. Nauk* 35, 4 (1980), 183.
22. A.M. Brunner, R.G. Burns & D. Solitar, Subgroup separability of free products of two free groups with cyclic amalgamation, *Contemp. Math.* 33, *Contributions to Group Theory* (1984), 90-115.
23. R.G. Burns, On the rank of the intersection of subgroups of a Fuchsian group, *Proc. Second Internat. Conf. Theory of Groups* (Canberra, 1973), Lecture Notes in Mathematics 372, Springer-Verlag, Berlin, Heidelberg, New York (1974), 165-187.
24. D.E. Cohen, Subgroups of HNN groups, *J. Austral. Math. Soc.* 17 (1974), 394-405.
25. P.M. Cohn, *Free Rings and their Relations*, Academic Press, London, New York (1971).
26. D.J. Collins & F. Levin, Automorphism groups of groups with a single defining relation, *Math. Z.*
27. D. Cooper, Automorphism groups of free groups having finitely generated fixed point sets, preprint.
28. M. Culler, Fixed points of automorphisms of free groups, to appear.
29. M. Culler & P.B. Shalen, Varieties of group representations and splittings of 3-manifolds, *Ann. of Math.* 117 (1983), 109-146.
30. M. Culler & K. Vogtmann, Moduli of graphs and automorphisms of free groups, to appear.
31. M. Dehn, Uber die Topologie des dreidimensionalen Raumes, *Math. Ann.* 69 (1910), 137-168.
32. M. Dehn, Uber unendliche diskontinuierliche Gruppen, *Math. Ann.* 72 (1912), 413-420.
33. E. Dyer & A. Vasquez, Some properties of two-dimensional Poincaré duality groups, *Algebra, Topology and Category Theory*, Academic Press Inc. (1976), 45-54.
34. J.L. Dyer, Separating conjugates in amalgamated free products and HNN extensions, *J. Austral. Math. Soc. Ser. A* 29 (1980), 35-51.
35. B. Eckmann & H. Muller, Poincaré duality groups of dimension two, *Comment. Math. Helv.* 55 (1980), 510-520.
36. M. Edjvet, On the "largeness" of one-relator groups, (1985), preprint.
37. R. Fricke & F. Klein, Vorlesungen uber die Theorie der Automorphen Funktionen. I: Die Gruppentheoretischen Grundlagen and II: Die funktionentheoretischen Ausfuhrungen und die Anwendungen, B.G. Teubner, Leipzig (1897), (1912); Johnson Reprint Co., New York, B.G. Teubner Verlagsgesellschaft, Stuttgart (1965).
38. S.M. Gersten, On fixed points of automorphisms of finitely generated free groups, *Bull. Amer. Math. Soc.* 8 (1983), 451-454.
39. S.M. Gersten, On Whitehead's algorithm, *Bull. Amer. Math. Soc.* 10 (1984), 281-285.
40. M. Gerstenhaber & O.S. Rothaus, The solution of sets of equations in groups, *Proc. Nat. Acad. Sci. U.S.A.* 48 (1962), 1531-1533.
41. M. Greendlinger, Dehn's algorithm for the word problem, *Comm. Pure Appl. Math.* 13 (1960), 67-83.
42. M. Greendlinger, On Dehn's algorithm for the conjugacy and word problems, with applications, *Comm. Pure Appl. Math.* 13 (1960), 641-677.
43. M. Gromov, Hyperbolic groups, in preparation.
44. K.W. Gruenberg, *Cohomological topics in group theory*, Lecture Notes in Mathematics 743, Springer-Verlag, Berlin-Heidelberg-New York, (1970).
45. W.J. Haboush, Reductive groups are geometrically reductive, *Ann. Math.* 102 (1975), 67-83.
46. R. Hartshorne, *Algebraic Geometry*, Graduate texts in Mathematics 52, Springer-Verlag, New York (1977).
47. G. Higman, The units of groups rings, *Proc. London Math. Soc.* (2) 46 (1940), 231-248.
48. P. Hill, S.J. Pride & A.D. Vella, Subgroups of small cancellation groups, *J. Reine Angew. Math.* 349 (1984), 24-54.

49. H. Hopf, Uber den Rand geschlossener Liescher Gruppen, *Comment. Math. Helv.* 13 (1940-1941), 119-143.

50. J. Howie, On locally indicable groups, *Math. Z.* 180 (1982), 445-461.

51. J. Howie, How to generalise one-relator group theory, *Proc. Alta Conf. Combinatorial Group Theory*, Ann. of Math. Stud., Princeton University Press, to appear.

52. A. Juhasz, The solution of the conjugacy problem for certain one relator groups I, Israel J. Math., to appear.

53. A. Juhasz, The solution of the conjugacy problem for certain groups, (1985), preprint.

54. D. Karrass, W. Magnus & D. Solitar, Elements of finite order in groups with a single defining relation, *Comm. Pure Appl. Math.* 13 (1960), 57-66.

55. A. Karrass, A. Pietrowski & D. Solitar, An improved subgroup theorem for HNN groups with some applications, *Canad. J. Math.* 26 (1974), 458-466.

56. A. Karrass & D. Solitar, The subgroups of two groups with an amalgamated subgroup, *Trans. Amer. Math. Soc.* 149 (1970), 227-255.

57. A. Karrass & D. Solitar, Subgroups of HNN groups and groups with a single defining relation, *Canad. J. Math.* 23 (1971), 627-643.

58. V. Krstic, Fixed subgroups of automorphisms of free by finite groups, (1985), preprint.

59. J.P. Labute, On the descending central series of groups with a single defining relation, *J. Algebra* 14 (1970), 16-23.

60. M. Lazard, Sur les groupes nilpotentes et les anneaux de Lie, *Ann. Ecole Norm. Sup.* 71 (1953), 101-190.

61. F. Levin, Solutions of equations over groups, *Bull. Amer. Math. Soc.* 68 (1962), 603-604.

62. J. Lewin & T. Lewin (=Taylor), An embedding of a torsion-free one-relator group in a field, *J. Algebra* 52 (1978), 39-77.

63. S. Lipschutz, Generalisation of Dehn's result on the conjugacy problem, *Proc. Amer. Math. Soc.* 17 (1966), 759-762.

64. A. Lubotsky & A. Magid, Varieties of representations of finitely generated groups, (1985), preprint.

65. R.C. Lyndon, Cohomology theory of groups with a single defining relation, *Ann. of Math.* (2) 52 (1950), 650-665.

66. R.C. Lyndon & P.E. Schupp, *Combinatorial group theory*, Springer-Verlag, Berlin-Heidelberg-New York (1977).

67. S. MacLane, *Homology*, Die Grundlehren der Mathematischen Wissenschaften 114, Academic Press, New York; Springer-Verlag, Berlin, Gottingen, Heidelberg (1963).

68. W. Magnus, Uber diskontinuierliche Gruppen mit einer definierenden Relation (Der Freiheitssatz.), *J. Reine Angew. Math.* 163 (1930), 141-165.

69. W. Magnus, Untersuchungen uber einige undendliche diskontinuierliche Gruppen, *Math. Ann.* 105 (1931), 52-74.

70. W. Magnus, Das Identitatsproblem fur Gruppen mit einer definierden Relation, *Math. Ann.* 106 (1932), 295-307.

71. W. Magnus, Uber freie Faktorgruppen und freie Untergruppen gegebener Gruppen, *Monatsch. Math. Phys.* 47 (1939), 307-313.

72. W. Magnus, Beziehungen zwischen Gruppen und Idealen in einem speziellen Ring, *Math. Ann.* 11 (1935), 259-280.

73. W. Magnus, Uber Gruppen und zugeordnete Liesche Ringe, *J. Reine Angew. Math.* 182 (1940), 142-149.

74. W. Magnus, The uses of 2 by 2 matrices in combinatorial group theory. A survey, *Resultate Math.* 4 (1981), 171-192.

75. W. Magnus, Two generator subgroups of PSL(2,ℂ), *Nachr. Akad. Wiss. Gottingen. II. Math. Phys. Kl.* 7 (1975), 81-94.

76. W. Magnus & B. Chandler, *The history of combinatorial group theory: A case study in the history of ideas*, Springer-Verlag, New York, Heidelberg, Berlin (1982).

77. W. Magnus, A. Karrass & D. Solitar, *Combinatorial group theory: Presentations of groups in terms of generators and relations*, Pure Appl. Math. 13, Interscience [John Wiley and Sons], New York, London, Sydney (1966).

78. A.I. Malcev, On faithful representations of infinite groups of matrices, *Amer. Math. Soc. Translations* (2) 45 (1965), 1-18.

79. W.S. Massey, *Algebraic Topology*, Harcourt Brace and World, Inc. New York (1967).

80. J. McCool, A presentation of the autormorphism group of a free group of finite rank, *J. London Math. Soc.* 8 (1974), 259-266.

81. J. McCool, On Nielsen's presentation of the automorphism group of a free group, *J. London Math. Soc.* 10 (1975), 265-270.
82. J. McCool, Some finitely presented subgroups of the automorphism group of a free group, *J. Algebra* 35 (1975), 205-213.
83. J. McCool & A. Pietrowski, On recognising certain one relation presentations, *Proc. Amer. Math. Soc.* 36 (1972), 31-33.
84. J. McCool & P.E. Schupp, On one relator groups and HNN extensions, *J. Austral. Math. Soc.* 16 (1973), 249-256.
85. S. Meskin, The isomorphism problem for a class of one-relator groups, *Math. Ann.* 217 (1975), 53-57.
86. D.I. Moldavanskii, Certain subgroups of groups with one defining relation, *Sibirsk. Mat. Zh.* 8 (1067), 1370-1384.
87. B.B. Newman, Some results on one relator groups, *Bull. Amer. Math. Soc.* 74 (1968), 568-571.
88. B.B. Newman, The soluble subgroups of a one-relator group with torsion, *J. Austral. Math. Soc.* 16 (1973), 278-285.
89. J. Nielsen, Die Isomorphismengruppe der freien Gruppen, *Math. Ann.* 91 (1924), 169-209.
90. A. Peitrowski, The isomorphism problem for one-relator groups with non-trivial centre, *Math. Z.* 136 (1974), 95-106.
91. S.J. Pride, The isomorphism problem for two-generator one-relator groups with torsion is solvable, *Trans. Amer. Math. Soc.* 227 (1977), 109-139.
92. S.J. Pride, Small cancellation conditions satisfied by one-relator groups, *Math. Z.* 184 (1983), 283-286.
93. R. Ree & N.S. Mendelsohn, Free subgroups of groups with a single defining relation, *Arch. Math.* 19 (1968), 577-580.
94. G. Rosenberger, Problem No. F2, *Homological group theory*, London Math. Soc. Lecture Notes 36 (1979).
95. G. Rosenberger, Uber Gruppen mit einer definierenden Relation, *Math. Z.* 155 (1977), 71-77.
96. P.E. Schupp, A survey of small cancellation theory, *Word Problems. Decision Problems and the Burnside Problem in Group Theory*, Studies in Logic and the Foundations of Mathematics 71, North-Holland, Amsterdam, London (1973).
97. P.E. Schupp, A strengthened Freiheitssatz, *Math. Ann.* 221 (1976), 73-80.
98. J-P. Serre, *Trees*, (Translated from the French by John Stillwell), Springer-Verlag, Berlin, Heidelberg, New York (1980).
99. J-P. Serre, *Cohomologie galoisienne*, Lecture Notes in Math. 5, Berlin, Heidelberg, New York (1964).
100. J.R. Stallings, On torsion-free groups with infinitely many ends, *Ann. of Math.* 88 (1968), 312-334.
101. J.R. Stallings, Homology and central series of groups, *J. Algebra* 2 (1965), 170-181.
102. N. Steenrod, *The topology of fibre bundles*, Princeton University Press (1951).
103. R. Stohr, Groups with one more generator than relators, *Math. Z.* 182 (1983), 45-47.
104. R.G. Swan, Groups of cohomological dimension one, *J. Algebra* 12 (1969), 585-610.
105. J. Tits, Free subgroups in linear groups, *J. Algebra* 20 (1972), 250-270.
106. B.A.F. Wehrfritz, Generalized free products of linear groups, *Proc. London Math. Soc.* (3) 27 (1973), 402-424.
107. E. Witt, Treue Darstellung liescher Ringe, *J. Reine Angew. Math.* 177 (1937), 152-160.
108. H. Zieschang, Algorithmen fur einfache Kurven und Flachen, II, *Math. Scand.* 25 (1969), 49-58.

SOME ALGORITHMS FOR COMPUTING WITH FINITE PERMUTATION GROUPS

Peter M Neumann
The Queen's College, Oxford, OX1 4AW, England

1. *INTRODUCTION*

Let Ω be a finite set of size n and G the subgroup of $\mathrm{Sym}(\Omega)$ generated by a collection $g_1,\ldots,g_k$ of explicitly known permutations. How do we compute the composition factors of G?

This is one of the many mathematical questions where theory and practice differ greatly. In theory we might work through the elements of G one by one seeking g such that the normal closure $N := \langle g^G \rangle$ is a non-trivial proper subgroup of G; if no such g exists then G is simple, otherwise we treat N and G/N in the same way until all the composition factors have been found. In practice this does not work. There are two reasons. The first is obvious - such a search will be absurdly slow even for permutation groups of quite small degree. The second is more subtle and more significant - we have at present no satisfactory way of handling a quotient group G/N as a permutation group except in certain special cases. Here is a simple example to show where the difficulty lies.

Example 1.1. Let $\Omega := \{1,2,\ldots,80\}$ and for $i := 0,1,\ldots,19$ let

$$x_i := (4i+1,\ 4i+2,\ 4i+3,\ 4i+4),$$
$$y_i := (4i+2,\ 4i+4).$$

The group X_i generated by x_i, y_i is dihedral of order 8 acting on $\{4i+1,\ldots,4i+4\}$. If

$$G := \langle x_0,\ldots,x_{19},\ y_0,\ldots,y_{19}\rangle$$

then $G = X_0 \times X_1 \times \ldots \times X_{19}$. As a permutation group of degree 80 with 40 generators this is well within the present-day range of fast computation. Now let

$$z_i := x_i^2 = (4i+1,\ 4i+3)(4i+2,\ 4i+4).$$

Then z_i generates the centre of X_i and if $Z := \langle z_0, z_1,\ldots,z_{19}\rangle$ then Z is the centre of G. Define

$$N := \langle z_0z_1, z_0z_2,\ldots,z_0z_{19}\rangle .$$

This is a subgroup of index 2 in Z and if $H := G/N$ then H is an extraspecial group of order 2^{41}. It is not hard to see that if $U \leqslant H$ and $|H:U| < 2^{21}$ then $Z/N \leqslant U$, and it follows that if H acts faithfully as a group of permutations of a set Γ then $|\Gamma| \geqslant 2^{21}$. But $2^{21} = 2{,}097{,}152$ and this is far too big for general-purpose computing on present-day machines. Of course, when machines become bigger and faster this particular extraspecial group H might come into range, but there are bigger versions of G and H: the direct product of m dihedral groups of order 8 has a natural permutation representation of degree 4m but it has an extraspecial quotient group which, if represented faithfully by permutations, requires degree $\geq 2^{m+1}$.

Although I chose to discuss the problem of computing composition factors of permutation groups to provide a focus for my course of lectures, the algorithms described below were originally designed in response to questions of John Cannon, who wanted to know how to compute minimal normal subgroups in primitive permutation groups. It was a happy afterthought that they could be used to solve the composition factors problem. Because of their genesis the context of the algorithms is general-purpose computing within a framework such as the CAYLEY system. What was asked for was a set of procedures that would cope well for a realistic range of n and k (recall that n was the degree of our group G and k was the number of its specified generators). In April 1983 we took $n \leq 3600$ and $k \leq 1000$ to be realistic for the CAYLEY system on present-day machines (although in his paper [2], p.488, Cannon suggests that n might go up to 10^4). In fact, I have tried to design my algorithms to survive future increases of computing power to permit calculation with permutation groups up to degree 10^5 or perhaps even 10^6.

The algorithms are constructed from a number of ingredients. Let (X,Γ) be any permutation group arising somewhere in our calculations, so that certain permutations of Γ that generate X are explicitly known. The following can be found extremely rapidly:

the fixed point set $\mathrm{Fix}_\Gamma(X)$ of X in Γ;

the orbit βX for any $\beta \in \Gamma$;

the action X^{Γ_1} of X on any X-invariant subset Γ_1;

then, if X is transitive on Γ, also

the congruence (X-invariant equivalence relation)

ρ generated by any pair (β,γ) (see Atkinson [1]);

the action X^{Γ_1} of X on Γ_1, where $\Gamma_1 := \Gamma/\rho$.

Using Atkinson's algorithm one may find minimal non-trivial congruences and maximal proper congruences very cheaply. Any more substantial computing with finite permutation groups depends on the notion of 'base' and 'strong generating set' introduced by Charles Sims [13]. His algorithms (see [13, 14,15]) in the improved versions worked out by J.S. Leon [9] will produce a strong generating set for X acting on Γ; then strong generating sets for subgroups of X and for actions X^{Γ_1}, where Γ_1 is an X-invariant subset of Γ or a quotient Γ/ρ of Γ by a congruence ρ, may be derived relatively quickly. This offers easy access to such items as

stabilisers X_β, X_{β_1,β_2},...;

the orders $|X|$, $|X_\beta|$,... in factorised form;

$\mathrm{Ker}(X^{\Gamma_1})$, the kernel of the action of X on Γ_1, where Γ_1 is an X-invariant subset of Γ;

$\mathrm{Ker}(X^{\Gamma_1})$, the kernel of the action of X on Γ_1, where Γ_1 is a quotient Γ/ρ.

In these last two situations, if $f: X \to X^{\Gamma_1}$ is the natural homomorphism then, for any explicitly known subgroup X_1 of X^{Γ_1}, the inverse image $f^{-1}(X_1)$ can be obtained very quickly as a permutation group on Γ. Whenever the notation seems clear enough I shall use $(X_1)^{-1}$ to denote this inverse image.

Beyond this what we need is a variety of components that are already available in one form or another (for example, either as explicit functions or as library routines in the CAYLEY system): procedures to calculate

the centraliser of an element;

the centraliser of a subgroup;

the intersection of subgroups;

the normal closure of an element;

the commutator subgroup;

Sylow subgroups;

the core of a subgroup.

These depend heavily on the availability of a base and strong generating set, and as a result they can be rather extravagant for groups of degree ≥ 1000. However, the algorithms that I shall describe should be hardly more expensive, and their range of applicability should grow directly as improvements to the individual components are discovered.

I shall write my algorithms in pidgin ALGOL, an ideal language for programming mathematicians. The following notation will be taken to be standard:

α is my favourite member of the set Ω on which G acts;

$|S|$ is the size of the set or group S;

X^{Γ} is the group of permutations of Γ induced by X;

$X_\beta := \{x \in X \mid \beta x = \beta\}$, the stabiliser of β; then $X_{\beta,\gamma} := (X_\beta)_\gamma$, etc;

$\mathrm{Ker}(X^{\Gamma}) := \{x \in X \mid \gamma x = \gamma \text{ for all } \gamma \in \Gamma\}$;

$\mathrm{Fix}_\Gamma(X) := \{\gamma \in \Gamma \mid \gamma x = \gamma \text{ for all } x \in X\}$;

$\mathrm{Orb}(X,\Gamma)$ is the set of X-orbits in Γ;

$Z(X) := \{z \in X \mid xz = zx \text{ for all } x \in X\}$, the centre of X;

$C_X(Y) := \{z \in X \mid yz = zy \text{ for all } y \in Y\}$, the centraliser of Y in X;

$N_X(Y) := \{x \in X \mid x^{-1}Yx = Y\}$, the normaliser of Y in X;

$\mathrm{Core}_X(Y) := \bigcap_{x \in X} x^{-1}Yx$;

$\langle Y \rangle$ is the subgroup of X generated by Y;

$\langle Y^X \rangle$ is the normal closure of Y in X;

$[x,y] := x^{-1}y^{-1}xy$, the commutator of x and y;

$X' := \langle [x,y] \mid x,y \in X \rangle$, the commutator subgroup of X;

$X, X', X'', \ldots, X^{(n)}, \ldots$ is the derived series of X;

$X^{(\infty)} := \cap X^{(n)}$, the soluble residual of X;

$Soc(X)$ is the product of the minimal non-trivial normal subgroups of X;

$O_p(X)$ is the maximal normal p-subgroup of X;

when X is a p-group, $\Omega_1(X) := \langle x \mid x^p = 1 \rangle$;

C_m is the cyclic group of order m;

S_m is the symmetric group of degree m - that is, it is $Sym(\Gamma)$ where $|\Gamma| = m$ but we have no real interest in this particular set Γ;

A_m is the alternating group of degree m.

Other notation will be introduced when it is needed.

The remainder of this paper is organised as follows. Since John Cannon's original questions were about minimal normal subgroups in primitive permutation groups we shall need a theoretical description of these. That is provided by the O'Nan-Scott Theorem, a version of which is expounded in §2. The algorithms are described in the six following sections -

§3: the EARNS routine is for finding abelian minimal normal subgroups in primitive permutation groups;

§4: the PCORE routine is for finding $O_p(G)$ in an arbitrary permutation group G;

§5: the SOCLE routine is for finding the socle of a primitive permutation group when this socle is non-abelian;

§6: the SOCLEBITS routine is for finding the simple direct factors of $Soc(G)$ when G is a primitive permutation group and $Soc(G)$ is non-abelian;

§7: the COMPFACTOR routine is a scheme for computing a list of the composition factors of an arbitrary permutation group G;

§8: the SIMPLETEST routine is a simple test for the simplicity of an arbitrary permutation group G.

Finally, §9 is a sort of up-market collection of footnotes concerning theory (computational complexity), practice and acknowledgements.

2. *THE O'NAN-SCOTT THEOREM*

Suppose that G is primitive on Ω. In purely group-theoretic terms this means that if $\alpha \in \Omega$ then the stabiliser G_α is a maximal proper subgroup of G. Also, G is core-free, that is, $Core_G(G_\alpha) = \{1\}$ because G acts faithfully on Ω. Let M be a minimal non-trivial normal subgroup of G. As is well-known (and easy to check) the M-orbits are blocks of imprimitivity for G and so M must be transitive on Ω; equivalently, $G = G_\alpha M$. Also, M is a direct power of some simple group S; that is, $M = X_1 \times X_2 \times \ldots \times X_d$, where $X_i \cong S$ for all i. The stabiliser G_α acts on M by conjugation and this action is faithful because its kernel $C_{G_\alpha}(M)$ is contained in $\mathrm{Core}_G(G_\alpha)$ and therefore is $\{1\}$. The stabiliser M_α, which is $G_\alpha \cap M$, is obviously a G_α-invariant subgroup of M: and the maximality of G_α in G implies that M_α is a maximal proper G_α-invariant subgroup of M.

If S is the cyclic group C_p of some prime order p, that is, if M is an elementary abelian p-group, then $M_\alpha = \{1\}$ (because $\mathrm{Core}_M(M_\alpha) = \{1\}$) and M must act regularly on Ω. The degree n is then p^d and G is the split extension of M by G_α. Notice that $M = C_G(M)$ (in fact, M is its own centraliser even in $\mathrm{Sym}(\Omega)$) and so M is the unique minimal (non-trivial) normal subgroup of G. If we identify M with the vector space $\mathbb{Z}_p^d$ then it becomes an irreducible $\mathbb{Z}_pG_\alpha$-module on which G_α acts faithfully; and, if we identify G_α with the appropriate group of invertible linear transformations of $\mathbb{Z}_p^d$ and identify Ω with $\mathbb{Z}_p^d$ by associating αx with x (for $x \in M = \mathbb{Z}_p^d$), then G becomes identified with the group of affine transformations

$$x \mapsto xh + u \quad (h \in G_\alpha,\ u \in \mathbb{Z}_p^d) \tag{*}$$

of $\mathbb{Z}_p^d$. Thus if $S \cong C_p$ then G is identifiable with a subgroup of the affine group $AGL(d,p)$ that contains the translation subgroup. I shall speak of G as being of *affine type*.

If S is non-abelian then $X_1, X_2, \ldots, X_d$ are the only minimal normal subgroups of M and therefore they must be permuted amongst themselves under conjugation by elements of G. Putting $\Sigma := \{1,2,\ldots,d\}$ we get an action of G, hence of G_α, on Σ. Since $G = G_\alpha M$ and M is in the kernel of this action, in fact $G^\Sigma = G_\alpha^\Sigma$. Furthermore, G (or G_α) is transitive on Σ because if $\{i_1,\ldots,i_c\}$ are permuted amongst themselves then $X_{i_1} \times \ldots \times X_{i_c}$ is a normal subgroup of G and, by minimality of M, must be the whole of M.

The information that we have collected up to this point is all "classical" in the sense that it was known and fully understood well over a hundred years ago. Much of Camille Jordan's doctoral dissertation (1860) represents a very early and rather ineffective attempt to go deeper (although Jordan, following Galois, used a definition of 'primitive' that is weaker than the modern one - in his early work a permutation group was 'primitive' if all its non-trivial normal subgroups were transitive). From time to time various mathematicians have added further little items of information but nothing truly memorable until a full and satisfactory description of finite primitive permutation groups in terms of their minimal normal subgroups was crystallised out and explicitly formulated by Michael O'Nan and Leonard Scott at the 1979 Santa Cruz Conference on Finite Groups (see [12]). Many versions of the O'Nan-Scott Theorem are now available. In the form in which we will find it most usable it is a taxonomy that classifies primitive permutation groups of finite degree into five types.

Type I: M *is abelian.* In this case G is of affine type as described above.

Type II: M *is non-abelian and* $C_G(M) \neq \{1\}$.

Let $M_1 := M$ and $M_2 := C_G(M)$. Then $\{1\} \neq M_2 \triangleleft G$ and so M_2 is transitive on Ω. The centraliser in $\mathrm{Sym}(\Omega)$ of a transitive group is transitive if and only if that group acts regularly. Therefore M_1 acts regularly on Ω and the same must be true of M_2. If N is any minimal normal subgroup of G and $N \neq M$ then $N \cap M = \{1\}$, so N centralises M and therefore $N \leq M_2$; it follows that $N = M_2$ since N has to be transitive and M_2 is regular on Ω. Thus M_2 is a minimal normal subgroup of G and M_1, M_2 are the *only* minimal normal subgroups. If $D := (M_1 \times M_2)_\alpha = G_\alpha \cap (M_1 \times M_2)$ then

$$D \cap M_1 = \{1\} \text{ and } DM_1 = M_1 \times M_2,$$

$$D \cap M_2 = \{1\} \text{ and } DM_2 = M_1 \times M_2,$$

whence

$$M_1 \cong (M_1 \times M_2)/M_2 \cong D \cong (M_1 \times M_2)/M_1 \cong M_2.$$

In fact, M_1, M_2 are isomorphic as G_α-operator groups. The subgroup D is a "diagonal" in $M_1 \times M_2$ and conjugation by elements of D induces the group of all inner automorphisms on M_1 and on M_2. We can identify Ω with M and G_α with a subgroup of its automorphism group $\mathrm{Aut}\,M$ that contains the group $\mathrm{Inn}\,M$ of all inner automorphisms, and then

$$G = \{x \mapsto x^h u \mid h \in G_\alpha,\ u \in M\}. \qquad (**)$$

Thus G becomes a subgroup of the group of all these "multiplicative affine" transformations of M, which used to be called the "holomorph" of M.

Types I and II are the cases where $C_G(M) \neq \{1\}$. Therefore from now on we treat groups in which $C_G(M) = \{1\}$. Then M is the unique minimal normal subgroup of G (in Philip Hall's terminology G is monolithic) and $M \leq G \leq \mathrm{Aut}M$. Let $\xi_i : M \to X_i$ be the projection homomorphism (for $i := 1,\ldots,d$) and consider the projections $M_\alpha \xi_i \leq X_i$. Since G_α normalises M_α and acts transitively on $\{X_1,\ldots,X_d\}$ by conjugation, there are just two possibilities: if $M_\alpha \xi_1 = X_1$ then $M_\alpha \xi_i = X_i$ for all i, in which case M_α is a subdirect product in M; if $M_\alpha \xi_1 < X_1$ then $M_\alpha \xi_i < X_i$ for all i.

Type III: *M is non-abelian,* $C_G(M) = \{1\}$ *and* M_α *is a subdirect product in* M. Being a subdirect product of finitely many simple groups M_α must be isomorphic to a direct product of some of them. Thus

$$M_\alpha = Y_1 \times \ldots \times Y_b,$$

where $Y_j \cong S$ for $j := 1,\ldots,b$; moreover, there is a partition $\Sigma = \{1,2,\ldots,d\} = \Sigma_1 \cup \ldots \cup \Sigma_b$ such that Y_j is a subdirect product (a "diagonal subgroup") in $\Pi\,\{X_i \mid i \in \Sigma_j\}$. Since G_α acts by conjugation to permute $Y_1,\ldots,Y_b$ among themselves it permutes the sets Σ_j among themselves, that is, $\{\Sigma_1,\ldots,\Sigma_b\}$ is a system of blocks of imprimitivity for G_α^Σ (and hence for G^Σ). Therefore the sets Σ_j all have the same size c and $d = bc$. If we number the direct factors X_i of M appropriately we may suppose that $\Sigma_j = \{i \mid 1 \leq i \leq d,\ i \equiv j \pmod b\}$. Notice that

$$n = |M:M_\alpha| = |S|^{d-b} = |S|^{b(c-1)}$$

and that $X_{b+1} \times X_{b+2} \times \ldots \times X_d$ acts regularly on Ω.

Type IV: M *is non-abelian*, $C_G(M) = \{1\}$, $d > 1$ *and* M_α *is not a subdirect power in* M. Let $Y_i := M_\alpha \xi_i$. Now G_α normalises $Y_1 \times Y_2 \times \ldots \times Y_d$, which is a proper subgroup of M that contains M_α, and therefore $M_\alpha = Y_1 \times Y_2 \times \ldots \times Y_d$. Let Γ_i be the coset space $(X_i : Y_i)$. Then Ω may naturally be identified as M-space with $\Gamma_1 \times \Gamma_2 \times \ldots \times \Gamma_d$. In fact, an isomorphism ϕ_i of S to X_i turns Γ_i into an S-space, and these isomorphisms can be so chosen that $\Gamma_1, \Gamma_2, \ldots, \Gamma_d$ are all isomorphic to one and the same S-space Γ: we choose elements $h_i \in G_\alpha$ such that $h_i^{-1} X_1 h_i = X_i$, we choose $\phi_1 : S \to X_1$ arbitrarily and we take ϕ_i to be the composite $\phi_1 \psi_i$ for $i := 2, \ldots, d$, where ψ_i is the map $X_1 \to X_i$ produced by conjugation by h_i. This S-space Γ will of course be faithful and transitive but unfortunately it need not be primitive. Nevertheless we have a useful description of G and Ω: if $m := |\Gamma|$ then $n = m^d$ and Ω is identifiable with Γ^d in such a way that M is correctly identified with S^d; furthermore, as we observed before, $S^d \leqslant G \leqslant \mathrm{Aut}(S^d)$.

Type V: M *is simple and* $C_G(M) = \{1\}$. In this case $S \leqslant G \leqslant \mathrm{Aut}\, S$. The action of S on Ω need not be primitive. Even so, knowing S we would expect to be able to say a great deal about G.

Example 2.1. The affine group AGL(d,p) itself is a primitive group of type I and degree p^d. Indeed, if H is any irreducible subgroup of GL(d,p) then we can use (*) to define a group

$$\{x \mapsto xh + u \mid h \in H,\ u \in \mathbb{Z}_p^d\}$$

of affine transformations of $\mathbb{Z}_p^d$ that is primitive and of course of affine type.

Example 2.2. Let S be any non-abelian simple group, take $\Omega := S$ and take

$$G := \{x \mapsto x^h u \mid h \in \mathrm{Inn} S,\ u \in S\}.$$

This group is primitive of type II. In fact $G \cong S \times S$ acting on S by the rule

$$x(y_1, y_2) = y_1^{-1} x y_2.$$

More generally, let $M := S \times S \times \ldots \times S$ (with d factors), let H be a subgroup of AutM that contains InnM and acts transitively on the set of simple direct factors of M, and let $\Omega := M$. If we define

$$G := \{x \mapsto x^h u \mid h \in H,\ u \in M\}$$

as in (**), then we have a typical primitive group of type II.

Example 2.3. Let T be a primitive group of permutations of the set $\{1, \ldots, c\}$, let S be any non-abelian simple group, and let G be the wreath product S Wr T (that is, the semidirect product $(X_1 \times \ldots \times X_c).T$, where $X_i \cong S$

for all i and T acts on the "base group" $X_1 \times \ldots \times X_c$ permuting the factors as it permutes $\{1,\ldots,c\}$). If Y is the diagonal $\{(x,\ldots,x) \mid x \in S\}$ in the base group and $H := Y \times T$ then, as one proves quite easily, H is a maximal subgroup of G. Consequently the coset space $\Omega := (G:H)$ is a primitive G-space. Here G is of type III with $d = c$ and $b = 1$.

Example 2.4. Let G be the twisted wreath product (B.H. Neumann [11]) of A_5 by A_6 where A_6 acts naturally on $\{1,\ldots,6\}$ and a stabiliser A_5 acts on the bottom factor A_5 as the group of inner automorphisms. Thus G is a semidirect product $A_5^6.A_6$ in which the six factors A_5 in the base group are permuted in the natural way by the "top group" A_6. It is not hard to show that the top group is maximal in this twisted wreath product so if Ω is the coset space $(G:A_6)$ then G acts as a primitive group of degree 60^6 on Ω. This is an example of a group of type IV in which S happens to act regularly on Γ.

Example 2.5. Let Ω be the set of flags (in this case, incident point-line pairs) in the projective plane PG(2,2), and let G be the group of all collineations and dualities of PG(2,2). Thus G is isomorphic to Aut(SL(3,2)) (and SL(3,2) is the simple group of order 168). It acts naturally on Ω as a primitive permutation group of degree 21 and type V. Its socle SL(3,2) is imprimitive; there are two different SL(3,2)-congruences, each having seven blocks of imprimitivity of size 3. In the one congruence flags are equivalent if and only if they have their point in common; in the other, flags are equivalent if their lines are the same.

Example 2.6. Let G_1 be a permutation group on a set Ω_1 of size n_1 and let U be a permutation group on the set $\{1,\ldots,b\}$. The wreath product $G_1 \operatorname{Wr} U$ acts naturally on the set Ω_1^b : if we think of Ω_1^b as $\Omega_1 \times \Omega_2 \times \ldots \times \Omega_b$ then the base group $G_1 \times G_2 \times \ldots \times G_b$ in the wreath product acts in the natural way and the top group U acts to permute the factors. This *product action* of the wreath product G can easily be proved to be primitive as long as G_1 is primitive and not regular on Ω_1 and U is transitive on $\{1,\ldots,b\}$. If we take G_1 to be a non-cyclic primitive group of type I then G will again be of affine type. If G_1 is primitive of type II then G is of type II. If G_1 is a primitive group of type III then G will again be of type III; thus from Example 2.3 we can construct rather more typical (though not yet completely general) examples of groups of type III with b, c arbitrary and $d = bc$. If we take G_1 to be of type IV or type V (and in the latter case if $b > 1$) then G will be of type IV.

The theory can be pushed substantially further to analyse general primitive permutation groups in terms of wreath products in their product actions. But what I have described of the O'Nan-Scott Theorem will be quite enough for our present purposes.

3. *THE EARNS ROUTINE*

We will suppose now that G is known to be primitive and that a strong generating set has already been computed (and also that G is not alternating or symmetric - quick tests will have decided this point before a strong generating set was calculated). Our first concern is to decide whether or not G is of affine type, and if it is, to find its elementary abelian regular normal subgroup (hereinafter known as earns). That is what the EARNS routine is for.

There is an obvious preliminary step which, however, takes negligible computing power. We first test whether n is a prime-power. If not then G is not of affine type. Let us suppose that this test has already been done and that *n is known to be p^d where p is a prime number.* Notice that in the range $n \leq 3600$ we have $d < 12$ and even if in our more ambitious moments we dream of computing with groups of degree up to a million, we will still have $d < 20$. Our problem is to find $O_p(G)$. If $O_p(G) = \{1\}$ then G is not of affine type, otherwise $O_p(G)$ is the earns we are seeking. Now $O_p(G)$ could be computed by finding a Sylow p-subgroup P of G and calculating its core. The difficulty with this (I am told) is that finding P is very expensive, impracticably so when $|P| = p^m$ with $m \geq 20$ or when n exceeds about 500. And there certainly are some groups that might arise which are well outside this range. For example, the affine group AGL(10,2) has degree 1024 and Sylow 2-subgroups of order 2^{55}, and the wreath product $S_{32}\,\mathrm{Wr}\,S_2$ in its product action of degree 1024 has Sylow 2-subgroups of order 2^{63}. Therefore we proceed with more care.

Step 1. Test whether or not G is a Frobenius group and if not find α,β such that $G_{\alpha,\beta} \neq \{1\}$. If the test gives a positive answer then G is of affine type and its earns may be computed very quickly:

```
            Begin
(1)         if G_α = {1} then
               begin
(1.1)          M := G;
(1.2)          goto line (15);
               end;
(2)         if G_α,β = {1} for all β ∈ Ω - {α} then
               begin
(2.1)          choose a ∈ Z(G_α) - {1};
(2.2)          choose x ∈ G - G_α;
(2.3)          b := x^-1 ax;  c := [a,b];
(2.4)          M := ⟨c^G⟩;
(2.5)          goto line (15);
               end;
(3)         choose β ∈ Ω - {α} such that G_α,β ≠ {1};
(to be continued).
```

Justification. If $G_\alpha = \{1\}$ then G is regular on Ω and since it is primitive it is cyclic of prime order and equal to its earns. Suppose therefore that $G_\alpha \neq \{1\}$. If $G_{\alpha,\beta} = \{1\}$ for all $\beta \in \Omega - \{\alpha\}$ then G is (by definition) a Frobenius group. By Frobenius' theorem G certainly has an earns M and our task is to find it. We exploit the following fact (see, for example, [7], Satz V.8.18, p.506):

Lemma 3.1. *If* H *is a Frobenius complement then* $Z(H) \neq \{1\}$.

Thus at line (2.1) we know that $Z(G_\alpha) \neq \{1\}$ and we can find an element a very quickly. Now a has α as its unique fixed point and so b has αx, which is different from α, as its unique fixed point. Therefore b moves α and so a, b cannot commute, that is, $c \neq 1$. On the other hand, a is central modulo M, whence $[a,b] \in M$, that is $c \in M - \{1\}$. Consequently $M = \langle c^G \rangle$ as claimed in lines (2.4) and (2.5).

Step 2. We now have $\alpha, \beta \in \Omega$ such that $\alpha \neq \beta$ and $G_{\alpha,\beta} \neq \{1\}$. The computation continues:

(4) $\Gamma := \mathrm{Fix}_\Omega(G_{\alpha,\beta})$;

(5) if $|\Gamma|$ is not a power of p then goto line (14);

(6) $C := C_G(G_{\alpha,\beta})$; $P := O_p(C^\Gamma)$;

(7) if P is not transitive on Γ then goto line (14);

(to be continued further).

The motivation for calculating Γ, C and P is given by the following fact.

Lemma 3.2. *Let* G *be a permutation group that has a regular normal subgroup* M. *If* $X \leq G$, $\Gamma := \mathrm{Fix}(X)$ *and* $C := C_G(X)$, *then* C *is transitive on* Γ *and* $(C \cap M)^\Gamma$ *is a regular normal subgroup of* C^Γ.

Proof. Let $\gamma_1, \gamma_2 \in \Gamma$ and let u be the member of M such that $\gamma_1 u = \gamma_2$. If $x \in X$ then $\gamma_1 x^{-1} u = \gamma_1 u = \gamma_2 = \gamma_2 x^{-1}$, and so $\gamma_1 x^{-1} u x = \gamma_2$. But $x^{-1}ux \in M$ since $M \trianglelefteq G$ and so, since M acts regularly, $x^{-1}ux = u$. As x was an arbitrary member of X we have $u \in C$, that is, $u \in C \cap M$. Thus $C \cap M$ is transitive, therefore regular, on Γ, as required.

In our computation, if G does have an earns M then $C \cap M$ will be an elementary abelian p-group, so $|\Gamma|$ will be a power of p and $O_p(C^\Gamma)$ will be transitive on Γ. Thus if $|\Gamma|$ is not a power of p then G is not of affine type. Likewise, if $|\Gamma|$ is a power of p but P is not transitive on Γ then G cannot be of affine type. In practice the simple numerical test at line (5) is decisive in a large proportion of all cases.

At line (6) the algorithm requires the computation of $O_p(C^\Gamma)$. This could be done by calling the PCORE procedure described in the next section of this paper. That procedure calls EARNS, but, as it applies EARNS to groups that are no larger than its own input, and as EARNS calls PCORE to deal with groups of strictly smaller degree, these references back and forth are permissible under the rules for recursion. As a matter of fact, however, C^Γ is a rather small group because the stabiliser of the two points α, β in C^Γ is $\{1\}$ (that is, the pair α, β is a base for C^Γ), and so $|C^\Gamma| \leq |\Gamma|(|\Gamma| - 1)$, and Γ will usually be quite a small set. Consequently there is no need to call PCORE because much cruder methods will supply $O_p(C^\Gamma)$ quickly enough. It is possible that it might be worth adjusting line (3) in Step 1 so as to choose β to minimise $|\Gamma|$ but this is hardly necessary and could possibly conflict with an adjustment that I shall suggest later.

Step 3. If $|\Gamma|$ does turn out to be a power of p and P turns out to be transitive on Γ then the previous step will not have decided the matter. But if an earns M exists then $C \cap M \trianglelefteq O_p(C)$ and so $C \cap M \cap Z(O_p(C)) \neq \{1\}$, and we could find M by searching through $Z(O_p(C))$ for fixed-point-free elements of order p and testing each one to see if its normal closure in G acts regularly on Ω. Since $Z(O_p(C))$ should be quite a small group (see Lemma 3.4 below) the number of elements that have to be tested in the search should not be outrageous. Even so, these tactics can be improved.

Consider the natural homomorphism $f : C \to C^\Gamma$. Clearly,

$$\mathrm{Ker}(f) = C \cap G_{\alpha,\beta} = Z(G_{\alpha,\beta}) \leq Z(C).$$

Let Q be the p-primary subgroup of Ker(f), that is, of $Z(G_{\alpha,\beta})$, let $S := Z(P) = Z(O_p(C^\Gamma))$ and let R be the Sylow p-subgroup of $f^{-1}(S)$. Notice that $f^{-1}(S)$ is nilpotent (of class ≤ 2) and so R is uniquely determined. Furthermore, $Q \leq R$ and $S = R^\Gamma \cong R/Q$.

Lemma 3.3. *Suppose that* G *has a regular normal subgroup* M, *and let* $D := C \cap M$. *Then*

(i) $R \leq D \times Q$ *and* $R \cap D \neq \{1\}$,

(ii) $[R,C] \leq D$.

Proof. Since D^Γ is a normal p-subgroup of C^Γ it is contained in P, and since D^Γ is regular it is its own centraliser in C^Γ. Therefore $S = Z(P) \leq D^\Gamma$ and $f^{-1}(S) \leq f^{-1}(D^\Gamma)$. But $D \cap \mathrm{Ker}(f) = \{1\}$ and so

$$f^{-1}(D^\Gamma) = D \times \mathrm{Ker}(f) = D \times Z(G_{\alpha,\beta}).$$

Since R is the Sylow p-subgroup of $f^{-1}(S)$ it is contained in $D \times Q$, which is the p-primary subgroup of $D \times Z(G_{\alpha,\beta})$. Now $Q \leq R \leq D \times Q$, so $R = (R \cap D) \times Q$. Therefore $R^\Gamma = (R \cap D)^\Gamma = S \cap D^\Gamma$. Since D^Γ is a non-trivial normal subgroup of P its intersection with Z(P) is not trivial. That is, $S \cap D^\Gamma \neq \{1\}$ and so $R \cap D \neq \{1\}$. This proves (1).

In part (ii) the notation [R,C] stands for $\langle [x,y] \mid x \in R,\ y \in C \rangle$. We have $[R,C] \leq [D \times Q, C] \leq [D,C]$ since Q is central in C. But $[D,C] \leq D$ since $D \trianglelefteq C$. Therefore $[R,C] \leq D$ as promised.

Guided by this discussion we complete the algorithm as follows (recall that $(S)^{-1}$ is shorthand for $f^{-1}(S)$):

```
(8)       Q := p-primary subgroup of Z(G_{α,β});
(9)       R := p-primary subgroup of (Z(P))^{-1};
(10)      if R ≰ Z(C) then
             begin
(10.1)       choose x ∈ R, y ∈ C such that
             z := [x,y] ≠ 1;
(10.2)       M := ⟨z^G⟩;
(10.3)       if M is regular on Ω then goto line (15)
             else goto line (14);
             end;
```

```
(11)      if |Q| > p^(d-1) then goto line (14);
(12)      Q0 := Ω1(Q); R0 := Ω1(R); choose y ∈ R0 - Q0;
(13)      for z ∈ yQ0 do
             begin
(13.1)       M := ⟨z^G⟩;
(13.2)       if M is regular on Ω then goto line (15);
             end;
(14)      return to the outside world with the news that G
          is not of affine type;
(15)      return to the outside world with the news that M
          is the earns of G;
          End.
```

Justification. The calculation of Q in line (8) may be done very quickly; similarly, to calculate R in line (9) one simply needs to choose elements $x_1,\dots,x_s$ of C whose f-images generate Z(P), calculate their p-parts $y_1,\dots,y_s$, and take R to be $\langle y_1,\dots,y_s,Q\rangle$. Line (10) and its subsidiary lines are justified by Lemma 3.3 (ii). These lines are not a necessary part of the calculation but they are effective in a high proportion of cases and then they save the relatively expensive search at line (13). The bound used in line (11) is justified as follows.

Lemma 3.4. *Let* H *be any group, let* A *be an abelian* p*-subgroup of order* q*, let* $C := C_H(A)$ *and let* $m := |H:C|$. *If* $O_p(H) = \{1\}$ *then* $q \leq m$.
Proof. We use induction on m. If $m = 1$ then $A \leq O_p(H)$ and so the assertion certainly holds. Suppose therefore that $m > 1$, that the assertion is known to hold for smaller values of m, and that $O_p(H) = \{1\}$.

If the conjugates of A in H all centralise A then they centralise each other. In this case the normal closure of A is a normal p-subgroup of H and therefore $q = 1 < m$. Suppose therefore that there is a conjugate B of A which does not centralise A. Without loss of generality we may also suppose that

$$|A \cap C_H(B)| \geq |B \cap C_H(A)|.$$

Let $A^* := A \cap C_H(B)$ and $C^* := C_H(A^*)$. Since $A^* \leq A$ we have $C \leq C^*$, and since $A^* \leq C_H(B)$ we have $B \leq C^*$. Therefore, since $B \not\leq C$, we have $|H:C^*| < |H:C|$ and, by inductive hypothesis, $|A^*| \leq |H:C^*|$. Also,

$$|C^*:C| \geq |B:B \cap C| = |B|.|B \cap C_H(A)|^{-1}$$
$$\geq |A|.|A \cap C_H(B)|^{-1} = |A:A^*|.$$

Consequently,

$$q = |A| = |A:A^*|.|A^*| \leq |C^*:C|.|H:C^*| = |H:C| = m,$$

as required.

Corollary 3.5. *In the situation of* Lemma 3.3 *we have* $|Q| \leq p^{d-1}$.
Proof. Let $H := G_\alpha$. Then H acts faithfully and irreducibly on M, and so $O_p(H) = \{1\}$. Therefore $|Q| \leq |H:C_H(Q)| \leq |H:H_\beta| \leq n-1$, whence $|Q| \leq p^{d-1}$.

As a matter of fact, $|Q|$ is very much smaller than p^{d-1} for any primitive permutation group of degree p^d; thus line (11) never operates and the search at line (13) is much shorter than Corollary 3.5 suggests. I have, however, been able to find only a rather long and unsatisfying proof depending on the classification of the finite simple groups. The inclusion of line (11) seems a small price to pay to circumvent this.

It may possibly be worth spending what little time it would take at line (3) in Step (1) choosing β judiciously. If at that point one were to minimise either $|\mathrm{Fix}(G_{\alpha,\beta})|$ or $|\Omega_1(\text{p-component}(Z(G_{\alpha,\beta})))|$ then one might make useful savings in later parts of the calculation, particularly in the search at line (13). Whether it is worth doing this or not is a question that should be settled experimentally.

4. *THE PCORE ROUTINE*

The EARNS routine can be used to find $O_p(G)$ in case G is known to be primitive. It is useful to be able to find $O_p(G)$ in general however, and the PCORE routine is proposed as an algorithm to do this. It is heavily based on recursion in that it calls itself on many occasions; it also calls EARNS, which at line (6) may call PCORE. At each such call, however, it will be clear that the new input is a group of strictly smaller degree or order. There are several points at which the routine calls for the computation of $O_p(N)$ where N is a proper normal subgroup of G. It would be permissible, since N is strictly smaller than G, to use recursion directly, but to do so would usually be uneconomic since N may have many orbits (think of 512 orbits, each of length 4) and computing $O_p(N)$ would, in effect, require most of the algorithm to be performed separately for each orbit. Accordingly we shall proceed differently at these points.

If Γ is a block of imprimitivity for G in Ω a subset $\{x_1,\ldots,x_s\}$ of G will be called a 'transversal' for Γ in G if $\Gamma x_1,\ldots,\Gamma x_s$ are all different and $\Omega = \Gamma x_1 \cup \ldots \cup \Gamma x_s$. It takes negligible computing time to find a transversal for Γ in G.

Lemma 4.1. *Suppose that* G *is transitive on* Ω *and that* $N \triangleleft G$. *Let* Γ *be an* N-*orbit,* T *a transversal for* Γ *in* G *and* $K := (O_p(N^\Gamma))^{-1}$. *Then*
$O_p(N) = \bigcap_{x\in T} x^{-1}Kx$.

Proof. Recall that $(O_p(N^\Gamma))^{-1}$ is shorthand for the complete inverse image in N of the subgroup $O_p(N^\Gamma)$ of N^Γ. Let $P := \bigcap_{x\in T} x^{-1}Kx$. Certainly $O_p(N)^\Gamma \leq O_p(N^\Gamma)$, so $O_p(N) \leq K$ and, since $O_p(N) \leq G$, therefore $O_p(N) \leq P$. On the other hand, $P^{\Gamma x} = x^{-1}(P^\Gamma)x$ and this is a subgroup of $x^{-1}K^\Gamma x$, which is a p-group. Since P is a subdirect product of the groups $P^{\Gamma x}$ for $x \in T$, it is itself a p-group, and since $P \triangleleft N$ we have $P \leq O_p(N)$. Therefore $O_p(N) = P$, as required.

Here, then, is PCORE. Its input consists of a prime number p and an arbitrary subgroup G of Sym(Ω), for which a strong generating set has already been calculated.

Step 1. This first step is designed to reduce the calculation to the transitive case:

```
          Begin
(1)       if G = {1} then
             begin
(1.1)        K := {1}; goto line (24);
             end;
(2)       Ω1 := αG;
(3)       if Ω1 ≠ Ω then
             begin
(3.1)        Ω2 := Ω - Ω1;
(3.2)        K1 := (Op(G^Ω1))^-1; K2 := (Op(G^Ω2))^-1;
(3.3)        K := K1 ∩ K2; goto line (24);
             end;
(to be continued).
```

Justification. At line (3.2) the recursion is permissible because Ω_1 and Ω_2 are proper subsets of Ω. Clearly $O_p(G)^{\Omega_1} \leq O_p(G^{\Omega_1})$ and so $O_p(G) \leq K_1$; similarly $O_p(G) \leq K_2$, and so $O_p(G) \leq K$. On the other hand $K \trianglelefteq G$ and K is a p-group since it is a subdirect product of K^{Ω_1} and K^{Ω_2}, both of which are p-groups; therefore $K \leq O_p(G)$. Thus $K = O_p(G)$ and line (3.3) is correct.

Step 2. We suppose from now on that $\Omega_1 = \Omega$, that is, that G is transitive on Ω. The computation continues:

```
(4)       find a minimal non-trivial congruence ρ1;
(5)       if ρ1 = Ω × Ω then
             begin
(5.1)        if n is not a power of p then
                begin
(5.1.1)         K := {1}; goto line (24);
                end;
(5.2)        call EARNS;
(5.3)        if G has an earns M then K := M else K := {1};
(5.4)        goto line (24);
             end;
(to be continued further).
```

Justification. Line (4) can be accomplished using Atkinson's algorithm. If ρ_1 turns out to be the universal relation then G is primitive on Ω. In this case if n is not a power of p then $O_p(G)$ must be trivial, while if n is a power of p then $O_p(G)$ will be the earns if there is one, {1} otherwise.

Step 3. We now assume that G is imprimitive so that our minimal non-trivial congruence ρ_1 is proper. The algorithm continues:

```
(6)      Ω1 := Ω/ρ1;  N1 := Ker(G^Ω1);
(7)      K1 := (O_p(G^Ω1))^-1;
(8)      if N1 is a p-group then
            begin
(8.1)       K := K1; goto line (24);
            end;
(9)      if K1 ≠ G then
            begin
(9.1)       Γ := αK1; T := a transversal for Γ in G;
(9.2)       K := (O_p(K1^Γ))^-1; for t ∈ T do K := K ∩ t^-1 Kt;
(9.3)       goto line (24);
            end;
(10)     if there exists a non-trivial congruence ρ2
         that does not contain ρ1 then
            begin
(10.1)      Ω2 := Ω/ρ2;
(10.2)      K := (O_p(G^Ω2))^-1; goto line (24);
            end;
(to be continued further).
```

Justification. Recursive calls are permissible at lines (7) and (10.2) because Ω_1 and Ω_2 are strictly smaller than Ω, and at line (9.2) because K_1 is strictly smaller than G. At line (8), if N_1 is a p-group (in particular, if $N_1 = \{1\}$, which means that G is faithful on Ω_1) then $N_1 \leq O_p(G)$ and $O_p(G)/N_1 = O_p(G/N_1)$, which is why $K_1 = O_p(G)$. Clearly $O_p(G) = O_p(K_1)$ and so lines (9) to (9.3) are justified by Lemma 4.1. At line (10.2) we certainly have $O_p(G) \leq K$; on the other hand, if we have come beyond line (9) then G^{Ω_1} is a p-group and so both K^{Ω_1} and K^{Ω_2} are p-groups, it follows that K is a p-group because K is a subdirect product of K^{Ω_1} and K^{Ω_2} since $\rho_1 \cap \rho_2$ is the trivial congruence; thus $K = O_p(G)$, as claimed.

Step 4. If we come past lines (8), (9) and (10) of the program then we know that G is transitive but imprimitive on Ω, with ρ_1 as its unique minimal non-trivial congruence, that G^{Ω_1} is a p-group and that N_1 is not a p-group. Let $Z_1 := O_p(N_1)$.

Lemma 4.2. *If* $Z_1 = \{1\}$ *then* $O_p(G) = \{1\}$.

Proof. Let $P := O_p(G)$ and suppose that $Z_1 = \{1\}$ whereas $P \neq \{1\}$. Then $N_1 \cap P = \{1\}$ and so P, N_1 centralise each other. Let ρ be equivalence modulo P. Then ρ is a non-trivial G-congruence, so $\rho_1 \leq \rho$ and N_1 fixes every ρ-class, that is, every P-orbit, set-wise. If Γ is any P-orbit then $N_1^\Gamma \leq C_{Sym(\Gamma)}(P^\Gamma)$ and this centraliser is a p-group. Therefore N_1, which is a subdirect product of the groups N_1^Γ as Γ ranges over the P-orbits, would be a p-group: and we know that this is not so. Thus if $Z_1 = \{1\}$ then $O_p(G) = \{1\}$.

Lemma 4.3. *If $Z_1 \neq \{1\}$ then Z_1 is transitive on each ρ_1-class and Z_1 is an elementary abelian* p-*group. Moreover, if $\{1\} \neq Z \leq Z_1$ and $Z \trianglelefteq G$ then* $C_G(Z) \leq O_p(G)$.

Proof. Since Z_1-orbits are non-trivial blocks of imprimitivity for G and are contained in ρ_1-classes, by minimality of ρ_1 the Z_1-orbits are the ρ_1-classes. By the same argument $\Omega_1(Z(Z_1))$ is transitive on each ρ_1-class. It follows that for each ρ_1-class Γ we have $Z_1^\Gamma = (\Omega_1(Z(Z_1)))^\Gamma$ (because a transitive abelian subgroup of $\mathrm{Sym}(\Gamma)$ is self-centralising) and so Z_1^Γ is an elementary abelian p-group. Therefore Z_1, which is a subdirect product of the groups Z_1^Γ, is an elementary abelian p-group.

Now suppose that $\{1\} \neq Z \leq Z_1$ and $Z \trianglelefteq G$. Then, as above, Z must be transitive on each ρ_1-class and so $C_{N_1}(Z)$ is a subdirect product of the groups Z^Γ as Γ ranges over ρ_1-classes. Therefore $C_{N_1}(Z)$ is an elementary abelian p-group and so $C_{N_1}(Z) = Z_1$. Also, $C_G(Z)/C_{N_1}(Z) \cong C_G(Z)^{\Omega_1}$ and this is a p-group since G^{Ω_1} is a p-group. Consequently $C_G(Z)$ is a p-group and, since $C_G(Z) \trianglelefteq G$, we have $C_G(Z) \leq O_p(G)$.

Lemma 4.3 will be used later. For now we use Lemma 4.2 and continue the algorithm as follows:

```
(11)        Γ := αN₁; T := a transversal for Γ in G;
(12)        Z₁ := (O_p(N₁^Γ))^(-1); for t ∈ T do Z₁ := Z₁ ∩ t^(-1)Z₁t;
(13)        if Z₁ = {1} then
               begin
(13.1)         K := {1}; goto line (24);
               end;
(14)        find a maximal proper congruence ρ₂;
(15)        Ω₂ := Ω/ρ₂; N₂ := Ker(G^Ω₂);
(16)        Γ := αN₂; T := a transversal for Γ in G;
(17)        Z₂ := (O_p(N₂^Γ))^(-1); for t ∈ T do Z₂ := Z₂ ∩ t^(-1)Z₂t;
(18)        if αZ₂ ≠ ρ₂(α) then
               begin
(18.1)         ρ* := equivalence modulo Z₂;
(18.2)         for congruences ρ such that
               ρ* ≤ ρ and |ρ:ρ*| = p do
                  begin
(18.2.1)          N := Ker(G^(Ω/ρ));
(18.2.2)          Γ := αN; T := a transversal for Γ in G;
(18.2.3)          K := (O_p(N^Γ))^(-1); for t ∈ T do K := K ∩ t^(-1)Kt;
(18.2.4)          if K ≠ Z₂ then goto line (24);
                  end;
(18.3)         K := Z₂; goto line (24);
               end;
(to be continued further).
```

Justification. By Lemma 4.1 lines (11), (12) produce $Z_1 := O_p(N_1)$ and then lines (13), (13.1) are justified by Lemma 4.2. The search for a maximal proper congruence ρ_2 can be done very rapidly at line (14) by Atkinson's algorithm. Notice that, since G^{Ω_1} is a p-group, $|\Omega_2| = p$ and $G/N_2 \cong C_p$. Therefore either $O_p(G) \leq N_2$ or $|O_p(G) : O_p(N_2)| = p$: that is, Z_2, which is $O_p(N_2)$ calculated according to Lemma 4.1 by lines (16) and (17), either is $O_p(G)$ or is a subgroup of index p in $O_p(G)$. Let ρ^* be equivalence modulo Z_2 and let ρ^{**} be equivalence modulo $O_p(G)$. Of course, if $Z_2 = O_p(G)$ then $\rho^* = \rho^{**}$, but if $Z_2 < O_p(G)$ then $\rho^{**} \not\leq \rho_2$ and so ρ^{**} must contain ρ^* with index p (that is, each ρ^{**}-class is a union of p different ρ^*-classes). Therefore if $\rho^* < \rho_2$, that is, if Z_2 is not transitive on the ρ_2-classes, then we can use recursion to find $O_p(G)$: we search through the congruences ρ that contain ρ^* with index p; since ρ is then a proper congruence $N := \mathrm{Ker}(G^{\Omega/\rho})$ is an intransitive normal subgroup of G and lines (18.2.2), (18.2.3) produce $O_p(N)$ by Lemma 4.1; since $Z_2 \leq N$, if $O_p(N) \neq Z_2$ then $O_p(N) = O_p(G)$, which explains line (18.2.4); if $O_p(N) = Z_2$ for all the congruences ρ that contain ρ^* with index p then $O_p(G) = Z_2$, which explains line (18.3).

The congruences that contain ρ^* with index p may be identified with minimal congruences in Ω/ρ^* and Atkinson's algorithm seeks them out very quickly indeed. The number of them is at most $(|\Omega/\rho^*| - 1)/(p-1)$, which is less than $n/p(p-1)$. If there turn out to be anywhere like this number of them then $|\rho^*|$ will be small, so N will be small and the calculation of $O_p(N)$ at lines (18.2.2), (18.2.3) should be quick. Therefore the search in lines (18.1) to (18.3) is not likely to be too expensive.

Step 5. If the computation comes through past line (18) then Z_2 (which, recall, is $O_p(N_2)$) is transitive on ρ_2-classes. The algorithm is completed like this:

```
(19)     find x0 ∈ G - N2;   x := p-part of x0;
(20)     X := ⟨Z2,x⟩;   Y := C_G(X);
(21)     if  Y ≰ Z2 then
             begin
(21.1)       K := ⟨Z2,Y⟩; goto line (24);
             end;
(22)     K := Z2;
(23)     for z ∈ (Y ∩ Z1) - {1} do
             begin
(23.1)       Z := ⟨z^G⟩;
(23.2)       K := ⟨K, C_G(Z)⟩;
             end;
(24)     return to the outside world with the news that O_p(G) = K;
         End.
```

Justification. Computing x at line (19) takes negligible time; then x acts as a p-cycle on Ω_2 and so X is a transitive p-subgroup of G. Therefore its centraliser Y is semi-regular on Ω and is a p-group. Moreover,

$Y = C_G(X) \leq C_G(Z_1)$ and $C_G(Z_1) \leq O_p(G)$ by Lemma 4.3; thus $Y \leq O_p(G)$. If $Y \nleq Z_2$ then $Z_2 < \langle Z_2, Y \rangle \leq O_p(G)$ and so, since $|O_p(G) : Z_2| \leq p$, we have $O_p(G) = \langle Z_2, Y \rangle$. This explains lines (21) and (21.1). If $z \in (Y \cap Z_1) - \{1\}$ and $Z := \langle z^G \rangle$ then $C_G(Z) \leq O_p(G)$ by Lemma 4.3, and so, after line (23.2) we certainly have $K \leq O_p(G)$. We need to prove the reverse inclusion. Let P be a Sylow p-subgroup of G that contains X. Certainly $Z(P) \leq C_G(X) = Y$. Now $Z(O_p(G)) \cap Z_1 \cap Z(P) \neq \{1\}$ and so $Z(O_p(G)) \cap Z_1 \cap Y \neq \{1\}$. Therefore there exists $z \in (Z_1 \cap Y) - \{1\}$ such that $z \in Z(O_p(G))$ and so $O_p(G) \leq C_G(\langle z^G \rangle)$. Consequently $O_p(G) \leq K$ and so $O_p(G) = K$ after the search at lines (23) to (23.2) has been completed. The cost of that search should be quite acceptable because $Y \cap Z_1$, which is semi-regular on Ω, is quite a small group.

5. *THE SOCLE ROUTINE*

Suppose now that G is known to be primitive and not of affine type. The SOCLE routine is designed to find its minimal normal subgroups. As in §2, let M be a minimal normal subgroup $X_1 \times X_2 \times \ldots \times X_d$, where $X_1, X_2, \ldots, X_d$ are all isomorphic to a certain non-abelian simple group S. Let

$$K := \bigcap_{i:=1}^{d} N_G(X_i) = \mathrm{Ker}(G^\Sigma),$$

where $\Sigma = \{1,2,\ldots,d\}$ as in §2. Recall that if G is of type III, IV or V in the classification of §2 then M = Soc(G). If G is of type II then

$$\mathrm{Soc}(G) = M_1 \times M_2 = X_1 \times \ldots \times X_d \times X_{d+1} \times \ldots \times X_{2d},$$

where $M_1 = M$ and $M_2 = X_{d+1} \times \ldots \times X_{2d}$. In any case, K is the kernel of the permutation representation of G acting by conjugation on the set of *all* simple direct factors of its socle.

Lemma 5.1. *The quotient* K/Soc(G) *is soluble.*
This follows immediately from the truth of the "Schreier Conjecture" according to which Aut(S)/Inn(S) (which we think of as Aut(S)/S) is soluble. For, since $C_G(\mathrm{Soc}(G)) = \{1\}$, we have

$$\mathrm{Soc}(G) = \prod X_i \leq K \leq \prod \mathrm{Aut}(X_i),$$

and so K/Soc(G) is isomorphic to a subgroup of $\prod \mathrm{Aut}(X_i)/X_i$. In fact, by examining the list of finite simple groups with care one convinces oneself that Aut(S)/S is soluble of derived length ≤ 3, from which it follows that $\mathrm{Soc}(G) = K'''$.

For groups of moderate degree it is almost always the case that G/K is soluble and that therefore Soc(G) is $G^{(\infty)}$, the last term of the derived series of G:

Lemma 5.2. *If* $n < 60^4 = 12{,}960{,}000$ *and* n *cannot be expressed in the form* $m_0^{d_0}$ *with* $m_0 \geq 5$ *and* $d_0 \geq 5$ *then* G/K *is soluble. In particular, if* $n < 3125$ *then* G/K *is soluble.*
Proof. Suppose that G/K is insoluble. Then $d \geq 5$ and so G is certainly not of type V. If G is of type II then $n = |S|^d \geq 60^5$. If G is of type III then

$n = |S|^{b(c-1)}$. Since G^{Σ} is insoluble and has a system of imprimitivity consisting of b blocks of size c, either $b \geq 5$ or $c \geq 5$. Therefore $n = |S|^{b(c-1)} \geq 60^4$. If G is of type IV then $n = m^d$ where m is the degree of a faithful permutation representation of S and so $m \geq 5$. Thus $n \geq 5^5$ and if $n < 60^4$ then n must be expressible in the form $m_0^{d_0}$ with $m_0 \geq 5$ and $d_0 \geq 5$.

For the reader's convenience and to put the lemma into perspective, here is a table of the powers of 2, 3, 5, 6, 7 that lie below 1,000,000:

1	2	3	5	6	7
2	4	9	25	36	49
3	8	27	125	216	343
4	16	81	625	1296	2401
5	32	243	3125	7776	16807
6	64	729	15625	46656	117649
7	128	2187	78125	279936	823543
8	256	6561	390625		
9	512	19683			
10	1024	59049			
11	2048	177147			
12	4096	531441			
13	8192				
14	16384				
15	32768				
16	65536				
17	131072				
18	262144				
19	524288				

In practice Lemma 5.2 means that if $n \leq 3600$ then $\mathrm{Soc}(G) = G^{(\infty)}$ unless $n = 3125$; if $n \leq 10^5$ then $\mathrm{Soc}(G) = G^{(\infty)}$ except perhaps when n is one of 5^5, 5^6, 5^7, 6^5, 6^6, 7^5, 8^5, 9^5, 10^5; and in the range $n \leq 10^6$ the only further exceptional degrees that have to be considered are 5^8, 6^7, 7^6, 7^7, 8^6, 9^6, 10^6, 11^5, 12^5, 13^5, 14^5 and 15^5.

To deal with these exceptional cases we use variants of the following very general observation.

Lemma 5.3. *Let X be a finite group that has no non-trivial abelian normal subgroups. If T is a Sylow 2-subgroup and* $N := \langle Z(T)^X \rangle$ *then* $\mathrm{Soc}(X) \leq N \leq K$, *where K is the intersection of the normalisers of the simple direct factors of* Soc(X).

Proof. Let $X_1, X_2, \ldots, X_s$ be the simple direct factors of Soc(X). Since X_i is subnormal in X the intersection $T_i := X_i \cap T$ is a Sylow 2-subgroup of X_i, and $T_i \neq \{1\}$ by the Feit-Thompson theorem. If $x \in T - K$ then x acts non-trivially on $\{X_1, X_2, \ldots, X_s\}$ by conjugation and therefore it also acts non-trivially on $\{T_1, T_2, \ldots, T_s\}$; thus x cannot lie in Z(T). Consequently $Z(T) \leq K$ and so $N = \langle Z(T)^X \rangle \leq K$. Also, if M is any minimal (non-trivial) normal subgroup of X then $M \cap T$ is a non-trivial normal subgroup of T and so $M \cap Z(T) \neq \{1\}$, from which it follows that $M \leqslant \langle Z(T)^X \rangle$. Thus $Soc(X) \leqslant N \leqslant K$ as the lemma states.

Although this observation may certainly be used to calculate Soc(G) (and indeed, has been so used [4], p.167), it is not very well suited for general-purpose computing. The group $S_{32} \operatorname{Wr} S_2$ in its product representation of degree 1024 has Sylow 2-subgroups of order 2^{63} and so a computation to find one is not likely to succeed. Of course, by Lemma 5.2, we could find the socle of this particular group by calculating the last term of its derived series, but there are other groups, such as $S_5 \operatorname{Wr} S_5$ in its product representation of degree 3125, whose Sylow 2-groups have order 2^{18}, or $S_8 \operatorname{Wr} S_5$ in its product representation of degree 32,768, whose Sylow 2-groups have order 2^{38}, which are not covered by Lemma 5.2 and for which a computation to find a Sylow 2-subgroup is likely to be prohibitively expensive. To reduce the cost we shall work with other prime numbers.

Lemma 5.4. *Let* p *be a prime divisor of* $|S|$, *let* P *be a Sylow* p*-subgroup of* G *and let* $N := \langle Z(P)^G \rangle$. *Then* $Soc(G) \leqslant N \leqslant K$.
The proof is the same as that of Lemma 5.3 except that the Feit-Thompson theorem is of course not needed.

On the face of it there is still a difficulty that stands in the way if we wish to use Lemma 5.4 for computational purposes. Since we do not yet know S we are unlikely to have the set of prime divisors of $|S|$ available. Of course we know *some* of these primes because every prime divisor of n must divide $|S|$. But the Sylow p-subgroups for a prime p that divides n may be very large, as in the cases of $S_{32} \operatorname{Wr} S_2$ and $S_8 \operatorname{Wr} S_5$ in their product representations mentioned above. Accordingly we turn to the largest prime that divides $|G|$.

Lemma 5.5. *Let* p *be the largest prime number that divides* $|G|$, *let* P *be a Sylow* p*-subgroup of* G *and let* $N := \langle Z(P)^G \rangle$. *If* $n < 78{,}125$ *then* $N = Soc(G)$.
Proof. If p does not divide $|K|$ then $p > 7$ because at least three different primes divide $|S|$ and p is larger than all of them; moreover, p then divides $|G/K|$. It follows that $d \geqslant p$ and then that $n \geqslant 5^7$. Therefore if $n < 5^7 = 78{,}125$ then p divides $|K|$. Examination of the list of simple groups convinces one that the largest prime dividing $|Aut(S)|$ does not divide $|Aut(S)/S|$, and therefore p divides $|S|$ but not $|K/Soc(G)|$. Now the argument of Lemma 5.2 proves that if $n < 5^7$ then $N = Soc(G)$.

If at some future time the condition $n < 78{,}125$ becomes irksome then one might use the following in place of Lemma 5.5.

Lemma 5.6. *Let* p *be the largest prime number such that* p^2 *divides* $|G|$, *let* P *be a Sylow* p*-subgroup of* G *and let* $N := \langle Z(P)^G \rangle$. *If* $n < 5^{14} = 6{,}103{,}515{,}625$ *then* $Soc(G) = N^{(\infty)}$
Proof. If G is of type V then (assuming the "Schreier Conjecture") there is nothing to prove. Suppose therefore that G is of type II, type III or type IV, that is, that Soc(G) is not simple, and let p_0 be the largest prime divisor of $|S|$. Then certainly p_0^2 divides $|G|$. If $p = p_0$ then $Soc(G) \leqslant N \leqslant K$

by Lemma 5.4. Otherwise $p > p_0$. The prime divisors of $|Aut(S)/S|$ are smaller than p_0 and so in this case p does not divide $|K|$. Therefore p^2 divides $|G/K|$ and, since G/K has a faithful permutation representation of degree d, we must have $d \geq 2p \geq 14$ and $n \geq 5^{14}$. Thus if $n < 5^{14}$ then $Soc(G) \leq N \leq K$ and so $Soc(G) = N^{(\infty)}$ by Lemma 5.1.

It is perhaps worth observing that if $n \leq 2^{25}$ then in fact $N = Soc(G)$ in this lemma. But the proof requires careful examination of the automorphisms of groups of Lie type and the small piece of extra information gleaned hardly seems worth the effort.

Guided by these lemmas we design the SOCLE routine as follows. The input must be a permutation group G of degree n that is known to be primitive and not of affine type, and for which a strong generating set is available.

Version 1 (appropriate if $n < 78,125$):

Begin

(1) if n is not of form $m_0^{d_0}$ with $m_0 \geq 5$ and $d_0 \geq 5$ then

begin

(1.1) $L := G^{(\infty)}$; goto line (6);

end;

(2) $p :=$ largest prime divisor of $|G|$;

(3) $P :=$ a Sylow p-subgroup of G;

(4) $Z := Z(P)$;

(5) $L := \langle Z^G \rangle$;

(6) return to the outside world with the news that $Soc(G) = L$;

End.

Here line (1) is justified by Lemma 5.2; it is not a necessary part of the program, but is effective in almost all cases and then it saves the search for P which is relatively expensive. Lines (2) to (6) are justified by Lemma 5.5.

Version 2 (appropriate if $n < 6,103,515,625$):

Begin

(1) if $n < 12,960,000$ and n is not of form $m_0^{d_0}$ with $m_0 \geq 5$ and $d_0 \geq 5$ then

begin

(1.1) $L := G^{(\infty)}$; goto line (6);

end;

(2) $p :=$ largest prime number such that p^2 divides $|G|$;

(3) $P :=$ a Sylow p-subgroup of G;

(4) $Z := Z(P)$;

(5) $L := \langle Z^G \rangle^{(\infty)}$;

(6) return to the outside world with the news that $Soc(G) = L$;

End.

Here line (1) is justified by Lemma 5.2; lines (2) to (6) are justified by Lemma 5.6.

In both versions of SOCLE the slowest part ofthe calculation will usually be the search for a Sylow p-subgroup at line (3). This can, however, be speeded up:

Version 3 (a modification of Version 1 or Version 2):

. . . .

(3) find a p-element $z \in G - \{1\}$ such that p does not divide $|G:C_G(z)|$;

(4) $L := \langle z^G \rangle$; ($L := \langle z^G \rangle^{(\infty)}$; in Version 2);

(5) if L is regular on Ω then $L := \langle L, C_G(L) \rangle$;

(6) return to the outside world with the news that $Soc(G) = L$;

End.

This works because z at line (3) is an element of the centre of some Sylow p-subgroup and so, by Lemmas 5.5 or 5.6 and 5.1, $L \leqslant Soc(G)$. Therefore if G is of type III, IV or V then $L = Soc(G)$. If G is of type II then L could be one of the two minimal normal subgroups of G, but then L would be regular on Ω and $Soc(G) = L \times C_G(L)$. Version 3 represents an improvement over Versions 1 and 2 because one of the earlier steps in the search for a Sylow subgroup, at least in one possible algorithm, is to find a p-element $z \in G - \{1\}$ whose centraliser contains a Sylow p-subgroup of G. Thus Version 3 by-passes the rest of the Sylow subgroup search.

6. *THE SOCLEBITS ROUTINE*

Suppose now that G is primitive on Ω, not of affine type, and that we have computed its socle L. Our next problem is to find the simple direct factors $X_1, \ldots, X_d$ (or $X_1, \ldots, X_d, X_{d+1}, \ldots, X_{2d}$ if G is of type II) of L. We do this as follows:

Begin

(1) find a maximal proper L-congruence ρ on Ω;

(2) $\Gamma := \Omega/\rho$; $Y := Ker(L^\Gamma)$; $X := C_L(Y)$;

(3) if $|X| \neq |\Gamma|^2$ then goto line (7);

(4) choose $x \in X - \{1\}$;

(5) $Z := Z(C_X(x))$; $Z_0 :=$ p-component of Z for some prime p dividing $|Z|$; $Z_1 := \Omega_1(Z_0)$;

```
(6)        for z ∈ Z_1 - {1} do
             begin
(6.1)          X_1 := ⟨z^X⟩;
(6.2)          if X_1 ≠ X then
                 begin
(6.2.1)            compute the G-conjugates X_1,...,X_d of X_1;
(6.2.2)            compute the G-conjugates X_{d+1},...,X_{2d} of C_X(X_1);
(6.2.3)            return to the outside world with the news (that G is of
                   type II, its minimal normal subgroups are X_1 ×...× X_d and
                   X_{d+1} ×...× X_{2d}, and) that the simple direct factors of L
                   are X_1,...,X_{2d};
                 end;
(7)        compute the G-conjugates X_1,...,X_d of X;
(8)        return to the outside world with the news that the simple direct
           factors of L are X_1,...,X_d;
           End.
```

Justification. The search for ρ may be accomplished very quickly using Atkinson's algorithm at line (1). Since L is a direct product of non-abelian simple groups, $L = Y \times C_L(Y)$ for any normal subgroup Y of L, in particular for $Ker(L^\Gamma)$. Thus $L = X \times Y$ at line (2). Now L^Γ is primitive, $L^\Gamma = X^\Gamma$, and X acts faithfully on Γ. Since X is a direct product of simple groups and is faithful and primitive on Γ the O'Nan-Scott Theorem tells us that either X is simple or X is of type II and has two simple factors (as in Example 2.2). In the latter case each of the simple factors acts regularly on Γ and so $|X| = |\Gamma|^2$. Thus if $|X| \neq |\Gamma|^2$ then X must be one of the simple direct factors of L and the others are its G-conjugates. This explains line (3).

Suppose that $X = A \times B$ where A and B are simple groups. If $x \in X - \{1\}$, say $x = ab$ with $a \in A$, $b \in B$, then $C_X(x) = C_A(a) \times C_B(b)$. Therefore if

$$Z_A := \Omega_1(\text{p-component } (Z(C_A(a))),$$

and

$$Z_B := \Omega_1(\text{p-component } (Z(C_B(b))),$$

then $Z_1 = Z_A \times Z_B$ and at least one of Z_A, Z_B will be non-trivial. A search through Z_1 must therefore produce a non-trivial element z of A or of B, and $\langle z^X \rangle$ will then be A or B respectively. This explains lines (6.2) to (6.2.3). If $\langle z^X \rangle = X$ for all $z \in Z_1 - \{1\}$ then X is simple, line (6.2) is never operative and after the search at line (6) we come through to line (7) as we should. (As a matter of fact, if $|X| = |\Gamma|^2$ then X is not simple and line (6.2) will operate at some stage.) The point of computing Ω_1(p-component $(Z(C_X(x)))$) at line 5 is that this will be a very small group and so the search at line (6) is quite short.

7. *THE COMPFACTORS ROUTINE*

We observed in §1 that if N is a normal subgroup of our permutation group G it is not always possible to compute economically in the quotient group G/N. For certain special types of normal subgroup N there is, however, no problem:

(i) if $N = \mathrm{Ker}(G^{\Omega_1})$ where Ω_1 is a G-invariant subset of Ω then G/N is just G^{Ω_1};

(ii) if $N = \mathrm{Ker}(G^{\Omega_1})$ where Ω_1 is Ω/ρ for a G-congruence ρ then again G/N is just G^{Ω_1};

(iii) if N acts regularly on Ω, as, for example, when G is primitive of affine type and N is its minimal normal subgroup, then G_α is a complement for N in G and so G/N may be identified with G_α acting on $\Omega - \{\alpha\}$ (or, of course, on Ω);

(iv) if G is primitive and not of affine type and N is the group K of §5, then G/N is just G^Σ, where $\Sigma = \{1,2,\ldots,d\}$ as in §2.

Such normal subgroups as these turn out to be sufficient for our purposes.

COMPFACTORS will make heavy use of recursion to compute a list of the composition factors of G. Let $\ell(X)$ denote the composition length of a finite group X. In order to estimate how deep the recursion dives and how long the list of composition factors in the output may be we need

Lemma 7.1. *If* $t := |\mathrm{Orb}(G,\Omega)|$ *then* $\ell(G) \leq \frac{4}{3}(n - t)$.

Proof. It seems simplest to reduce this lemma to the soluble case and so we first prove that every finite group G has a soluble subgroup G* such that $\ell(G) \leq \ell(G^*)$. We use induction on $|G|$. If G is soluble (and, in particular, if $G = \{1\}$) we take $G^* := G$, as we must. Suppose now that G is insoluble. Let $H := G^{(\infty)}$ and let K be a normal subgroup of G maximal subject to being properly contained in H. The chief factor H/K of G is a direct product of groups $X_1,\ldots,X_d$, each isomorphic to one and the same simple group S. Let p be a prime number that divides $|S|$ and let P/K be a Sylow p-subgroup of H/K. Notice that, as $K \trianglelefteq P$ and P/K is a direct product of d non-trivial groups, we have

$$\ell(H) = d + \ell(K) \leq \ell(P/K) + \ell(K) = \ell(P).$$

Now let $N := N_G(P)$. By the Frattini Lemma we have $G = N.H$ and so $\ell(G/H) = \ell(N/(N \cap H))$. Since $\ell(N/(N\cap H)) = \ell(N/P) - \ell((N\cap H)/P) \leq \ell(N/P)$ it follows that

$$\ell(G) = \ell(G/H) + \ell(H) \leq \ell(N/P) + \ell(P) = \ell(N).$$

The group N is a proper subgroup of G and, by inductive hypothesis, there is a soluble subgroup G* of N with $\ell(N) \leq \ell(G^*)$. Then G* is a soluble subgroup of G such that $\ell(G) \leq \ell(G^*)$.

To prove the lemma we can suppose now that our subgroup G of $\mathrm{Sym}(\Omega)$ is soluble. We use induction on n: the assertion is certainly true if n is 1, so our inductive hypothesis is that it is true for groups of smaller degree than G. If G is intransitive and Ω_1 is an orbit of length n_1, and $K := \mathrm{Ker}(G^{\Omega_1})$, then

$$\ell(G) \leqslant \ell(G^{\Omega_1}) + \ell(K) \leqslant \tfrac{4}{3}(n_1 - 1) + \tfrac{4}{3}((n - n_1) - (t - 1))$$

by inductive hypothesis; and therefore $\ell(G) \leqslant \frac{4}{3}(n - t)$ as required. Suppose then that G is transitive but imprimitive, let ρ be a non-trivial proper congruence, and let $\Omega_1 := \Omega/\rho$, $K := \mathrm{Ker}(G^{\Omega_1})$. If $n_1 := |\Omega_1|$ then K has at least n_1 orbits in Ω. Therefore

$$\ell(G) = \ell(G^{\Omega_1}) + \ell(K) \leqslant \tfrac{4}{3}(n_1 - 1) + \tfrac{4}{3}(n - n_1)$$

by inductive hypothesis; that is, $\ell(G) \leqslant \frac{4}{3}(n - 1)$.

Suppose finally that G is primitive on Ω. Since G is soluble it is of affine type. If its degree is p^d then $\ell(G) = d + \ell(H)$, where $H := G_\alpha$ and this is a soluble irreducible subgroup of GL(d,p). Using crude estimates we find that

$$\ell(H) \leqslant \log_2|H| < d^2 \log_2 p = d \log_2 n$$

and so $\ell(G) < d(1 + \log_2 n) \leqslant \frac{4}{3}(n - 1)$ if $n \geqslant 16$. The primitive soluble groups with $n < 16$ can be examined by degrees. The most extravagant ones that arise are GL(2,3) of degree 9, whose composition length is 7, well below the stipulated upper bound, and S_4, whose composition length is 4, equal to the bound. Thus $\ell(G) \leqslant \frac{4}{3}(n - t)$ in all cases.

The bound given in Lemma 7.1 is achieved by many groups (but only by soluble groups). For, if (G_1,Ω_1) and (G_2,Ω_2) achieve the bound then so does $(G_1 \times G_2, \Omega_1 \dot{\cup} \Omega_2)$ and so also does $(G_1 \mathrm{Wr}\, G_2, \Omega_1 \times \Omega_2)$ provided that G_1 is transitive on Ω_1. For example, the wreath power $\mathrm{Wr}^k S_4$ of degree 4^k achieves the bound for all k.

Here now is the COMPFACTORS routine. Its input should be a permutation group G for which a strong generating set is known. The output will be a list CF(G) of the composition factors of G in the order in which they occur in at least one descending composition series of G. If LIST1 and LIST2 are lists then LIST1 + LIST2 will denote the list obtained by appending LIST2 to LIST1.

```
          Begin
(1)       CF := ∅;
(2)       if G = {1} then goto line (15);
(3)       Ω1 := αG;
(4)       if Ω1 ≠ Ω then
              begin
(4.1)         N := Ker(G^Ω1);
```

(4.2) $CF := CF(G^{\Omega_1}) + CF(N^{\Omega-\Omega_1})$; goto line (15);

end;

(to be continued).

At each stage of the calculation CF is the list of composition factors already found, which explains lines (1) and (2). At line (4), if Ω_1 is a proper subset of Ω then $CF(G^{\Omega_1})$ and $CF(N^{\Omega-\Omega_1})$ may be obtained by recursion as line (4.2) requires. If we come past line (4) then G is transitive on Ω and we continue:

(5) find a minimal non-trivial congruence ρ;

(6) if $\rho \neq \Omega \times \Omega$ then

begin

(6.1) $\Omega_1 := \Omega/\rho$; $N := Ker(G^{\Omega_1})$;

(6.2) $CF := CF(G^{\Omega_1}) + CF(N^{\Omega})$; goto line (15);

end;

(to be continued further).

As usual, line (5) is accomplished by Atkinson's algorithm. If ρ is not the universal relation then $|\Omega_1| < \frac{1}{2}|\Omega|$ and N is a proper normal subgroup of G, so the recursion in line (6.2) is permissible. If we come past line (6) then ρ is the universal relation on Ω and so G is primitive. Therefore the algorithms of §§3,5,6 may be used:

(7) call EARNS;

(8) if G is of affine type and its earns has order p^d then

begin

(8.1) $CF := CF(G_\alpha)$ + (d cyclic groups C_p); goto line (15);

end;

(9) call SOCLE to compute the socle L of G;

(10) call SOCLEBITS to compute the simple direct factors $X_1,\ldots,X_d$ (or $X_1,\ldots,X_{2d}$ if G turns out to be of type II) of L and list those in $LIST_1$;

(11) $\Sigma := \{1,2,\ldots,d\}$; $K := Ker(G^{\Sigma})$;

(to be continued further).

At line (8.1) the stabiliser G_α is strictly smaller than G and so recursion is permissible. If we come past line (8) then G is primitive and not of affine type, so the SOCLE and SOCLEBITS routines are applicable. After line (11) we have (or can compute) the composition factors of G/K because this is the group G^{Σ}, and we have the composition factors of L. We do not have a permutation representation of K/L available but fortunately we do not need one because this is a soluble group. Accordingly the last part of the algorithm is:

```
(12)      LIST_2 := ∅;
(13)      while K ≠ L do
            begin
(13.1)      N := K';
(13.2)      factorise |K:N| as p_1 p_2 ... p_s, where p_1, p_2, ..., p_s are prime
            numbers;
(13.3)      LIST_2 := LIST_2 + (C_{p_1}, C_{p_2}, ..., C_{p_s});
(13.4)      K := N;
            end;
(14)      CF := CF(G^Σ) + LIST_2 + LIST_1;
(15)      publish CF as the list of composition factors of G;
          End.
```

Here lines (12) to (13.4) calculate $LIST_2$ as the list of composition factors of the soluble group K/L, and then line (14) puts the composition factors of G/K, K/L, L together as the list of composition factors of G.

There remains the problem of identifying the simple groups that appear in the output. By the Classification Theorem for finite simple groups they are determined modulo rare ambiguities by their orders, and these ambiguities can be resolved without much difficulty by examining suitable subgroups. I understand from John Cannon that he has devised and implemented an algorithm (part of which was contributed by Richard Lyons) for solving this problem.

8. *THE SIMPLETEST ROUTINE*

One can use COMPFACTORS to test for simplicity of G. But if that is all one wants then a very much faster test can be devised using components from our various routines. Here is an example, appropriate for groups G of degree $n < 78,125$, in which a strong generating set is given. I have called it SIMPLETEST:

```
          Begin
(1)       if G = {1} then return to the outside world and say so;
(2)       Ω := Ω - Fix_Ω(G);   Ω_1 ∈ Orb(G,Ω);
(3)       if Ker(G^{Ω_1}) ≠ {1} then goto line (20);
(4)       (Ω,G,n) := (Ω_1, G^{Ω_1}, |Ω_1|);
(5)       find a maximal proper G-congruence ρ on Ω;
(6)       Ω_1 := Ω/ρ;
```

```
(7)       if Ker(G^Ω1) ≠ {1} then goto line (20);
(8)       (Ω,G,n) := (Ω1,G^Ω1,|Ω1|);
(to be continued).
```

Justification. At line (2) Ω_1 is a non-trivial G-orbit and so $G/\mathrm{Ker}(G^{\Omega_1})$ is non-trivial. Therefore if $\mathrm{Ker}(G^{\Omega_1}) \neq \{1\}$ then G is not simple, as line (3) says. If we come past line (3) then G is faithful on Ω_1 and so after line (4) we have G transitive on Ω. Line (5) is done with Atkinson's algorithm; then at line (6) Ω_1 is again a non-trivial G-space and so if $\mathrm{Ker}(G^{\Omega_1}) \neq \{1\}$ then G is not simple, as line (7) says. If we come past line 7 then G is faithful on Ω_1 and so after line (8) we have G primitive on Ω.

Note that as it is written the procedure modifies the originally given permutation representation of G. If it is required that that should be preserved then the program needs to be slightly altered.

Next we test for a cyclic group of prime order acting regularly, and then reduce to the case where G is perfect. To do this we work with the generators $g_1,\ldots,g_k$ of G, of which we assume that $g_1 \neq 1$:

```
 (9)      for i := 2 to k do
              begin
 (9.1)        x := [g1,gi];
 (9.2)        if x ≠ 1 then goto line (11);
              end;
(10)      goto line (19);
(11)      if ⟨x^G⟩ ≠ G then goto line (20);
(to be continued further).
```

Justification. We come through to line (10) if and only if $[g_1,g_i] = 1$ for all i, that is, $g_1 \in Z(G)$. Since a primitive group whose centre is non-trivial is cyclic of prime order we then know that G is simple, as line (10) says. If at line (9.2) we find a non-trivial commutator then we test at line (11) whether its normal closure is the whole of G. If not then G is not simple, as line (11) says; if so then we pass beyond line (11) with the knowledge that G is perfect.

It might be worth including another few lines somewhere in the early parts of the program, namely:

```
(a)       for i := 1 to k do
              begin
(a.a)         if gi is an odd permutation then
                 begin
(a.a.a)          if gi^2 ≠ 1 then goto line (20);
```

```
(a.a.b)         for j := 1 to k do
                    begin
(a.a.b.a)           if g_j ≠ 1 and g_j ≠ g_i then goto line (20);
                    end;
(a.a.c)         goto line (19);
                end;
            end.
```

If all the generators of G are even permutations then nothing happens, but if one of the generators is odd then these lines test whether or not G is cyclic of order 2. If not then it is not simple, as lines (a.a.a) and (a.a.b.a) say; if so, then it is simple as line (a.a.c) says. This is the sort of filter that could usefully be built in even before a strong generating set was computed. It is extremely cheap, and in those cases when it is decisive it saves a considerable amount of relatively expensive computing.

From now on we assume that G is primitive on Ω and that it is a perfect group. The computation continues:

```
(12)      call EARNS;
(13)      if G is of affine type then goto line (20);
(14)      if  n ∈ {5^5,5^6,6^5,6^6,7^5,8^5,9^5} then
              begin
(14.1)        p := largest prime divisor of |G|;
(14.2)        find a p-element z ∈ G - {1} such that p does not divide
              |G : C_G(z)|;
(14.3)        if ⟨z^G⟩ ≠ G then goto line (20);
              end;
(15)      if |G| ≠ n^2 then goto line (19);
(16)      choose x ∈ G - {1};
(17)      Z := Z(C_G(x)); Z_0 := p-component of Z for some prime p dividing
          |Z|; Z_1 := Ω_1(Z_0);
(18)      for z ∈ Z_1 - {1} do
              begin
(18.1)        if ⟨z^G⟩ ≠ G then goto line (20);
              end;
(19)      return to the outside world with the news that G is simple;
(20)      return to the outside world with the news that G is not simple;
          End.
```

Justification. A group of affine type is simple if and only if it is cyclic of prime order, which, by the time we get to lines (12) and (13), G is not. Therefore if G is of affine type then G is not simple. Lines (14) - (14.3) are adapted from Lines (1) - (4) of the SOCLE routine (Version 3) and are justified by Lemmas 5.2 and 5.5. If we come through to line (15) then we know that G is its own socle, that is, it is a direct product of isomorphic simple groups. Either there are two factors and G is of type II (as in Example 2.2) or G is of type V and is simple. Lines (15) - (18.1) are adapted from lines (3) - (6.2) of the SOCLEBITS routine and are justified as they are.

In this algorithm the only costly calculation is at line (14.2). Even that, however, is considerably cheaper than computing a whole Sylow subgroup of G, and besides, that line does not operate if $n < 3125$. Therefore I believe that SIMPLETEST promotes 'Test whether simple' from the expensive category in John Cannon's list ([2], p.489) to medium, and perhaps even to the cheaper end of medium.

9. *COMMENTARY*

Theory. Although the algorithms that I have described were intended for practical purposes it seems worth examining them briefly for their computational complexity. Most of the elementary constituents of the programs are known to be computable in polynomially bounded time (that is, in time bounded by $f(n+k)$ where $f(t)$ is some polynomial) but there are two significant exceptions. Apparently it is not known whether intersections $X \cap Y$ and centralisers $C_G(X)$ can always be computed in polynomially bounded time (see Hoffman [6]). In [8] Kantor proves the existence of a polynomial-time algorithm to compute $O_p(G)$ but his argument depends on the Classification Theorem for finite simple groups. Therefore it may be of some interest that the following observation does not depend on that theorem.

Theorem 9.1. *The* EARNS *and* PCORE *routines are polynomial-time algorithms.* The proof, which I shall only sketch, depends on the following two facts.

Lemma 9.2. *If* $Y \trianglelefteq X \leqslant \mathrm{Sym}(\Omega)$ *then* $C_X(Y)$ *can be computed in polynomially bounded time.*

Proof. First $C_{\mathrm{Sym}(\Omega)}(Y)$ can be obtained in polynomially bounded time (see [6], p.259), and then, since X normalises $C_{\mathrm{Sym}(\Omega)}(Y)$, also $X \cap C_{\mathrm{Sym}(\Omega)}(Y)$, which is $C_X(Y)$, can be obtained in polynomially bounded time (see [6], p.251).

Lemma 9.3. *Let* r *be a fixed positive integer. If* $X \leqslant \mathrm{Sym}(\Omega)$, *if* $\Delta \subseteq \Omega$ *and* $|\Delta| \leqslant r$, *and if* Y *is the pointwise stabiliser* $X_{(\Delta)}$, *then* $C_X(Y)$ *can be computed in polynomially bounded time.*

Proof. Let $N := N_X(Y)$. We first show that N can be obtained in polynomially bounded time (the degree of a polynomial bound depends on r, however). Let $\Gamma := \mathrm{Fix}_\Omega(Y)$. Then

$$N = \{x \in X \mid x^{-1}Yx = Y\}$$

$= \{x \in X \mid X_{(\Delta x)} = Y\}$

$= \{x \in X \mid \Delta x \subseteq \Gamma\}$.

Let $x_1, \ldots, x_m$ be a set of coset representatives for Y in X (such as may be computed in polynomially bounded time by the Schreier-Sims method) and let $S := \{x_i \mid \Delta x_i \subseteq \Gamma\}$. For a given element x of X the condition $\Delta x \subseteq \Gamma$ can be tested in time bounded by $O(rn)$. Since $m \leq n^r$ the set S can be obtained in polynomially bounded time and since $N = \cup\{Hx_i \mid x_i \in S\}$, therefore also N can be found in polynomially bounded time.

Now $C_X(Y) = C_N(Y)$ and so Lemma 9.2 tells us, since $Y \trianglelefteq N$, that $C_X(Y)$ can be obtained in polynomially bounded time.

Sketch proof of Theorem 9.1. We examine EARNS first. Step 1 can all be accomplished in polynomially bounded time: lines (1) to (1.2) are obviously all right, the groups $G_{\alpha,\beta}$ can be tested for triviality in polynomially bounded time by Sims' (or simpler) algorithms, and if they are all {1} then G is a Frobenius group and $|G_\alpha| \leq n-1$. In Step 2 it is line (6) that needs analysis. By Lemma 9.3 above $C_G(G_{\alpha,\beta})$ can be obtained in polynomially bounded time; and the calculation of $O_p(C^\Gamma)$ can also be achieved in polynomially bounded time because $|C^\Gamma| \leq |\Gamma|(|\Gamma| - 1) < n^2$ as was pointed out on p.29. In Step 3 lines (8), (9), (10) are all right because the centre can be found in polynomially bounded time and the p-component of a group that is known to be nilpotent can be found in linearly bounded time. Line (10) and its dependants are not necessary to the algorithm (but can certainly be executed in polynomially bounded time), line (11) presents no problem and if we come through beyond it then the search at line (13) requires examination of fewer than n cases. The normal closure at line (13.1) (and also at line (10.2)) can be computed in polynomially bounded time as was proved by Furst, Hopcroft and Luks [5] (see [6], p.264). Therefore EARNS is a polynomial-time algorithm.

The PCORE routine refers back to itself at lines (3.2), (7), (9.2), (10.2), (12), (17), (18.2.3), and each time the permutation group which is offered as its new input is either a proper quotient, or a quotient of a proper normal subgroup of G, or it is a permutation representation of G of strictly smaller degree. It follows from Lemma 7.1 that the depth of recursion required is at most $\frac{4}{3}(n-1)$, and, as a matter of fact, once the problem is reduced to computation in transitive groups the depth of recursion can be bounded by $(\log_2 n)^2$. Therefore it is sufficient to show that each line of the algorithm (other than calls back to PCORE) can be executed in polynomially bounded time.

In Step 1 the only doubtful line is (3.3). But the intersection $K_1 \cap K_2$ can be obtained in polynomially bounded time by an algorithm of Luks (see [6], p.250) because K_1 and K_2 are normal subgroups of G. In Step 2 line (4) can be done in polynomially bounded time by Atkinson's algorithm, we have already seen that EARNS at line (5.2) is a polynomial-time computation, and the other lines are clearly all right. Whenever Lemma 4.1 is used (as at lines (9.1), (9.2) in Step 3) it is a polynomial-time computation: the transversal T can be found in linearly bounded time (given that we have a strong generating set for G available) and then Luks' algorithm (see [6], p.250) yields $\bigcap_{t \in T} t^{-1}Kt$ in polynomially bounded time because the groups $t^{-1}Kt$ are normal in N, hence normalise each other. Line (10) is completed in polynomially bounded time by Atkinson's algorithm, and all the remaining lines in Step 3 are obviously all right. In Step 4 lines (11) and (12), (16) and (17), and (18.2.2) and (18.2.3) implement Lemma 4.1 and are therefore all right; in lines (14) and (18.2) the relevant congruences can be found in polynomial time by Atkinson's algorithm; at line (18.2) the number of cases to be examined is much less than n; all other lines are obviously all right. In Step 5 line (19) can be done by computing a suitable power of one of the generators of G, which can be done in linearly bounded time. At line (20) the group X is a transitive p-group, so $C_{Sym(\Omega)}(X)$ is a p-group and therefore $C_G(X)$ can be obtained in polynomially bounded time (see [6], pp.251,259). At line (23) the intersection $Y \cap Z_1$ is all right since $Z_1 \leqslant G$ (see [6], p.250) and then $|Y \cap Z_1| < n$ since this group is semi-regular on Ω, so there are fewer than n elements z to be examined. Finally, $\langle z^G \rangle$ can be obtained in polynomially bounded time ([5]; see [6], p.264) and then so can $C_G(Z)$ by Lemma 9.2. Therefore PCORE is a polynomial-time algorithm, as the theorem asserts.

As it stands the SOCLE routine is designed to work only for a limited range of values of n and it does not make sense to ask if it is a polynomial-time algorithm. But of course we could use Lemma 5.3 instead of Lemmas 5.5 and 5.6, and simply compute as follows:

```
        Begin
(1)     T := a Sylow 2-subgroup of G;
(2)     Z := Z(T);
(3)     N := ⟨Z^G⟩;
(4)     L := N^(∞);
(5)     return to the outside world with the news that Soc(G) = L;
        End.
```

Though crude, this certainly is a polynomial-time algorithm: line (1) by Kantor's theorem [8], line (2) by, for example, Lemma 9.2 above (see [6], p.260), lines (3) and (4) by theorems of Furst, Hopcroft and Luks [5] (see [6], pp.264,269). However, both Kantor's proof and the proof that Soc(G) = $N^{(\infty)}$ use the Classification Theorem for finite simple groups. In fact, Kantor's proof uses the theorem due to Luks that the composition factors of G can be computed in polynomial time. Therefore my methods certainly do not give an independent proof of that theorem.

The SOCLEBITS routine also works in polynomially bounded time, and no deep theorems are needed to prove this. The calculation of $C_L(Y)$ in line (2) is all right by Lemma 9.2, and after line (3) any potentially slow calculations are taking place in a group of order $\leqslant n^2$.

The COMPFACTORS and SIMPLETEST routines are also polynomial-time algorithms (provided that they are adjusted to cope with permutation groups of arbitrary degree as SOCLE has to be). But again, the proof depends ultimately on the Classification Theorem for finite simple groups to justify lines (9) and (12)-(14) of COMPFACTORS and lines (14)-(14.3) of SIMPLETEST. I imagine that it is at a similar point that the polynomial-time algorithm for finding composition factors due to Luks (quoted in [10], p.63 and [8]) requires the Classification Theorem.

Practice. Although the algorithms described in §§3-8 are intended for practical use, they have not yet all been properly tested. The EARNS routine was implemented by John Cannon in April 1983 and is available in the CAYLEY library under the name ERNIES. The PCORE routine is quite new and untested. I had begun its design in May 1983 but did not then see how to finish it, and it is only now, some weeks after proposing its completion as an open problem in my last conference lecture, that I have been able to deal with my earlier difficulties. The basic ideas of SOCLE and SOCLEBITS have been tested and found to work by John Cannon. Therefore I believe that these routines, and also COMPFACTORS and SIMPLETEST, which are constructed from them, should work satisfactorily in practice. Nevertheless, a substantial amount of testing is still needed.

Acknowledgements. Many of the ideas in this paper were hatched during the London Mathematical Society Symposium on Computational Group Theory held in Durham from 30 July to 9 August 1982, and during a visit that I paid to the University of Sydney from 20 March to 20 April 1983. I am grateful to the organisers of that conference for inviting me to come there - as a stranger in paradise - and to the Mathematics Department of the University of Sydney for its genial hospitality. I also offer my warm thanks to Colin Campbell and Edmund Robertson for inviting me to organise this material at last and expound it at their conference Groups - St Andrews 1985. Most of all I am grateful to John Cannon. He aroused my interest by asking three questions, the answers to which are the content of §3, §5 and §6, he gave me tutorials about availability and costing of various facilities in the CAYLEY system, and he has tested some of the ideas.

REFERENCES

1. M.D. Atkinson, An algorithm for finding the blocks of a permutation group, *Math. Comp.* 29 (1975), 911-913.
2. John J. Cannon, Effective procedures for the recognition of primitive groups, in *The Santa-Cruz conference on finite groups*, Amer. Math. Soc. Proc. Symp. Pure Math. 37 (1980), 487-493.
3. John J. Cannon, *A language for group theory*, preprint, University of Sydney (1982).
4. John J. Cannon, An introduction to the group theory language, Cayley, in *Computational group theory* (edited by M.D. Atkinson), Academic Press (1984), 145-183.
5. M. Furst, J.E. Hopcroft & E. Luks, Polynomial-time algorithms for permutation groups, *Proc. 21st I.E.E.E. Symp. on Foundations of Comp. Sci.* (1980), 36-41.

6. Christoph M. Hoffman, *Group-theoretic algorithms and graph isomorphism*, Lecture Notes in Comp. Sci., Vol. 136, Springer-Verlag (1982).
7. B. Huppert, *Endliche Gruppen I*, Springer-Verlag (1967).
8. William M. Kantor, Sylow's theorem, in polynomial time, *J. Comp. Syst. Sci.* 30 (1985), 359-394.
9. Jeffrey S. Leon, On an algorithm for finding a base and a strong generating set for a group given by generating permutations, *Math. Comp.* 35 (1980), 941-974.
10. E. Luks, Isomorphism of graphs of bounded valence can be tested in polynomial time, *J. Comp. Syst. Sci.* 25 (1982), 42-65.
11. B.H. Neumann, Twisted wreath products of groups, *Archiv der Math.* 14 (1963), 1-6.
12. Leonard L. Scott, Representations in characteristic p, in *The Santa-Cruz conference on finite groups*, Amer. Math. Soc. Proc. Symp. Pure Math. 37 (1980), 319-331.
13. Charles C. Sims, Determining the conjugacy classes of a permutation group, in *Computers in algebra and number theory*, SIAM-AMS Procs, Vol. IV, Amer. Math. Soc. (1971), 191-195.
14. Charles C. Sims, Computation with permutation groups, in *Proc. 2nd Symp. on Symbolic and Algebraic Manipulation, Los Angeles* 1971 (edited by S.R. Petrick), Ass. for Comp. Machinery, New York (1971), 23-28.
15. Charles C. Sims, Some group-theoretic algorithms, in *Topics in Algebra* (edited by M.F. Newman), Lecture Notes in Math., Vol. 697, Springer-Verlag (1978), 108-124.

Note added 17 February 1986. The following recent references, relevant particularly to the considerations of §9, have been drawn to my attention by W.M. Kantor and E.M. Luks.

16. L. Babai, W.M. Kantor & E.M. Luks, Computational complexity and the classification of finite simple groups, *Proc. 24th I.E.E.E. Symp. Found. Comp. Sci.* (1983), 162-171.
17. William M. Kantor, Algorithms for computing in permutation groups, University of Oregon (1986), preprint.
18. Eugene M. Luks, Computing the composition factors of a permutation group in polynomial time, University of Oregon (1985), preprint.

FIVE LECTURES ON GROUP RINGS

J.E. Roseblade
Jesus College, Cambridge CB5 8BL, England.

LECTURE 1

Let G be a group. The integral group ring $\mathbb{Z}G$ consists of the formal sums

$$\xi = n_1x_1 + \dots + n_rx_r$$

with integers $n_1,\dots,n_r$ and group elements $x_1,\dots,x_r$; they add and multiply in an obvious way. In this first talk G will be the factor group Γ/M of a soluble group Γ by an Abelian normal subgroup A, and I shall try to explain how knowing about $\mathbb{Z}G$ helps to understand about Γ. For γ in Γ the induced automorphism $\mu \to \mu^\gamma$ ($\mu \in M$) of M depends only on the projection $x = M\gamma$ in G, so there is an induced action $\mu \to \mu^x$ of G on M; this extends to one of $\mathbb{Z}G$ on M:

$$\mu^\xi = (\mu^{n_1})^{x_1}(\mu^{n_2})^{n_2} \dots (\mu^{n_r})^{x_r}.$$

Thus M is a $\mathbb{Z}G$-module, written multiplicatively. To know about Γ we must know about $\mathbb{Z}G$-modules: there is no escape, for given a $\mathbb{Z}G$-module M there is always a group Γ with $M \lhd \Gamma$ and $G = \Gamma/M$. The split extension $G \ltimes M$ (consisting of the pairs $x\mu$ with $(x\mu)(y\nu) = xy.\mu^y\nu$) is always one, but frequently there will be more — depending on $H^2(G,M)$. At a very basic level, therefore, we cannot avoid studying group rings. Naturally, ring theorists' main interest in them is as examples of rings; but we have a wider perspective. Much of the work on group rings over the last few decades (and certainly my own work) has been motivated by problems in group theory. In this first lecture I shall mainly and briefly talk about how certain results of ring theory translate into results on groups. This is mostly for those who are new to group rings. In the remaining lectures I hope to do something less hackneyed: survey and discuss parts of the subject that I find particularly attractive.

A simple translation from multiplicative to additive notation is what is usually needed, see Fig. 1. Thus, for example, *Φ is nilpotent if and only if* H *is nilpotent and* $Mh^n = 0$ *for some* n.

Let me illustrate this by referring to three papers of Philip Hall in 1954, 1959 and 1961, *in as far as they relate to metabelian groups.* We take $M \geq \Gamma'$ and assume, then, that G is Abelian. Hall noticed that when Γ is finitely generated

(1) Γ *has* Max-n;

(2) *Chief factors of* Γ *are finite and* Γ *is residually finite;*

(3) *Fratt*(Γ) *is nilpotent.*

Let $\Gamma = \langle \gamma_1, \ldots, \gamma_m \rangle$, so that $G = \langle x_1, \ldots, x_m \rangle$ with $x_i = M\gamma_i$. Then $\mathbb{Z}G$ here is a finitely generated commutative ring (generated by the 2m elements $x_1^{\pm 1}, \ldots, x_m^{\pm 1}$); the three results correspond to classical theorems about such rings.

(1) corresponds to Hilbert's Basis Theorem, from which it follows that finitely generated $\mathbb{Z}G$-modules have the ascending chain condition. Since Γ' is the normal closure of the finitely many commutators $[\gamma_i, \gamma_j]$, M is finitely generated and so has the ascending chain condition on normal subgroups of Γ; that is enough.

From the translation, chief factors below M correspond to simple sections of M. The 'weak' Nullstellensatz asserts these are finite. This is the first half of (2). For the second half, which I don't want to discuss, the Artin-Rees lemma can be brought in to play.

For (3) the 'strong' Nullstellansatz is needed. Let Φ stand for Fratt(Γ). We need $Mh^n = 0$ for some n. The classical theorem says this is true if $Mh \leq N$ for every maximal submodule N of M. Translating back, we need $[M, \Phi] \leq N$. By (2) M/N is finite; by (2) for Γ/N, therefore, M/N survives in a finite image of Γ. We may now appeal to Frattini [1885].

Hall proved (1) for finitely generated Abelian by polycyclic groups, by showing $\mathbb{Z}G$ is right-Noetherian if G is polycyclic. He proved (2) and (3) for finitely generated Abelian by nilpotent groups, and others extended them to finitely generated Abelian by polycyclic groups. I think it's fair to say that the applications to groups are made in very much the same way as in the metabelian case once the corresponding group ring theorems have been proved, and that much of classical commutative algebra has now been discussed for polycyclic group rings. I shall mention some of these, without very much detail, later on.

Let G be a polycyclic group, so that there exists $1 = G_0 \lhd G_1 \lhd \ldots \lhd G_n = G$ with each G_i/G_{i-1} cyclic. Polycyclic groups generalize finitely generated Abelian ones via finitely generated nilpotent groups; they are the only soluble groups that can be embedded in matrix groups over the integers. A good source of examples comes from number theory.

Fig. 1

	$\times$	$+$	
	μ^{ξ}	$\Sigma\, n_i(\mu x_i) = \mu^{\xi}$	
	μ^{Γ}	$\mu\mathbb{Z}G$	
$N \leq M$	$N \lhd \Gamma$	N a submodule	
$\Phi \lhd \Gamma$	$[\mu, \gamma]$	$\mu(x - 1)$	$x = M\gamma$
$H = M\Phi/M$	$[M, \Phi] = 1$	$Mh = 0$	$h = \langle {}^{+}x - 1 \mid x \in H \rangle$
	$[M, \underbrace{\Phi, \ldots, \Phi}_{n}]$	Mh^n	

If K is a number field, I the ring of integers and U the unit group then I is finitely generated as an additive group, by Dedekind, and U is finitely generated as a multiplicative one, by Dirichlet. The split extension $U \ltimes I$ is polycyclic, as are all its subgroups. The more interesting parts of polycyclic groups have the flavour of these. A very good place to learn about them is Dan Segal's book [1983].

Suppose $G_i = G_{i-1}\langle x_i \rangle$, then $G = \langle x_1 \rangle \langle x_2 \rangle \ldots \langle x_n \rangle$ and the elements of G are of the form $x_1^{r1} x_2^{r2} \ldots x_n^{rn}$. They look like monomials in $x_1, \ldots, x_n$ and their inverses; but the 'variables' need not commute and the expression is not usually unique. The group rings of G look sufficiently like finitely generated commutative rings for us to expect some similarities of behaviour. This is what has now been firmly established. But it is not just the superficial resemblance that counts, for P.H. Kropholler [1984] has shown that any soluble finitely generated minimax group is a product of finitely many cyclic groups; their group rings are not so well behaved. For instance they only have the maximum condition on right ideals if they are polycyclic.

It is the poly-cyclic nature of polycyclic groups that endows their group rings with allure.

Put $R = \mathbb{Z}G$, $S = \mathbb{Z}G_{n-1}$ and write x for x_n; suppose G/G_{n-1} is infinite. Then $R = \langle S, x \rangle$ with $S = S^x$, and there is a direct sum decomposition of R as an S-module: $R = \sum_{n=-\infty}^{+\infty} x^n S$. Multiplication is now modified by the action of x on S: for a,b in $\langle x \rangle$ and σ, τ in S,

$$(a\sigma)(b\tau) = ab\sigma^b\tau;$$

just as in the split extension of groups, indeed R is a split extension $\langle x \rangle \ltimes S$, a special sort of crossed product. We can expect that a great deal will have to do with the structure of S as a ring with the operator x. Many theorems, of course, have been proved by induction on n. For example, that R has Max-r is easily seen this way. So also are the various results to do with 'generic flatness' (see, for example, Hall [1959], J.E. Roseblade [1973], J.C. McConnell [1982], K.A. Brown [1982]); but the more interesting ones have come from taking a rather different point of view.

Let A be an Abelian normal subgroup of G and set S, now, equal to $\mathbb{Z}A$. The conjugation action of G on A extends by linearity to one of G on S. The properties of such rings S viewed as G-rings are the ones that give the key information about $\mathbb{Z}G$. I don't at all want to describe this key information, because it is written up by Roseblade [1978] and D.S. Passman [1984], but I do want to say something of the questions about these G-rings, and the results and methods.

LECTURE 2

In Hall's work [1959] on simple modules for finitely generated nilpotent groups and mine [1973] for polycyclic groups, much depended on analysing the structure of a finitely generated module when viewed as a module for an Abelian normal subgroup. Throughout most of this lecture R will denote kG and S will stand for kA with A an Abelian normal subgroup of G. For simplicity I shall assume A is free and k is a field. Let me remind you first of a result stemming from Hall's work.

(1) *Let* G *be polycyclic and* M *a finitely generated* R*-module. Suppose* A *is in the centre of* G. *There exists a non-zero element* λ *of* S *so that* $M[1/\lambda]$ *is a free* $S[1/\lambda]$*-module.*

To avoid trivialities we should suppose M is a torsion-free S-module, so $M[1/\lambda]$ is non-zero.

Now suppose M is simple as an R-module and consider a maximal ideal L of S (we write $L \lhd\!\!\cdot\, S$). If λ is not in L then L survives in $S[1/\lambda]$, i.e. $L[1/\lambda] < S[1/\lambda]$. Hence, and because $M[1/\lambda]$ is free, we have $ML < M$. Because S is central in R, ML is a submodule and because M is simple, we deduce that $ML = 0$.

Thus, either $ML = 0$ for some $L \lhd\!\!\cdot\, S$ (which is a situation where further deductions can be made), or else λ lies in every maximal ideal of S. The Strong Nullstellensatz describes $\bigcap_{L \lhd\!\cdot\, S} L$ as the nilpotent radical.

Considering central A was enough for Hall, because he mainly dealt with nilpotent groups. In the polycyclic case one has to go further.

(2) *Let* G *be polycyclic and* M *a finitely generated* R*-module. There exists a non-zero element* λ *of* S *such that* $M[1/\lambda^G]$ *is free over* $S[1/\lambda^G]$.

Here, of course, $S[1/\lambda^G]$ is short for $S[1/\lambda^x;\ x \in G]$: a whole conjugacy class of elements has to be inverted to achieve freeness.

Recently McConnell [1982] and Brown [1982] have extended these results, to cope with non-Noetherian situations. Let $R = \langle S, G \rangle$ be a ring generated by a commutative domain S and a group G with $S = S^G$ and $G/G \cap S$ polycyclic. Brown shows that the same conclusions as in (2) hold for finitely generated R-modules M. The novelty here is that S is not required to satisfy a chain condition. This result has applications to questions of primitivity of group rings of nilpotent groups.

(1) and (2) are proved by induction on the length of a cyclic chain for G; the proofs of Brown and McConnell, of course use the cyclic length of $G/G \cap S$; but a clever idea called vector *generic flatness* is exploited.

For $L \lhd\!\!\cdot\, S$ we write $°L$ for $\bigcap_{x \in G} L^x$. Just as in the central case, we deduce from (2) that if M is a simple R-module then either $M°L = 0$ for some $L \lhd\!\!\cdot\, S$ or every maximal ideal of S contains some conjugate of λ. Hall's problem on simple modules therefore inevitably leads to these two questions:

(a) *What* λ *in* S *conjugate into every maximal ideal of* S*?*

(b) *Given* $L \lhd\!\!\cdot\, S$, *what is* $°L$*?*

These are evidently both concerned with orbits of maximal ideals of S under the acting group $\Gamma = G/C_G(A)$. From now on Γ will denote a subgroup of Aut A. Let me say something of the results before turning to the methods. A distinction has to be drawn between the fields involved. Either k^* is periodic (which happens if and only if k is absolutely algebraic of prime characteristic) or k^* contains free subgroups of any finite rank. The first are sufficiently typified by finite fields; these in any case are the ones that have most to do with integral group rings.

Suppose first then that k is finite and that $L \triangleleft\!\!\cdot\, S$. Then S/L is also finite by the weak Nullstellensatz and so $L^{\dagger} = \mathrm{Ker}(A \to S/L)$ has finite index in A. It follows at once that the orbits are all finite. With orbits all finite we may expect there to be some stable maximal ideals. If $L = L^{\Gamma}$ then Γ induces k-automorphisms of S/L, so that $\Gamma/C_{\Gamma}(S/L)$ is cyclic. We may expect to get many stable maximal ideals if Γ itself is cyclic. Indeed the first result that is at all helpful with question (a) is given in my 1973 paper:

(3) *If* k *is finite and* $\Gamma = \langle \gamma \rangle$ *then* $\displaystyle\bigcap_{L = L^{\gamma} \triangleleft\!\cdot\, S} L = 0.$

Evidently this answers (a) when Γ is cyclic:

(4) *If* k *is finite and* Γ *cyclic then no non-zero element of* S *can conjugate into every maximal ideal.*

B.A.F. Wehrfritz [1979] took up the question of stable maximal ideals. He showed that (3) does not extend to acting groups Γ that are free Abelian of rank 2, but that it may be extended with cyclic Γ in a different direction. He proved

(5) *Suppose* k *is finite and* $\Gamma = \langle \gamma \rangle$. *If* $P = P^{\gamma} \triangleleft_p S$ *then there exists* m *such that* $\displaystyle\bigcap_{P \leq L = L^{\gamma^m} \triangleleft\!\cdot\, S} L = P.$

$\triangleleft_p$ means 'is a prime ideal of'.

Wehrfritz showed that m could be taken to be 1 in certain special situations, but, as far as I know, it is unknown whether m can be taken to be 1 in general. (5) depends on deep results that I shall mention in Lecture 3.

(4) certainly extends away from a cyclic acting group, and I'll say more of that in due course.

Naturally one way of tackling the questions and proving the theorems is by induction, say on the composition length of $A \otimes_{\mathbb{Z}} \mathbb{Q}$ as a $\mathbb{Q}\Gamma$-module, but we have only to reflect for a moment to see that this would be unnecessarily complicated. Consider question (a) for example, where we might now expect that the answer is 'only 0'. If Γ_0 is of finite index in Γ and T is a transversal to the cosets of Γ_0 in Γ then $\mu = \prod_{t \in T} \lambda^t$ conjugates into every maximal ideal under the action of Γ_0 if λ does under Γ, and if μ is zero then so is λ. So we allow ourselves to take as basic not just an irreducible situation, but what D.R. Farkas and R.L. Snider [1984], in their transatlantic way, have called a 'super irreducible' situation: when no subgroup of finite index in Γ acts reducibly on $A \otimes_{\mathbb{Z}} \mathbb{Q}$.

We say that A is a *plinth* for Γ if Γ and all its subgroups of finite index act irreducibly on $A \otimes_{\mathbb{Z}} \mathbb{Q}$.

Most important for all questions in this area is a theorem of G.M. Bergman [1971], which I shall discuss more fully later on. One way of stating it is:

(6) *Suppose* A *is a plinth for* Γ. *If* X *is a non-zero* Γ*-invariant ideal of* S *then* $\dim_k(S/X)$ *is finite.*

This is one of the vital ingredients in a proof of:

(7) *Suppose* k *is finite,* G *polycyclic and* A *a plinth for* G. *If* λ *conjugates into every maximal ideal of* S *then* $\lambda = 0$.

The other ingredient is due to Passman: *if* A *is a plinth for the Abelian group* Γ *then there is some* γ *in* Γ *with* A *a plinth for* $\langle\gamma\rangle$. Because of Mal'cev's theorem the acting group in (7) can be supposed Abelian. In each Γ-orbit of γ-stable maximal ideals of S, take one that contains λ. Their intersection has infinite co-dimension by (3) and is γ-stable, so by (6) it must be zero.

(7) was extended by D.L. Harper in 1980. He proved it for any field k by a method that used (7) but did not involve any result like (3) for fields k that do not have k* periodic.

Let me conclude this talk by saying just a word or two about this other sort of field. As I remarked above, when k* is not periodic A embeds in k*. For any embedding $\phi : A \hookrightarrow k^*$ a maximal ideal $L = \mathrm{Ker}(kA \to k)$ is defined that tends to have a large orbit under Γ. Evidently $L^\dagger$ is trivial here. If $L = L^\gamma$ then $a - \phi(a)$ and $a^\gamma - \phi(a)$ are both in L, so that $a - a^\gamma$ is as well. Hence $a = a^\gamma$; this holds for all a in A, so that $\gamma = 1$. Thus the L^γ are in (1-1) correspondence with the γ. A rather difficult result that I'll mention later ((3) in Lecture 3) will show that $^\circ L$ will be zero unless there are elements other than 1 in A that lie in a finite orbit under Γ.

Incidentally, (7) and Harper's result remain true for any free Abelian A of finite rank (cf. Brown [1981]).

LECTURE 3

To-day I shall talk about the important paper by Bergman [1971] and some of the work related to it. There is a discussion in Passman [1977] and in Roseblade [1978] but both these accounts manage to conceal its true beauty. I should like to say enough about the paper to encourage you to look at it again, as R. Bieri and R. Strebel have done in their investigations on finitely presented groups. A recent account is given by Strebel [1984].

Bergman didn't state his theorem the way I did yesterday; what he said was this:

(1) *Let* $I \triangleleft kA$ *and suppose* $I = I^\Gamma$. *If* $0 < I$ *and* $\dim_k(kA/I)$ *is infinite, then some subgroup of finite index in* Γ *stabilizes a subgroup* B *of* A *with* B *and* A/B *infinite.*

Most of his paper doesn't mention the acting group at all. There is no restriction on Γ, but A is free Abelian of finite rank. I shall say something of work done without this restriction later on.

Naturally when $I = I^\Gamma$, the finitely many primes minimal over I are permuted by Γ. Nothing is lost by replacing I by one of these also of infinite co-dimension and Γ by its stabilizer; henceforth, then, we shall consider a prime $P = P^\Gamma$ with $0 < P$ and kA/P of infinite dimension over k. This last statement is the same as saying that the (ring-theoretic) dimension dim kA/P is positive. Let F be the field of fractions of kA/P and write n for the rank of A. The dimension of kA/P is the transcendence degree [F:k] of F over k; let it be r, so that our assumptions are that $0 < r < n$.

To prove (1) we have to associate a finite system of isolated subgroups of A with P in a natural way, for then Γ will have a subgroup of finite index that stabilizes them. The way to do this is to consider valuations of F, and nothing is lost by considering real valuations. These, you will recall, are homomorphisms v from F* to $\mathbb{R}$ that satisfy

$$v(\alpha+\beta) \geq \mathrm{Min}\{v(\alpha), v(\beta)\}.$$

We shall assume throughout that the valuations are trivial on k. We usually write $v(0) = \infty$ and think of v as defined on F. Each of these v induces a homomorphism $v_A : A \to \mathbb{R}$ with kernel A_v an isolated subgroup of A. Γ acts on the valuations and hence on the A_v. Bergman's theorem therefore follows from

(2) *There are some* A_v, *but only finitely many, that have co-rank* r.

To produce A_v of co-rank r is very easy. Let $a_1, \ldots, a_r$ be elements of A giving rise to a transcendence base of F over k and let C be the subgroup they generate. Then kC embeds in F and F is finite over k(C). [It is easy to see that C may actually be chosen so that kA/P is integral over kC. Although we do not need this here is a proof: If $0 \neq \lambda = \Sigma \lambda_a a$ is in P, there is a subgroup B of A with A/B infinite cyclic and the Ba, a in sup(λ), all different. If we write $A = B\langle c\rangle$ then λ takes the form $\Sigma \lambda_a b_a c^{n_a}$ with $a = b_a c^{n_a}$. Since the n_a are all different and the $\lambda_a b_a$ are all units of kB, it follows that P+c and $P+c^{-1}$ are both integral over P + kB/P. The remark now follows by induction on dim(S/P).] We now embed C in $\mathbb{R}$ by a map $\phi : C \to \mathbb{R}$, say, and define

$$\phi(\lambda_1 c_1 + \ldots + \lambda_r c_r) = \underset{1 \leq i \leq r}{\mathrm{Min}} \phi(c_i).$$

This gives a real valuation on k(C) that extends to one, v say, on F. The value group of v is at worst a finite extension of φ(C); thus A_v has co-rank r.

As for the finite number of these A_v, let me tell you how that may be proved. The fact about valuations that we need is that $v(\alpha_1 + \alpha_2 + \ldots + \alpha_m)$ equals $\mathrm{Min}_i\, v(\alpha_i)$ if the minimum is achieved only once. Now for $\lambda = \Sigma \lambda_a a$ in kA we write $s(\lambda) = \{a \mid \lambda_a \neq 0\}$ for the support of λ. If λ is in P then $v(P+\lambda) = \infty$, so the minimum of the $v(P+a)$, $a \in s(\lambda)$, must be achieved twice. Hence there are different elements a,b of s(λ) with $v_A(ab^{-1}) = 0$. A_v therefore satisfies a rather special property in relation to P. Just temporarily say an isolated subgroup B of A is a P-*subgroup* if for every non-zero λ in P there is some pair $a \neq b$ of elements of s(λ) with $ab^{-1} \in B$. An easy combinatorial argument (in Passman [1977]) shows that there are only a finite number of minimal P-subgroups. These, of course, would do just as well as those in (2) to prove (1); but as a matter of fact the A_v of co-rank r are a subset of the minimal P-subgroups. This is easily seen: if B is a P-subgroup, let C be a direct complement of B in A. Evidently kC embeds in kA/P.

It follows that C must have rank at most r, and B co-rank at most r.

This proves Bergman's theorem, but we could hardly say that what we have so far said helps very much with understanding what the Γ-stable primes of kA look like. Perhaps the best theorem generalising Bergman's is one that does help significantly. I am afraid in some recent papers it has been called Roseblade's theorem, so let me put the record straight by saying again what I said in my 1978 paper: the theorem is joint work with D.C. Brewster. It is best stated for faithful primes of kA. A prime P is *faithful* if $P^{\dagger} = \mathrm{Ker}(A \to kA/P)$ is trivial. Most questions about primes of kA can be reduced to ones about faithful ones merely by proceeding to $A/P^{\dagger}$. Of course one has to allow a slightly wider class of groups A, i.e. the finitely generated ones. So now let A be finitely generated and Γ a group of automorphisms of it. It saves a good deal of writing if we use a single word to mean lying in a finite orbit under the action of Γ; I use *orbital*. The orbital elements of A are then those that are fixed by some subgroup of finite index in Γ; they form a subgroup which I denote by O. The theorem is

(3) *If* $P = P^{\Gamma}$ *is a faithful prime of* kA *then* $P = (P \cap kO)kA$.

When A is a plinth for Γ then either $O = A$ and Γ is finite, or else $O = 1$ and a non-zero stable prime P cannot be faithful, so the group $P^{\dagger}$ will be infinite and stabilized by Γ. (3) is therefore a generalization of Bergman's theorem. I don't want to go into the details of the proof; they depend on a careful consideration of the valuations of F.

Because Γ induces a finite group of automorphisms of O, theorem (3) shows that stable primes are no harder to deal with than they would be if Γ were finite.

There is another way of stating (3) that has the flavour of the fundamental theorem of Galois theory. When P is faithful, A embeds in kA/P and we may as well think of kA/P as $k[A]$ and F then as $k(A)$. It is an easy exercise to show that $P = (P \cap kO)kA$ is equivalent with $[k(A) : k(O)] = r(A/O)$. It is also not too hard to show that $k(O)$ consists of all the orbital elements of $k(A)$. Once this has been done the theorem becomes:

(4) *Let* $k(A)$ *be a field generated by* k *and* A. *If* O *is the subgroup of orbital elements of* A, *then* $k(O)$ *is the subfield of orbital elements of* $k(A)$ *and* $[k(A) : k(O)] = r(A/O)$.

This has applications to prime ideals of polycyclic group algebras. If $P \lhd_p kG$ and $A \lhd G$ then $P \cap kA$ is stable under G. It is not necessarily prime, but it is G-prime in an obvious sense, and as such it is an intersection of a finite orbit of primes; theorem (3) can be applied to these. Theorem (3) is also the key to Wehrfritz's result that I mentioned at (5) yesterday: the γ^m involved there is chosen to centralize the subgroup O.

Theorems like (3) have also been proved for Abelian groups of finite torsion-free rank. I think it was Brewster who first considered these in 1975. His results were not put into his Ph.D. dissertation [1976] and remained unpublished. Let A be a torsion-free Abelian group of finite rank acted on by a group Γ. If P is a faithful Γ-invariant prime of kA then $P = (P \cap kO)kA$ just as before if k has characteristic zero; but if the characteristic is a prime p, adjustments have to be made: O must be replaced by the subgroup of A generated by the orbital subgroups that (viewed additively) embed in $\mathbb{Z}[1/p]$. To see that this cannot be avoided consider a group A that is the direct sum of two copies of $\mathbb{Z}[1/p]$, written multiplicatively. There is a free subgroup B of rank 2 such that $A = \bigcup_{m=1}^{\infty} B^{\gamma^m}$, with γ the automorphism $\alpha \to \alpha^{1/p}$ of A. Of course $B < B^{\gamma} < B^{\gamma^2} < \dots$. With $\Gamma = \langle \gamma \rangle$, the subgroup O is trivial and yet, when k has characteristic p, any non-zero faithful prime Q of kB satisfies $Q < Q^{\gamma} < Q^{\gamma^2} < \dots$ and gives rise to a faithful prime $P = \bigcup Q^{\gamma^m}$ of kA. Such Q exist since B has rank 2.

Theorems on stable primes in the finite rank situation are not so relevant to soluble groups of finite rank as those in the finitely generated case are to polycyclic groups. It is really the Γ-primes which are of most interest. C.J.B. Brookes [1985], using similar methods, has proved a theorem of the same type concerned with these. If I is an ideal of kA he defines the *standardizer* of I in Γ to be the subgroup of Γ consisting of all γ such that $I \cap kB = I^{\gamma} \cap kB$ for some finitely generated subgroup B (depending on γ) having the same rank as A. If P is a faithful prime of kA whose standardizer has finite index in Γ then Brookes shows $P = (P \cap kO)kA$ as before. The relevance of this is that every ideal stabilized by Γ has at least one minimal overprime whose standardizer is of finite index. Although it follows from Brookes' theorem that if a faithful prime has its standardizer of finite index then it is actually orbital, this is not necessarily the case for non-faithful primes P, for $P^{\dagger}$ is not necessarily normalized by any very sensible sort of subgroup of Γ and we cannot proceed, as one does with orbital primes, to $A/P^{\dagger}$. Brookes' theorem has applications to soluble groups of finite rank; his main application is that if k* is not periodic and if G is a soluble group of finite rank with no non-trivial finite conjugacy classes, then kG is primitive.

LECTURE 4

I should like now to consider more carefully the situation of Bergman's theorem. We have a free Abelian group A of finite rank n, say, a field k and a prime ideal P of kA. The field of fractions of kA/P is denoted by F. We considered real valuations v on F and the kernels of the induced maps $v_A : a \to v(P + a)$. These v_A all belong to $\mathrm{Hom}(A, \mathbb{R})$, the real characters of A and the relevant ones were all non-zero.

$\mathrm{Hom}(A, \mathbb{R})$ is a real vector space of dimension n and can therefore be made into a topological space. We do this in a concrete way by viewing $\mathbb{R}^n$ as an inner product space in the usual way: $\langle x, y \rangle = \sum_i x_i y_i$. The $\mathbb{Z}$-lattice on the standard basis of $\mathbb{R}^n$ will be denoted by $\mathbb{Z}^n$. We fix a basis $a_1, \dots, a_n$ of A and identify $x = (x_1, x_2, \dots, x_n)$ in $\mathbb{R}^n$ with the character determined by $a_i \to x_i$ for each i. All concepts defined in $\mathbb{R}^n$ that are invariant under $GL_n(\mathbb{Z})$ can thus be transferred to $\mathrm{Hom}(A, \mathbb{R})$. Notice at once that if χ is a character identified with x then $\chi(a_1^{m_1} \dots a_n^{m_n})$ equals $\sum_i m_i x_i$ which is $\langle \chi, a \rangle$, if A is thought of as identified with $\mathbb{Z}^n$.

For χ, Ψ in $\mathrm{Hom}(A, \mathbb{R})$, say $\chi \sim \Psi$ if and only if χ is a positive multiple of Ψ and write $[\chi]$ for the equivalence class of χ. We put S(A) for the set of all $[\chi]$ with $\chi \neq 0$; it is homeomorphic with the unit (n-1)-sphere

S^{n-1}. Bieri and Strebel have called S(A) the *character sphere* of A; it is where all the action takes place.

For $n = 2$, S(A) is the unit circle. It is perhaps helpful to think geometrically. A character χ can be thought of as giving a direction. If $a \in A$ then $\chi(a) > 0$ means 'a lies in the direction χ'; $\chi(a) = 0$ means 'a is perpendicular to χ'; $\chi(a) < 0$ means 'a lies in the opposite direction from χ'. In the diagram below x corresponds with $a_1^{m_1}a_2^{m_2} \to m_1 + 2m_2$; $x/|x| = (1/\sqrt{5}, 2/\sqrt{5}) = u$ is on the circle; [u] is a discrete point in the sense that it corresponds to a discrete character: one of the form $A \twoheadrightarrow \mathbb{Z} \hookrightarrow \mathbb{R}$.

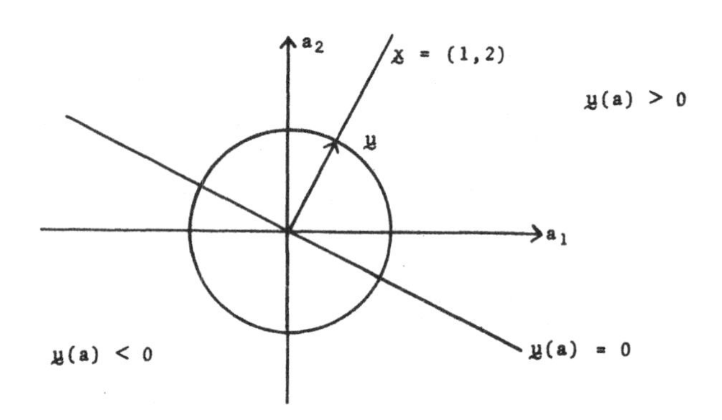

A given character χ may be extended in many different ways to a valuation, not of F perhaps but certainly to the field of fractions of kA. Most important for us is its extension using the minimum formula. Define $\bar{\chi} : kA \to \mathbb{R} \cup \{\infty\}$ by $\bar{\chi}(0) = \infty$ and

$$\bar{\chi}\left(\sum_a \lambda_a a\right) = \underset{\lambda_a \neq 0}{\operatorname{Min}} \ \chi(a), \quad \text{if} \ \sum_a \lambda_a a \neq 0;$$

and extend to the field of fractions in the obvious way.

Valuations v of F that do not have A in their kernel induce a non-zero character v_A and hence a point $[v_A]$ of S(A); but exactly the same can be said of valuations on the field of fractions of kA/Q for any prime Q, and we must take account of those where $Q \geq P$. To do this it is best to extend the meaning of valuation and say that v is a real valuation of kA/P if it is a multiplicative map to $\mathbb{R} \cup \{\infty\}$ with $v(\alpha + \beta) \geq \operatorname{Min}(v(\alpha), v(\beta))$, with $v(0) = \infty$ and $v(1) = 0$; this allows for the prime ideal $v^{-1}(\infty)$ to be greater than zero. We write $B = B(P)$ for the subset of S(A) consisting of all the $[v_A]$ induced by these (generalized) real valuations of kA/P.

Now for an element λ of kA, define

$$\lambda^* = \left\{ [\chi] \in S(A) \ \middle| \ \underset{a \in s(\lambda)}{\operatorname{Min}} \ \chi(a) \text{ is achieved twice} \right\},$$

and put $C = C(P)$ for the subset of all $[\chi]$ of S(A) that lie in λ^* for every non-zero λ in P. The point we used in the last lecture was that B is a subset of C. In his paper, Bergman proved

(1) $B = C$

Here I shall sketch a proof that $C \leqslant B$. Let $[\chi] \in C$. Write $A_\chi = \{a \in A | \chi(a) \geqslant 0\}$ and $A_\chi^+ = \{a \in A | \chi(a) > 0\}$. We put $T = (P + kA_\chi^+)/P$ and $R = (P + kA_\chi)/P$. Since no element of P is $1 + \alpha$ with α in kA_χ^+, it follows that T is a proper ideal of R and embeds, therefore, in a maximal ideal T_0 of R. By the fundamental theorem of valuation theory, there exists a valuation v of F with $v(R) \geqslant 0$ and for r in R, $v(r) > 0$ if and only if $r \in T_0$. An easy argument shows that v_A and χ induce the same pre-order in A, so that $[\chi] = [v_A]$.

This isn't quite right, for the v might not be a real valuation. The difficulty is illusory because every valuation can be approximated by a real one; the precise sense in which this can be done need not concern us, all we want is:

(2) *Let* v *be any valuation on* F *with* v_A *non-zero. There is a real valuation* v' *of* kA/P *inducing the same pre-order as* v *on* A.

It is this subset $B = C$ of $S(A)$ that Bergman was most concerned with. Without going into details just let me say that if $r = \dim(kA/P)$ with $0 < r < n$ then he proved that the subset lies in a finite union of 'rational' $(r-1)$-dimensional great subspheres of $S(A)$ and has non-empty intersection with every $(n-r)$-dimensional great hemisphere. These correspond (dually) with the statements I made yesterday about the P-subgroups, but have a much more geometrical appeal. Bergman makes the suggestion that the geometry could be further investigated so as to elucidate the variety defined by P. As far as I know this has been taken up only by Robert Bieri and his associates. There are interesting papers on the subject by Bieri and J.R.J. Groves [1984, 1986]. Bieri and Strebel's work [1980, 1981] on finite presentability uses very much the same set.

I'll finish this lecture off by considering a special case. We suppose that P is generated by a single element $\mu = \sum \mu_a a \neq 0$. In this special case we have:

(3) $B(P) = \mu^*$

For, of course, $B(P) = C(P) \leqslant \mu^*$; to see the converse we write each element ρ of kA as $\rho_0 + \rho_1$, where, if $\rho = \sum \rho_a a$, then ρ_0 is the sum of the $\rho_a a$ with $\chi(a) = \overline{\chi}(\rho)$ and $\rho_1 = \rho - \rho_0$ satisfies $\chi(\rho_1) > \overline{\chi}(\rho)$. Of course $[\chi] \in C(P)$ if and only if for each non-zero ρ in P then ρ_0 is not a single term $\rho_a a$. Now it is easy to see that $(\mu\tau)_0 = \mu_0\tau_0$; if $[\chi] \in \mu^*$ then μ_0 is not a single term and therefore not a unit of kA; it follows that $(\mu\tau)_0$ is not a single term either, so that $[\chi] \in (\mu\tau)^*$.

Bergman suggested that B(P), known by (1) to be $\bigcap_{0 \neq \mu \in P} \mu^*$, was always the intersection of a finite number of μ^*, with μ in P. (3) shows this to be true if P is principal. Bieri and Groves [1984] showed it is always true.

We shall get a feel for the geometry if we consider μ^* more carefully. For an element c of A we write 0_c for the subset of all $[\chi]$ on S(A) with $\chi(c) > 0$: this is an open hemisphere; its complement 0_c^- is a closed one.

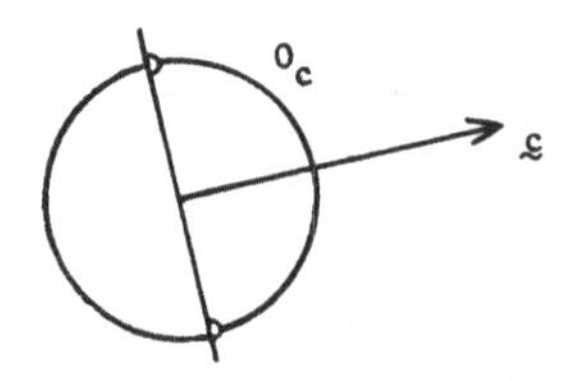

Write S = S(μ). Following Strebel [1984] we say that an element a in S is a *corner* of μ if there is a character χ with $\chi(a) < \chi(b)$ for all b different from a in S. Thus a is a corner if and only if

$$C_a = \{[\chi] \in S(A) \mid \chi(a) < \chi(b) \text{ for all b in } S\backslash\{a\}\}$$

is non-empty. Notice that $\chi(a) < \chi(b)$ if and only if $\chi(ba^{-1}) > 0$, so that $C_a = \bigcap_{a \neq b} 0_{ba^{-1}}$. Clearly the points not in μ^* are just those that are in some C_a for some corner a, so $\mu^* = \bigcap_{\text{corners } a} C_a^-$ is the intersection of the complements. Thus $\mu^* = \bigcap_{\text{corners } a} \bigcup_{a \neq b \in S} 0_{ba^{-1}}$, which is the union of a finite number of convex polyhedra.

Let me draw an example. Take A = $\langle a,b \rangle$ and $\mu = a + b + a^{-1}b + a^2b + b^2$. The diagram of S has a node at (m,n) if a^mb^n is in S.

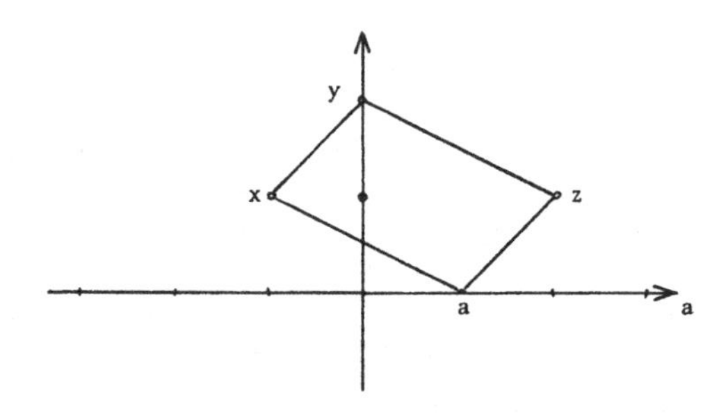

The corners are a, $a^{-1}b$, b^2, a^2b; b is not a corner.

Consider C_a.

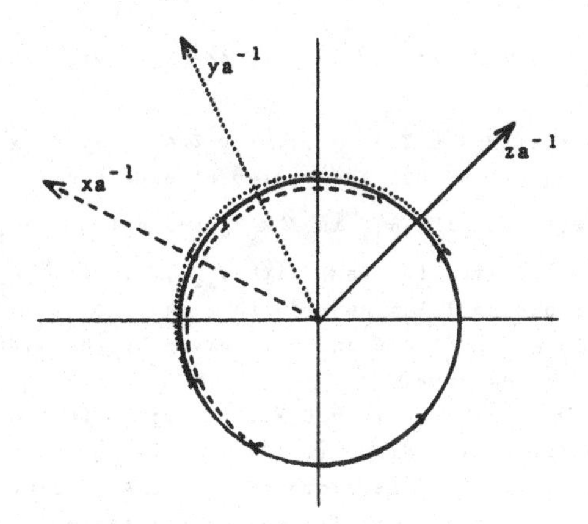

Here $0_{xa^{-1}} \cap 0_{za^{-1}} \subseteq 0_{ya^{-1}}$; C_a is the region ▬▬▬ , which is open at both ends, and C_a^- looks like this:

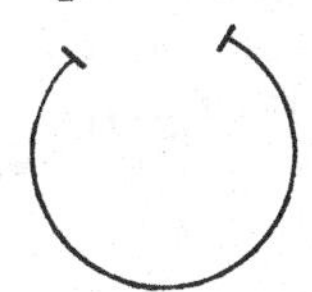

μ^* in this case consists of 4 points as shown:

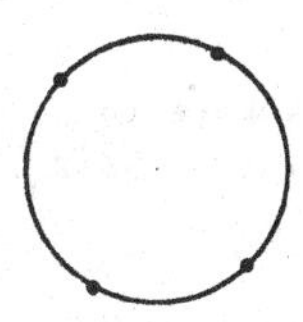

LECTURE 5

I'd like to finish off by talking briefly about the Bieri and Strebel 'geometric invariant' for a module over a finitely generated Abelian group.

Let Γ be a finitely generated metabelian group and M an Abelian normal subgroup of Γ with Abelian quotient group A. We know M is finitely generated as a $\mathbb{Z}A$-module. For $\chi \in \mathrm{Hom}(A,\mathbb{R})$ define A_χ, as before, to be $\{a \in A | \chi(a) > 0\}$. Define Σ_M on $S(A)$ to be

$$\{[\chi] \mid M \text{ is finitely generated over } \mathbb{Z}A_\chi\}.$$

This is the geometric invariant introduced by Bieri and Strebel [1980]. It is defined whether or not A is free, but for the purposes of this talk I shall suppose that A is free. Write C(M) for the centralizer in $\mathbb{Z}A$ of M, that is the set of elements λ of $\mathbb{Z}A$ with $\lambda \equiv 1$ (Ann M).

There is a useful criterion for $[\chi]$ to be in Σ_M:

(1) *Let* $0 \neq \chi \in \mathrm{Hom}(A,\mathbb{R})$. $[\chi] \in \Sigma_M$ *if and only if there exists* λ *in* $C(M)$ *with* $\bar{\chi}(\lambda) > 0$.

Proof. Suppose $[\chi] \in \Sigma_M$, so that $M = u_1\mathbb{Z}A_\chi + \ldots + u_m\mathbb{Z}A_\chi$. Since χ is non-zero there is an element a with $\chi(a) < 0$. The system of equations $u_i a = \sum_{j=1}^{m} u_j\alpha_{ji}$ $(1 \leq i \leq m)$ with all the α_{ji} in $\mathbb{Z}A_\chi$ gives $\det(a - (\alpha_{ji})) \in \mathrm{Ann}\,M$. We see $a^m + \sigma_{m-1}a^{m-1} + \ldots$ lies in $\mathrm{Ann}\,M$ with $\sigma_{m-1},\ldots,$ in $\mathbb{Z}A_\chi$. If we multiply through by a^{-m} we deduce $1-\lambda \in \mathrm{Ann}\,M$ with λ in $a^{-1}\mathbb{Z}A_\chi + a^{-2}\mathbb{Z}A_\chi + \ldots$. Since $\chi(a^{-1}) > 0$ and $\bar{\chi}$ is given by the minimum formula it follows that $\bar{\chi}(\lambda)$ is positive.

Conversely if $\bar{\chi}(\lambda) > 0$ for some λ in $C(M)$ then for u in M and a in A we have $ua = ua\lambda^r$ for all positive r. Since $\bar{\chi}(a\lambda^r) = \chi(a) + r\bar{\chi}(\lambda)$, it is clear that $a\lambda^r \in \mathbb{Z}A_\chi$ for all large r. Therefore $u\mathbb{Z}A = u\mathbb{Z}A_\chi$. Since M is finitely generated over $\mathbb{Z}A$, it must also be finitely generated over $\mathbb{Z}A_\chi$.

The elementary properties of the geometric invariant stem directly from (1). Modules with the same centralizer evidently have the same invariant, so $\Sigma_M = \Sigma_{\mathbb{Z}A/\mathrm{Ann}\,M}$. A simple argument with centralizers shows that if N is a submodule of M then $\Sigma_M = \Sigma_N \cap \Sigma_{M/N}$, whilst for ideals I and J of $\mathbb{Z}A$, $\Sigma_{\mathbb{Z}A/IJ}$ equals $\Sigma_{\mathbb{Z}A/I} \cap \Sigma_{\mathbb{Z}A/J}$; in particular $\Sigma_{\mathbb{Z}A/I}$ equals $\Sigma_{\mathbb{Z}A/I^m}$ for all m. Now the radical of I has some positive power in I and is the intersection of the finitely many minimal primes above I, so it's easy to see

(2) *If* $P_1,\ldots,P_r$ *are the minimal primes above* $\mathrm{Ann}\,M$ *then* Σ_M *equals* $\Sigma_{\mathbb{Z}A/P_1} \cap \ldots \cap \Sigma_{\mathbb{Z}A/P_r}$.

Thus it is just the cyclic prime modules whose invariants have to be calculated. Bieri and Strebel give the following description of Σ_M. 'Real valuation' is used in the same sense as before.

(3) *Let* $0 \neq \chi \in \mathrm{Hom}(A,\mathbb{R})$. *Then* $[\chi] \in \Sigma_M^-$ *if and only if there is a prime* P *minimal over* $\mathrm{Ann}\,M$ *and a real valuation* v *of* $\mathbb{Z}A/P$ *inducing* χ.

If $[\chi] \notin \Sigma_M$ we can by (2) find P with $[\chi] \notin \Sigma_{\mathbb{Z}A/P}$. By (1) no element of $1 + P$ $(= C(\mathbb{Z}A/P))$ is positive under $\bar{\chi}$. Hence $P + \mathbb{Z}A_\chi^+ < P + \mathbb{Z}A_\chi$. We were in this position in the last lecture whilst proving (1) and the proof can now proceed in the same way.

Conversely let $\lambda \in C(M)$; so that $P + \lambda = P + 1$. Suppose v is a real valuation of $\mathbb{Z}A/P$ with $\chi = v_A$. If $\lambda = \Sigma\,\lambda_a a$ then

$$0 = v(P+1) = v(P+\lambda) \geq \min_{a \in s(\lambda)} v(P+\lambda_a a) = \min_{a \in s(\lambda)} v(P+a) = \bar{v}_A(\lambda).$$

Hence $\bar{v}_A(\lambda) \leq 0$ and (1) shows $[v_A] \notin \Sigma_M$.

An immediate consequence is:

(4) *If all the discrete points of* $S(A)$ *are in* Σ_M *then* M *is finitely generated.*

For if $\mathbb{Z}A/P_i$ is not finitely generated over $\mathbb{Z}$ we can find a discrete valuation v with v_A non-zero.

(3) shows that Σ_M^- is very similar to the subsets of S(A) that we discussed in Lecture 4. The only difference we need note between then and now is that the units of $\mathbb{Z}A$ are ± the group elements. So, for example, using arguments similar to those used for proving (3) of Lecture 4, we can say

(5) *If* $P = \mu\mathbb{Z}A$ *then*

$$\Sigma_M = \cup \{C_a \mid a \text{ a corner of } \mu \text{ and } \mu_a = \pm 1\}.$$

Now the marvellous thing about Bieri and Strebel is that they discovered a beautiful connection between finite presentability and this geometric invariant. Let me state it as:

(6) *If* $S(A) = \Sigma_M \cup -\Sigma_M$ *then any extension of* M *by* A *is finitely presented and conversely.*

Here $-\Sigma_M$ is the antipodal set; in other words the one corresponding to the 'opposite' module M*, where the action on M is modified by the involution $a \to a^{-1}$ $(a \in A)$. I have nothing like enough time even to hint at how they prove it. Suffice to mention that in proving finite presentability from $S(A) = \Sigma_M \cup \Sigma_{M^*}$ it is the compactness of S(A) that plays a vital role. For λ in $\mathbb{Z}A$ we may set $O_\lambda = \{[\chi] \in S(A) \mid \bar{\chi}(\lambda) > 0\}$. This is the intersection of a finite number of open hemispheres: the O_a $(a \in s(\lambda))$: and so is open. When $S(A) = \Sigma_M \cup \Sigma_{M^*}$ we get from (1) an open cover of S(A) consisting of the sets O_λ, $\lambda \in C(M) \cup C(M^*)$. Compactness is used to pick a finite subset Λ of $C(M) \cup C(M^*)$ so that $S(A) = \bigcup_{\lambda \in \Lambda} O_\lambda$. If $\lambda = \lambda_1 a_1 + \ldots + \lambda_m a_m$ is in $\Lambda \cap C(M)$ relations of the form $u = u^{\lambda_1 a_1} u^{\lambda_2 a_2} \ldots u^{\lambda_m a_m}$, and relations of the form $u = u^{\lambda_1 a_1^{-1}} u^{\lambda_2 a_2^{-1}} \ldots u^{\lambda_m a_m^{-1}}$ if $\lambda \in C(M^*)$, turn out to be significant.

The only way I can give you any indication of this result perhaps is to give some examples.

(i) $\Sigma_{\mathbb{Z}A} = \emptyset$. Here $C(\mathbb{Z}A) = 1$ and $\bar{\chi}(1) = 0$ for all χ.

This example corresponds to the wreath product C_∞ wr A; as a module the base group is $\mathbb{Z}A$. The group is not finitely presented unless A is trivial.

(ii) Here we take $A = \langle a \rangle$ to be infinite cyclic. S(A) is $[+1] \cup [-1]$.

(a) $M = \mathbb{Z}\langle a \rangle/(a-6)$. Here $6a^{-1}$ is in C(M) and is positive under -1, but no element of C(M) is positive under +1. Therefore $\Sigma_M = [-1]$ and $\Sigma_{M^*} = [+1]$. The split extension $\langle a \rangle \ltimes M = \langle a, b \mid b^a = b^6 \rangle$ is finitely presented. M here is $\mathbb{Z}[1/6]$ and a multiplies by 6.

(b) $M = \mathbb{Z}\langle a \rangle/(2a-3)$. Here again $M = \mathbb{Z}[1/6]$, but a multiplies by 3/2. No extension of M by $\langle a \rangle$ is finitely presented. Take $1 + (2a-3)(\lambda_m a^m + \ldots + \lambda_n a^n)$ in C(M). Under +1, this has value the least power of a actually occurring; if this is positive then m cannot be negative, nor positive so $m = 0$ and $1 - 3\lambda_m$ must vanish. Since this is impossible, $[+1] \notin \Sigma_M$. Similarly $[-1] \notin \Sigma_M$. Hence $\Sigma_M = \Sigma_{M*} = \emptyset$.

(iii) Let me finish by mentioning the famous example of a finitely presented metabelian group Γ with Γ' of infinite rank found by G. Baumslag [1972] and V.N. Remeslennikov [1973]. One way of viewing it is

$$\Gamma = \langle x,a,b \mid [a,b] = 1,\ [x,x^{a^n}] = 1,\ n = 0,\pm1,\pm2,\ldots,x^b = xx^a \rangle.$$

Here $A = \langle a,b \rangle$ is free of rank 2 and $M = \langle x^\Gamma \rangle$ is $\mathbb{Z}\langle a,b\rangle/(1+a-b)$. Using what we learnt in Lecture 4, with $\mu = 1+a-b$, the diagram of μ is

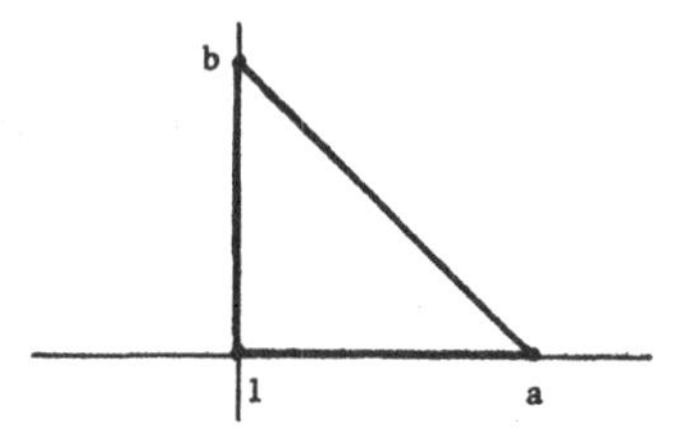

Σ_M is very easily seen to be

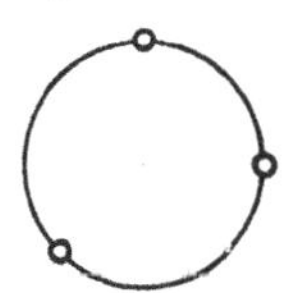

Σ_M^- is 3 isolated points. Evidently $\Sigma_M \cup -\Sigma_M$ is $S(A)$, illustrating the fact that Γ is finitely presentable.

REFERENCES

Baumslag, G. (1972), A finitely presented metabelian group with a free abelian derived group of infinite rank, *Proc. Amer. Math. Soc.* 35, 61-62.

Bergman, G.M. (1971), The logarithmic limit-set of an algebraic variety, *Trans. Amer. Math. Soc.* 157, 459-469.

Bieri, R., Groves, J.R.J. (1984), On the geometry of the set of characters induced by valuations, *J. reine angew. Math.* 347, 168-195.

Bieri, R., Groves, J.R.J. (1986), A rigidity property for the set of all characters induced by valuations, *Trans. Amer. Math. Soc.*, to appear.

Bieri, R., Strebel, R. (1980), Valuations and finitely presented metabelian groups, *Proc. London Math. Soc.* (3) 41, 439-464.

Bieri, R., Strebel, R. (1981), A geometric invariant for modules over an abelian group, *J. reine angew. Math.* 322, 170-189.

Brewster, D.C. (1976), *The maximum condition on ideals of the group ring*, Ph.D. dissertation, Cambridge University.

Brookes, C.J.B. (1985), Ideals in group rings of soluble groups of finite rank, *Math. Proc. Cambridge Philos. Soc.* 97, 27-49.

Brown, K.A. (1981), Modules over polycyclic groups have many irreducible images, *Glasgow Math. J.* 22, 141-150.

Brown, K.A. (1982), The Nullstellensatz for certain group rings, *J. London Math. Soc.* (2) 26, 385-396.

Farkas, D.R., Snider, R.L. (1984), Arithmeticity of stabilizers of ideals in group rings, *Invent. Math.* 75, 75-82.

Frattini, G. (1885), Intorno alla generazione dei gruppi di operazioni, *Atti R. Accad. dei Lincei, Rendiconti* (IV) 1, 281-285.
Hall, P. (1954), Finiteness conditions for soluble groups, *Proc. London Math. Soc.* (3) 4, 419-436.
Hall, P. (1959), On the finiteness of certain soluble groups, *Proc. London Math. Soc.* (3) 9, 595-622.
Hall, P. (1961), The Frattini subgroups of finitely generated groups, *Proc. London Math. Soc.* (3) 11, 327-352.
Harper, D.L. (1980), Primitivity in representations of polycyclic groups, *Math. Proc. Cambridge Philos. Soc.* 88, 15-31.
Kropholler, P.H. (1984), On finitely generated soluble groups with no large wreath product sections, *Proc. London Math. Soc.* (3) 49, 155-169.
McConnell, J.C. (1982), The Nullstellensatz and Jacobson properties for rings of differential operators, *J. London Math. Soc.* (2) 26, 37-42.
Passman, D.S. (1977), *The algebraic structure of group rings*, Wiley-Interscience, New York.
Passman, D.S. (1984), Group rings of polycyclic groups, in *Group theory: essays for Philip Hall* (eds. K.W. Gruenberg and J.E. Roseblade), Academic Press, London.
Remeslennikov, V.N. (1973), On finitely presented groups, *Proc. Fourth All-Union Symposium on the Theory of Groups*, Novosibirsk, 164-169.
Roseblade, J.E. (1973), Group rings of polycyclic groups, *J. Pure Appl. Algebra* 3, 307-328.
Roseblade, J.E. (1978), Prime ideals in group rings of polycyclic groups, *Proc. London Math. Soc.* (3) 36, 385-447, (Corrigenda, *ibid* 38 (1979), 216-218).
Segal, D. (1983), *Polycyclic groups*, Cambridge Tracts in Mathematics, 82, Cambridge University Press.
Strebel, R. (1984), Finitely presented soluble groups, in *Group theory: essays for Philip Hall* (eds. K.W. Gruenberg and J.E. Roseblade), Academic Press, London.
Wehrfritz, B.A.F. (1979), Invariant maximal ideals of commutative rings, *J. Algebra* 56, 472-480.

BUILDINGS AND GROUP AMALGAMATIONS

Jacques Tits
Collège de France, Paris, France

The following pages are but a brief summary of five lectures held at the conference "Groups - St Andrews 1985". We recall basic definitions, state the main results and sketch some applications. Further motivation, detailed proofs and more examples will be published elsewhere. Some extra information can already be found in [Ti8], [Ti9].

1. *DEFINITIONS. PRELIMINARIES*

1.1 *Amalgamated sums*

Let $\mathfrak{X}$ be a set of groups and Φ a set of homomorphisms $\phi: A_\phi \to B_\phi$, with $A_\phi, B_\phi \in \mathfrak{X}$. The *inductive limit* of the system $(\mathfrak{X}, \Phi)$, also called the *amalgamated sum* of $\mathfrak{X}$ relative to Φ, is a group X endowed with a set of homomorphisms $\psi_A: A \to X$, for all $A \in \mathfrak{X}$, such that, for all $\phi \in \Phi$, $\psi_{A_\phi} = \psi_{B_\phi} \circ \phi$, the system $(X, (\psi_A))$ being *universal* for that property, in the usual sense. This characterizes $(X, (\psi_A))$ up to unique isomorphism. One can also view X as the group defined by the following presentation: the set of generators is the disjoint union of the elements of $\mathfrak{X}$ and the relations are those provided by the multiplication tables of the elements of $\mathfrak{X}$ plus all relations of the form $\phi(a) = a$ for $\phi \in \Phi$ and $a \in A_\phi$.

As is well known, a crucial problem with amalgamated sums is to give conditions under which the group X "does not collapse" or, more exactingly, conditions for the injectivity of the maps ψ_A. Theorem 1 of Section 2.4 provides such a condition in a special but interesting situation.

Exercise. We denote by Fr_{21} (= Frobenius group of order 21) the semi-direct product $\mathbb{Z}/3\mathbb{Z} \ltimes \mathbb{Z}/7\mathbb{Z}$, where the first group operates non-trivially on the second. Let $X_{\{0,1\}}$, $X_{\{1,2\}}$ be two copies of Fr_{21}, let $X_{\{0,2\}}$ be the direct product of two copies of $\mathbb{Z}/3\mathbb{Z}$, for $i \in \{0,1,2\}$ let X_i denote a cyclic group of order 3 and, for any two distinct elements i, j of $\{0,1,2\}$, let $\phi_{ij}: X_i \to X_{\{i,j\}}$ be an injection. Set $\mathfrak{X} = \{X_{\{i,j\}}, X_i \mid i,j \in \{0,1,2\}\}$ and $\Phi = \{\phi_{ij} \mid i,j \in \{0,1,2\}\}$. Prove that, if $\phi_{ij}(X_i) \neq \phi_{ji}(X_j)$ for all i, j, the amalgamated sum of $\mathfrak{X}$ relative to Φ is isomorphic to $\mathbb{Z}/3\mathbb{Z}$ or $\{1\}$, and determine for which choice of the ϕ_{ij} the first answer holds. What happens if $\phi_{ij}(X_i) = \phi_{ji}(X_j)$ for some i,j?

1.2 *Chamber systems* ([Ti5], [Ka3])

A *chamber system* over a set I consists of a set $\mathbf{C}$ and a system $(\mathcal{P}_i)_{i \in I}$ of partitions of $\mathbf{C}$, indexed by I. The elements of $\mathbf{C}$ are called *chambers* and two chambers belonging to the same element of $\mathfrak{P}_i$ are said to be i-*adjacent*. By abuse of notation, the chamber system $(\mathfrak{C}, (\mathfrak{P}_i)_{i \in I})$ will often simply be denoted by $\mathfrak{C}$.

Examples. (a) A "numbered complex" (over I) is a simplicial complex together with a map $\tau: V \to I$ of the set V of its vertices into the set of reference I, such that the restriction of τ to any simplex of the complex is injective. The image of a simplex by τ is called the *type* of the simplex. The *chambers* of such a complex are defined as the simplices of type I and two chambers are said to be i-*adjacent* if their faces of type $I - \{i\}$ coincide. Thus, to every numbered complex Δ, we have associated a chamber system $\mathfrak{C} = \mathfrak{C}(\Delta)$. Most chamber systems met in applications so far arise in that way. Suppose that Δ has the following property:

(C) every simplex A is contained in a chamber and, if card $(I - \tau(A)) \geq 2$, the "adjacency graph" of all chambers containing A is connected.

Then, clearly, $\mathfrak{C}(\Delta)$ determines Δ: if $\mathfrak{C}(\Delta) = \mathfrak{C}(\Delta')$ with Δ' also satisfying (C), the complexes Δ and Δ' are canonically isomorphic. In that case, we set $\Delta = \Delta(\mathfrak{C})$.

(b) The non-empty linear subspaces of a projective space Π are the vertices of a numbered complex $\Delta(\Pi)$ whose simplices are the sets consisting of pairwise incident subspaces (*flags*), the type of a subspace being its rank (projective dimension plus one). This complex, called the flag complex of Π, has property (C). The associated chamber system $\mathfrak{C}(\Pi) = \mathfrak{C}(\Delta(\Pi))$ has the maximal flags of Π as chambers.

(c) Let G be a group, B a subgroup and $(P_i)_{i \in I}$ a system of subgroups of G containing B. We denote by $\mathfrak{C}(G; B; (P_i)_{i \in I})$ the chamber system $\mathfrak{C}(G/B; (\mathfrak{P}_i)_{i \in I}))$, where $\mathfrak{P}_i$ is the partition of G/B whose elements are the fibres of the canonical map $G/B \to G/P_i$. Important special cases are the following.

Let $\mathbf{G}$ be an algebraic semi-simple group over a field K, let $\mathbf{B}$ denote a minimal parabolic K-subgroup of $\mathbf{G}$ and let $\mathbf{P}_i$ ($i \in I$) be the parabolic K-subgroups containing $\mathbf{B}$ properly and minimal with that property. Set $G = \mathbf{G}(K)$, $B = \mathbf{B}(K)$, $P_i = \mathbf{P}_i(K)$ and $\mathfrak{C} = \mathfrak{C}(G; B; (P_i)_{i \in I})$. This is the chamber system associated (cf. Example (a)) with the so-called *spherical building* Δ of $\mathbf{G}$ over K, whose simplices "are" the parabolic K-subgroups of $\mathbf{G}$ (cf. [Ti2], 5.2); since that building has property (C), it is characterized by the condition $\mathfrak{C}(\Delta) = \mathfrak{C}$.

Let K be a locally compact local field and let $\mathbf{G}$ be an algebraic almost simple group over K which, for the commodity of the exposition, we assume to be simply connected. The group $G = \mathbf{G}(K)$ has a natural topology making it a locally compact group. All maximal pro-p-subgroups of G, where p is the characteristic of the residue field, are conjugate to each other. Let B be the normalizer of one of them and let P_i ($i \in I$) denote the subgroups of G containing B properly and minimal with that property. Here again, $\mathbf{C} = \mathbf{C}(G;\ B;\ (P_i)_{i \in I})$ is the chamber system associated with a numbered complex satisfying condition (C) (hence characterized by $\mathbf{C}$), called the *affine* building of G over K. The stabilizers in G of the vertices of that building are the maximal compact subgroups of G. (Cf. [BT], [Ti4].)

1.3 *Coxeter chamber systems*

A *Coxeter matrix* is a symmetric matrix $M = (m_{ij})_{i,j \in I}$ whose entries are integers or ∞, and such that $m_{ij} = 1$ or ≥ 2 according as $i = j$ or $i \neq j$. As is customary, we represent such a matrix by a graph whose vertices are the elements of I and whose edges are the pairs $\{i,j\}$ with $m_{ij} \geq 3$, the number m_{ij} being written on the edge $\{i,j\}$ whenever it is ≥ 4 (here, we shall not make use of the convention $\overset{4}{\bullet\!-\!\bullet} = \bullet\!=\!\bullet$).

A *Coxeter system* of type M is a group $W = W(M)$ ("the" Coxeter group of type M) together with a generating system $(s_i)_{i \in I}$ such that the relations $(s_i s_j)^{m_{ij}} = 1$ for $m_{ij} \neq \infty$ form a presentation of W. The chamber system $\mathbf{C}(M) = \mathbf{C}(W;\ \{1\};\ (\langle s_i \rangle)_{i \in I})$ is called "the" *Coxeter chamber system* of type M; as one can see, its chambers are the elements of W. "The" *Coxeter complex* $\Delta(M)$ of type M, defined in [Ti2], can be characterized by the relation $\mathbf{C}(\Delta(M)) = \mathbf{C}(M)$, together with property (C).

Examples. (a) The Coxeter complex of type $\bullet\!-\!\bullet\!-\!\bullet$ (resp. $\bullet\!-\!\bullet\!\overset{4}{-}\!\bullet$, resp. $\bullet\!-\!\bullet\!\overset{5}{-}\!\bullet$, resp. $\bullet\!-\!\bullet\!\overset{6}{-}\!\bullet$) is the barycentric subdivision of a tetrahedron (resp. a cube, resp. an icosahedron, resp. a paving of the Euclidean plane by regular hexagons).

(b) The Coxeter complex of type $\tilde{A}_2 = \triangle$ is shown by the following picture. The chambers are the triangles of the drawn subdivision of the plane, and the three types of adjacencies are materialized by the three types of edges:

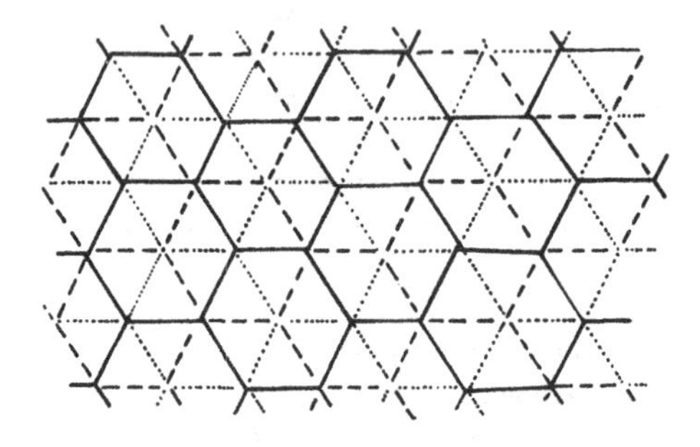

1.4 *Buildings*

Let M be a Coxeter matrix. Following [Ti5], we give the name of *buildings of type* M to all complexes called "weak buildings" of type M in [Ti2], and also to the associated chamber systems. Since the complexes in question satisfy condition (C), the two viewpoints - simplicial complexes and chamber systems - are equivalent. We do not recall here the definitions of [Ti2] and [Ti5], but we shall give a characterization of buildings which will be sufficient for our purpose and which the reader may take as the definition if he wishes to.

Let $L(I)$ denote the free monoid generated by I. Its elements are called *I-words*. An I-word $\underline{i} = i_1 \ldots i_m$ is said to be *reduced* (with respect to M) if the element $s(\underline{i}) = s_{i_1} \ldots s_{i_m}$ of $W(M)$ is not the product of less than m factors belonging to $\{s_i \mid i \in I\}$. A *gallery* of type $\underline{i}$ in a chamber system over I is a sequence of chambers $(C_0, \ldots, C_m)$ such that C_{h-1} and C_h are i_h-adjacent and distinct for all h, $1 \leq h \leq m$. The following two results are essentially proved in [Ti5]; the first one is the characterization of buildings announced above.

(1.4.1) *A chamber system* $(\mathcal{C}, (\mathcal{P}_i)_{i \in I})$ *is a building of type* M *if and only if* $\mathcal{C}$ *is not empty, each element of each* $\mathcal{P}_i$ *contains at least two chambers, and there exists a "distance"* $\delta : \mathcal{C} \times \mathcal{C} \to W(M)$ *such that, for* $C, C' \in \mathcal{C}$ *and any reduced I-word* $\underline{i}$*, the relation* $\delta(C,C') = s(\underline{i})$ *is equivalent to the existence of a gallery of type* $\underline{i}$ *joining* C *and* C'.

(1.4.2) *If* $\underline{i}$ *is a reduced I-word, the gallery of type* $\underline{i}$ *joining two chambers at distance* $s(\underline{i})$ *in a building is unique.*

Examples of buildings. (a) The spherical and affine buildings of Example 1.2(c) are indeed buildings in the above sense. This reflects (partially) the so-called "BN-pair property" of the groups involved and, in particular, the existence of a Bruhat decomposition.

(b) Let $M = A_n = \circ\!-\!\circ\!-\!\circ \;\ldots\; \circ\!-\!\circ$ (with n vertices). Then the buildings of type M are nothing else but the flag complexes of n-dimensional projective spaces (cf. Example 1.2(b)). Here, in the definition of projective spaces, lines are required to have at least *two* (not three) points.

(c) Let $M = \begin{pmatrix} 1 & \infty \\ \infty & 1 \end{pmatrix} = \circ\overset{\infty}{-\!-}\circ$. Then, a building of type M is a tree without dangling edge (i.e. without vertex of order 1), the chambers being the edges of the tree.

2. *CHAMBER-TRANSITIVE AUTOMORPHISM GROUPS OF BUILDINGS*

We suppose a Coxeter matrix $M = (m_{ij})_{i,j \in I}$ is given once and for all, and, for $J \subset I$, we denote by M_J the principal minor $(m_{ij})_{i,j \in J}$.

2.1 *Preliminary remarks*

Let $(\mathcal{C}, (\mathcal{P}_i)_{i \in I})$ be a chamber system. By an *automorphism* of such a system, we shall always mean a permutation of $\mathcal{C}$ preserving *each* $\mathcal{P}_i$. Let G be an automorphism group of the system, transitive on $\mathcal{C}$ (we also say: chamber-transitive); choose a chamber C, let B be the stabilizer of C and P_i the stabilizer of the element of $\mathcal{P}_i$ containing B. Then, it is clear that

$(\mathbf{C}, (\mathbf{P}_i)_{i \in I}) = (G; B; (P_i)_{i \in I})$.

Conversely, if G, B, P_i for $i \in I$ are as in Example 1.2(c), the group G acts on G/B in the usual way and induces a chamber-transitive automorphism group of the chamber system $\mathbf{C}(G; B; (P_i)_{i \in I})$.

2.2 *When is the chamber system* $\mathbf{C}(G; B; (P_i)_{i \in I})$ *a building?*

The answer to that question is given by Proposition 1 below, which is a trivial consequence of (1.4.1) and (1.4.2). When the system *is* a building, the proposition also provides a - so to speak - "half-canonical" decomposition of the elements of G as products of elements of the P_i's, generalizing in a straightforward manner a decomposition considered by R. Steinberg ([St], §8). Let G, B, $(P_i)_{i \in I}$ be as in 1.2(c). For $i \in I$, set $P_i^* = P_i - B$ and let R_i denote a set of representatives of P_i^*/B in P_i^* and, for any I-word $\underline{i} = i_1 \dots i_m$, set $P_{\underline{i}}^* = P_{i_1}^* \dots P_{i_m}^*$.

Proposition 1. *The chamber system* $\mathbf{C}(G; B; (P_i)_{i \in I})$ *is a building of type* M *if and only if the following conditions are fulfilled:*

(i) *for* $i \in I$, $P_i^* \neq \emptyset$;

(ii) *if* $\underline{i} \in L(I)$ *is reduced (with respect to* M*),* $P_{\underline{i}}^*$ *depends only on the element* $w = s(\underline{i})$ *of* $W(M)$*; we denote it by* P_w^*;

(iii) *the group* G *is the disjoint union of the sets* P_w^*, *for* $w \in W(M)$.

When they are satisfied, the product map $R_{i_1} \times \dots \times R_{i_m} \times B \to P_{\underline{i}}^*$, *for* $\underline{i} = i_1 \dots i_m$ *reduced, is injective.*

2.3 *The group* G *as an amalgamated sum*

Suppose that $\mathbf{C}(G; B; (P_i)_{i \in I})$ is a building of type M. For any subset J of I, let P_J denote the subgroup of G generated by all P_j for $j \in J$. The following assertion is an easy consequence of Proposition 1 and of known properties of Coxeter groups (to be found in [Bo] or [Ti2]).

Proposition 2. *The group* G *is the amalgamated sum of the system of groups* B, P_i, $P_{\{i,j\}}$ *for* $m_{ij} \neq \infty$, *relative to the inclusion maps.*

2.4 *An existence theorem*

The theorem which we are going to state is, in a weak sense, a converse to Proposition 2.

A subset J of I is called *spherical* if the Coxeter group $W(M_J)$ is finite. Let there be given:

for any spherical set $J \subset I$ of cardinality ≤ 3, a group P_J;

for any pair (J,J') of such sets, with $J \subset J'$, an injective homomorphism $\phi_{JJ'} : P_J \to P_{J'}$, those homomorphisms satisfying the conditions $\phi_{JJ} = \text{id.}$ and $\phi_{JJ''} = \phi_{J'J''} \circ \phi_{JJ'}$ (whenever the two members are defined).

We shall often identify the group P_J with its image by $\phi_{JJ'}$.

Let $\mathcal{J}$ be the set of all spherical subsets of I of cardinality ≤ 2 (i.e. the set of all subsets of I of the form $\emptyset$, $\{i\}$ or $\{i,j\}$ with $m_{ij} \neq \infty$). For $K \subset I$, let G_K denote the amalgamated sum of the system of groups $(P_J | J \in \mathcal{J}, J \subset K)$ relative to the set of homomorphisms $(\phi_{JJ'} | J \subset J' \subset K, J' \in \mathcal{J})$. Set $P_\emptyset = B$, $G_I = G$, and assume $P_i \neq B$ for all $i \in I$. Observe that, by the above convention, $B \subset P_j \subset P_J$ whenever P_J is defined and $j \in J$.

Theorem 1. *Suppose that for all spherical subsets* J *of* I *of cardinality* ≤ 3, $\mathbf{C}(P_J; B; (P_j)_{j \in J})$ *is a building of type* M_J. *Then, for all* $K \subset I$, *the canonical map* $G_K \to G$ *is injective and, if we identify each* $P_j = G_j$ *with its canonical image in* G, $\mathbf{C}(G; B; (P_i)_{i \in I})$ *is a building of type* M.

Observe that the groups P_J for card $J = 3$ play a role only through their *existence*, and that, for such J, the canonical homomorphism $G_J \to P_J$ is an isomorphism, by Proposition 2.

When $M = \begin{pmatrix} 1 & \infty \\ \infty & 1 \end{pmatrix}$, $\mathbf{C}(G; B; (P_i)_{i \in I})$ is a tree (cf. 1.4(c)) and Theorem 1 combined with Proposition 1 reduces to well-known structure theorems for amalgamated sums of two groups (cf. [Se], Chap. I, théorèmes 1 et 7).

2.5 *Homotopy of* I-*words*

The proof of Theorem 1 combines two classical concepts of "reduced words", occurring respectively in the theory of Coxeter groups and the theory of free groups, and a notion of "homotopy of words", which we shall first illustrate on the simplest example, that of I-words.

For any $i,j \in I$ with $i \neq j$, let $\underline{m}(i,j)$ denote the I-word ijiji... of length m_{ij}. We define an *elementary homotopy* of I-words (with respect to the given Coxeter matrix M) as the replacement, in a given I-word, of a section of the form $\underline{m}(i,j)$ by the corresponding $\underline{m}(j,i)$. A *homotopy* is the result of a succession of elementary homotopies. Clearly, if two I-words $\underline{i}$ and $\underline{i}'$ are homotopic, we have $s(\underline{i}) = s(\underline{i}')$ (with the notation s of 1.4). It is an important property of Coxeter groups that the converse holds for reduced words (cf. [Ti1]):

Lemma 1. *Two reduced* I-*words* $\underline{i}$ *and* $\underline{i}'$ *are homotopic if and only if* $s(\underline{i}) = s(\underline{i}')$.

The proposition which we are now going to state is responsible for the role played by the spherical sets of cardinality 3 in Theorem 1. For $w \in W(M)$, let Γ_w denote the graph whose vertices are the reduced I-words $\underline{i}$ such that $s(\underline{i}) = w$, and whose edges are the elementary homotopies between such words. We shall be interested in two types of closed paths in Γ_w, namely

(A) $ff_1f', \ldots, ff_mf', ff_1f'$

where f, f' are given I-words and $f_1, \ldots, f_m, f_1$ is a closed path in the graph $\Gamma_{w'}$ for some $w' \in \langle s_i, s_j, s_k \rangle$ with $\{i,j,k\}$ spherical, and

(B) $f\underline{m}(i,j)f'\underline{m}(k,\ell)f''$, $f\underline{m}(j,i)f'\underline{m}(k,\ell)f''$, $f\underline{m}(j,i)f'\underline{m}(\ell,k)f''$,
$f\underline{m}(i,j)f'\underline{m}(\ell,k)f''$, $f\underline{m}(i,j)f'\underline{m}(k,\ell)f''$

where f, f', f" are given I-words, $i,j,k,\ell \in I$, $i \neq j$ and $k \neq \ell$.

Proposition 3. *The fundamental group of* Γ_W *with a given base point* * *is generated by all loops of the form* pqp^{-1}, *where* p *is a path with origin* * *and* q *is a closed path of type* (A) *or* (B).

This is Proposition 4 of [Ti5]. One can probably reformulate that result more conceptually, in terms of a suitably defined "second homotopy group" of the Coxeter chamber system of type M.

2.6 *A sketch of proof of Theorem* 1

In order to describe the idea of that proof, we first give the basic definitions and then state two lemmas which summarize the main steps of the argument.

We keep the notation and assumptions of the theorem (see 2.4). For any I-word $\underline{i} = (i_1,\ldots,i_m)$, a sequence $(p_1,\ldots,p_m, b)$ with $p_j \in P^*_{i_j}$ and $b \in B$ will be called a *word of type* $\underline{i}$. The I-words are special cases of these in the following sense: if we take for P_J the subgroup $\langle s_j \mid j \in J \rangle$ of $W = W(M)$ (notation of 1.3), so that, in particular, $B = \{1\}$, $P_i = \langle s_i \rangle$ and $P^*_i = \{s_i\}$, a word is a sequence of the form $(s_{i_1},\ldots,s_{i_m},1)$, and the "type map" establishes a one-to-one correspondence between words (in this special situation) and I-words. We are going to extend from the I-words to the words in general the notions of "elementary homotopy" and "homotopy" but beforehand, we have to define a simpler operation, the "equivalence" of words, which in the case of I-words reduces to the identity.

An *equivalence* (of words) is the replacement of $(p_1,\ldots,p_m,b)$ by a word of the form $(p_1b_1, b_1^{-1}p_2b_2,\ldots,b_{m-1}^{-1}p_mb_m, b_m^{-1}b)$, where $b_1,\ldots,b_m \in B$;

an *elementary homotopy* is the replacement in $(p_1,\ldots,p_m,b)$ of a section $(p_{s+1},\ldots,p_{s+m_{ij}})$ of type $\underline{m}(i,j)$ by a section $(p'_{s+1},\ldots,p'_{s+m_{ij}})$ of type $\underline{m}(j,i)$ such that the products $p_{s+1}\cdots p_{s+m_{ij}}$ and $p'_{s+1}\cdots p'_{s+m_{ij}}$ *in the group* $P_{\{i,j\}}$ are equal (as before we identify the groups P_i and P_j with their images in $P_{\{i,j\}}$ by the given maps $\phi_{i,\{i,j\}}$ and $\phi_{j,\{i,j\}}$);

a *homotopy* is the result of an arbitrary succession of equivalences and elementary homotopies.

We denote by $[p_1,\ldots,p_m, b]$ the homotopy class of the word $(p_1,\ldots,p_m,b)$, by $\mathfrak{G}$ the set of all homotopy classes of words, by $v(p_1,\ldots,p_m,b)$ the "value" of the word $(p_1,\ldots,p_m,b)$ in the group $G = G_I$ (cf. 2.4), that is, the product of the canonical images of $p_1,\ldots,p_m,b$ in G, and by $\bar{v}$ the map $[p_1,\ldots,p_m, b] \to v(p_1,\ldots,p_m,b)$ of $\mathfrak{G}$ in G (this map exists since, clearly, two homotopic words have the same value in G). If $\mathcal{C}(G, B, (P_i)_{i \in I})$ is to be a building of type M, the last assertion of Proposition 1 implies that, for any *reduced* word $i_1\ldots i_m$, the map v must *inject* the product $R_{i_1} \times \ldots \times R_{i_m} \times B$ into G. The following lemma, a major step of the proof of Theorem 1, expresses that this is indeed the case.

Lemma 2. *Two homotopic words of the same reduced type are equivalent.*

We just outline the proof, showing where Proposition 3 and the hypothesis concerning spherical sets of cardinality ≤ 3 come into play. Let $x_0,\ldots,x_n$ be a sequence of words of reduced types $\underline{i}_0,\ldots,\underline{i}_{n-1}$, $\underline{i}_n = \underline{i}_0$ such that, for all $i \in \{1,\ldots,n\}$, x_i is deduced from x_{i-1} by an equivalence or an elementary homotopy. One must show that x_0 and x_n are equivalent. But the sequence $\underline{i}_0,\ldots,\underline{i}_n$ is a closed path (with possible redundancies) in the graph Γ_w of 2.5, for $w = s(\underline{i}_0) = s(\underline{i}_1) = \ldots$. By Proposition 3 (and "lifting homotopies" of I-words to words by means of Proposition 1 applied to chamber systems of the form $\mathfrak{C}(P_{\{i,j\}}; B; P_i, P_j)$), one is then reduced to considering the case where the sequence $\underline{i}_0,\ldots,\underline{i}_n$ has one of the types (A) or (B) of 2.5. Type (B) is essentially trivial. Type (A) is also easy to handle, using the existence of the group P_J, for J the spherical subset of cardinality ≤ 3 involved in the definition of type (A), and applying the last assertion of Proposition 1 to the building $\mathfrak{C}(P_J; B; (P_j)_{j\in J})$.

Let us now prove that Theorem 1 is a consequence of the above Lemma and the following one.

Lemma 3. *There exist consistent right actions of the groups* B, P_i $(i \in I)$ *and* $P_{\{i,j\}}$ $(i,j \in I,\ m_{ij} \neq \infty)$ *on the set* $\mathfrak{G}$ *of homotopy classes of words of reduced types such that, for any such word* $(p_1,\ldots,p_m,b)$, *one has* $[p_1,\ldots,p_m,b] = [p_1,\ldots,p_{m-1},1].(p_m b)$.

Indeed, supposing Lemma 3 established, we see that the amalgamated sum $G = G_I$ operates on $\mathfrak{G}$ on the right in such a way that $[p_1,\ldots,p_m,b] = [1].\bar{v}([p_1,\ldots,p_m,b])$. Thus, the map $\bar{v} : \mathfrak{G} \to G$ is a bijection (its inverse being $g \mapsto [1].g$), by means of which we identify G and $\mathfrak{G}$. In view of Lemma 2, it follows that the canonical maps $P_i \to G$ are injective and we identify the P_i's with their images in G.

In order to show that $\mathfrak{C}(G; B; (P_i)_{i\in I})$ is a building (last assertion of Theorem 1), we must verify the conditions (i), (ii), (iii) of Proposition 1. Condition (i) is satisfied by assumption ($P_i \neq B$). Condition (ii) is clear by Lemma 1 and the existence of "lifting of homotopies" (from I-words to words) already used in the proof of Lemma 2. Finally, if $g \in G$ is identified with $[p_1,\ldots,p_m,b] \in \mathfrak{G}$, then $g = p_1\ldots p_m b$, and (iii) follows in view of the fact that, if two reduced I-words $\underline{i}$ and $\underline{i}'$ are the types of two homotopic words, one has $s(\underline{i}) = s(\underline{i}')$.

By the result just established, $\mathfrak{C}(G_K; B; (P_i)_{i\in K})$ is a building for all $K \subset I$. Since the canonical map $G_K \to G_I$ obviously maps any product of the form $P_{i_1}\ldots P_{i_m}$ "computed in G_K" into "the same product computed in G", its injectivity now follows from Proposition 1 (iii) and the injectivity of the canonical homomorphism $W(M_K) \to W(M)$ (cf. e.g. [Ti1], Corollaire 6). This proves Theorem 1, granting Lemma 3.

The proof of Lemma 3, which uses Proposition 1 applied to the buildings $\mathfrak{C}(P_{\{i,j\}}; B; P_i, P_j)$ and known properties of Coxeter systems (to be found in [Bo], see in particular Chap. 4, §1, Exercise 3, or in [Ti2], §2), is fairly easy, but long, and will be omitted.

3. *EXAMPLES*

3.1 *Some buildings of type* $\tilde{A}_2$

Let $q = 2$ or 8. Consider the field $\mathbb{F}_{q^3}$ as a 3-dimensional vector space over $\mathbb{F}_q$ and let Π denote its projective plane "at infinity". Let A be the full automorphism group of $\mathbb{F}_{q^3}$; it is a cyclic group of order q+1 (a fact which is true for no other value of q!). The semi-direct product $\mathbb{F}^{\times}_{q^3} \rtimes A$ operates on $\mathbb{F}_{q^3}$ and induces on Π a collineation group $F \cong (\mathbb{Z}/(q^2+q+1)\mathbb{Z}) \rtimes (\mathbb{Z}/(q+1)\mathbb{Z})$ (a Frobenius group of order $(q^2+q+1)(q+1)$), which is easily shown to be simply transitive on the flags (incident point-line pairs). Consequently, if H, H' are two distinct cyclic subgroups of order q+1 of F, then $\mathbf{C}(F; \{1\}; H, H')$ is the flag complex of Π, hence, by Example 1.4(b), a building of type A_2 = ⊢—⊣.

Now, consider three copies of F, called $X_{\{0,1\}}$, $X_{\{1,2\}}$, $X_{\{2,0\}}$ and, in $X_{\{i,j\}}$, choose two distinct cyclic subgroups of order q+1, X_i^j and X_j^i. For any value of the index i, "identify" the groups X_i^j and X_i^k (with $\{i,j,k\} = \{0,1,2\}$), denote by X_i the group resulting from that identification and consider the amalgamated sum G of the system $(X_i, X_{\{i,j\}} \mid i,j \in \{0,1,2\})$ (relative to the inclusion maps). Theorem 1 applied to this situation shows that

the canonical maps $X_{\{i,j\}} \to G$ *are injective and* $\Delta = \mathbf{C}(G; \{1\}; (X_i)_{i=0,1,2})$ *is a building of type* $\tilde{A}_2$ (cf. 1.3(b)).

In particular, G is an infinite group (in contrast with the result of the exercise in 1.1).

So far, we have said nothing about the way the groups X_i^j and X_i^k are identified, and that is not immaterial. Indeed, for any two distinct $i,j \in \{0,1,2\}$, let f_i^j denote the element of X_i^j which conjugates every element of the (q^2+q+1)-cycle of $X_{\{i,j\}}$ into its square. Then, f_i^k can be identified with any power $(f_i^j)^a$ of f_i^j, provided that a is a unit of the ring $\mathbb{Z}/(q+1)\mathbb{Z}$. If $q = 2$ (resp. 8), we obtain in that way two (resp. six) *essentially different* amalgamations of X_i^j and X_i^k. Let us say that the index i is *polarized* if $a = 1$. If all three indices 0, 1, 2 are polarized, we say that the amalgamation is *totally polarized*. If $q = 2$, there are four non-isomorphic amalgamations, according to the number (= 0,1,2,3) of polarized indices (the existence of four different ways of amalgamating three copies of Fr_{21} has been first pointed out by M. Ronan). If $q = 8$, one checks that the total number of cases, up to isomorphism, is equal to 44.

Thus, we have defined 4 + 44 *a priori* different buildings of type $\tilde{A}_2$. It is natural to ask which of them are affine buildings of algebraic simple groups (necessarily of relative type A_2) over local fields (cf. 1.2(c)). The answer is as follows:

(i) in the totally polarized case, Δ is the affine building of the group $SL_3(\mathbb{F}_q((y)))$ (cf. [KMW2] and [Me], §2);
(ii) if $q = 2$ and all three indices are non-polarized, Δ is the affine building of $SL_3(\mathbb{Q}_2)$ (cf. [KMW1]);

(iii) in the remaining $2+43$ cases, Δ is not the affine building of an algebraic simple group over a local field (cf. [Ti8], [Ti9]).

When Δ is the affine building of an algebraic group $\mathbf{G}$ over a local field K (cases (i) and (ii) above), our group G is a discrete cocompact subgroup of the locally compact group $\mathbf{G}(K)$. The question arises then: is it arithmetic (or rather S-arithmetic in the standard terminology) and, if so, which arithmetic group is it? In case (ii), Margulis' arithmeticity theorem ([Ma], [Ti3]) implies *a priori* an affirmative answer to the first question, but the arithmeticity of G is also apparent from the explicit embedding $G \to PSL_3(\mathbb{Q}_2)$ given in [KMW1]: it turns out that G is the group of integral points of a certain special unitary group over $\mathbb{Z}[\frac{1}{2}, \sqrt{-7}]$ (cf. [KMW1], [Ti8], [Ti9]). In case (i), Margulis' theorem is not (yet) available and [KMW2], [Me] only give generating matrices for the subgroup G of $P\Gamma L_3(\mathbb{F}_q((y)))$, but the arithmeticity of G is shown by the following description.

(For $q=8$, that result was obtained after the conference. In finding the arithmetic description of G given here, I was much helped by the knowledge of [KMW2] and [Me], §2; I am grateful to Thomas Meixner for communicating to me the relevant formulas out of [Me] long before that memoir was completely written.)

Let ξ' be a generator of the field extension $\mathbb{F}_{q^3}/\mathbb{F}_q$ of trace zero, set $\xi = \xi'^{-1}$ and let t and n denote the trace and norm of ξ in $\mathbb{F}_q$. Thus,

$$\xi^3 - t\xi^2 - n = 0, \tag{1}$$

$$\xi^{q^2} + \xi^q + \xi = t, \tag{2}$$

$$\xi^{q^2} \cdot \xi^q \cdot \xi = n. \tag{3}$$

(We use plus and minus signs, even though the ground field has characteristic 2, for reasons which will become clear at the end of this subsection 3.1.) Over the purely transcendental extension $\mathbb{F}_q(y)$, consider the cyclic division algebra D of dimension 9 generated by $\mathbb{F}_{q^3}(y)$ and an element σ such that $\sigma^3 = y-1$, $\sigma y = y\sigma$ and $\sigma x \sigma^{-1} = x^q$ for all $x \in \mathbb{F}_{q^3}$. Set $\tau = \xi\sigma(\sigma+1) \in D$. One has

$$^{\tau}\sigma = \xi\sigma\xi^{-1} = \xi^{1-q} \cdot \sigma \tag{4}$$

and, by a straightforward computation, using (1), (2), (3), (4),

$$\tau^3 = ny(y-1). \tag{5}$$

Let Aut D denote the group of all *algebra automorphisms* of D and $\mathrm{Aut}_y D$ the group of all *ring automorphisms fixing* y. Thus, Aut D is a subgroup of $\mathrm{Aut}_y D$ of index 1 or 3 according as $q=2$ or 8. The group $A = \mathrm{Aut}(\mathbb{F}_{q^3})$, of order q+1, operates on D in the obvious way (fixing σ and y), and we shall also denote by A its image in $\mathrm{Aut}_y D$. For $a \in A$, we have

$$a(\tau) = a(\xi) \cdot \xi^{-1} \cdot \tau. \tag{6}$$

For every element or subset X of D, let $\overline{X}$ denote its canonical image in the group of inner automorphisms of D. Observe that $\overline{\tau}^3 = 1$. In $\mathrm{Aut}_y D$, we consider three copies of F, namely $A \cdot \overline{\mathbb{F}_{q^3}^{\times}}$ and its conjugates by $\overline{\tau}$ and $\overline{\tau}^2$. Those groups intersect pairwise along cyclic groups of order $q+1$, namely A and its conjugates by $\overline{\tau}$ and $\overline{\tau}^2$. Their amalgamation in $\mathrm{Aut}_y D$ is totally polarized; indeed, it follows from (6) that if a is the generator of A which operates on $\mathbb{F}_{q^3}$ by $x \mapsto x^2$, then $\overline{\tau}_a = \overline{\xi}^{-1}.a$, an element of $A \cdot \overline{\mathbb{F}_{q^3}^{\times}}$ which also induces the automorphism $x \mapsto x^2$ of $\mathbb{F}_{q^3}$. Consequently, there is a homomorphism ι of the group G of case (i) above which maps the subgroups $X_{\{i,j\}}$ onto $A \cdot \overline{\mathbb{F}_{q^3}^{\times}}$ and its conjugates by $\overline{\tau}$ and $\overline{\tau}^2$.

Let Λ be the order in D over the ring $\mathbb{F}_q\,[y, y^{-1}, (y-1)^{-1}]$ generated by $\mathbb{F}_{q^3}$ and σ. We denote by $PGL(\Lambda)$ (resp. $P\Gamma L_y(\Lambda)$) the stabilizer of Λ in $\mathrm{Aut}\, D$ (resp. $\mathrm{Aut}_y D$). The group $PGL(\Lambda)$ is an arithmetic subgroup of $PGL(D) = \mathrm{Aut}\, D$. Clearly, $\iota(G) \subset P\Gamma L_y(\Lambda)$.

Since $y-1$ has a cubic root in $\mathbb{F}_q((y))$, the tensor product $D \otimes \mathbb{F}_q((y))$ is a full matrix algebra $M_3(\mathbb{F}_q((y)))$. Thus $\mathrm{Aut}_y D$ is naturally embedded in $P\Gamma L_3(\mathbb{F}_q((y)))$, and it acts on the affine building, call it Δ', of $SL_3(\mathbb{F}_q((y)))$. Now, it can be verified that the three groups $\iota(X_{\{i,j\}})$, explicitly described above, are the stabilizers in $P\Gamma L_y(\Lambda)$ of the three vertices of a chamber of Δ' (viewed as a simplicial complex: cf. 1.4) and that each one of those groups permutes transitively and regularly the chambers containing the corresponding vertex. From those facts and Proposition 2, one deduces, by simple arguments left to the reader, that

ι is injective, which allows us to identify G and $\iota(G)$;

$\Delta' = \Delta$ (this is the result of Köhler-Meixner-Wester and of Meixner stated above: cf. (i));

$G \cap PGL(\Lambda)$ has index 1 or 3 in G and index 3 in $PGL(\Lambda)$; in particular, it is arithmetic.

The above result can be generalized to any pair consisting of a field k (replacing $\mathbb{F}_q$) and a cyclic extension of degree 3 (replacing $\mathbb{F}_{q^3}$), except that the group one constructs is no longer chamber-transitive but only edge-transitive on the affine building of $SL_3(k((y)))$.

3.2 *A building of type* $\widetilde{D}_4$

Let us consider a split group H of type D_4 over $\mathbb{F}_2$ "rigidified" in the usual way by the choice of a Borel subgroup and a maximal torus containing it (the torus in question being defined over $\mathbb{F}_2$ but viewed as an algebraic group, because its group of rational points would not be of much use!). The vertices of the Dynkin diagram being labelled as follows:

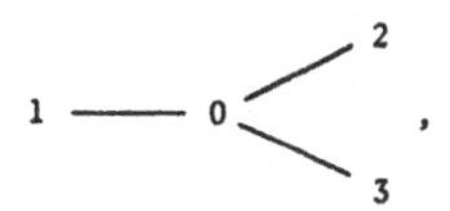

we denote the positive roots by 0, 1, 2, 3, 01, 02, 03, 012, 013, 023, 0123, 00123, with the obvious convention. For any positive root a, let $e_a = e_0, e_1, \ldots, e_{00123}$ (resp. $f_a = f_0, f_1, \ldots, f_{00123}$) be the non-neutral rational element of the one-parameter root group corresponding to a (resp. -a). For any subset J of {0,1,2,3}, let Q_J denote the group generated by all e_a and all f_j for $j \in J$ (group of $\mathbb{F}_2$-rational points of the standard parabolic subgroup of type J). The symmetric group $\mathfrak{S}_3 = \mathfrak{S}(\{1,2,3\})$ operates on H in the obvious way. The following lemma is verified by a straightforward computation.

Lemma 4. (i) *The map*

$$e_0 \longmapsto e_0 e_{012} e_{013} e_{00123}, \qquad f_0 \longmapsto f_0 e_1 e_{012} e_{013},$$
$$e_1 \longmapsto e_1,$$
$$e_2 \longmapsto e_2 e_1 e_{0123} e_{00123}, \qquad f_2 \longmapsto f_2 e_1 e_{01} e_{00123},$$
$$e_3 \longmapsto e_3 e_1 e_{0123} e_{00123}, \qquad f_3 \longmapsto f_3 e_1 e_{01} e_{00123},$$

extends to an automorphism t *of* Q_{023}.

(ii) *The automorphism* t *has order 2 and commutes with the transposition* (2,3) *(acting on* Q_{023}*).*

(iii) *The automorphism* $(1,2).t|_{Q_{03}}$ *of* Q_{03} *has order* 3.

Now, let us consider a system of groups P_J, for $J \subset \{0,1,2,3,4\}$ and card $J \leq 3$, and a system of injective homomorphisms $\phi_{JJ'} : P_J \to P_{J'}$, for $J \subset J'$, on which the symmetric group $\mathfrak{S}_4 = \mathfrak{S}(\{1,2,3,4\})$ acts in such a way that the following properties hold:

(a) if $s \in \mathfrak{S}_4$, then $s.P_J = P_{s(J)}$ and ${}^s\phi_{JJ'} = \phi_{s(J),s(J')}$;

(b) for $J \in \underline{J} = \{\{0\}, \{3\}, \{0,3\}, \{2,3\}, \{0,2,3\}, \{1,2,3\}\}$, one has $P_J = Q_J$, and if $J \subset J'$ and $J, J' \in \underline{J}$, $\phi_{JJ'} : Q_J \to Q_{J'}$ is the inclusion map (inside H);

(c) an element of $\mathfrak{S}_3$ stabilizing $J \in \underline{J}$ has the same action on Q_J (via the action of $\mathfrak{S}_3$ on H) as on P_J (via the canonical inclusion of $\mathfrak{S}_3$ in $\mathfrak{S}_4$ and the given action of $\mathfrak{S}_4$ on the system of groups $P_{J'}$);

(d) the permutation (1,4) acts on $P_{023} = Q_{023}$ through t.

The existence and uniqueness of such a system readily follows from the assertions (ii) and (iii) of the above Lemma: in order to define P_J for any given $J \subset \{0,1,2,3,4\}$ with card $J \leq 3$, choose $s \in \mathfrak{S}_4$ such that $s^{-1}(J) = J' \in \underline{J}$, take for P_J a copy of $Q_{J'}$, written "formally" ${}^sQ_{J'}$ and, for different choices of s, identify the various ${}^sQ_{J'}$ by means of (c) and (d).

Now, Theorem 1 implies that *if* G *is the amalgamated sum of the system* $(P_i, P_{\{i,j\}} \mid i,j \in \{0,1,2,3,4\},\ i \neq j)$ *with respect to the maps* $\phi_{i,\{i,j\}}$, *then* $\Delta = \mathcal{C}(G; P_\emptyset; (P_i)_{i=0,1,2,3,4})$ *is a building of type*

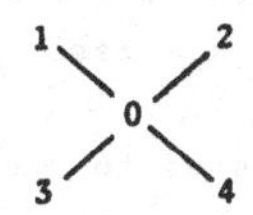

By a general classification theorem for buildings of affine type (cf. [Ti7]), Δ must be the affine building of an algebraic simple group over a local field; in fact, it is the affine building of $SO_8(q,\mathbb{Q}_2)$, where $q = \sum_{h=1}^{8} x_i^2$. Then, by Margulis' theorem, G must be an arithmetic group; in fact, it is the derived group of $SO_8(q,\mathbb{Z}[\frac{1}{2}])$. These identifications of Δ and G easily follow from results of W. Kantor [Ka1]. The above formulas have been obtained using Kantor's description of G; our purpose here was only to illustrate Theorem 1, but Kantor's approach certainly gives a better hold on the group G than ours (except for the fact that the existence of an automorphism group $\mathfrak{S}_4$ is more directly clear in our construction).

Observe that the existence of the above example is related to the known fact (cf. [JP], p. 327) that H^1 of $SO_6(\mathbb{F}_2)$ with coefficients in the standard module is not trivial. Indeed, the group $Q_{\cup 23}$ is the semi-direct product of $SO_6(\mathbb{F}_2)$ by its six-dimensional representation space, and the automorphism t maps the canonical section (Levi subgroup) onto a non-conjugate one.

3.3 *A classification theorem*

One might believe that the examples given above are only special instances of a very general situation. This is not true as is shown by the following theorem, due to W. Kantor, R. Liebler and the author [KLT]:

Theorem 2. *Let* $\mathbf{G}$ *be a simple algebraic group of relative rank* ≥ 2 *over a locally compact local field* K. *Suppose that the affine building* Δ *of* $\mathbf{G}$ *over* K *possesses an automorphism group* G *which permutes transitively the chambers of* Δ *and such that the stabilizer of a chamber in* G *is finite (in other words, that* G *is a discrete subgroup of* Aut Δ*). Then, one of the following holds:*

$K = \mathbb{Q}_2$ *and* $\mathbf{G}$ *is split of type* A_2, C_2, G_2, A_3, B_3 *or* D_4;

$K = \mathbb{Q}_2$, $\mathbf{G}$ *is quasi-split of type* A_3 *and it splits over* $\mathbb{Q}_2(\sqrt{-1})$ *or* $\mathbb{Q}_2(\sqrt{-3})$;

$K = \mathbb{Q}_3$ *and* $\mathbf{G}$ *is split of type* C_2 *or quasi-split of type* A_3, *splitting over* $\mathbb{Q}_3(\sqrt{-3})$;

$K = \mathbb{F}_2((y))$ *or* $\mathbb{F}_8((y))$, *and* $\mathbf{G}$ *is split of type* A_2.

In all those cases, a group G meeting our requirements does exist (cf. [Ka1], [Ka2], [KMW1], [KMW2], [MW1], [MW2], [M]). A complete description of all existing G has been obtained by W. Kantor: in each case, there is a small finite number of, and most of the time only one, conjugacy class(es) of such groups G in Aut Δ. All those groups can of course, in principle, be constructed by means of Theorem 1, in a similar way as the examples treated in 3.1 and 3.2, but I did not make the exercise in all cases (for the split A_3 over $\mathbb{Q}_2$, see [Ti8], 3.3).

It should be mentioned that a result closely related to the above one has been obtained by F. Timmesfeld ([Tim]). His hypotheses are considerably more general - he does not restrict himself to buildings of affine type nor, in the affine case, to buildings of algebraic groups - but in the case of affine buildings associated to algebraic groups, his result is less complete than ours. The methods of proof are totally different.

3.4 *Compact Lie groups and "unitary" Kac-Moody groups*

Let $\mathbf{G}$ be a semi-simple group of relative rank ≥ 4 over a field K, let $\mathbf{B}$ be a minimal parabolic K-subgroup of $\mathbf{G}$ and let $\mathbf{P}_i$ $(i \in I)$ denote the parabolic K-subgroups containing property $\mathbf{B}$ and minimal with that property. Set $G = \mathbf{G}(K)$, $B = \mathbf{B}(K)$ and $P_i = \mathbf{P}_i(K)$. We have seen (1.2(c)) that $\mathbf{C}(G; B; (P_i)_{i \in I})$ is a building. In principle, therefore, Theorem 1 enables one to prove the existence of the group G, or rather, to deduce it from the existence and properties of groups of smaller rank. One may question the utility of this procedure since, for the groups under consideration, other existence proofs are available. But besides the fact that, for such important objects, new and especially simple existence proofs are always instructive (think of what is involved if one wants to show, from scratch, the existence of the group E_8 over some field, say), a point of the method is that it extends - always in principle! - to Kac-Moody groups, a wide generalization of split semi-simple groups (cf. [Ti6] and the literature of that paper).

Now, in all cases, and especially that of Kac-Moody groups, this method comes up against a difficulty in that the group B, an essential ingredient of the construction, is a "complicated" group, whose existence is hardly simpler to establish than that of G itself. Here are three possible ways to get around that obstacle. For simplicity, we assume that we are in the case of Kac-Moody groups (including, as we have seen, that of *split* semi-simple groups).

First, one can try to prove the existence of the "chamber system G/B" by a kind of geometric analogue of Theorem 1 (the group G being recovered afterwards as an automorphism group of the building). Thus, the main difficulty, i.e. the group B, is so to speak factored out! Recent work of M. Ronan and the author seems to indicate that such a thing is feasible.

Another way out would consist in proving the existence of B by an amalgamation method similar to that used for G. Let B' be a "Borel subgroup" opposite to B (in general, B' is *not conjugate* to B: cf. [Ti6], §4). The set G/B' is the set of chambers of a building on which G, and hence B, operates. The action of B is of course not chamber-transitive, but a suitable generalization of Theorem 1 still applies. I hope to come back to this in a forthcoming paper.

The third, and simplest, method which will now be described in somewhat greater detail, does not apply directly to Kac-Moody groups as defined in [Ti6], but rather to the "unitary forms" of complex Kac-Moody groups. Compact forms of complex semi-simple Lie groups are well-known objects; for Kac-Moody groups over $\mathbb{C}$, "unitary forms" - which are *not* compact groups! - have been introduced by H. Garland ([Ga]) in the affine case, and by V. Kac and D. Peterson (cf. e.g. [KP]) in general. It so happens that, by an "Iwasawa-like" theorem, such a "unitary form" - call it G - acts chamber-transitively on the building of the complex group. Thus,

§2 applies. In particular, Proposition 2 and Proposition 1 provide a presentation of G and a "canonical decomposition" for its elements, due to R. Steinberg ([St]) in the finite dimensional case and extended to the general situation by Kac and Peterson (*loc. cit.*). What makes Theorem 1 easily applicable here is that the stabilizer of a chamber in G is no longer a "complicated" group but just a torus! Let us then proceed and outline the construction and existence proof of the "unitary forms" of complex Kac-Moody groups based on that remark.

Adopting the general framework of [Ti6], §2, we suppose given a free abelian group Λ of finite rank, whose $\mathbb{Z}$-dual is denoted $\Lambda^\vee$, and two finite sets $\Phi = (\alpha_i)_{i\in I} \subset \Lambda$, $\Phi^\vee = (\alpha_i^\vee)_{i\in I} \subset \Lambda^\vee$, indexed by the same set I, such that the matrix $A = (A_{ij})$ with $A_{ij} = \alpha_i^\vee(\alpha_j)$ is a *generalized Cartan matrix*, i.e. that $A_{ii} = 2$, $A_{ij} \leq 0$ if $i \neq j$ and $A_{ij} = 0$ if and only if $A_{ji} = 0$. To those data, we associate a Coxeter matrix $M = (m_{ij})$ defined as follows: $m_{ii} = 1$ and, for $i \neq j$, $m_{ij} = 2, 3, 4, 6$ or ∞ according as $A_{ij}A_{ji} = 0, 1, 2, 3$ or ≥ 4. (In the most familiar case of a compact Lie group G, Λ is the character group of a maximal torus, Φ is a basis of the root system, $\Phi^\vee$ is the corresponding set of coroots and W(M) is the Weyl group.) Our subgroup $P_\emptyset$ will be Hom $(\Lambda, \mathbb{R}/\mathbb{Z})$ and we call it T (a torus) in order to avoid the misleading notation B. Similarly, the groups which play the role of our P_J will now be denoted by L_J ("Levi subgroups"). They are defined as follows: for J spherical of cardinality ≤ 3, L_J is "the" compact Lie group having T as a maximal torus, such that $(\alpha_j)_{j\in J}$ is a basis of the root system of L_J with respect to T and $\alpha_j^\vee$ is the coroot corresponding to α_j.

Let us briefly comment on that definition. For the sake of brevity, we have assumed knowledge of the general structure theory of compact Lie groups. But since all semi-simple groups involved have rank ≤ 3, the above description can be replaced by a simple explicit construction in each case. More precisely, representing by T_J' the neutral component of the intersection of the kernels of the α_j's for $j \in J$, we can write $L_J = (\tilde{L}_J \times T_J')/Z$, $T = (\tilde{T}_J \times T_J')/Z$, where $\tilde{L}_J$ is a simply connected compact group (of rank card J), $\tilde{T}_J$ is a maximal torus of $\tilde{L}_J$ and the quotient map $\kappa : \tilde{T}_J \times T_J' \to T$ induces the inclusion map of T_J' in T. Thus, in order to describe L_J, it suffices, in each case, to give $\tilde{L}_J$ and the homomorphism $\kappa' : \tilde{T}_J \to T$ induced by κ; indeed, Z is then the kernel of the map $\kappa = \kappa' \times \mathrm{id}|_{T_J'}$ and $L_J = (\tilde{L}_J \times T_J')/Z$. Suppose for instance that M_J has type A_ℓ for $\ell = \text{card } J = 1, 2$ or 3. Then, one takes for $\tilde{L}_J$ the (complex) special unitary group $SU_{\ell+1}$, for $\tilde{T}_J$ the group of diagonal unitary matrices of determinant 1 and the homomorphism κ' is given by

$$\mathrm{diag}(e(\theta_1), e(-\theta_1+\theta_2), e(-\theta_2+\theta_3), e(\theta_3)) \longmapsto$$
$$\alpha_1^\vee(\theta_1)\alpha_2^\vee(\theta_2)\alpha_3^\vee(\theta_3) \qquad (e(\theta) = \exp 2\pi i\theta)$$

when $\ell = 3$ and by similar but simpler formulas when $\ell = 1$ or 2; here, the elements of J are labelled 1, 2, ... in a "natural order" (the graph representing M_J is a "string") and the elements of $\Lambda^\vee$ are interpreted as homomorphisms of $\mathbb{R}/\mathbb{Z}$ in T. The case where M_J is a "direct sum" of Coxeter matrices of types A_ℓ, which suffices to handle all matrices A whose non-diagonal coefficients are 0 or -1 (e.g. matrices of type E_8!), is a straightforward generalization of the above one, the group L_J being now a direct product of special unitary groups. In the general situation, besides the complex unitary groups, one must consider also groups of type Spin 5, Spin 7, G_2 and $SU_3(\mathbb{H})$ (quaternionic special unitary group in three variables). In all cases, explicit formulas for κ' are easily supplied.

Now, we proceed with our application of Theorem 1. For every ordered pair of distinct elements i, j or I, with $m_{ij} \neq \infty$, let us choose a homomorphism $\phi_{ij} : L_i \to L_{\{i,j\}}$ inducing the identity on T; this homomorphism is easily seen to exist and to be unique up to multiplication on the right by an inner automorphism of L_i corresponding to an element of T. Moreover, since T is the product of the center of L_i and the center of L_j, the pair (ϕ_{ij}, ϕ_{ji}) is unique up to multiplication on the left by an inner automorphism of $L_{\{i,j\}}$ corresponding to an element of T. Consequently;

(3.4.1) *The system* $(T, L_i, L_{\{i,j\}}, \phi_{ij} | i,j \in I,\ i \neq j,\ m_{ij} \neq \infty)$ *is unique (i.e. depends only on* Λ, Φ, $\Phi^\vee$*) up to isomorphism.*
Let G be the amalgamated sum of that system. We claim that Theorem 1 implies:

Proposition 4. *The canonical homomorphisms* $L_i \to G$ *are injective and if we identify* T *and the groups* L_i *with their images by those homomorphisms the chamber system* $\mathfrak{C}(G;\ T;\ (L_i)_{i \in I})$ *is a building of type* M.

In order to show that Theorem 1 does apply here, we must

(a) exhibit a system of injective maps $\phi_{JJ'} : L_J \to L_{J'}$ satisfying the conditions of 2.4 (we say that the system is *consistent*), such that $\phi_{\emptyset J} : T \to L_J$ is the inclusion map for all J and that $\phi_{i,\{i,j\}} = \phi_{ij}$ for $i,j \in I$ with $i \neq j$ and $m_{ij} \neq \infty$;

(b) verify that, modulo the identifications resulting from the choice of the ϕ_{jJ}'s, $\mathfrak{C}(L_J;\ T;\ (L_j)_{j \in J})$ is a building of type M_J for all spherical sets J of cardinality ≤ 3.

For (a), let us consider a given spherical set J'. If card J' ≤ 2, the maps $\phi_{JJ'}$ are already prescribed, so assume card J' = 3. One verifies that there exists a system of embeddings $\phi'_{JJ'} : L_J \to L_{J'}$, for $J \subset J'$, extending the inclusion map $T \to L_{J'}$ and such that, if one identifies the groups L_j $(j \in J')$ with their images by the $\phi'_{jJ'}$, then $\mathfrak{C}(L_{J'};\ T;\ (L_j)_{j \in J'})$ is a building of type $M_{J'}$. (One can either give a uniform argument using the general theory of compact Lie groups, or just check the assertion "by hand", for each one of the few groups $L_{J'}$ that can occur; as before, if all non-diagonal coefficients of A are 0 or -1, one only has eventually to deal with special unitary groups.) For $i,j \in J'$ with $i \neq j$, set $\phi'_{ij} = \phi'^{-1}_{\{i,j\},J'} \circ \phi'_{iJ'}$. By (3.4.1), the systems $\mathcal{L}_{J'} = (L_i, L_{\{i,j\}}, \phi_{ij} | i,j \in J',\ i \neq j)$ and $\mathcal{L}'_{J'} = (L_i, L_{\{i,j\}}, \phi'_{ij} | i,j \in J',\ i \neq j)$ are isomorphic. On the other hand, Proposition 2 implies that $L_{J'}$ "is" the amalgamated sum of $\mathcal{L}'_{J'}$. Consequently, there is no loss of generality in assuming that L_J "is" the amalgamated sum of the system $\mathcal{L}_{J'}$. Now, we take for $\phi_{iJ'}$ and $\phi_{\{i,j\},J'}$ the canonical maps $L_i \to L_{J'}$ and $L_{\{i,j\}} \to L_{J'}$.

This achieves (a). As for (b), it is part of the above argument if card $J = 3$, and it consists of a straightforward checking if card $J \leq 2$.

Thus, Proposition 3 is proved and we have associated a "unitary form" to any system $(\Lambda, \Phi, \Phi^\vee)$ satisfying the conditions stated above. If $\Phi^\vee$ is a basis of $\Lambda^\vee$, this is nothing else but the "unitary form" defined by V. Kac and D. Peterson [KP] via the representation theory of the Kac-Moody Lie algebra corresponding to A; indeed, *loc. cit.* contains a presentation of the group in question which is nothing else but the presentation we have taken for definition.

Let us conclude with a few remarks concerning the topological structure of G. Here, the sets R_i of 2.2 can be chosen to be 2-dimensional discs (inside the compact Lie groups L_i). Therefore, if w is an element of length ℓ of the Coxeter group W(M), the set P_w^* of 2.2, which we shall now rather call L_w^*, "is" a manifold, product of the torus T and a 2ℓ-dimensional disc. It is an interesting exercise to prove that the topological structure of L_w^* defined in that way does not depend on the choices made (in particular the choice of the reduced word representing w) and that the manifolds L_w^* can be glued together in such a way as to make a topological group out of G. (In the Kac-Peterson approach, these facts are straightforward consequences of the definition.) Suppose now that W is finite. Then, it has a unique longest element w_0, of length say ℓ_0, therefore G is a manifold of dimension $2\ell_0 + \dim T$ in which $L_{w_0}^*$ is open and dense. It is well-known that if $\underline{i}_0 = i_1 \ldots i_{\ell_0}$ is any reduced word such that $w_0 = s(\underline{i}_0)$, then every element $w \in W$ can be written as $w = s(\underline{i})$, where $\underline{i}$ is a reduced subword of $\underline{i}_0$ (i.e. a reduced word obtained by deleting some terms in $\underline{i}_0$). Therefore, $G = L_{i_1} \ldots L_{i_{\ell_0}}$ (by Proposition 1), and G is a *compact group*.

REFERENCES

[Bo] N. Bourbaki, *Groupes et algèbres de Lie*, Chap. 4, 5, 6, Actu. Sci. Ind. 1337, Hermann, Paris (1968).

[BT] F. Bruhat & J. Tits, Groupes réductifs sur un corps local, I. Données radicielles valuées, *Inst. Hautes Etudes Sci. Publ. Math.* 41 (1972), 5-251; II. Schémas en groupes, existence d'une donnée radicielle valuée, *Ibid.* 60 (1983), 5-184.

[Ga] H. Garland, The arithmetic theory of loop groups, *Inst. Hautes Etudes Sci. Publ. Math.* 52 (1980), 5-136.

[JP] W. Jones & B. Parshall, On the 1-cohomology of finite groups of Lie type, *Proc. Conf. Finite Groups, Park City, Utah*, 1975, ed. W. Scott and F. Gross, Academic Press (1976), 313-328.

[KP] V. Kac & D. Peterson, Defining relations of certain infinite dimensional groups, Colloque Elie Cartan, Lyon, June 1984, *Astérisque*, to appear.

[Ka1] W. Kantor, Some exceptional 2-adic buildings, *J. Algebra* 92 (1985), 208-223.

[Ka2] W. Kantor, Some locally finite flag-transitive buildings, *European J. Combin.*, to appear.

[Ka3] W. Kantor, Generalized polygons, SCABs and GABs, *C.I.M.E. Conference on Buildings and the Geometry of Diagrams*, Como, September 1984, Springer Lecture Notes, to appear.

[KLT] W. Kantor, R. Liebler & J. Tits, Locally finite chamber-transitive automorphism groups of classical affine buildings, in preparation.

[KMW1] P. Köhler, T. Meixner & M. Wester, The 2-adic affine building of type A_2 and its finite projections, *J. Combin. Theory Ser. A* 38 (1985), 203-209.

[KMW2] P. Köhler, T. Meixner & M. Wester, The affine building of type A_2 over a local field of characteristic two, *Arch. Math.* 42 (1984), 400-407.

[Ma] G. Margulis, Diskretnye gruppy dviženii mnogoobrazii nepoložitel'noi krivizny (Discrete motion groups of manifolds with non-positive curvature), *Proc. Internat. Congress Math., Vancouver*, 1974, Vol. 2 (1975), 21-34.

[Me] Th. Meixner, *Klassische Tits Kammersysteme mit einer transitiven Automorphismengruppe*, Habilitationsschrift, Giessen (1985).

[MW1] Th. Meixner & M. Wester, Some locally finite buildings derived from Kantor's 2-adic groups, *Comm. Algebra*, to appear.

[MW2] Th. Meixner & M. Wester, A locally finite building of type BC_2 over Q_3, (1985), preprint.

[Se] J.-P. Serre, Arbres, amalgames, SL_2, *Astérisque* 46, Soc. Math. Fr. (1977).

[St] R. Steinberg, *Lectures on Chevalley groups*, Yale University, 1967, notes by J. Faulkner and R. Wilson (1968).

[Tim] F. Timmesfeld, Locally finite classical Tits chamber systems of large order, *Invent. Math.*, to appear.

[Ti1] J. Tits, Le problème des mots dans les groupes de Coxeter, *Simposia Matematica, Ist. Naz. Alta Mat.*, 1 (1968), 175-185.

[Ti2] J. Tits, *Buildings of spherical type and finite BN-pairs*, Springer Lecture Notes in Math. 386 (1974).

[Ti3] J. Tits, Travaux de Margulis sur les sous-groupes discrets des groupes de Lie, *Sém. Bourbaki*, exposé 482, février 1976, Springer Lecture Notes in Math. 567 (1977), 174-190.

[Ti4] J. Tits, Reductive groups over local fields, *Proc. Sympos. Pure Math.* 33 (Summer Institute on Group Representations and Automorphic Forms, Corvallis, 1977) Vol. 1 (1979), 29-69.

[Ti5] J. Tits, A local approach to buildings, *The geometric vein, The Coxeter Festschrift*, Springer-Verlag (1981), 519-547.

[Ti6] J. Tits, Groups and group functors attached to Kac-Moody data, *Arbeitstagung Bonn* 1984, Springer Lecture Notes in Math. 1111 (1985), 193-223.

[Ti7] J. Tits, Immeubles de type affine, *C.I.M.E. Conference on Buildings and Geometry of Diagrams*, Como, September 1984, Springer Lecture Notes, to appear.

[Ti8] J. Tits, Résumé du cours 1984-1985, *Annuaire du Collège de France*, 85e année, (1985), 93-111.

[Ti9] J. Tits, *Lectures on buildings and arithmetic groups*, Yale University, 1984, notes by R. Scaramuzzi and D. White, (1986).

FINITE PRESENTABILITY OF S-ARITHMETIC GROUPS

Herbert Abels
University of Bielefeld, 48 Bielefeld, West Germany

This is a survey of the recent solution of the problem indicated by the title [Abels 1985]. I shall first explain the concepts mentioned in the title, then state the problem - which obviously is, how the two concepts are related, i.e. which S-arithmetic groups have a finite presentation -, then give the positive results and the negative examples which marked the path to the solution of the problem, then state the solution and finally try to give an idea of the proof. There are also concluding remarks about the higher finiteness conditions F_n, $n \geq 3$.

Let Γ be a group. It is called *finitely presentable* or sometimes *finitely presented* if there is an exact sequence of groups $N \rightarrowtail F \twoheadrightarrow \Gamma$, where F is a finitely generated free group with finite basis X say and N is a subgroup finitely generated as a normal subgroup, say $N = \langle R \rangle^F$, R finite $\subset F$. The pair X, R is then called a *finite presentation* of Γ and one writes $\Gamma = \langle X;R \rangle$. Why is one interested in finite presentations of groups? Well, a finite presentation of Γ gives a complete description of the group in finite terms. A reason for a topologist to look at finitely presented groups might be that a group Γ has a finite presentation iff it is the fundamental group of a compact manifold.

The problem of which groups are finitely presented is unsolved. At this point one has to mention that Bieri and Strebel gave a necessary condition for a solvable group to be finitely presented [Bieri-Strebel 1980]. For a formulation of this condition in our framework see below.

One can ask which groups of a certain class of groups have a finite presentation. One particularly important class of groups is the class of arithmetic and more generally S-arithmetic groups whose definition I shall recall now.

Let k be a finite extension field of $\mathbb{Q}$ (I only talk about the characteristic zero case. For the case of positive characteristic see [Behr 1985] which includes a report on the current state of the art for reductive groups). Let G be a linear algebraic group defined over k. So G is a subgroup of the general linear group GL_n with entries from some big field, e.g. $\mathbb{C}$, and G is given as the set of zeros of a set of polynomials in the variables $X_{11},\ldots,X_{nn}$ and $(\det X)^{-1}$, $X = (X_{ij}) \in GL_n$, with coefficients in k. For the sake of easier exposition let us take $k = \mathbb{Q}$. The *arithmetic group* corresponding to G is by definition

$$G(\mathbb{Z}) = G(\mathbb{Q}) \cap GL_n(\mathbb{Z}),$$

the group of points of G with integer coefficients (and of course determinant a unit in $\mathbb{Z}$).

There is the following definitive result for arithmetic groups.

Theorem [Borel, Harish-Chandra 1962]. *$G(\mathbb{Z})$ is finitely presented.*

Reduction theory is essentially used in the proof of this theorem and the following result: *$G(\mathbb{Z})$ has a torsion free subgroup of finite index which has a finite Eilenberg-MacLane complex* ([Raghunathan 1968], cf. [Borel, Serre 1973]). This is a much stronger finiteness theorem.

Now let S be a finite set of prime numbers, $S = \{p_1,\dots,p_s\}$, and put $\mathbb{Z}_S = \mathbb{Z}[1/p_1,\dots,1/p_s]$. The *S-arithmetic group* corresponding to the linear algebraic group G defined over $\mathbb{Q}$ is

$$\Gamma = G(\mathbb{Z}_S) = G(\mathbb{Q}) \cap GL_n(\mathbb{Z}_S).$$

Problem. *For which G and S is the S-arithmetic group Γ finitely presented?*

There was the following partial answer:

Theorem (Behr 1967). *If G is reductive, Γ is finitely presented for every S.*

There is a corresponding stronger finiteness theorem, ([Serre 1970, Theorem 4 p.124], cf. [Borel, Serre 1976]): *For any reductive linear algebraic group defined over $\mathbb{Q}$, the S-arithmetic group $G(\mathbb{Z}_S)$ has a torsion free subgroup of finite index which has a finite Eilenberg-MacLane complex.*

Let me explain when an algebraic group is called reductive. A linear map g of a vector space V into itself is called *unipotent* if $g - \mathrm{Id}$ is nilpotent. A subgroup of GL(V) is called unipotent if every one of its elements is unipotent. Every linear algebraic group G has a unique biggest unipotent normal subgroup, called the *unipotent radical* of G and denoted $\mathrm{rad}_u(G)$. A group is called *reductive* if $\mathrm{rad}_u(G) = \{e\}$. The general linear group GL_n is reductive, for example. The analogous problem of which S-arithmetic groups are finitely generated was solved in [Borel, Tits 1965]. I shall describe the solution below, at least for a special case.

After Behr's result one is tempted to look at unipotent groups and ask which S-arithmetic groups for unipotent groups are finitely generated or finitely presented. The answer is: none - except if $S = \emptyset$ - since every such group maps onto the following group.

Example. Let $G_2 = \left\{\begin{pmatrix} 1 & * \\ 0 & 1 \end{pmatrix}\right\}$, then $G_2(\mathbb{Z}_S)$ is isomorphic to the additive group $\mathbb{Z}_S$ hence not finitely generated unless $S = \emptyset$.

So for triangular groups of matrices there must be something non trivial on the diagonal and it had better act vigorously on the unipotent radical if Γ stands a chance to be even finitely generated. The precise statement is the result of Borel and Tits given below. We need some preparatory material to state it. First we pass to the completion.

Theorem [Kneser 1964]. *$G(\mathbb{Z}_S)$ is finitely generated (finitely presented) iff $G(\mathbb{Q}_p)$ is compactly generated (compactly presented) for every $p \in S$.*

The right hand side needs explanation. The p-adic completion $\mathbb{Q}_p$ of $\mathbb{Q}$ is a locally compact topological field, so the set $G(\mathbb{Q}_p)$ of zeros of polynomials with rational coefficients is a closed subgroup of $GL_n(\mathbb{Q}_p)$, hence a locally compact topological group. So it makes sense to say that it has a compact set of generators. For the definition of compact presentability just replace "finite" by "compact" in the definition of finite presentability above. Here $F = FX$ is the free topological group on the compact set X. One can rephrase this avoiding the concept of free topological group as follows.

Lemma. *Let* G *be a locally compact topological group.* G *has a compact presentation iff* G *can be presented as* $\langle X;R \rangle$ *where* X *is a compact subset of* G *and the set* R *of words of the free group* FX *has bounded length.*

Actually the bound can then be reduced to 3, see [Abels 1972].

Kneser's theorem has two advantageous features, firstly one has to look at one prime number at a time only and secondly one deals with algebraic groups over local fields, where a lot more about the structure is known (for an application see the next theorem). Kneser's theorem also has a disadvantageous feature: there is only little known about compact presentability.

A linear algebraic group over a field K is called a K-*split torus* if it is isomorphic over K to a product of copies of GL_1's. A linear algebraic group G is called K-*split solvable* if it is a semidirect product of a K-split torus T and a unipotent normal subgroup U, $G = T \ltimes U$.

Theorem [Borel, Tits 1965]. *Every linear algebraic group* G *over a local field* K *of characteristic zero contains a maximal* K-*split solvable subgroup* H *and* $G(K)/H(K)$ *is compact.*

Since G(K) is compactly generated (compactly presented) iff H(K) is, our initial problem is reduced to the:

Problem. *Let* G *be a* K-*split solvable group,* $K = \mathbb{Q}_p$. *When is* G(K) *compactly presented?*

The analogous problem for compact generation was solved in [Borel, Tits 1965].

Theorem. *Let* $G = T \ltimes U$ *be a* K-*split solvable group,* $K = \mathbb{Q}_p$. *Then* G(K) *is compactly generated iff the representation of* T *on* $\mathfrak{u}^{ab}$ *– the abelianized Lie algebra of* U *– given by the adjoint representation fixes no non-zero vector of* $\mathfrak{u}^{ab}$.

This statement makes precise what we meant by saying that T acts vigorously on U. So for

$$G_3 = \left\{ \begin{pmatrix} 1 & * & * \\ 0 & * & * \\ 0 & 0 & 1 \end{pmatrix} \right\}$$

the S-arithmetic group $G_3(\mathbb{Z}_S)$ is finitely generated, since $G_3(\mathbb{Q}_p)$ is compactly generated (both of which can also be verified directly). Is this group finitely presented resp. compactly presented? The answer is no by the following theorem due essentially to Bieri and Strebel.

Let T be a K-split torus and let ρ be a representation of T on a vector space V, both V and ρ defined over K. The simplest representations are the one dimensional representations $T \to GL_1$, also called characters. The characters form a free abelian group $X(T)$ of rank $\dim T$. The representation theory of a split torus T is very easy: Every representation space V of T splits into one dimensional ones. So if we define for every character $\chi \in X(T)$ the weight space

$$V^\chi = \{v \in V \mid \rho(t)v = \chi(t)v \text{ for every } t \in T\}$$

we have

$$V = \bigoplus_{\chi \in X(T)} V^\chi.$$

The set of weights of the representation ρ of T on V is defined as $\{\chi \in X(T) \mid V^\chi \neq 0\}$. We regard $X(T)$ as a lattice in $X(T) \otimes_{\mathbb{Z}} \mathbb{R}$, so all concepts of real linear geometry make sense. A representation of T on V is called *tame* if for any two weights the point 0 is not on the closed segment joining them.

Theorem. *Let* $G = T \ltimes U$ *be a* K*-split solvable group,* $K = \mathbb{Q}_p$. *If* $G(K)$ *has a compact presentation then the representation of* T *on* $\mathfrak{u}^{ab}$ *is tame.*

This theorem is a version for topological groups of a more general theorem of Bieri and Strebel [Bieri, Strebel 1978, 1980] for discrete groups (which also yields directly that $G_3(\mathbb{Z}_S)$ is not finitely presented, $S \neq \emptyset$).

For G_3 as above the set of weights of T on $\mathfrak{u}^{ab}$ consists of exactly two points $\pm\chi$, $\chi \in X(T)$, as is seen by direct inspection. So $G_3(\mathbb{Q}_p)$ has no compact presentation.

Is the necessary condition "tameness" of the theorem above sufficient? That the answer is no follows from the following example.

Example. Define

$$G_4 = \left\{\begin{bmatrix} 1 & * & * & * \\ 0 & * & * & * \\ 0 & 0 & * & * \\ 0 & 0 & 0 & 1 \end{bmatrix}\right\}.$$

Then $G_4(\mathbb{Q}_p)$ has a compact presentation. This is implied by the main theorem below or by the finite presentability of $\Gamma = G_4(\mathbb{Z}[1/p])$, see [Abels 1979]. The example Γ is interesting in group theory because it answers several questions negatively, the most prominent of which is P. Hall's [Hall 1954]: Is every homomorphic image of a finitely presented solvable group finitely presented? Equivalently, is every normal subgroup of a finitely presented solvable group finitely generated as a normal subgroup? In Γ even the center is not finitely generated, since isomorphic to $\mathbb{Z}[1/p]$.

The center of G_4 is

$$Z = \left\{\begin{bmatrix} 1 & 0 & 0 & * \\ & 1 & 0 & 0 \\ 0 & & 1 & 0 \\ & & & 1 \end{bmatrix}\right\}.$$

The linear algebraic group G_4/Z furnishes a counter-example to the conjecture that tameness is sufficient for compact presentability, because $(G_4/Z)(\mathbb{Q}_p) = G_4(\mathbb{Q}_p)/Z(\mathbb{Q}_p)$ has no compact presentation by the following lemma, but G_4/Z satisfies tameness, in fact the representations of T on $\mathfrak{u}^{ab}$ are the same for G_4 and G_4/Z. So there is no way to decide compact presentability from just looking at the representation of T on $\mathfrak{u}^{ab}$.

Lemma. *Let* $A \rightarrowtail B \twoheadrightarrow C$ *be an exact sequence of locally compact topological groups. If* B *is compactly generated and* C *is compactly presented, then* A *is compactly generated as a normal subgroup of* B. *In particular, if furthermore* A *is central in* B, *then* A *is compactly generated as a group.*

Why is $(G_4/Z)(\mathbb{Q}_p) = (T \ltimes U)(\mathbb{Q}_p)$ not compactly presented? Because it has a non trivial central extension which induces an isomorphism on the metabelianizations. So one has to forbid this. A rephrasing of the condition that no such extension exists is the condition (2) of the following theorem. So the two conditions of the theorem are necessary. The main result is that they are also sufficient.

Main theorem [Abels 1985]. *Let* $G = T \ltimes U$ *be a* K*-split solvable group,* $K = \mathbb{Q}_p$. *Then* G(K) *has a compact presentation iff the following two conditions hold.*

(1) *The representation of* T *on* $\mathfrak{u}^{ab}$ *is tame.*

(2) *The representation of* T *on* $H_2(\mathfrak{u})$ *fixes no element* $\neq 0$.

Here $H_2(\mathfrak{u})$ is the second Lie algebra homology of the Lie algebra $\mathfrak{u}$ over K with coefficients in the trivial $\mathfrak{u}$-module K. It can be computed by the Koszul complex.

Before giving some idea of the proof let me summarize the solution of the initial problem - for which G and S is the corresponding S-arithmetic group Γ finitely presented - in the form of a recipe. For every $p \in S$ find a maximal $\mathbb{Q}_p$-split solvable subgroup of $G(\mathbb{Q}_p)$ and apply the main theorem for it. In case $G(\mathbb{Q}_p)$ contains a maximal torus which is $\mathbb{Q}_p$-split, there is a direct criterion for compact presentability, avoiding the search for a maximal $\mathbb{Q}_p$-split solvable subgroup [Abels 1985, Theorem 6.4.3].

In the proof of the main theorem one first studies the case $\dim T = 1$. For this case the result is: G has a compact presentation iff all the weights of T on $\mathfrak{u}^{ab}$ lie on one side of zero in $X(T) \otimes_{\mathbb{Z}} \mathbb{R}$. So in contrast with the case $\dim T > 1$, the question of compact presentability is decided by the representation of T on $\mathfrak{u}^{ab}$ for $\dim T = 1$. This result for $\dim T = 1$ rather easily implies necessity of (1) for general T. Its proof is modelled after [Bieri, Strebel 1978].

For the general case we first pass to a discrete subgroup Q of T(K) such that T(K)/Q is compact, hence Q is finitely generated abelian of rank $\dim T$. Put $U(K) =: N$. For every $t \in Q$ define the subgroup

$$N_t := \{u \in N \mid t^n u t^{-n} \text{ converges to } e \text{ for } n \to \infty\}$$

of N "contracted by t". Then N_t is the group of K-points of a linear algebraic subgroup of U whose Lie algebra is $\oplus\, \mathfrak{u}^{\chi}$, $\chi(t) \in p\mathbb{Z}_p$. The result above for $\dim T = 1$ immediately implies that $Q \ltimes N_t$ has a compact presentation for every $t \in Q$. So one is led to study the free product of the (finitely many) groups $Q \ltimes N_t$ amalgamated along their intersections $H := \coprod\limits_{\cap} (Q \ltimes N_t) = Q \ltimes M$, where $M := \coprod\limits_{\cap} N_t$. It turns out by a long and intricate computation that M is nilpotent if (1) holds. If we turn H into a locally compact topological group $H_1 = Q \ltimes M_1$ locally isomorphic to G(K) by imposing a compact set of further relations we obtain a central extension $M_1 \to N$ with the following properties: $M^{ab} \to N^{ab}$ is an isomorphism, the kernel of $M_1 \to N$ is a p-divisible discrete group and Q acts on this kernel with finite orbits. So either $H_1 \to G(K)$ is an isomorphism - and hence G(K) has a compact presentation by construction of H_1 - or there is no compact presentation of G(K) at all. So the final task is to relate this central extension of the topological group N with central extensions of the Lie algebra $\mathfrak{n}$ and hence with $H_2(\mathfrak{n})$ in order to obtain sufficiency of (1) and (2).

Concluding remarks concerning higher finiteness conditions. A discrete group Γ is called of type FP_n if the trivial $\mathbb{Z}\Gamma$-module $\mathbb{Z}$ has a projective resolution $\ldots \to P_i \to P_{i-1} \to \ldots \to P_o \to \mathbb{Z}$ with finitely generated $\mathbb{Z}\Gamma$-modules P_i for $i \leq n$. The group Γ is called of type F_n if there is an Eilenberg-MacLane complex for Γ with finite n-skeleton. The three conditions FP_1, F_1 and finitely generated are equivalent. The conditions F_2 and finitely presented are equivalent and imply FP_2. For $n > 2$ a group is of type F_n iff it is finitely presented and of type FP_n ([Wall]). It is an open problem to give necessary and sufficient conditions for an S-arithmetic group Γ to be of type F_n. The work of Åberg implies that for a $\mathbb{Q}$-split solvable group $G = T \ltimes U$ a necessary condition for $\Gamma = G(\mathbb{Z}[1/p])$ to be of type F_n is that the representation of T on $\mathfrak{u}^{ab}$ is n-tame, i.e. the convex hull of any n weights of T on $\mathfrak{u}^{ab}$ does not contain 0. This is not sufficient, of course.

The following sequence of S-arithmetic groups includes our earlier examples. Let G_n be the algebraic subgroup of GL_n defined by $G_n = \{X = (X_{ij}) \mid X_{11} = X_{nn} = 1, X_{ij} = 0 \text{ for } i > j\}$. Then $G_n(\mathbb{Z}[1/p])$ is of type FP_{n-2}, not of type FP_{n-1} and finitely presented for $n \geq 4$ ([Holz, Abels-Brown, Brown, Bieri]).

REFERENCES

H. Abels, Kompakt definierbare topologische Gruppen, *Math. Ann.* 197 (1972), 221-233.

H. Abels, An example of a finitely presented solvable group, in: *Homological Group Theory*, Proc. Durham 1977, ed. C.T.C. Wall, London MS Lecture Notes 36 (1979), 205-211.

H. Abels, Finite presentability of S-arithmetic groups. Compact presentability of solvable groups, Lecture Notes in Mathematics, Springer, to appear.

H. Abels & K.S. Brown, Finiteness properties of solvable S-arithmetic groups: An example, *J. Pure Appl. Algebra,* to appear.

H. Åberg, Bieri-Strebel valuations (of finite rank), *Proc. London Math. Soc.,* to appear.

H. Behr, Über die endliche Definierbarkeit verallgemeinerter Einheitengruppen. II, *Invent. Math.* 4 (1967), 265-274.

H. Behr, Finite presentability of arithmetic groups over global function fields, Frankfurt (1985), preprint.

R. Bieri, A connection between the integral homology and the centre of a rational linear group, *Math. Z.* 170 (1980), 263-266.

R. Bieri & R. Strebel, Almost finitely presented soluble groups, *Comment. Math. Helv.* 53 (1978), 258-278.

R. Bieri & R. Strebel, Valuations and finitely presented metabelian groups, *Proc. London Math. Soc.* (3) 41 (1980), 439-464.
A. Borel & Harish-Chandra, Arithmetic subgroups of algebraic groups, *Ann. of Math.* 75 (1962), 485-535.
A. Borel & J.P. Serre, Corners and arithmetic groups, *Comment. Math. Helv.* 48 (1973), 436-491.
A. Borel & J.P. Serre, Cohomologie d'immeubles et de groupes S-arithmétiques, *Topology* 15 (1976), 211-232.
A. Borel & J. Tits, Groupes réductifs, *Inst. Hautes Etudes Sci. Publ. Math.* 27 (1965), 55-152.
P. Hall, Finiteness conditions for soluble groups, *Proc. London Math. Soc.* (3) 4 (1954), 419-436.
S. Holz, *Endliche Identifizierbarkeit von Gruppen,* Thesis Bielefeld (1985).
M. Kneser, Erzeugende und Relationen verallgemeinerter Einheitengruppen, *J. Reine Angew. Math.* 214/215 (1964), 345-349.
M.S. Raghunathan, A note on quotients of real algebraic groups by arithmetic subgroups, *Invent. Math.* 4 (1968), 318-335.
J.P. Serre, Cohomologie des groupes discrets, in: Prospects in Mathematics, *Ann. of Math. Stud.* 70 (1971), 77-169.
C.T.C. Wall, Finiteness conditions for CW-Complexes, *Ann. of Math.* 81 (1965), 56-69.
C.T.C. Wall, Finiteness conditions for CW-complexes II, *Proc. Roy. Soc. London Ser.* A 295 (1966), 129-139.

ADDITIONAL REFERENCE

K.S. Brown, Finiteness properties of group, *J. Pure Appl. Algebra,* to appear.

EFFICIENT PRESENTATIONS OF $GL(2,\mathbb{Z})$ AND $PGL(2,\mathbb{Z})$

F. Rudolf Beyl
Portland State University, Portland, USA

Gerhard Rosenberger
University of Dortmund, Dortmund, West Germany

The efficient free presentations of the title are

$$GL(2,\mathbb{Z}) \cong \langle a,b : a^2 = ([b,a]b)^4 = ([b,a]ba)^2 = 1\rangle \quad (1)$$

$$PGL(2,\mathbb{Z}) \cong \langle a,b : a^2 = ([b,a]b)^2 = ([b,a]ba)^2 = 1\rangle \quad (2)$$

where a, b, and $[b,a] = bab^{-1}a^{-1}$ correspond to

$$\begin{pmatrix} 0 & 1 \\ 1 & 0 \end{pmatrix}, \quad \begin{pmatrix} 1 & -1 \\ 0 & 1 \end{pmatrix}, \quad \begin{pmatrix} 0 & -1 \\ 1 & 1 \end{pmatrix},$$

respectively. Our primary motivation is the possible application in the theory of Klein surfaces: May [M] showed that it is precisely the finite factor groups of $PGL(2,\mathbb{Z})$ with order greater than 6 that appear as the automorphism groups of compact Klein surfaces with non-empty boundary and genus $g \geq 2$ such that the group order is maximal, viz. $12(g-1)$. For groups G with finite abelianized group G_{ab}, P. Hall's Inequality [BT; IV.1.4] reads

$$\#(\text{relators}) \geq \#(\text{generators}) + \#(\text{generators of } H_2G) \quad (3)$$

for every free presentation of G, where H_2G denotes the Schur multiplicator of G. The asserted efficiency means equality in (3) for the presentations (1) and (2). By Swan's result [S], [BT; IV.1.10] efficiency cannot be expected in general, here it follows from our computation of the multiplicators:

$$H_2GL(2,\mathbb{Z}) \cong \mathbb{Z}/2\mathbb{Z} \text{ and } H_2PGL(2,\mathbb{Z}) \cong \mathbb{Z}/2\mathbb{Z}. \quad (4)$$

This proves that the presentations (1) and (2) are minimal in the strongest sense. Other 2-generator 3-relation presentations are already mentioned in [CM; (7.29)+(7.291)], but the 3-generator 4-relation presentation of $GL(2,\mathbb{Z})$ in [LS; p.25] is incorrect. (The relation $C^2 = 1$ is missing in the latter. If one adds $B^2 = AC = 1$ but not $C^2 = 1$, then C has order 3 rather than 2.)

Proof of (1). Since $GL(2,\mathbb{Z})$ is the semidirect product of $SL(2,\mathbb{Z}) \cong \langle x,y : x^2 = y^3, x^4 = 1\rangle$ and $\mathbb{Z}/2\mathbb{Z}$, we have

$$GL(2,\mathbb{Z}) \cong \langle x,y,t : x^2 = y^3, x^4 = t^2 = 1, txt^{-1} = x^{-1}, tyt^{-1} = y^{-1}\rangle,$$

where x, y, and t correspond to the matrices

$$\begin{pmatrix} 0 & -1 \\ 1 & 0 \end{pmatrix}, \quad \begin{pmatrix} 0 & -1 \\ 1 & 1 \end{pmatrix}, \quad \begin{pmatrix} 0 & 1 \\ 1 & 0 \end{pmatrix},$$

respectively. We add generators $a = t$ and $b = y^{-1}x$. Noting

$$[b,a] = y^{-1}xtx^{-1}yt^{-1} = y^{-1}tx^{-2}yt^{-1} = y^{-1}ty^{-2}t^{-1} = y,$$

we add the consequential relation $y = [b,a]$ and then eliminate t, x, y by $t = a$, $x = yb$, and $y = [b,a]$. We obtain the relations $([b,a]b)^2 = [b,a]^3$, $([b,a]b)^4 = 1$, $a^2 = 1$, $a[b,a]ba^{-1} = b^{-1}[a,b]$, and $a[b,a]a^{-1} = [a,b]$. The last relation obviously follows from $a^2 = 1$, we rewrite the fourth relation as $([b,a]ba)^2 = 1$. Again $a^2 = 1$ implies

$$([b,a]b)^2a([b,a]ba)^2a = [b,a]^3,$$

whence the first relation is a consequence of the others.

Proof of (2). Since the center of $GL(2,\mathbb{Z})$ is generated by $-I$, which corresponds to x^2, we obtain a presentation of $PGL(2,\mathbb{Z})$ from (1) by adding the relator $([b,a]b)^2$.

Proof of (3). We use the spectral sequence of the group extension

$$PSL(2,\mathbb{Z}) \rightarrowtail PGL(2,\mathbb{Z}) \twoheadrightarrow \mathbb{Z}/2\mathbb{Z}$$

in integral homology, cf. [B; No. 5.6], with

$$E^2_{pq} = H_p(\mathbb{Z}/2\mathbb{Z}, H_qPSL(2,\mathbb{Z}))$$

converging to $H_{p+q}PGL(2,\mathbb{Z})$, suitably filtered. Since $PSL(2,\mathbb{Z})$ has a 2-generator 2-relator presentation, (3) yields $H_2PSL(2,\mathbb{Z}) = 0$. Also $H_2(\mathbb{Z}/2\mathbb{Z}) = 0$, hence $E^2_{20} = E^2_{02} = 0$. The generator of $\mathbb{Z}/2\mathbb{Z}$ acts as multiplication by -1 on

$$H_1PSL(2,\mathbb{Z}) \cong PSL(2,\mathbb{Z})_{ab} \cong \mathbb{Z}/6\mathbb{Z},$$

due to $txt^{-1} = x^{-1}$ and $tyt^{-1} = y^{-1}$ in $GL(2,\mathbb{Z})$. Thus $E^2_{11} = H_1(\mathbb{Z}/2\mathbb{Z}, \mathbb{Z}/6\mathbb{Z}) \cong \mathbb{Z}/2\mathbb{Z}$. Since the group extension splits, $E^2_{30} = H_3(\mathbb{Z}/2\mathbb{Z})$ is a direct summand of $H_3PGL(2,\mathbb{Z})$ and must survive, so $d^2 : E^2_{30} \longrightarrow E^2_{11}$ vanishes. A similar argument, with $SL(2,\mathbb{Z}) \cong \langle x,y : x^4 = 1, x^2 = y^3\rangle$, yields $H_2GL(2,\mathbb{Z}) = \mathbb{Z}/2\mathbb{Z}$.

An analysis of the last proof shows that the induced mapping $H_2GL(2,\mathbb{Z}) \longrightarrow H_2PGL(2,\mathbb{Z})$ vanishes. Since the abelianized groups are isomorphic (vierer group), $GL(2,\mathbb{Z})$ is a representation group of $PGL(2,\mathbb{Z})$, cf. [BT; II.2.14]. We finally mention that any generator *pair* of $PGL(2,\mathbb{Z})$ is obtainable from our pair $\{a,b\}$ by a Nielsen transformation. The proof is a refinement of the arguments in [KR], but is rather technical.

REFERENCES

[B] F.R. Beyl, Abelian groups with a vanishing homology group, *J. Pure Appl. Algebra* 7 (1976), 175-193.

[BT] F.R. Beyl & J. Tappe, *Group Extensions, Representations, and the Schur Multiplicator*, Lecture Notes in Math. vol. 958, Springer-Verlag (1982).

[CM] H.S.M. Coxeter & W.O.J. Moser, *Generators and Relations of Discrete Groups*, 4th ed. Ergebnisse der Math. vol. 14, Springer-Verlag (1980).

[KR] R.N. Kalia & G. Rosenberger, Automorphisms of the Fuchsian groups of type (0;2,2,2,q;0), *Comm. Alg.* 6 (1978), 1115-1129.

[LS] R.C. Lyndon & P.E. Schupp, *Combinatorial Group Theory*, Ergebnisse der Math. vol. 89, Springer-Verlag (1977).

[M] C.L. May, Large automorphism groups of compact Klein surfaces with boundary, I, *Glasgow Math. J.* 18 (1977), 1-10.

[S] R.G. Swan, Minimal resolutions of finite groups, *Topology* 4 (1965), 193-208.

THE COMMUTATOR MAP

Rolf Brandl
Mathematisches Institut, University of Würzburg, Würzburg,
West Germany

The commutator map ρ_y, defined by $g \to [g,y] = g^{-1}y^{-1}gy$ of a group G into itself is not too well-understood. It can be represented by a directed graph $\Gamma(y)$ whose vertices are the elements of G and two vertices a and b are joined by a directed edge if and only if $[a,y] = b$. If G is a finite group then $\Gamma(y)$ has a number of connected components, each of which consists of a cycle (a so called Engel cycle) with various trees attached. It is recommended as an amusing exercise to draw $\Gamma(y)$ for $G = A_5$ and $y = (1\ 2\ 3)$.

For two elements x, y of a finite group G there exist a non-negative integer r and a positive integer d such that $[x,_r y] = [x,_{r+d} y]$. If r and d are chosen minimal with respect to this property, then $r = r(x,y)$ is called the (Engel-) depth of the cycle generated by x and y and $d = d(x,y)$ its diameter. Moreover $r(G) = \max\{r(x,y) \mid x,y \in G\}$ is called the depth of G and $d(G) = \mathrm{lcm}\ \{d(x,y) \mid x,y \in G\}$ its diameter.

These notions, with some obvious modifications, make sense for infinite groups. However, there is no recent progress in this area, and so we shall primarily be concerned with finite groups and refer the reader to [3] and [9] for more details.

1. *THE DIAMETERS*

The depth and the diameter have been determined for various types of groups. For an appropriate formulation we need to introduce some notation.

Definition. *Let* p *be a prime and let* $\ell > 2$ *be a positive integer coprime to* p. *Let* k *be an algebraically closed field of characteristic* p. *Then* $d_{p,\ell}$ *denotes the least common multiple of the multiplicative orders of* $-1+x$ *where* $x \neq 1$ *runs through the set of* ℓ*th roots of unity of* k.

Let $\mathfrak{A}_k$ be the class of all finite abelian groups of exponent dividing k. Then the first result reads as follows.

Theorem 1.1 ([10]). *Let* p *be a prime and let* $\ell > 2$ *be a positive integer coprime to* p.

(a) *If* $G \in \mathfrak{A}_p\mathfrak{A}_\ell$ *then* $r(G) \leq 2$ *and* $d(G)$ *divides* $d_{p,\ell}$. *Moreover, if an element of order* ℓ *in* G *acts faithfully on* $O_p(G)$ *then* $r(G) = 2$ *and* $d(G) = d_{p,\ell}$.

(b) *If* $G \in \mathfrak{A}_{p^\alpha}\mathfrak{A}_\ell$ *where* $\alpha > 1$ *then* $r(G) \leq 2$ *and* $d(G)$ *divides* $p^{\alpha-1}\,d_{p,\ell}$.

In general, part (b) cannot be improved as can be seen from the following example.

Example. Let $p = 1093$ and let $G \in \mathfrak{A}_{p^2}\mathfrak{A}_2$. Then $d(G)$ divides $d_{p,2}$. (The reason for this is that $(-2)^{p-1} \equiv 1 \bmod p^2$.)

The next result can be used to determine depth and diameter of a group in $\mathfrak{A}_p\mathfrak{A}_\ell\mathfrak{A}_2$.

Proposition 1.2. *Let $\ell > 3$ be odd and let $Q = \mathbb{D}_\ell$. Furthermore, let p be a prime not dividing ℓ, let N be a faithful and irreducible $\mathbb{F}_pQ$-module and let $G = NQ$ be the natural split extension. If δ denotes the least common multiple of $d_{p,\ell}$, $d_{p,2}$ and the multiplicative order of -2 modulo ℓ, then $r(G) = 2$ and $d(G) = \delta$ or $d(G) = p\cdot\delta$.*

Concerning the diameter, both cases can occur. If $\ell = 3$ then always $d(G) = p\cdot\delta$. However, if $\ell = 11$ and $p = 2$, we have $d(G) = \delta = d_{2,11}\ d_{11,2}$. Although most frequently the first case occurs, no general criterion is known.

The following provides a list of depth and diameter of some simple groups of the series $G = PSL(2,q)$. The cases $q = 5,7$ are due to J. Neubüser, $q = 8$ has been dealt with by Daniela B. Nikolova.

q	5	7	8	9	11	13
$\|G\|$	60	168	504	360	660	1092
$r(G)$	3	4	3	4	6	7
$d(G)$	60	168	126	120	1980	2184

In the above results, the diameters are closely related to the order of the group. If p denotes an odd prime then the diameter of the dihedral group $\mathbb{D}_p$ divides $p - 1$, which is the order of a chief factor subtracted by 1. For soluble groups this is typical.

Theorem 1.3. *Let G be a finite soluble group and let $R = \gamma_\infty(G)$. If p is a prime dividing $d(G)$ then either p divides the order of R or p divides $|H/K| - 1$ where H/K is an elementary abelian section of R.*

A corresponding result for nonsoluble groups does not seem to be at hand as can be seen from the following example that has been computed by the Aachen-Sydney GROUP system.

Example. Let M_{11} be the Mathieu group of degree 11. Then the elements $x = (1\ 2\ 11\ 10\ 6\ 4\ 3\ 8\ 7\ 5\ 9)$ and $y = (1\ 10\ 11)\ (2\ 9\ 8\ 4\ 3\ 5)\ (6\ 7)$ generate a cycle of length 13.

Let $\mathcal{F}_d$ be the class of all finite groups of diameter dividing d. It is easy to see that for $d \neq 2$ the dihedral group of order $2\cdot|(-2)^d - 1|$ has diameter d. From this one can prove that for $d_1 \neq 2 \neq d_2$ the classes $\mathcal{F}_{d_1}$ and $\mathcal{F}_{d_2}$ are equal if and only if $d_1 = d_2$. For $d = 2$, however, there is the following, somewhat surprising result.

Theorem 1.4 ([2]). $\mathcal{F}_1 = \mathcal{F}_2$.

There are various relations between the diameter of a group and nilpotency properties.

Theorem 1.5. (a) ([8]). *Let* G *be a finite soluble group of diameter* d. *If* d *is odd, let* $\varepsilon = 2$, *otherwise, let* $\varepsilon = 1$. *Then* $G^{\varepsilon d}$ *is nilpotent.*
(b) *There exists a function* ψ *such that any finite soluble group in* $\mathfrak{F}_d$ *has Fitting length* $\leq \psi(d)$.

Theorem 1.6 ([8]). *Let* G *be a finite group of diameter* d. *If* p *is a prime not dividing* $((-2)^d - 1) \cdot \prod (q^d - 1)$ *where the product runs over the primes* q *dividing* d, *then* G *is* p*-nilpotent.*

For some special values of d one has better results, for example:

Theorem 1.7. *Let* G *be a finite group of diameter* d.
(a) ([8]) *If* d *is odd then* G *is soluble and all subgroups of odd order of* G *are nilpotent.*
(b) *If* d *is a prime then* G *is metanilpotent.*

In general, groups G having odd diameter need not be metanilpotent as can be seen from the example following 1.2 and so there remains:

Question 1. *Is there a bound for the Fitting length of finite groups with odd diameter?*

It has now been proved that finite groups having odd diameter have Fitting length at most four.

The next problem is best motivated by the table exhibiting the diameters of some of the groups PSL(2,q).

Question 2. *Is* 1.5(a) *valid without the hypothesis that* G *is soluble. In particular, for a nonabelian finite simple group* G *is it true that* exp(G) *divides* d(G)?

The final result of this section deals with finite simple groups of a given diameter.

Theorem 1.8. *For any given* d *there are at most finitely many nonabelian finite simple groups of diameter* d.

2. *THE DEPTH*

In this paragraph we consider the class $\mathfrak{D}_r$ of all finite groups having depth $\leq r$. As a finite nilpotent group belongs to $\mathfrak{D}_r$ if and only if it satisfies the rth Engel condition, the depth, in a sense, is a generalization of the Engel length. As in the theory of Engel conditions, groups of depth ≤ 2 are quite well understood.

Theorem 2.1 ([6]). *Finite groups of depth* 1 *are abelian.*

There are various types of groups of depth 2. If π denotes a set of primes then every group in $\mathfrak{A}_\pi \mathfrak{A}_{\pi'}$ has depth ≤ 2. Moreover, every Frobenius group with abelian kernel and complement of odd order has depth 2. More examples of Fitting length 3 may be found in 1.2. The following collects some general information on groups in $\mathfrak{D}_2$.

Theorem 2.2 ([4]). *Let* G *be a finite group of depth* 2. *Then* G/F(G) *is supersoluble, metabelian and has abelian Sylow* p-*subgroups for all odd primes* p. *Moreover,* $\ell_p(G) \leq 1$ *for* $p \neq 2$ *and* $\ell_2(G^2) \leq 1$.

In particular, $\mathfrak{D}_2$ is contained in the variety of all (2-Engel)-by-metabelian groups and so all finite groups of depth 2 satisfy a nontrivial law. It is an open question whether a corresponding statement is valid in general.

Question 3. *Is it true that for any given* r *the class* $\mathfrak{D}_r$ *satisfies a non-trivial law?*

An affirmative answer to Question 3 would yield a positive solution of:

Question 3'. *Is it true that for given* r *the class* $\mathfrak{D}_r$ *contains at most finitely many nonabelian simple groups?*

One result in this context is

Theorem 2.3 ([5]). *Let* $q \geq 4$ *be a prime power. Then* PSL(2,q) *has depth* 3 *if and only if* q = 4,5 *or* 8.

There is some computational evidence that the groups mentioned in 2.3 are the only finite simple groups of depth 3, for example, the smallest Suzuki group Sz(8) has depth at least 11.

Another question of considerable interest is: whether the depth bounds the structure constants of a soluble group. For example, 2.2 implies that finite groups of depth are soluble of derived length 4 (it is not known at present, whether this bound can be replaced by 3). In general, however, the derived length is not bounded by the depth as, for example, every group of exponent 4 has depth ≤ 4, but there are such groups of arbitrary derived length. Using extensions of elementary abelian p-groups by certain groups of exponent 4 it was shown in [4] that in general, the depth of a soluble group G neither bounds the derived length nor the ranks of the chief factors of G/F(G).

Question 4. *Is there a function* f *having the property that any finite soluble group of depth* r *has Fitting length* $\leq f(r)$?

3. *CONNECTED COMPONENTS*

The number of connected components of the graph $\Gamma(y)$ will be denoted by $\zeta(y)$. If G is finite then $\zeta(y)$ is equal to the number of different cycles of $\Gamma(y)$. Now $\Gamma(y)$ always has a component containing the identity of G and this is the only component if and only if y is a right Engel element in G. Hence if G is finite then $\zeta(y) = 1$ if and only if $y \in F(G)$. Setting $\zeta(G) = \max\{\zeta(y) \mid y \in G\}$, we infer that $\zeta(G) = 1$ if and only if G is nilpotent. So, in a sense, $\zeta(G)$ measures, how far G is away from being nilpotent. One result in this spirit is:

Theorem 3.1. *Let* G *be a finite group.*
(a) *If* $\zeta(G) < 4$ *then* G' *is nilpotent.*
(b) *If* $\zeta(G) < 6$ *then* G *is soluble.*

As $\zeta(S_4) = 5$ and $\zeta(A_5) = 7$, the bounds given are best possible.

It may be remarked that questions concerning the number of connected components rather quickly lead to problems in number theory. For example, if p is an odd prime then $\zeta(\mathbb{D}_p) = 2$ if and only if -2 is a primitive root modulo p.

One can also consider the depth of a certain component of G, for example the one containing the identity element. This is equivalent to considering the following conditional identity (in the sense of [1])

(V - r): If for some $x,y \in G$, $k \in \mathbb{N}$, we have $[x,_k y] = 1$, then $[x,_r y] = 1$.

It is easy to prove that a finite group satisfying (V - 1) is abelian. The corresponding statement is not true in the infinite case as, for example, every free group satisfies (V - 1). In particular, this property is not inherited by quotients.

Question 5. *Is the class of all finite groups satisfying* (V - r) *closed with respect to forming quotients?*

The above conditional identities can be used to derive an "odd" characterization of the first Janko group.

Theorem 3.2. *Let* G *be a finite nonabelian simple group. Then the following are equivalent:*
(i) G *satisfies* (V - 2);
(ii) $G \cong PSL(2,q)$ *for* $q = 2^f$ *or* $q \equiv \pm 3 \bmod 8$ *or* $G \cong J_1$.

REFERENCES

1. R. Baer, Principal factors, maximal subgroups and conditional identities, *Ill. J. Math.* 13 (1969), 1-52.
2. R. Brandl, Engel cycles in finite groups, *Arch. Math.* 41 (1983), 97-102.
3. R. Brandl, Infinite soluble groups with Engel cycles; a finiteness condition, *Math. Z.* 182 (1983), 259-264.
4. R. Brandl, On groups with small Engel depth, *Bull. Austral. Math. Soc.* 28 (1983), 101-110.
5. R. Brandl & Daniela B. Nikolova, Simple groups of small Engel depth, *Bull. Austral. Math. Soc.*, to appear.
6. N.D. Gupta, Some group-laws equivalent to the commutative law, *Arch. Math.* 17 (1966), 97-102.
7. N.D. Gupta, Groups with Engel-like conditions, *Arch. Math.* 17 (1966), 193-199.
8. N.D. Gupta & H. Heineken, Groups with a two-variable commutator identity, *Math. Z.* 95 (1967), 276-287.
9. Daniela B. Nikolova, On a class of group identities, *Serdika* 9 (1983), 189-197.
10. Daniela B. Nikolova, Laws in metabelian varieties of groups of type $\mathfrak{A}_k\mathfrak{A}_\ell$, *Serdika* 9 (1983), 235-243.
11. Daniela B. Nikolova, Calculation of some commutator laws in alternating groups, *Serdika* 10 (1984).
12. Daniela B. Nikolova, Solubility of finite groups with a two-variable commutator identity, *Serdika* 10 (1984), 59-63.
13. Daniela B. Nikolova, Groups with a two-variable commutator identity, *Doklady BAN* 36 (1983), 721-724.

POLYNOMIAL FUNCTIONS AND REPRESENTATIONS

H.K. Farahat
University of Calgary, Alberta T2N 1N4, Canada

1. Several authors have discussed polynomial functions on real euclidean space (see [1] and references listed therein), apparently without realizing the connection of their work to group representation theory. The following theorem is a generalization of some results of that sort. The proof is deliberately self-contained in the interests of wider communication.

Theorem. *Let* T *be the group of translations of* R^n. *Let* H *be a group of linear transformations of* R^n *such that:*
(1) H *has no proper subgroup of finite index;*
(2) *every element of* R^n *has the form* $\eta x - x$ *for some* $\eta \in H$, $x \in R^n$.
Let $G = HT = \{\eta\tau : \eta \in H, \tau \in T\}$. *If the* G*-translates of a continuous complex valued function* f *on* R^n *span a finite-dimensional vector space, then* f *is a polynomial function.*

Proof. Observe that G is a group of mappings of R^n to R^n and that T is a normal subgroup of G. Indeed, if for $x \in R^n$, t_x denotes the translation of R^n by x, that is

$$t_x(y) = y + x \qquad (y \in R^n),$$

and if $\eta \in H$, then since η is linear,

$$(\eta t_x \eta^{-1})(y) = \eta(\eta^{-1}(y) + x) = y + \eta(x) = t_{\eta(x)}(y),$$

hence

$$\eta t_x \eta^{-1} = t_{\eta(x)} \qquad (\eta \in H,\ x \in R^n).$$

Now if $\sigma \in G$ and f is a continuous function on R^n then $f\sigma$ is a mapping, of R^n to $\mathbb{C}$, which we refer to as the *translate of* f *by* σ. We are assuming that the vector space V spanned by $\{f\sigma : \sigma \in G\}$ is finite-dimensional. Choose a basis $\{e_1, \dots, e_m\}$ of V and write

$$e_i\sigma = \sum_{j=1}^{m} M_{ij}(\sigma)e_j \qquad (\sigma \in G),$$

so that $M(\sigma)$ is an $m \times m$ complex matrix-valued function of $\sigma \in G$. The map $\sigma \in G \to M(\sigma) \in GL_m(\mathbb{C})$ is a *representation:*

$$M(\sigma\sigma') = M(\sigma)M(\sigma') \qquad (\sigma, \sigma' \in G).$$

Observe that the restriction of any M_{ij} to the translation subgroup T is necessarily continuous. For every translation τ is continuous and hence each $e_i\sigma$ is also continuous on R^n. Accordingly the mapping

$$x \in R^n \to M(t_x) \in GL_m(\mathbb{C})$$

is a continuous complex matrix representation of the group $(R^n,+)$. It follows from Section 2 below that there exist matrices $C_1,\ldots,C_n$ such that $C_iC_j = C_jC_i$ (all i,j) and

$$M(t_x) = \exp(x_1C_1 + \ldots + x_nC_n)$$

for all $x = (x_1,\ldots,x_n) \in R^n$.

On the other hand we shall use (1) and (2) above to deduce that $M(t_x)$ is always unipotent, that is the sum of the identity matrix and a nilpotent matrix. Equivalently, we shall prove that 1 is the only eigenvalue of $M(t_x)$ for each $x \in R^n$. Let $V_1 \subset V_2 \subset \ldots \subset V_k = V$ be a composition series for V as a representation space for G. This means that $V_1, V_2, \ldots, V_k$ are subspaces of V stable under the action of G and that each factor space V_j/V_{j-1} $(1 \leq j \leq k)$ is irreducible as a G-space ($V_0 = \{0\}$). It follows from Section 3 below that the linear transformation induced on any V_j/V_{j-1} by any $\sigma \in T$ is the identity. Hence the linear transformation induced on V by any $\sigma \in T$ is unipotent. But, of course, $M(\sigma)$ is the matrix for this linear transformation with respect to the basis $\{e_1,\ldots,e_m\}$. Consequently, each $M(t_x)$ $(x \in R^n)$ is a unipotent matrix.

Now if α_i is any eigenvalue of C_i then $\exp(x_i\alpha_i)$ is an eigenvalue of $\exp(x_iC_i)$; consequently we have $\exp(x_i\alpha_i) = 1$ for all $x_i \in R$. Differentiate with respect to x_i to deduce that $\alpha_1 = 0,\ldots,\alpha_n = 0$, and hence that $X = x_1C_1 + \ldots + x_mC_m$ is nilpotent, in fact that $X^m = 0$. Then

$$M(t_x) = \sum_{r=0}^{m-1} X^r/r! \qquad (x \in R^n)$$

has entries which are *polynomial functions* on R^n. Now, for any i and any $x \in R^n$ we have

$$e_i(x) = e_it_x(0) = \sum_{j=1}^{m} M_{ij}(t_x)e_j(0),$$

which shows that e_i is a polynomial function on R^n. Hence every element of V, and in particular f itself, is a polynomial function.

2. The group $(R^n,+)$ is one of the standard and well known Lie groups. Every continuous homomorphism of Lie groups is necessarily differentiable and this fact severely restricts the nature of continuous representations. Books on Lie groups and their representations deal with these matters. Since R^n is such a familiar object we provide here a proof from first principles of the required facts:

Proposition. *Every complex continuous matrix representation of* $(R^n,+)$ *has the form*

$$A(x) = \exp(x_1C_1 + \ldots + x_nC_n) \qquad (x \in R^n)$$

with $C_1,\ldots,C_n$ *a commutative set of matrices.*

Proof. If A is any matrix representation of $(R^n,+)$ then

$$A(x) = A((x_1,0,\dots,0) + \dots + (0,0,\dots,x_n))$$
$$= A(x_1,\dots,0) \dots A(0,0,\dots,x_n)$$
$$= A_1(x_1) \dots A_n(x_n), \text{ say,}$$

so that $A_1,\dots,A_n$ are representations of $(R,+)$ which are mutually commutative. Therefore it is sufficient to prove the proposition for the case $n = 1$.

Suppose then that A is a continuous matrix representation of $(R,+)$. We set

$$A_\xi(x) = \int_x^{x+\xi} A(t)dt \qquad (x \in R,\ \xi \text{ small real}).$$

If ξ is small enough then $A_\xi(x)$ will be close enough to $A(x)$ to ensure its invertibility. But

$$A_\xi(x+y) = \int_{x+y}^{x+y+\xi} A(t)dt = \int_x^{x+\xi} A(y+v)dv = \int_x^{x+\xi} A(y)A(v)dv = A(y)A_\xi(x),$$

and so

$$A(y) = A_\xi(x+y)A_\xi(x)^{-1}.$$

As we have $A'(x) = A(x+\xi) - A(\xi)$ (by the fundamental theorem of Calculus), both $A_\xi(x+y)$ and $A_\xi(x)$ are differentiable functions of y, hence A *is differentiable.*

Now the identity $A(x+y) = A(x)A(y)$ $(x,y \in R^n)$ yields, upon differentiation with respect to y:

$$A'(x+y) = A(x)A'(y).$$

Now put $y = 0$ to get $A'(x) = A(x)A'(0)$ whence $A(x) = \exp(xA'(0))$ as required.

3. The result needed in Section 1 is a simple application of the so-called Clifford Theory in group representations. To save the reader the trouble of learning this theory, a complete proof of what is needed is provided here. Let $G = HT$ as in Section 1 with T a normal abelian subgroup of G and H satisfying properties (1) and (2) of the theorem enunciated there. Let W be an irreducible complex G-space. We want to show that $\tau w = w$ for every $\tau \in T$ and every $w \in W$.

Since T is commutative, every complex irreducible T-space is 1-dimensional. Let W_0 be any such T-subspace of W. For any $\eta \in H$, ηW_0 is also such a T-subspace and $\sum_\eta \eta W_0$ is easily seen to be a non-zero G-subspace of W. As the latter is irreducible we must have $W = \sum_\eta \eta W_0$, a sum of irreducible T-spaces. Collecting T-isomorphic summands we find that W is a *direct sum* of a finite number of T-subspaces, say $W = W_1 + \dots + W_k$ where T acts by "scalar multiplication" on each W_i. This means that for each i,

$$\tau(w) = \chi_i(\tau)w, \quad \text{for } \tau \in T,\ w \in W_i,$$

where $\chi_1,\dots,\chi_k$ are distinct complex representations of T of degree 1. Furthermore, for any $\eta \in H$, $W_i \to \eta W_i$ is a permutation of the finite set $\{W_1,\dots,W_k\}$, and for any j there exists η such that $\eta W_1 = W_j$. Now it follows that

$$\{\eta \in H: \ \eta W_1 = W_1\}$$

is a subgroup of H of index k exactly. As H has no proper subgroup of finite index we conclude that $k = 1$; and that $W = W_1$. We now have

$$\tau(w) = \chi_1(\tau)w \quad \text{for} \quad w \in W, \ \tau \in T.$$

However, by property (2) of H, $\tau(0) = \eta x - x$ for some $x \in R^n$ and so

$$\tau = t_{\tau(0)} = t_{\eta x - x} = t_{\eta x} t_x^{-1} = \eta t_x \eta^{-1} \cdot t_x^{-1},$$

from which it follows that, for $w \in W$,

$$\tau(w) = \eta t_x \eta^{-1} \cdot t_x^{-1}(w) = \chi_1(t_x)^{-1} \cdot \eta t_x \eta^{-1}(w) = \chi_1(t_x)^{-1} \cdot \eta \cdot \chi_1(t_x)\eta^{-1}(w) = w,$$

as required.

4. *Examples.* There are many examples of groups H satisfying requirements (1) and (2).

Consider firstly the proper orthogonal group $O_n^+(R)$. Given any proper orthogonal transformation η and any positive integer r there exists an orthogonal transformation λ such that $\eta = \lambda^r$, as may be seen by, for example, expressing η as a product of commuting plane rotations. Thus $O_n^+(R)$ admits "root extraction". Because of this $O_n^+(R)$ has no proper subgroups of finite index. If K were such a subgroup of index s then $\lambda^{s!} \in K$ for every orthogonal transformation λ, and hence $\eta \in K$ for every $\eta \in O_n^+(R)$. This establishes property (1); and property (2) also holds if $n \geq 2$. To see this choose an orthonormal basis $u_1,\ldots,u_n$ for R^n such that a given $y \in R^n$ is a multiple of u_1. Take η to be the orthogonal transformation given by $\eta(z_1u_1 + \ldots + z_nu_n) = -z_1u_1 - z_2u_2 + z_3u_3 + \ldots + z_nu_n$, and check that $\eta x - x = y$ where $x = -\frac{1}{2}y$.

For a second example, take H to be the group of all linear transformations of the form $x \in R^n \to (\lambda_1x_1,\ldots,\lambda_nx_n)$ where $\lambda_1,\ldots,\lambda_n$ are positive real constants. Conditions (1) and (2) are easily verified in this case.

These examples show that Theorem 1 includes the results discussed by P.G. Laird and R. McCann [1] by other methods. The present paper was inspired by a seminar of P.G. Laird at the University of Calgary.

REFERENCES

1. P.G. Laird & R. McCann, On some characterizations of polynomials, *Amer. Math. Monthly* 91 (1984), 114-116.

ON QUESTIONS OF BRAUER AND FEIT

Pamela A. Ferguson
University of Miami, Coral Gables, FL 33124, USA

All groups considered in this paper are finite. Let G be a group and χ be a complex character of G. Then $Q(\chi)$ denotes the field generated by the values $\chi(x)$, $x \in G$. If m is a positive integer we denote by Q_m the field of m-th roots of 1 over Q. We define $f(\chi)$ to be the smallest positive integer m such that $Q(\chi) \subseteq Q_m$. Then $f(\chi)$ is a positive integer, $f(\chi) \not\equiv 2 \pmod 4$ and $f(\chi)$ divides the exponent of G.

Brauer [2, Theorem 2] proved the following theorem.

Theorem 1. *Let* χ *be an irreducible character of* G. *Let* $\Omega = Q(\chi)$ *and let* Γ *be the Galois group of* Ω *over* Q. *If* $\theta_1, \theta_2, \ldots, \theta_n$ *is a system of* n *elements of* Γ, *we have one of the following two cases.*

(1) *There exists an element* $g \in G$ *such that* $\theta_i \circ \chi(g) \neq \chi(g)$ *for* $i = 1, \ldots, n$.

(2) *There exists a product* $\theta_\alpha \theta_\beta \ldots \theta_\rho$ *with* $1 \leq \alpha < \beta < \ldots < \rho \leq n$ *with an odd number of factors which leaves* χ *invariant.*

It is a direct consequence of Theorem 1 that if $p^a q^b | f(\chi)$ where $p^a q^b$ is the product of two powers of primes, then G has an element of order $p^a q^b$. Presumably, this led Brauer to pose Problem A in [3, Problem 41].

Problem A. *Let* p,q,r *be three distinct primes and suppose that* G *(is a group) and has an irreducible character* χ *for which there exist elements* α, β, γ *of* G *such that* $\chi(\alpha)$, $\chi(\beta)$, $\chi(\gamma)$ *are irrational and* $\chi(\alpha) \in Q(\sqrt{\pm qr})$, $\chi(\beta) \in Q(\sqrt{\pm pr})$, $\chi(\gamma) \in Q(\sqrt{\pm pq})$. *Does* G *contain elements of order* pqr?

The techniques used in the proof of Theorem 1 above show that if $p^a q^b r^c | f(\chi)$, r^c a power of a prime, and G has no element of order $p^a q^b r^c$, the certain field extensions are quadratic and $p^a q^b r^c$ is not divisible by the square of an odd prime (see [7, Proposition 1], for a more precise statement). Hence, a slight generalization of Brauer's is the following:

Problem B. *If* G *is a group,* χ *an irreducible character of* G, $m = p^a q^b r^c$ *is the product of at most three powers of primes, and* $m | f(\chi)$, *then does* G *have an element of order* m?

Walter Feit generalized Problem B in [4, p.41] by removing the condition on the number of primes dividing m. In particular he posed the following:

Problem C. *Let* χ *be an irreducible complex character of a finite group, then does* G *contain an element of order* $f(\chi)$?

Gow [10] proved the answer to Problem C is yes if G has odd order. Amit and Chillag [1] generalized this to the case that G is solvable.

A. Turull and the author [5] showed the answer is also yes if G is π-solvable where π is the set of primes dividing $\chi(1)$. We have also obtained affirmative answers to Problems A-C in more general situations (see [6] and [7]). Our results will be briefly discussed.

In the case that G is π-solvable for π the set of primes dividing $\chi(1)$, the proof of the affirmative answer to Problem C is based on the following idea. Let $f(\chi) = \pi_{i=1}^{n} p_i^{a_i}$ where the p_i are distinct primes. Let $p_i^{r_i}$ be the order of a Sylow p_i-subgroup of G for $i = 1,\ldots,n$. Let Ω denote the field of $|G|$-th roots of unity over Q and Ω_i the subfield of $(|G|/p_i^{r_i-a_i+1})$st roots of unity over Q. Let θ_i be a generator of $\mathrm{Gal}(\Omega/\Omega_i)$, the Galois group of Ω over Ω_i, if p_i is odd and $\theta_i \in \mathrm{Gal}(\Omega/\Omega_i)$ such that $\theta_i \circ \chi \neq \chi$ if $p_i = 2$. Let $S = \{\theta_1,\ldots,\theta_n\}$. It follows from Theorem 1 that G contains an element of order $f(\chi)$ if no product β of an odd number of distinct elements in S leaves χ invariant. Using unique factorizations of χ into appropriate products of characters due to Gajendragadkar [9] and Isaacs [11], we were able to show no such β existed.

In [6], we wished to generalize the ideas described in the previous paragraph. Hence, suitable "unique" factorizations of χ into a product of characters were necessary. In [8], prime characters were defined and it was shown ([8, Theorem A]) that any quasi-primitive irreducible character could be written essentially uniquely as a product of an admissible set of prime characters in a Schur extension of G. By combining this with a condition on some characters ψ of central extensions of non-abelian simple groups involved in G with $\psi(1)|\chi(1)$, an affirmative answer was obtained to Feit's Problem C for a larger class of groups (see [6, Theorem 2.4]). The previously known affirmative answers to Feit's question follow from [6, Theorem 2.4]. Also by applying this theorem, the following are obtained in [6].

Theorem. *Let* G *be a finite group all of whose non-abelian simple composition factors are minimal simple groups. Let* χ *be any irreducible character of* G. *Then* G *contains an element of order* $f(\chi)$.

Theorem.. *Let* G *be a finite group all of whose composition factors are N-groups. Suppose that* χ *is irreducible and* m *is some positive integer such that* $m|f(\chi)$. *Assume at least one of the following holds:*

(i) $36 \nmid \chi(1)$;

(ii) A_7 *is not involved in* G, *(as a composition factor if* χ *is quasi-primitive;*

(iii) $2 \nmid m$;

(iv) $3 \nmid m$;

(v) $5 \nmid m$;

(vi) $G' = G$ *and* χ *is a quasi-primitive.*

Then G *contains an element of order* m.

In [7], we addressed Problems A and B and were able to attain some affirmative answers. We reduced Problem B for $\chi(1)$ odd to the case where G is a perfect central extension of a simple group, i.e. we obtained the following:

Theorem. *Assume* χ *is an irreducible character of odd degree of the group* G *and* $m = \pi_{i=1}^{3} p_i^{c_i}$ *divides* $f(\chi)$ *where the* p_i *are distinct primes and* $(p_1p_2,2) = 1$. *Assume that whenever* $\psi \in Irr(S)$ *with* $\psi(1)|\chi(1)$ *and* S *is a perfect central extension of a non-abelian simple group involved in* G *with* $p_1p_2p_3^d|f(\psi)$ *where* $d = 1$ *if* p_3 *is odd and* $d \geq 2$ *if* $p_3 = 2$, *then* S *contains an element of order* $p_1p_2p_3^d$. *Then* G *contains an element of order* m.

Likewise we reduce Problem B for $\chi(1)$ arbitrary, χ quasi-primitive, G perfect and p,q,r odd to the case where G is a perfect central extension of a simple group (see [7, Theorem 4]). We investigated various families of simple groups and showed that the answer to Problem B is yes for every one of their perfect central extension. In fact we do not know of an example where the answer to Problem B is no. Here is an example of what we obtained.

Theorem. *Assume* χ *is an irreducible character of odd degree of* G *and* $\chi_{|M}$ *is homogeneous whenever* M *is a characteristic subgroup of* G. *Let* $m = \pi_{i=1}^{3} p_i^{c_i}$ *where the* p_i *are distinct primes and* $m|f(\chi)$. *If every non-abelian composition factor of* G *is an* N*-group, a sporadic group or* A_n *for* $n \geq 5$, *then* G *contains an element of order* m.

REFERENCES

1. G. Amit & D. Chillag, On a Question of Feit, *Proc. Amer. Math. Soc.* (to appear).
2. R. Brauer, A note on theorems of Burnside and Blichfeldt, *Proc. Amer. Math. Soc.* 15 (1964), 31-34.
3. R. Brauer, *Representations of Finite Groups*, Lectures on Modern Mathematics, Vol. I, edited by T. Saaty, Wiley & Sons (1963).
4. W. Feit, *Characters of Finite Groups*, W.A. Benjamin, New York (1967).
5. P. Ferguson and A. Turull, On a Question of Feit, *Proc. Amer. Math. Soc.* (to appear).
6. P. Ferguson, Factorizations of characters and a question of Feit, *J. Algebra* (submitted).
7. P. Ferguson, On a Question of Brauer, *Arkiv der Math.* (submitted).
8. P. Ferguson, Prime characters and factorizations of quasi-primitive characters, *Math. Zeit.* (submitted).
9. D. Gajendragadkar, A characteristic class of characters of finite π-separable groups, *J. Algebra* 59 (1979), 237-259.
10. R. Gow, Character values of groups of odd order and a question of Feit, *J. Algebra* 68 (1981), 75-78.
11. I.M. Isaacs, Primitive characters, normal subgroups, and M-groups, *Math. Z.* 177 (1981), 267-284.

THE PICARD GROUP AND THE MODULAR GROUP

Benjamin Fine
University of California, Santa Barbara, CA 93106, USA and
Fairfield University, Fairfield, CT 06430, USA

1. The *classical modular group* M is $PSL_2(\mathbb{Z})$ the 2×2 projective special linear group with integral entries. The modular group has been one of the most extensively studied single groups (see references). The reason for this is that M arises in so many different contexts - number theory, group theory, automorphic function theory, Riemann surfaces and elsewhere. Each of these disciplines claims the modular group for its own and looks at it in a slightly different manner. Related to M is its cousin the *Picard group* Γ which is $PSL_2(\mathbb{Z}[i])$, the projective special linear group with Gaussian integer entries. Γ has also been extensively studied (references) although the work more recent. Group theoretically Γ is quite similar to M [10]. However Γ and M differ greatly in their action on the complex plane. Whereas M is a Fuchsian group, Γ is nowhere discontinuous in $\mathbb{C}$ and therefore has no Fuchsian subgroups of finite index [23].

What we will do in this survey is compare and contrast Γ and M in four different areas - group theoretical structure, general subgroup structure, congruence subgroups and Fuchsian subgroups. As a general rule of thumb group theoretical properties of M will have close analogs in Γ. As might be expected given the number theoretical similarities of $\mathbb{Z}$ and $\mathbb{Z}[i]$, the closer the group theoretical property reflects the underlying number theory (i.e., congruence subgroups) the closer the analog. Properties involving discontinuity however must be considerably revised.

A bit of history. The structure of the modular group was known to Fricke-Klein in the 1890's [22]. The algebraic properties which we will be stressing were developed in large part in the 1960's. For this we rely heavily on the book of Morris Newman [32] who is also responsible for a large chunk of these results. The Picard group was mentioned in Fricke-Klein where a presentation was derived. However the major work on it has been done since 1970. Many of the results that we quote on the subgroup lattice and the number theory have not yet appeared and proofs will be published elsewhere.

2. *GROUP THEORETICAL STRUCTURE*

First some notation and conventions. We will refer to the modular group as M (this is not standard). Since $PSL_2(\mathbb{Z}) = SL_2(\mathbb{Z})/\{I,-I\}$ the elements of M can be considered as $\pm\begin{pmatrix} a & b \\ c & d \end{pmatrix}$ with $ad - bc = 1$, $a,b,c,d \in \mathbb{Z}$. Elements of M can also be considered as linear fractional transformations $z' = az + b/cz + d$, $ad - bc = 1$. We will use either interpretation depending on which is most convenient. The Picard group will be referred to as Γ. As in M the elements of Γ will be considered either as $\pm\begin{pmatrix} \alpha & \beta \\ \gamma & \delta \end{pmatrix}$ with $\alpha\delta - \beta\gamma = 1$ and $\alpha,\beta,\gamma,\delta \in \mathbb{Z}[i]$ or as linear fractional transformations with Gaussian integer coefficients. However since the matrix $\begin{pmatrix} i & 0 \\ 0 & -i \end{pmatrix} \in SL_2\mathbb{Z}[i]$ the transformation $z' = -z$ is in Γ and we must allow the linear fractional maps in Γ to have determinant ± 1 - that is $z' = \frac{\alpha z + \beta}{\gamma z + \beta}$ with $\alpha\delta - \beta\gamma = \pm 1$ and $\alpha, \beta, \gamma, \delta \in \mathbb{Z}[i]$.

For later reference we identify the following standard transformations

$$a : z' = -\frac{1}{z}, \quad t : z' = z + 1, \quad u : z' = z + i, \quad \ell : z' = -z.$$

Notice all four are in Γ and a, t are also in M. We also define in M

$$x = a;\ z' = -\frac{1}{z}, \quad y = at : z' = -\frac{1}{z+1}.$$

As a first step in obtaining the structure of these groups we need presentations. The standard presentation for M was given in Fricke-Klein [22]. (Magnus [28] mentions that its structure was in a sense known to Gauss.) Specifically a, t above generate M and the only relations are $a^2 = (at)^3 = 1$. By simple Tietze transformations, M is also generated by x, y with relations $x^2 = y^3 = 1$. That is:

Theorem M1. $M = \langle a,t;\ a^2 = (at)^3 = 1\rangle = \langle x,y;\ x^2 = y^3 = 1\rangle$.

From the above it is obvious that M is a free product of a cyclic group of order 2 and a cyclic group of order 3. Much of M's algebraic nature follows from this fact.

Theorem M2. $M \cong \langle x\rangle * \langle y\rangle \cong \mathbb{Z}_2 * \mathbb{Z}_3$.

There are several different proofs of the above theorem reflecting the discipline from which one is looking at M. A very direct matrix proof is given in [32] and in [27]. A second proof involves constructing a fundamental domain for M in the upper half-plane and then deducing a presentation of M from standard Fuchsian group techniques (see [23] or [28] for a description). Finally the structure of M can be determined from an algebraic method of P.M. Cohn on $SL_2(R)$ for discretely normed rings [5]. We will say more about this later.

From Theorem M2 we obtain several immediate and well-known facts about M. Since elements of finite order in free products are conjugates to elements in the factors we get only 1 conjugacy class in order 2 (namely (x)) and 2 classes in order 3 ((y) and (y^2)). Further since the normal closures of {x} and {y} have finite index then a normal subgroup with torsion must have finite index. Finally as a direct result of the Kurosh subgroup theorem we get that torsion-free subgroups must actually be free and further since there are torsion-free subgroups of finite index (M' for example) then there are free subgroups of finite index. We compile these:

Theorem M3. (1) *There is* 1 *conjugacy class in* M *in order* 2 *and* 2 *conjugacy classes in order* 3.

(2) *Every normal subgroup with torsion has finite index - in fact the index is* 1, 2 *or* 3.

(3) *Every torsion-free subgroup is free.*

(4) *There are free subgroups of finite index.*

Recall that a group G is SQ-universal if every countable group can be embedded as a subgroup of a quotient of G. As a corollary of (4) we get that M is SQ-universal (although there are more direct ways). We will see that Γ is also SQ-universal.

(5) M *is SQ-universal.*

Now let us look at Γ. A presentation for Γ was given in Fricke-Klein [22]. In particular:

Theorem P1 ([22]).

$$\Gamma = \langle a,\ell,t,u;\ a^2 = \ell^2 = (a\ell)^2 = (t\ell)^2 = (u\ell)^2 = (ua\ell)^3 = (at)^3 = [t,u] = 1\rangle .$$

Somewhat later Sansone ([36], [28]) using a geometric method of Bianchi derived another presentation:

Theorem P'1 ([28]). $\Gamma = \langle A,B,C,D;\ A^3 = B^2 = C^3 = D^2 = (AC)^2 = (AD)^2 = (BD)^2 = (BC)^2 = 1\rangle$ *where here*

$$A : z' = -\frac{1}{z+1},\quad B : z' = -\frac{1}{z},\quad C : z' = \frac{1}{z+i},\quad D : z' = \frac{1}{z}.$$

(This is actually a modification of what Sansone actually gave.)

From Sansone's presentation it was observed by Fine [8], and independently by Karrass-Solitar [21], that Γ is a free product with amalgamated subgroup with particularly simple factors. This was rederived more recently by Flöge [17].

Theorem P2. *Γ is a free product with amalgamation $\Gamma \cong G_1 *_M G_2$ where*

$$G_1 \cong S_3 \underset{\mathbb{Z}_3}{*} A_4,\quad G_2 \cong S_3 \underset{\mathbb{Z}_2}{*} D_2$$

and the amalgamated subgroup is the modular group M.

Somewhat later a different type of product structure for Γ was given by Brunner, Lee and Wielenberg [4]. This is a polygonal product which was introduced by Karrass, Pietrowski and Solitar [21]. (See the latter for terminology). In particular Γ is a *quadrangular product of spherical triangle groups*. That is Γ can be described by the picture

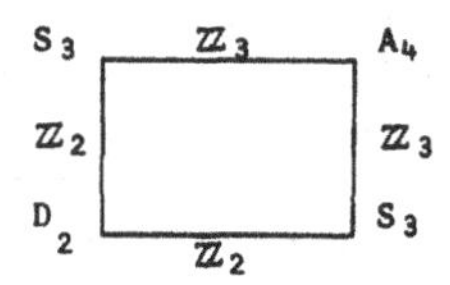

where the corner vertex groups are amalgamated via the groups indicated on the edges.

Before we use these decompositions some more history is in order. If n is a positive square-free integer and $\mathcal{O}_n$ are the integers in $Q(\sqrt{-n})$ the groups $\Gamma_n = PSL_2(\mathcal{O}_n)$ have been termed the *Bianchi groups*. In this context $\Gamma = \Gamma_1$. Although Fricke gave a presentation for Γ_1, this seems to be the only Bianchi group he does this for. Bianchi [2] found geometric regions for these groups. Real interest in these groups seems to have grown after Serre's work on groups acting on trees indicated that all these should have some sort of amalgam structure. Further the Bianchi groups are (together with $\mathbb{Z}$) the sole exceptions to the solution of the congruence subgroup problem for SL(2,R), R an algebraic number field. Swan [38] developed a comprehensive geometric method based on Bianchi's work to obtain presentations for Γ_n. About the same time P.M. Cohn developed algebraic methods to find presentations for Γ_n in the case where $\mathcal{O}_n$ is Euclidean [6], n = 1,2,3,7,11. Fine ([8], [9]) reduced these to finite presentations and showed that except for n = 3 they all were free products with amalgamations or HNN groups. This was also true for non-Euclidean $\mathcal{O}_n$'s for which explicit presentations were worked out. Flöge [17] extended some of these while Zimmert [40] showed that for large n's, Γ_n must have free quotients. Lubotzky [25], and Grunewald-Schwermer [18] proved that all have subgroups of finite index which have nonabelian free quotients.

Now how is Theorem M3 reflected on Γ? Since Γ is a free product with amalgamation its torsion elements are conjugate to torsion-elements in the factors. The factors are in turn free products with amalgamation of finite groups. From this we get:

Theorem P3 ([10]). *There are* 7 *conjugacy classes of torsion elements in* Γ - *the identity,* - 4 *in order* 2 *and* 2 *in order* 3. *Explicitly they are* $\{a\}$, $\{\ell\}$, $\{a\ell\}$, $\{ut\ell\}$, $\{at\}$, $\{ua\ell\}$.

By examining the normal closures of these elements we find that a normal subgroup of Γ with torsion must have finite index. A different method of Karrass-Pietrowski and Solitar using the polygonal product decomposition gives us that normal subgroups of infinite index must be infinitely generated. Then:

Theorem P4. (a) [10] *If* $N \triangleleft \Gamma$ *and* N *has torsion then* $|\Gamma:N|$ *is finite. In fact* $|\Gamma:N|$ *divides* 2, 4, 12 *or* 24.
(b) [20] *If* $N \triangleleft \Gamma$ *and* N *is f.g. then* $|\Gamma:N| < \infty$.

Results on free subgroups (Theorem M3 - (3), (4)) do not have direct analogs in Γ. In fact while Γ certainly has many free subgroups none are of finite index.

Theorem P5 ([15]). Γ *has no free subgroups of finite index. Every subgroup of finite index contains a free abelian subgroup of rank* 2.

However the torsion-free subgroups of Γ are in a sense almost free.

Theorem P6 ([15]). *If* G *is torsion-free in* Γ *then* G *is an* HNN *group whose base is a tree product with amalgamation of free groups.*

Brunner, Frame, Lee and Wielenberg [3] using the polygonal product structure of Γ have devised a method to classify the torsion-free subgroups of finite index. These subgroups correspond to fundamental groups of hyperbolic 3-manifolds. They carry out the classification for indexes up to 24. It was observed by Solitar [3] that torsion-free subgroups of Γ must have index divisible by 12. This corresponds to a result in M that says any torsion-free subgroup has index divisible by 6.

Theorem P7, M4 ([32], [3]). (1) *If* $G \subset \Gamma$, G *torsion-free then* $12 \mid |\Gamma:G|$.
(2) *If* $G \subset M$, G *torsion-free then* $6 \mid |M:G|$.

Finally Tretkoff-Fine [14] showed that Γ is SQ-universal by examining the structure of certain subgroups of finite index. A somewhat simpler proof of this is afforded by one of the subgroups of finite index classified by Brunner et.al. The *group of the Boromean rings* (see [3]) is represented faithfully by a subgroup of index 24 in Γ. This group has a free quotient of rank 2 and thus Γ is SQ-universal. These SQ-universal results have also been tremendously extended ([18], [25]).

3. *GENERAL SUBGROUP STRUCTURE*

In this section we will look at the structure of several classes of subgroups in M and Γ. In particular the abelian subgroups, the derived series and the series of powers (the series of subgroups G^n generated by n^{th} powers).

For the first two cases the situation in M is relatively simple. Abelian subgroups of free products are conjugate to subgroups of the factors. Since the factors are cyclic - all abelian subgroups are cyclic. (This also follows from general results on Fuchsian groups - again indicating that results on M can be arrived at from several different directions.) A straightforward computation shows that M' has index 6 and is free of rank 2. Thus the whole derived series consists of free groups and M'' (and therefore $M^{(n)}$ for all $n \geq 2$) has infinite index. So

Theorem M5 ([32]). (1) *An abelian subgroup of* M *is cyclic.*
(2) $|M:M'| = 6$ *and* M' *is free of rank* 2.
(3) $|M:M^{(n)}| = \infty \; \forall n > 1$ *and* $M^{(n)}$ *is free.*

The series of powers $\{M^n\}$ is rather direct if $6 \nmid n$. M^2 is the normal closure of y, M^3 is the normal closure of x and thus $|M:M^2| = 2$, $|M:M^3| = 3$. If n is even $6 \nmid n$ then $M^n = M^2$. If n is odd, $n \not\equiv 0 \pmod 3$ then $M^n = M$ and finally if n is odd $n \equiv 0 \pmod 3$ then $M^n = M^3$. If $6 \mid n$ the situation is much more complicated. M^6 has index 216 in M and is free of rank 37 [33]. In general the structure of M^{6n} is not known, precisely for $n > 1$, however for sufficiently large n, M^{6n} has infinite index [32].

There is a further nice connection between the series of powers and the derived series - that is $M^2 \cap M^3 = M'$. This will carry over to Γ.

Theorem M6. (1) $|M:M^2| = 2$, $|M:M^3| \equiv 3$.
(2) *If* $6 \nmid n$, $M^n = M$, M^2 *or* M^3 *depending on* $2 \nmid n$, $3 \nmid n$ *or* $2 \mid n$ *or* $2 \nmid n$, $3 \mid n$.
(3) $|M:M^6| = 216$, M^6 *is free of rank* 37.
(4) $M^2 \cap M^3 = M'$.

The situation in Γ is very similar. Γ is a free product with amalgamation so its abelian subgroups can be studied through this construction. However Γ is also a discrete (though not discontinuous) subgroup of $PSL_2(\mathbb{C})$ so its abelian subgroups are either cyclic, a Klein 4-group (D_2) or free abelian of rank 2. These all appear in Γ so they give a complete classification of its abelian subgroups. The same analysis can be arrived at using Moldavanskii's lemma on free products with amalgamation [26].

Theorem P8 ([15]). *An abelian subgroup of* Γ *is either cyclic* ($\mathbb{Z}_2, \mathbb{Z}_3, \mathbb{Z}$), *a Klein 4-group or free abelian of rank* 2.

The derived series in Γ is similar but more complicated than in M. The structure of the derived series is such that each term in the series is very symmetric. In particular:

Theorem P9 ([15]). (1) Γ' *has index* 4 *in* Γ. Γ' *is the free product with amalgamation*

$$\Gamma' \cong H_1 \underset{H}{*} H_2 \text{ with } H_1 \cong H_2 \text{ (and the amalgamation via this isomorphism)}$$

$$H_1 \cong H_2 \cong A_4 \underset{\mathbb{Z}_2}{*} A_4 \qquad H \cong \mathbb{Z}_3 * \mathbb{Z}_3.$$

(2) Γ'' *has index* 12 *(index* 3 *in* Γ'*) and*

$$\Gamma'' \cong (D_2 * D_2) \underset{H}{*} (D_2 * D_2) \text{ with } H \cong \mathbb{Z} * \mathbb{Z}.$$

(3) Γ''' *has index* 768 *(index* 64 *in* Γ''*).* Γ''' *is an* HNN *group. The base is a tree product with amalgamation of free groups each of rank* 9.

(4) $\Gamma^{(iv)}$ *and thus all higher derived groups have infinite index and all are torsion-free.*

From (4) we have that all of the derived series from Γ''' on are generalized free products of free groups. Whether they are actually free from some point on is still unknown.

The series of powers in Γ is even closer to that of M. By direct computation $\Gamma^2 = \Gamma'$ so $|\Gamma:\Gamma^2| = 4$. Further Γ^2 is the normal closure of $\{at, a\ell\}$. The cube powers generate Γ - that is $\Gamma^3 = \Gamma$. If n is odd, Γ^n is also Γ while if n is even $n \not\equiv 0 \pmod 6$ then $\Gamma^n = \Gamma^2$. Here again as in the modular group the multiples of 6 hold a special position. As of now for all $n > 1$ both the index $|\Gamma:\Gamma^{6n}|$ and the structure of Γ^{6n} are unknown.

Finally since $\Gamma^2 = \Gamma'$ and $\Gamma^3 = \Gamma$ we have $\Gamma^2 \cap \Gamma^3 = \Gamma'$ exactly equal to the relation in M. Summarizing all the above.

Theorem P10 ([15]). (1) $\Gamma^2 = \Gamma'$, $|\Gamma:\Gamma^2| = 4$, $\Gamma^2 = N(at, a\ell)$.

(2) $\Gamma^3 = \Gamma$, $\Gamma^n = \Gamma$ *if* $2 \nmid n$.

(3) $\Gamma^n = \Gamma^2$ *if* $2 | n$, $6 \nmid n$.

(4) $\Gamma^2 \cap \Gamma^3 = \Gamma'$.

4. *CONGRUENCE SUBGROUP LATTICE*

Number theoretic interest in these groups center for the most part on the congruence subgroups. If $n \in \mathbb{Z}$ the *principal congruence subgroup modulo* n in the modular group M is the normal subgroup consisting of those matrices $A \in SL_2(\mathbb{Z})$ for which $A \equiv \pm I \pmod n$ elementwise. We denote this by M(n). Equivalently M(n) is the kernel of the natural map $PSL_2(\mathbb{Z}) \to PSL_2(\mathbb{Z}/n)$. Finally M(n) can be said to consist of those linear fractional transformations $z' = az + b/cz + d$ with $a \equiv d \equiv \pm 1 \bmod n$, $b \equiv c \equiv 0 \bmod n$. (This last definition does not carry over to the Picard group so we will rely on the matrix definition. More on this later.)

Since $\mathbb{Z}/(n)$ is finite this implies that the principal congruence subgroups of all have finite index. A *congruence subgroup* of M is any subgroup of M which contains a principal congruence subgroup. For a general ring

R the *congruence subgroup property* is that every subgroup of finite index in SL(n,R) or PSL(n,R) is a congruence subgroup. Mennicke [30] proved this to be true for $SL(n,\mathbb{Z})$, $n \geq 3$ and Bass, Milnor and Serre [1] gave a positive solution to the congruence subgroup problem in a wide range of cases. However the congruence subgroup property is false in M and in the Bianchi groups in general. First we discuss the structure of the principal congruence subgroups.

Theorem M7 ([23]). *For all* $n \in \mathbb{Z}$, $M(n)$ *is a free group.*

There are two separate proofs of this. We will briefly discuss each of these since both carry over in a slightly different way to Γ.

First of all a trace argument shows that if $n > 2$ $M(n)$ must be torsion-free. Since torsion elements in M have trace 0 or 1, if $g = \begin{pmatrix} a & b \\ c & d \end{pmatrix} \in M(n)$ then $a + d \equiv 2 \pmod n$ and $a + d = 0$ or 1. But $2 \not\equiv 0, 1 \pmod n$ if $n \neq 2$. A little more work shows that M(2) is also torsion-free. Then M(n) is free by the Kurosh subgroup theorem.

A separate argument goes as follows. M(n) being a subgroup of M is also a Fuchsian group. The transformation $z' = z + n$ is in M(n) so M(n) contains parabolic maps. But torsion-free Fuchsian groups with parabolic maps are free [23].

In order to discuss the congruence subgroup property in M we need a theorem of Wohlfahrt. For any subgroup G of M we define the *level of* G as the least positive integer n such that G contains the normal closure of the parabolic element $\{t^n\}$.

For normal subgroups this is equivalent to being least positive integer n such that $t^n \in G$. Since $t^n \in M(n)$ and $t^k \notin M(n)$ is $k < n$ the principal congruence subgroup mod n also has level n. Wohlfahrt's theorem gives a criteria for determining a congruence subgroup.

Theorem M8 (Wohlfahrt [32]). *Let* G *be a subgroup of* M *of level* n. *Then* G *is a congruence subgroup if and only if* $G \supset M(n)$.

Using Wohlfahrt's theorem matrix examples of non-congruence subgroups can be given. In particular if M(2) is the principal congruence subgroup mod 2 then M(2) is free of rank 2. If A,B are the generators of M(2) let e(W;A), e(W;B) represent the exponent sum of any word W(A,B) in A and B. For any positive integer n let $M_n(2)$ be the subgroup of M(2) consisting of all words, W(A,B), with $e(W;A) \equiv e(W;B) \equiv 0 \pmod n$. Then $M_n(2)$ has finite index in M and an application of Wohlfahrt's theorem shows that $M_n(2)$ is not a congruence subgroup. The argument for $M_n(2)$ can be extended to any free subgroup of M.

Theorem M9 ([32]). *Let* G *be a free subgroup of* M *of level* n *and rank* r *with generators* $g_1,\ldots,g_r$. *Let* G_m *be the normal subgroup of* G *consisting of those words* $W(g_1,\ldots,g_r)$ *with* $e(W;g_1) \equiv e(W;g_2) \equiv 0 \bmod m$. *Then if* m *is divisible by a prime* p *greater than* $n+1$, G_m *is not a congruence subgroup.*

A final result gives some divisibility properties of the congruence subgroups. In M let $M(d)^*$ be the matrices in $PSL_2(\mathbb{Z})$ congruent to a scalar matrix modulo d. Thus $M(d) \subset M(d)^*$. Then:

Theorem M10 ([34]). *Suppose* G *is normal in* M. *If* $M(d)^* \subset G \subset M(n)^*$ *then* $G = M(kd)^*$, *for some* $k \in \mathbb{Z}$.

This has been shown to hold over many principal ideal domains.

The situation for congruence subgroups in Γ is quite similar. If α is a Gaussian integer we let $\Gamma(\alpha)$ be the principal congruence subgroup modulo the ideal (α). A trace argument shows that if $(\alpha) \neq (1+i)$ or (2) then $\Gamma(\alpha)$ is torsion-free. A computation using a result from Newman [32] shows that $|\Gamma:\Gamma(2)| = 48$ while theorem P6 shows that for normal subgroups with torsion $|\Gamma:N| \mid 24$. Therefore $\Gamma(2)$ is also torsion-free. Therefore $\Gamma(\alpha)$ is torsion-free for all $(\alpha) \neq (1+i)$. From Theorem P5 each $\Gamma(\alpha)$ must then be an HNN group whose base is a tree product of free groups. Since $|\Gamma:\Gamma(\alpha)|$ is finite for all α, we have, using results of Karrass-Solitar [19].

Theorem P11 ([11]). *If* $(\alpha) \neq (1+i)$, $\Gamma(\alpha)$ *is an HNN group with finitely generated free part and whose base is a tree product with amalgamation of finitely many free groups each of finite rank.*

Since the amalgamated subgroup in the generalized free product structure on Γ is M we have the following corollary.

Corollary. *In* $\Gamma(\alpha)$ *both the amalgamated subgroups in the base and the associated subgroups are conjugates of* $\Gamma(\alpha) \cap M$. ($M(\alpha)$ if $\alpha \in \mathbb{Z}$.)

The structure of the special case $\Gamma(1+i)$ can be determined by the Reidemeister-Shreier method.

Theorem P12 ([11]). $\Gamma(1+i)$ *has index* 6 *in* Γ. $\Gamma(1+i)$ *is a free product with amalgamation*

$$\Gamma(1+i) = H_1 \underset{U}{*} H_2.$$

The factors are

$$H_1 \cong D_2 \underset{\mathbb{Z}}{*} [\mathbb{Z}_2 * \mathbb{Z}_2] \underset{\mathbb{Z}}{*} [\mathbb{Z}_2 * \mathbb{Z}_2], \quad H_2 \cong \mathbb{Z}_2 * D_2$$

with amalgamated subgroup $U \cong \mathbb{Z} * \mathbb{Z}_2$.

A second class of congruence like subgroups in Γ are the $\overline{\Gamma(\alpha)}$ which we define as the group of linear transformations in Γ congruent to the identity transformation mod(α). In M, $\overline{M(n)} = M(n)$ for all integers n, however in Γ this is no longer the case. In particular $\Gamma(2) \neq \overline{\Gamma(2)}$. To see this notice that the transformation $\ell: z' = -z$ is congruent to the identity mod(2). However this transformation is the image of the matrix $\begin{pmatrix} i & 0 \\ 0 & -i \end{pmatrix}$ which is not congruent to the identity matrix mod(2). The subgroup $\overline{\Gamma(2)}$ is an interesting subgroup. Since $\ell \in \overline{\Gamma(2)}$ this subgroup has torsion. It can be shown [15] that $\overline{\Gamma(2)}$ is precisely $N(\ell)$, the normal closure of ℓ in Γ. Further this is the only normal subgroup with torsion of index 24 in Γ. A result in [15] shows that there are only 3 normal subgroups of index 24. We have:

Theorem P13 ([15]). (1) $\overline{\Gamma(2)}$ *has index* 24 *in* Γ. $\overline{\Gamma(2)} = N(\ell)$ *and* $\overline{\Gamma(2)}$ *has a complicated free product with amalgamation structure involving free products* *of Klein 4-groups.*

(2) $\overline{\Gamma(2)}$ *is the only normal subgroup with torsion of index* 24 *in* Γ

The decomposition of $\overline{\Gamma(2)}$ mentioned in (1) was used in [8] to obtain a subgroup of $\overline{\Gamma(2)}$ which provides a faithful representation of a surface group of genus 2.

Drillick [7] gave a technique to produce non-congruence subgroups of Γ. His method involved mapping Γ onto A_n. In the proof he shows that the only non-abelian groups that can appear in a composition series for $PSL_2(\mathbb{Z}[i]/(\alpha))$ are $PSL_2(\mathbb{Z}[i]/(\pi))$ where (π) is a prime in $\mathbb{Z}[i]$. He then shows that A_n is a quotient of Γ for infinitely many n's and then picks one which is isomorphic to no $PSL_2\mathbb{Z}[i]/(\pi)$. His method is patterned after an identical method for the modular group [28].

A general result by Lubotzky [25] gives examples of non-congruence subgroups in all the Bianchi groups. In the same paper he gives a separate proof of the result of Grunewald-Schwermer [18] that each of the Bianchi groups has subgroups of finite index which have non-abelian free quotients.

We take a slightly more matrix oriented approach to find non-congruence subgroups of Γ. First we extend Wolhfahrt's Theorem. For a subgroup G of finite index in Γ we define its *level* to be the least positive integer n such that $G \supset N(t^n,u^n)$. We then have:

Lemma. *Let* $G \subset \Gamma$ *of level* n. *If* $\alpha \in \mathbb{Z}[i]$ *and* $G \supset \Gamma(\alpha n)$ *then* $G \supset \Gamma(n)$.

Using this concept of level, Wohlfahrt's Theorem goes through for Γ unchanged.

Theorem P14 ([15]). *Let* $G \subset \Gamma$ *of level* n. *Then* G *is a congruence group if and only if* $G \supset \Gamma(n)$.

We note that the level of $\Gamma(n)$ is n if $n \in \mathbb{Z}$, while if $\alpha \notin \mathbb{Z}$, the level of $\Gamma(\alpha)$ is the smallest integer n such that $\alpha \,\big|\, |n|$.

Using this we can give matrix constructions of non-congruence subgroups on Γ. From Karrass-Solitar we know that every subroup of Γ is an HNN group [19], thus we can speak of the *rank of the free part* for any subgroup G of Γ. We then get:

Theorem P15 ([15]). *Let* $G \subseteq \Gamma$, $|\Gamma;G| < \infty$. *If rank of the free part of* $G > 2$ *then* G *contains a normal subgroup of finite index which is not a congruence subgroup.*

A formula of Karrass-Solitar determines the free part rank. If G_1,G_2 are the factors of Γ from Theorem P2, then:

Corollary P15 ([15]). *If* $|\Gamma:G| < \infty$, *and let* $d = |\Gamma:(G,M)| - |\Gamma:(G,G_1)| - |\Gamma:(G,G_2)| + 1$. *Then if* $d > 2$, G *contains a non-congruence subgroup of finite index, which is normal in* G.

Specific examples of Theorem P15 can be given.

Corollary P15 ([15]). $\Gamma(2)$ *contains a non-congruence subgroup.*

This is proven by computing the rank of the free part. It is nice in that it mirrors the modular group situation.

Many knot and link groups can be represented in Γ [3]. In particular the *group of the Boromeam rings* is faithfully represented as a torsion free normal subgroup of index 24 in Γ. Calling the image group B_1 we get:

Corollary P15 ([15]). *B_1 contains a non-congruence subgroup.*

This is the normal subgroup of smallest index which we can locate which contains a normal non-congruence subgroup in the manner constructed above.

Closely tied to the study of the congruence subgroups is the study of the general normal subgroup lattice. In M, Theorem M10 shows us that if $G \triangleleft M$ and G is wedged between the congruence-like subgroups $M(d)^*$ and $M(n)^*$ then $G = M(kd)^*$. Further results in M have led to a classification of normal subgroups by index, genus and parabolic class number. Several of these results are summarized in Theorem M11.

Theorem M11. (1) [32] *If $N(t^n)$ is the normal closure of the parabolic elements $\{t^n\}$ then $M(n) = N(t^n)$ if and only if $1 \leq n \leq 5$.*
(2) *For any $g \neq 1$ there are only finitely many normal subgroups of genus g (as Fuchsian groups) and finite index. For $g = 1$ there are infinitely many normal subgroups of M of genus 1 and of finite index.*
(3) [35] *Let $D(u)$ be the number of normal subgroups of index u in M. Then $D(u)$ becomes arbitrarily large and $D(u) = 0$ if $u = 12p$ with p a prime greater than 11.*

Thus from (3) there are gaps in the possible indices of normal subgroups of M - a phenomena even more pronounced in Γ.

Newman and Fine [15] investigated the corresponding normal subgroup lattice of the Picard group. This led to a complete classification for small indices together with results showing large gaps in the possible indices. The methods employed the use of principal series for possible quotients. Specifically:

Theorem P16 ([15]). (1) *There are exactly 3 normal subgroups of index 2 - $N(t)$, $N(u)$, $N(t^2,u^2,at,tu)$.*
(2) *Γ' is the only normal subgroup of index 4.*
(3) *$\Gamma(1+i)$ is the only normal subgroup of index 6.*
(4) *Γ'' is the only normal subgroup of index 12.*
(5) *There are 3 normal subgroups of index 24 and 6 of index 48. The above are the only normal subgroups of index ≤ 48.*

Theorem P17 ([15]). *If $\alpha \in Z[i]$ with $\alpha = a + bi$ then if $|\alpha| \leq 15$ we have $\Gamma(\alpha) = N(t^{|\alpha|}, u^{|\alpha|})$ if $a \in Z$ $(b = 0)$ and $\Gamma(\alpha) = N(t^{|\alpha|}, t^a u^{-b})$ if $b \neq 0$.*

Theorem P18 ([15]). (1) *There are no normal subgroups of index $12p^k$ where p is a prime $p \neq 2,3,5,11$.*
(2) *There are no normal subgroups of index 12pq if p,q are primes with $12p \not\equiv 1 \bmod q$ and $12q \not\equiv 1 \bmod p$.*

(3) *If* $G \triangleleft \Gamma$ *with* $|\Gamma:G| = 2^k 3$ *then* $G \subset \Gamma(1+i)$.

(4) *If* $G \triangleleft \Gamma$ *with* $|\Gamma:G| = 12m$ *with* $(m,6) = 1$ *and* Γ/G *solvable then* $G = \Gamma''$.

Another set of congruence like subgroups that have been fairly extensively studied in M are the groups $M_0(n) = \left\{\pm \begin{pmatrix} a & b \\ c & d \end{pmatrix} \in M,\ c \equiv 0 \bmod r\right\}$. Similarly $M^0(n)$ ([32], [24]). These groups have played a role in the study of genus properties of subgroups of M [24]. (Considered as Fuchsian groups.) The structure of the corresponding $\Gamma_0(\alpha)$ has not as yet been investigated.

5. *FUCHSIAN SUBGROUPS*

M is itself Fuchsian so every subgroup is Fuchsian. From Theorem M3 part (3) every torsion-free Fuchsian subgroup is free. Thus if ϕ_g is a Riemann surface group of genus g then there are no faithful representations of ϕ_g in M, Fuchsian or otherwise. The situation is quite different in Γ. Γ is nowhere discontinuous in $\mathbb{C}$ and therefore contains no Fuchsian subgroups of finite index [23]. On the other hand Maskit [29] and Mennicke [31] have given faithful Fuchsian representations of ϕ_g in Γ. An algebraic method to generate faithful representations of ϕ_g in Γ was given by Fine [16], but the images are non-Fuchsian. Thus a question of interest is how Fuchsian subgroups are actually embedded in Γ. In [13] it was shown that to obtain a Fuchsian subgroup F of Γ that is neither free nor a free product of cyclics, F must have a certain intersection property with the modular group. The results obtained are somewhat different if F has torsion or not. Specifically:

Theorem P19 ([13]). *Let* F *be a torsion-free Fuchsian subgroup of* Γ. *Then* F *is free unless it has at most cyclic intersection with all conjugates of* M *in* Γ *and non-trivial intersection with at least one conjugate of* M.

In particular the above theorem applies to Fuchsian representations of surface groups.

Corollary P19 ([13]). *Suppose* $F \subset \Gamma$, F *Fuchsian,* $F \cong \phi_g$ *then* F *has cyclic non-trivial intersection with some conjugate of* M *and cyclic (possibly trivial) intersection with all conjugates of* M.

The second part of the theorem follows from a lemma which is of interest in itself.

Lemma P19 ([13]). *If torsion-free Fuchsian* F *has non-cyclic intersection with some conjugate of* M *then* F *is contained in that conjugate.*

If F were Fuchsian and contained in a principal congruence subgroup it would be torsion-free. The following was conjectured in [11].

Conjecture. *If* $F \subset \Gamma(\alpha)$, *with* F *Fuchsian then* F *is free. A stronger version (without much evidence at present) would be* $F \subset \Gamma(\alpha)$, F *Fuchsian then* F *is conjugate to a subgroup of* M. (See note following references.)

The torsion situation is similar but requires some modification.

Theorem P20 ([13]). *Let* F *be a f.g. Fuchsian subgroup of* Γ. *Then* F *is either finite or a free product of cyclics if it has trivial intersection with all conjugates of* M *or non-cyclic, non-infinite dihedral intersection with some conjugate of* M.

Theorems P16 - P17 can be extended to the other Euclidean Bianchi groups [13]. Theorem P17 can be applied to representations of triangle groups $T(\ell,m,r)$ in Γ.

Corollary P20 ([13]). *Let* T *provide a faithful Fuchsian representation of a triangle group in* Γ. *If* $|T| = \infty$ *then* $|T|$ *has non-trivial intersection with some conjugate of* M *and at most cyclic intersection with all conjugates of* M.

In the modular group attention has been given to classifying normal subgroups by their genus [32]. Since a typical non-Fuchsian subgroup of Γ has no genus in the same sense an open question is how to extend this classification to Γ. Since ϕ_2 appears in Γ all genuses can be represented. However none are from normal subgroups.

Theorem P21 ([13]). Γ *contains no normal Fuchsian subgroups.*

This is a consequence of a general result [13] which shows that $PSL_2(R)$ will contain no normal Fuchsian subgroups if R is a non-real subring of $\mathbb{C}$.

Finally if C is a circle in the complex plane, the set of maps in Γ which fix C and its interior generate a Fuchsian subgroup P(C). If $P_N(C)$ is the normal closure of P(C) in Γ, the indices of $P_N(C)$ were investigated by Waldinger [39] and Fine [10]. As a corollary of Theorem P4.

Corollary P4. *If* C *and its interior are fixed by any elliptic map in* Γ *then* $|\Gamma:P_N(C)| \,\big|\, 24$.

6. *MISCELLANY*

We close by stating some questions concerning Γ which are in the same vein as the rest of the survey:

(1) *The structure of* $\Gamma_0(\alpha)$ *and* $\Gamma^0(\alpha)$ *for* $\alpha \in \mathbb{Z}[i]$.
(2) *The determination of* $|\Gamma:\Gamma^{6n}|$ *and the structure of* Γ^{6n} *(for* $n > 1$ *this is still open in* M*).*
(3) *An extension of the genus concept to* Γ *to allow a similar type of classification.*
(4) *The general relationship between* $\Gamma(n)$ *and* $N(t^n,u^n)$.

The final two questions were suggested by Lubotzky.

(5) *If* $H \subseteq M$ *of finite index does then there exist a subgroup* H' *of finite index in* Γ *such that* $H \supset H' \cap M$.

Lubotzky has a geometric proof for Γ. What is desired is an algebraic proof for Γ and a proof or counterexample for Γ_n in general. This can be reformulated in the following way. There is a short exact sequence for $\widehat{SL_2(\mathbb{Z})}$

$$1 \to \widehat{F_\omega} \to \widehat{SL_2(\mathbb{Z})} \to SL_2(\hat{\mathbb{Z}}) \to 1$$

where $\hat{G}$ stands for the pro-finite completion of G. The kernel $\hat{F}_\omega$ is a free pro-finite group. We then have a similar sequence for Γ and a diagram.

$$\begin{array}{ccccccccc} 1 & \to & F_\omega & \to & \widehat{SL_2(\mathbb{Z})} & \to & SL_2(\hat{\mathbb{Z}}) & \to & 1 \\ & & \downarrow\eta & & \downarrow & & \downarrow & & \\ 1 & \to & \ker & \to & \widehat{SL_2(\mathbb{Z}[i])} & \to & SL_2(\widehat{\mathbb{Z}[i]}) & \to & 1 \end{array}$$

Then

(1) What is the structure of the kernel above?

(2) Is the map η injective?

REFERENCES

This list of references is not intended to be complete. We list those that have been quoted in the survey. Reviews on Infinite Groups. Baumslag AMS, lists 36 papers alone in the section on congruence subgroups of the modular group - most done by Newman (and this is prior to 1970). A more complete set of references on the modular group is given in Newman's book (although this also dates prior to 1970).

1. H. Bass, J. Milnor & J.P. Serre, Solution of the congruence subgroups problem for SL_n ($n \geq 3$), SP_{2n}($n \geq 2$), *Inst. Hautes Etudes Sci. Publ. Math.* 33 (1967), 59-137.
2. L. Bianchi, Sui Gruppi de sostituzioni lineari con coefficienti appartenenti a corpi quadratici imaginari, *Math. Ann.* 40 (1892), 332-412.
3. A.M. Brunner, M. Frame, Y.W. Lee & N. Wielenberg, Classifying torsion-free subgroups of the Picard Group, *Trans. Amer. Math. Soc.* 282 (1984), 205-235.
4. A.M. Brunner, Y.W. Lee & N. Wielenberg, Polyhedral groups and graph amalgamation products (to appear).
5. P.M. Cohn, On the structure of GL_2 of a ring, *Inst. Hautes Etudes Sci. Publ. Math.* 30 (1966), 5-53.
6. P.M. Cohn, A presentation of SL_2 for Euclidean Quadratic Imaginary Number Fields, *Mathematika* 15 (1968), 156-163.
7. A. Drillick, *The Picard Group*, Ph.D. Thesis, N.Y.U. (1971).
8. B. Fine, The structure of $PSL_2(R)$, *Ann. Math. Studies* 79 (1974), 145-170.
9. B. Fine, The HNN and generalized free product structure of certain linear groups, *Bull. Amer. Math. Soc.* 81 (1975), 413-416.
10. B. Fine, Fuchsian subgroups of the Picard Group, *Canad. J. Math.* 28 (1976), 481-486.
11. B. Fine, Congruence subgroups of the Picard group, *Canad. J. Math.* 32 (1980), 1474-1481.
12. B. Fine, Groups whose torsion-free subgroups are free, *Bull. Acad. Sin.* 12 (1984), 31-36.
13. B. Fine, Fuchsian embeddings in the Bianchi Groups (to appear).
14. B. Fine & M. Tretkoff, The SQ-universality of certain arithmetically defined groups, *J. London Math. Soc.* 13 (1976), 64-68.
15. B. Fine & M. Newman, The number theory of the Picard Group (in preparation).
16. B. Fine, Representations of surface groups in the Picard group (to appear).
17. D. Flöge, Zür struktur der PSL_2 über einegen imaginar quadratischen Zahlringen, *Math. Z.* (1983), 255-279.
18. F. Grunewald & J. Schwermer, Free non-abelian quotients of SL_2 over orders of imaginary quadratic number fields, *J. Algebra* 69 (1981), 162-175.
19. A. Karrass & D. Solitar, The subgroups of a free product of two groups with an amalgamated subgroup, *Trans. Amer. Math. Soc.* 150 (1970), 227-255.
20. A. Karrass & D. Solitar, Subgroups of HNN groups with one defining relation, *Canad. J. Math.* 23 (1971), 627-643.

21. A. Karrass, A. Pietrowski & D. Solitar, The subgroups of a polygonal product (to appear).
22. F. Klein & R. Fricke, *Vorlesungen über die Theorie die Modulfunktionen*, Teubner, Leipzig (1890).
23. J. Lehner, *Discontinuous groups and automorphic functions*, Math. Surveys, No. 8, Amer. Math. Soc. (1964).
24. J. Lehner & M. Newman, Weierstrass points of $\Gamma_0(n)$, *Ann. Math.* 79 (1964), 360-368.
25. A. Lubotzky, Free quotients and the congruence kernel of SL_2, *J. Algebra* 77 (1982), 411-418.
26. R. Lyndon & P. Schupp, *Combinatorial group theory*, Springer-Verlag (1977).
27. W. Magnus, A. Karrass & D. Solitar, *Combinatorial group theory*, Wiley-Interscience, New York (1966).
28. W. Magnus, *Non-Euclidean tesselations and their groups*, Academic Press (1974).
29. B. Maskit, On a class of Kleinian groups, *Ann. Acad. Sci. Fenn. AI Math.* (1969), 1-3.
30. J. Mennicke, Finite Factor Groups of the Unimodular Group, *Ann. Math.* 81 (1965), 31-37.
31. J. Mennick, A note on regular coverings of closed orientable surfaces, *Proc. Glasgow Math. Assoc.* 5 (1969), 49-66.
32. M. Newman, *Integer Matrices*, Academic Press, New York (1972).
33. M. Newman, Normal congruence subgroups of the modular group, *Amer. J. Math.* 85 (1963), 419-427.
34. M. Newman, Free subgroups and normal subgroups of the modular group, *Illinois J. Math.* 8 (1964), 262-265.
35. M. Newman, A complete description of normal subgroups of genus one of the modular group, *Amer. J. Math.* 86 (1964), 17-24.
36. G. Sansone, I sotto gruppi del gruppo di Picard, *Rend. Arc. Mat. Palermo* 47 (1923), 273-333.
37. J.P. Serre, *Trees*, Springer-Verlag, New York (1980).
38. R.G. Swan, Generators and relations for certain special linear groups, *Adv. in Math.* 6 (1971), 1-77.
39. H. Waldinger, On the subgroups of the Picard groups, *Proc. Amer. Math. Soc.* 16 (1965), 1375-1378.
40. R. Zimmert, Zür SL_2 der ganzen Zahlen eines imaginar-quadratischen Zahlkorpers, *Inventiones Math.* 19 (1973), 73-82.

Note added in proof. Colin Maclachlan has subsequently shown that the conjecture following Lemma P19 is false.

FACTORS GROUPS OF THE LOWER CENTRAL SERIES OF FREE PRODUCTS OF FINITELY GENERATED ABELIAN GROUPS

Anthony M. Gaglione
U.S. Naval Academy, Annapolis, Maryland 21402, USA
Hermann V. Waldinger
Polytechnic University, Brooklyn, New York 11201, USA

1. *INTRODUCTION*

This paper deals with the factor groups of the lower central series of a certain class of groups to be defined below. For a free group F of finite rank, the structure of the factor groups $\overline{F}_n = F_n/F_{n+1}$ where F_n denotes the nth subgroup of the lower central series of F is well known [5,9].

The lower central series is important in several applications. For example, such quotient groups arise in the study of grammars [10] and are also important in Burnside's problem [7,8]. This being the case, it seems desirable to determine the structure of these factor groups for other classes of non-free groups.

Here we shall study the factor groups of the lower central series arising from groups G through the use of basic commutators. We shall assume that G is a free product of finitely many groups, G(i), and that every G(i) is a finitely generated abelian group. As such this paper continues and generalizes the work of Dark [2], Struik [11,12] and the authors [15,16,17]. We will give an algorithm for finding bases of the factor groups of the lower central series of our groups, G, thus extending the results of Theorem 2.1 of [3]. This algorithm is based on the "representation algorithm" of [17].

Finally, it is the intent of the authors that this paper be as self-contained and accessible as possible. With this idea in mind, the only items which are assumed are certain elementary facts about the lower central series which may be found in Chapter 5 of [9] or Chapters 10-12 of [5] and a theorem on free generators of the commutator subgroup of G, i.e., Theorem 4.1. We omit the proof of Theorem 4.1 (see [4]) because it involves technical details (e.g., the Kurosh subgroup theorem) that are too far removed from the techniques used throughout this paper.

2. *GROUP THEORETICAL FOUNDATIONS*

We start by giving some notations, definitions, known results and immediate consequences of these results. These things will be used throughout the paper. Let G be a group. Let a,b ∈ G. Then the element A of G given by

$$A = (a,b) = a^{-1}b^{-1}ab \tag{2.1}$$

is the commutator of a and b. We will write $a = A^L$ and $b = A^R$. Also the lower central series of G, i.e.,

$$G_1 \supseteq G_2 \supseteq \cdots \supseteq G_m \supseteq \cdots \tag{2.2}$$

is the sequence of subgroups defined inductively as follows $G_1 = G$. For

$n > 1$, G_n is the group generated by all commutators (a, b_{n-1}) where $a \in G$ and $b_{n-1} \in G_{n-1}$.

We will say that the element c of G has weight n in G if $c \in G_n$ but $c \notin G_{n+1}$. This will be denoted by writing $W_G(c) = n$. For each of notation, the weight of a commutator (a,b) in G, $W_G((a,b))$, will be written $W_G(a,b)$. It is evident that $a \in G_n$ implies that if $W_G(a)$ is defined then $W_G(a) \geq n$. We will also often abbreviate the commutator $(\ldots((a_1,a_2),a_3),\ldots,a_n)$ by $(a_1,a_2,a_3,\ldots,a_n)$.

The following properties of the lower central series are well known [5,9]: If $(a,b) \neq 1$ and $W_G(a,b)$ is defined, then

$$W_G(a,b) \geq W_G(a) + W_G(b). \tag{2.3a}$$

If $W_G(a_i) = n_1$ and $W_G(b_j) = n_2$, then

$$\left(\prod_i a_i^{\alpha_i}, \prod_j b_j^{\beta_j}\right) \equiv \prod_{i,j} (a_i,b_j)^{\alpha_i\beta_j} \bmod G_{n_1+n_2+1}. \tag{2.3b}$$

If $W_G(a) = n_1$, $W_G(b) = n_2$, $a \equiv c \bmod G_{n_1+1}$ and $b \equiv d \bmod G_{n_2+1}$, then

$$(a,b) \equiv (c,d) \bmod G_{n_1+n_2+1}. \tag{2.3c}$$

The Jacobi identity

$$(a,b,c)(b,c,a)(c,a,b) \equiv 1 \bmod G_{W+1} \tag{2.3d}$$

where $W = W_G(a) + W_G(b) + W_G(c)$.

We now proceed to define basic commutator according to a natural linear ordering given in [14] (i.e., basic commutators of the same weight are ordered lexicographically with respect to each other). We will need the properties of this ordering in our investigation of a group G which is a homomorphic image of the free group F of finite rank r.

Definition 2.1. *The basic commutators of weight one in* F *are the free generators of* F *in the order* $c_1 < c_2 < \ldots < c_r$. *Having defined and ordered basic commutators of weight less than* n, *we use them to define and order the basic commutators of weight* n. *The basic commutators of weight* n *in* F *are* $c_m = (c_i,c_j)$ *where* c_i *and* c_j *are basic commutators such that*
(i) $W_F(c_i) + W_F(c_j) = n$;
(ii) $c_i > c_j$;
(iii) *if* $c_i = (c_s,c_t)$, *then* $c_j \geq c_t$.
Let $c_{m_1} = (c_{i_1},c_{j_1})$ *and* $c_{m_2} = (c_{i_2},c_{j_2})$ *be such that* $W_F(c_{m_1}) = W_F(c_{m_2})$. *Then we order* $c_{m_1} > c_{m_2}$ *if* $c_{i_1} > c_{i_2}$, *or* $c_{i_1} = c_{i_2}$ *but* $c_{j_1} > c_{j_2}$. *A basic commutator of weight* n *is greater than any of smaller weight. Having ordered all basic commutators, we assume that their subscripts are chosen so that* c_i *is the ith basic commutator.* (In this definition, we are using the word weight in agreement with its general meaning since it can be shown that a basic commutator of weight n in F is in F_n but not in F_{n+1} [9]. Moreover for $W_F(c)$ we will write $W(c)$ in the rest of this paper.)

To proceed we introduce an auxiliary definition.

Definition 2.2. *Let the group* G *have presentation* $G = \langle c_1, c_2, \ldots, c_r; s_1, \ldots, s_t \rangle$. (Then G is the factor group F/N, where N is the normal closure of the subgroup of F generated by the words $s_1, s_2, \ldots, s_t$. In particular when $t = 1$ and $s_1 = 1$ then $G = F$.) *Let the basic commutator* c_m *be the element of* F *of Definition 2.1 as well as its image in* G *under the homomorphism* $F \to G = F/N$; *we shall, however, always mean by the weight of* c_m, $W(c_m)$, *the weight of* c_m *in* F; i.e. *the number of Definition 2.1. The element* $a \in G_n$ *is said to be basic commutator representable (b.c.-representable) if*

$$a \equiv c_{i_1}^{\varepsilon_1} c_{i_2}^{\varepsilon_2} \ldots c_{i_h}^{\varepsilon_h} \bmod G_{n+1} \tag{2.4}$$

where the c_{i_j} *are elements of* G *as well as basic commutators of weight* n, $c_{i_1} < c_{i_2} < \ldots < c_{i_h}$ *if* $h > 1$ *and* $\varepsilon_1, \ldots, \varepsilon_h$ *are nonzero exponents. The product on the right hand side of* (2.4) *will be called the basic commutator representation of* a *(b.c. rep. of* a*).*

Before going further, we make the following important remark:

Remark 2.1. If $a \in F$ $(a \neq 1)$ and $\bar{a}$ is its image under the homomorphism $F \to G = F/N$, then $W(a) \leq W_G(\bar{a})$ by the definition of weight when $W_G(\bar{a})$ is defined.

The name basic commutator is appropriate in the sense of the following well known theorem [5,9].

Theorem 2.1. *Every group* F_{n+1} *is a normal subgroup of* F *and every factor group* $\bar{F}_n = F_n/F_{n+1}$ *is a free abelian group of finite rank. The basic commutators of weight* $n (n \geq 1)$ *are mapped into a basis of* $\bar{F}_n$ *(under the homomorphism* $F_n \to \bar{F}_n$*) such that every nontrivial element of* F *has a unique weight and a unique b.c. rep. Moreover if* a *and* b *are distinct basic commutators then*

$$W(a,b) = W(a) + W(b). \tag{2.5}$$

By the definition of the lower central series, we obtain at once the following corollary for the group presented in Definition 2.2.

Corollary 2.1. *Every* G_{n+1} *is a normal subgroup of* G *and every factor group* $\bar{G}_n = G_n/G_{n+1}$ *is a finitely generated abelian group. The basic commutators of weight* n *in* F *are mapped into generators of* $\bar{G}_n$ *(under the homomorphisms* $F_n \to G_n \to \bar{G}_n$*) such that every element of weight* $n > 0$ *in* G *is b.c.-representable.*

To compute a b.c. rep. of a group element, we make use of the well known "collection process" [5,6] which is discussed in Section 3(b).

3. *PREREQUISITES FROM THE COMMUTATOR CALCULUS*

3(a) *The Ordering Theorem and Maximal Components*

For the previously stated properties of basic commutators, our natural linear order is not required. It is, however, order preserving under commutation. Before we can state this result precisely, we need two preliminary definitions.

Definition 3.1. *Let* $a \in F$ *be nontrivial. The maximal component of* a, $M(a)$, *is the largest commutator in the b.c. rep. of* a, *i.e. in* (2.4) $M(a) = c_{i_h}$. *Also again for ease of notation, the maximal component of a commutator* $M((a,b))$ *will be written* $M(a,b)$.

Definition 3.2. *Let* $a,b \in F$. *The inequalities* $a > b$ *and* $a \geqslant b$ *will mean that* $M(a) > M(b)$ *and* $M(a) \geqslant M(b)$, *respectively.*

The following ordering theorem is of importance in this paper.

Theorem 3.1. *Let* a, b *and* c *be basic commutators in* F *such that* $b > c$, $a \neq b$ *and* $a \neq c$. *Then* $(a,b) > (a,c)$.

The proof of this theorem will be accomplished easily once we have established the following three auxiliary lemmas on the basic commutators a, b and c.

Lemma 3.1. *Let* $a > b > c$ *then* $(a,b) > (a,c)$.

Lemma 3.2. *Let* $a > b > c$ *then* $(a,c) > (b,c)$.

Lemma 3.3. *Let* $a > b$ *and* $a = (\alpha,\beta)$ *for* $W(a) > 1$. *Then the following formula holds for obtaining the maximal component of* (a,b):

$$M(a,b) = \begin{cases} (a,b) \text{ for } W(a) = 1 \text{ or } W(a) > 1 \text{ and } \beta \leqslant b \\ (M(\alpha,b),\beta) \text{ for } W(a) > 1 \text{ and } b < \beta. \end{cases}$$

We shall prove these lemmas simultaneously by induction on the place of the basic commutator a in the ordering of Definition 2.1. We note first that Lemma 3.3 holds trivially for $W(a) = 1$.

When F has only two generators ($r = 2$), we begin the induction with $a = c_3 = (c_2,c_1)$ and so $W(a) = 2$. Then we must take $b = c_2$ and $c = c_1$ so that $M(a,b) = ((c_2,c_1),c_2) > ((c_2,c_1),c_1) = M(a,c) > (c_2,c_1) = M(b,c)$. This verifies our lemmas for $a = c_3$ when $r = 2$.

When $r > 2$, however, we must start with $W(a) = 1$. Let us take for a the generator c_{j_1}, $b = c_{j_2}$ and $c = c_{j_3}$ with $r \geqslant j_1 > j_2 > j_3 \geqslant 1$. Then $(a,b) = (c_{j_1},c_{j_2}) > (c_{j_1},c_{j_3}) = (a,c) > (c_{j_2},c_{j_3}) = (b,c)$ by Definition 2.1. Thus our lemmas hold for $a = c_{j_1}$.

Let us now suppose that we have proven our lemmas for every $a = c_j$ with $3 \leqslant j < i$ and $W(c_i) > 1$. We shall establish them below for $a = c_i = (\alpha,\beta)$.

Proof of Lemma 3.3. When $\beta \leqslant b < a$ then $(a,b) = (\alpha,\beta,b)$ is a basic commutator and thus $M(a,b) = (a,b)$. For the more interesting case $b < \beta$ we take as a point of departure the Jacobi identity (see (2.3d)) $(\alpha,\beta,b)(\beta,b,\alpha)(b,\alpha,\beta) \equiv 1 \bmod F_{W(\alpha)+W(\beta)+W(b)+1}$, to which we apply (2.3b) to write (a,b) in the form:

$$(a,b) = (\alpha,\beta,b) \equiv (\alpha,b,\beta)(\beta,b,\alpha)^{-1} \bmod F_{W(a)+W(b)+1}. \tag{3.1}$$

(In the future, we will refer to this computation as application of (2.3d) and (2.3b).) By Theorem 2.1, we have the unique b.c. rep's.

$$(\alpha,b) \equiv \prod_{s=1}^{m} c_{i_s}^{e_s} \bmod F_{W(\alpha)+W(b)+1},$$
$$(\beta,b) \equiv \prod_{t=1}^{n} c_{j_t}^{f_t} \bmod F_{W(\beta)+W(b)+1}. \tag{3.2}$$

Combining (3.1) and (3.2), we find by means of (2.3b) and (2.3c) that

$$((\alpha,b),\beta) \equiv \prod_{s=1}^{m} (c_{i_s},\beta)^{e_s} \bmod F_{W(a)+W(b)+1}, \tag{3.3a}$$

$$((\beta,b),\alpha) \equiv \prod_{t=1}^{n} (c_{j_t},\alpha)^{f_t} \bmod F_{W(a)+W(b)+1}, \tag{3.3b}$$

and so

$$(a,b) \equiv \prod_{s=1}^{m} \prod_{t=1}^{n} (c_{i_s},\beta)^{e_s} (c_{j_t},\alpha)^{-f_t} \bmod F_{W(a)+W(b)+1}. \tag{3.3c}$$

According to our subscript convention (see Definition 2.1) and the definition of b.c. rep. (see (2.4)), the basic commutators in (3.2) are ordered so that (i) $i_m = \max\{i_1,i_2,\ldots,i_m\}$ and $j_n = \max\{j_1,j_2,\ldots,j_n\}$, (ii) $c_{i_m} = M(\alpha,b)$ and $c_{j_n} = M(\beta,b)$. Since $\alpha < a = c_i$, we may apply the induction hypothesis on Lemma 3.1 to $\alpha > \beta > b$ to find that $a = (\alpha,\beta) > (\alpha,b)$, i.e.

$$c_{i_m} < a. \tag{3.4}$$

Next we note that $W(c_{i_m}) = W(c_{i_s}) > W(\alpha) \geq W(\beta)$ and thus $a > c_{i_m} > c_{i_s} > \alpha > \beta$ for $s = 1,\ldots,m-1$. So we may apply the induction hypothesis on Lemma 3.2 to find from $c_{i_m} > c_{i_s} > \beta$ that $(c_{i_m},\beta) > (c_{i_s},\beta)$. Therefore $M((\alpha,b),\beta) = M(c_{i_m},\beta) = M(M(\alpha,b),\beta)$ according to (3.3a).

If $W(\alpha) = 1$, we note that (α,b,β) is a basic commutator, since we are assuming $b < \beta$ so that $M(\alpha,b,\beta) = (\alpha,b,\beta)$. But if $W(\alpha) > 1$ then we may apply the induction hypothesis on Lemma 3.3 to c_{i_m} because $c_{i_m} < a$ by (3.4). It says that $c_{i_m} = M(\alpha,b) = (\alpha,b)$ if $\alpha^R \leq b$ or $c_{i_m} = M(\alpha,b) = (M(\alpha^L,b),\alpha^R)$ if $b < \alpha^R$; thus if we let $\delta = c_{i_m}^R$, either $\delta = b$ or $\delta = \alpha^R$. Since $a = (\alpha,\beta)$ is basic, $\alpha^R \leq \beta$ and thus $\delta \leq \beta$ in any case. Hence (c_{i_m},β) is a basic commutator and $M(c_{i_m},\beta) = (c_{i_m},\beta)$ under all circumstances.

Next we must examine (β,b,α) as given in (3.3b). First suppose that $W(\alpha) > W(\beta) + W(b)$. In this case $\alpha > c_{j_t} > \beta$ for $t = 1,2,\ldots,n$. It follows from equation (2.5) in Theorem 2.1 that every commutator (α,c_{j_t}) has $W(\alpha,c_{j_t}) = W(a) + W(b)$ and is clearly basic. But the basic commutator (c_{i_m},β) has the same weight (by (2.5)) and is greater than every (α,c_{j_t}) since $W(\alpha) < W(c_{i_m})$ so that $c_{i_m} > \alpha$. It is then clear from (3.3c) that

$$M(a,b) = (c_{i_m},\beta) = (M(\alpha,b),\beta). \tag{3.5}$$

The second case we treat is $W(\alpha) = W(\beta) + W(b)$. But then all the $(\alpha_j,c_{j_t}) \neq 1$ are either basic or inverses of basic commutators. We note that $W(\alpha) = W(c_{j_t}) < W(c_{i_m})$ and $W(\alpha) + W(c_{j_t}) = W(c_{i_m}) + W(\beta)$ again imply that $(\alpha,c_{j_t}) < (c_{i_m},\beta)$ for all t and so (3.5) holds.

Our third case is $W(\alpha) < W(\beta) + W(b)$. We first show that

$$a = (\alpha,\beta) > M(\beta,b) = c_{j_n}. \tag{3.6}$$

Since $a > \alpha > \beta > b$, we may apply the induction hypothesis to $\alpha > \beta > b$. Lemma 3.1 then implies that $a = (\alpha,\beta) > (\beta,b)$ and Lemma 3.2 implies that $(\alpha,b) > (\beta,b)$. Combining these, we obtain (3.6). Having established (3.6) we may apply the induction hypothesis on Lemma 3.2 to $c_{j_n} > c_{j_t} > \alpha$ to find that $(c_{j_n},\alpha) > (c_{j_t},\alpha)$ for $t = 1,\ldots,n-1$.

Finally we note that if $W(\beta) = 1$, then $c_{j_n} = M(\beta,b) = (\beta,b)$; so that $M(c_{j_n},\alpha) = (c_{j_n},\alpha)$ since $\alpha > \beta > b$. If $W(\beta) > 1$, we write $c_{j_n} = M(\beta,b) = (\lambda,\mu)$. Since $= a^R < a$, we may apply the induction hypothesis on Lemma 3.3 to finding $M(\beta,b) = c_{j_n}$. This gives $M(\beta,b) = (\beta,b)$ if $\beta^R \le b$ or $M(\beta,b) = (M(\beta^L,b),\beta^R)$ if $b < \beta^R$; thus either $\mu = b$ or $\mu = \beta^R$. In any case, $\mu < \beta$ and so $(c_{j_n},\alpha) = (\lambda,\mu,\alpha)$ is a basic commutator (note that $\beta < \alpha$ since $a = (\alpha,\beta)$ is basic). This means that $M(c_{j_n},\alpha) = (c_{j_n},\alpha)$ under all circumstances. But $W(c_{j_n},\alpha) = W(c_{i_m},\beta)$ by (2.5) and applying the induction hypothesis on Lemma 3.2 to $a > \alpha > \beta > b$, we find that $c_{i_m} = M(\alpha,b) > c_{j_n} = M(\beta,b)$. Thus we have $(c_{j_n},\alpha) < (c_{i_m},\beta)$ and we again arrive at (3.5) from (3.3c). We have now proven under the inductive hypothesis on Lemmas 3.1, 3.2 and 3.3 that Lemma 3.3 holds for $a = c_i$.

Proof of Lemma 3.1. Consider the sequence $M(a,c_{i-1})$, $M(a,c_{i-2}),\ldots,M(a,c_k) = M(a,\beta)$, $M(a,c_{k-1}),\ldots,M(a,c_1)$, where we write c_k for the basic commutator β. Because of Lemma 3.3, it is identical with (a,c_{i-1}), $(a,c_{i-2}),\ldots,(a,c_k) = ((\alpha,c_k),c_k)$, $(M(\alpha,c_{k-1}),c_k),\ldots,(M(\alpha,c_1),c_k)$. For Lemma 3.1, it is evidently sufficient to show that

$$(a,c_{j+1}) > (a,c_j) \text{ for } k \le j \le i-2, \tag{3.7}$$

and also that

$$(M(\alpha,c_{v+1}),c_k) > (M(\alpha,c_v),c_k) \text{ for } 1 \le v \le k-1. \tag{3.8}$$

Since $W(c_{j+1}) \ge W(c_j)$, $W(a,c_{j+1}) \ge W(a,c_j)$ by (2.5) and hence (3.7) follows by Definition 2.1. Since $a > \alpha > \beta \ge c_{v+1} > c_v$, we can apply our inductive hypothesis on Lemma 3.1 to obtain $(\alpha,c_{v+1}) > (\alpha,c_v)$; hence, again (3.8) follows by Definition 2.1. This completes the proof of Lemma 3.1 for $a = c_i$.

It remains to examine Lemma 3.2.

Proof of Lemma 3.2. When $W(a) > W(b)$ then $W(M(a,c)) > W(M(b,c))$ by (2.5) and therefore $(a,c) > (b,c)$. It is thus sufficient to proceed under the hypothesis that $W(a) = W(b) \ge 2$ from here on. We shall write $a = (\alpha_1,\beta_1)$ and $b = (\alpha_2,\beta_2)$ and consider four cases:

1. $\beta_1 \le c$, $\beta_2 \le c$.
2. $c < \beta_1$, $\beta_2 \le c$.
3. $\beta_1 \le c$, $c < \beta_2$.
4. $c < \beta_1$, $c < \beta_2$.

1. Here (a,c) and (b,c) are basic commutators. Since $a > b$ and $W(a,c) = W(b,c)$ by (2.5), $(a,c) > (b,c)$ by definition of our ordering (Definition 2.1).

2. From Lemma 3.3, we find that

$$M(a,c) = (M(\alpha_1,c),\beta_1),\ M(b,c) = ((\alpha_2,\beta_2),c).$$

We note that $\beta_2 \leqslant c < \beta_1$ in this case. Since $a > \alpha_1$ we already know from Lemma 3.1 that $(\alpha_1,c) \geqslant (\alpha_1,\beta_2)$. But $a > b$ and $W(a) = W(b)$, hence $\alpha_1 \geqslant \alpha_2$ by Definition 2.1. Thus $(\alpha_1,\beta_2) \geqslant (\alpha_2,\beta_2)$ by the inductive hypothesis on Lemma 3.2. Combining the above, we find $(\alpha_1,c) \geqslant (\alpha_2,\beta_2)$. Since $W(M(\alpha_1,c),\beta_1) = W(M(a,c)) = W(M(b,c)) = W(\alpha_2,\beta_2,c)$ by (2.5), we find that $M(a,c) = (M(\alpha_1,c),\beta_1) > ((\alpha_2,\beta_2),c) = M(b,c)$ by Definition 2.1.

3. From Lemma 3.3, we have

$$M(a,c) = ((\alpha_1,\beta_1),c),\ M(b,c) = (M(\alpha_2,c),\beta_2).$$

We note that $a > \alpha_1 \geqslant \alpha_2 > \beta_2 > c$ since $a > b$. Hence by the inductive hypothesis on Lemma 3.1 we find that $b = (\alpha_2,\beta_2) > (\alpha_2,c)$ and so $a = (\alpha_1,\beta_1) > b > (\alpha_2,c)$. Furthermore, using (2.5), $W(\alpha_1,\beta_1,c) = W(\alpha_1) + W(\beta_1) + W(c) = W(\alpha_2) + W(\beta_2) + W(c) = W(M(\alpha_2,c),\beta_2)$, since $W(a) = W(b)$. Thus $(a,c) > (b,c)$ by Definition 2.1.

4. Applying Lemma 3.3 one last time yields here

$$M(a,c) = (M(\alpha_1,c),\beta_1),\ M(b,c) = (M(\alpha_2,c),\beta_2).$$

Since $a > \alpha_1 \geqslant \alpha_2$, we may apply the induction hypothesis on Lemma 3.2 to find that $(\alpha_1,c) \geqslant (\alpha_2,c)$ where equality holds only for $\alpha_1 = \alpha_2$. But when $\alpha_1 = \alpha_2$, then $\beta_1 > \beta_2$ since $a > b$. Since $W(a) = W(b)$, $(a,c) > (b,c)$ under all circumstances by Definition 2.1. This concludes our examination of Lemma 3.2.

We have now shown that the hypothesis that Lemmas 3.1, 3.2, and 3.3 hold for every $a = c_j$ with $3 \leqslant j < i$ and $W(c_i) > 1$ implies that they must also hold for $a = c_i$. Since Lemmas 3.1-3.3, have been established, we are now ready for the:

Proof of Theorem 3.1. Since $b > c$, $a \neq b$, and $a \neq c$ there are three possible cases: Case (i): $a > b > c$. Here the theorem is identical to Lemma 3.1. Case (ii): $b > a > c$. Then $M(a,b) = M(b,a) > M(b,c)$ by Lemma 3.1. Also $(b,c) > (a,c)$ by Lemma 3.2. Hence $(a,b) > (a,c)$. Case (iii): $b > c > a$. Here the theorem is identical to Lemma 3.2. This completes the proof of Theorem 3.1.

To continue, we require:

Lemma 3.4. *Let* $a,b \in F$ *with* $M(a) \neq M(b)$. *Then* $M(a,b) = M(M(a),M(b))$.
Proof. By Theorem 2.1, a and b have unique b.c. rep's

$$a \equiv \prod_i c_i^{\varepsilon_i} \bmod F_{W(a)+1} \quad \text{and} \quad b \equiv \prod_j c_j^{\gamma_j} \bmod F_{W(b)+1}.$$

So that by (2.3b), $(a,b) \equiv \prod_{i,j} (c_i,c_j)^{\varepsilon_i\gamma_j} \bmod F_{W(a)+W(b)+1}$. Now letting $\Pi = \prod_{i,j} (c_i,c_j)^{\varepsilon_i\gamma_j}$, clearly $M(a,b) = M(\Pi)$. But from Theorem 3.1, $M(\Pi) = M(M(a),M(b))$.

It is evident from Lemma 3.4 that Theorem 3.1 has the alternate, more general formulation:

Corollary 3.1. *Let* $a,b,c \in F$ *such that* $b > c$, $a \neq 1$, $M(a) \neq M(b)$, $M(a) \neq M(c)$. *Then* $(a,b) > (a,c)$.

To apply Theorem 3.1 we shall need more machinery. We shall introduce for every basic commutator c its "regular sequence" [c], i.e. $[c] = [d_1,d_2,\ldots,d_h]$.

Definition 3.3. *The regular sequence of* c, [c], *consists of* c *only when* $W(c) = 1$ (i.e. $c = c_i$, $1 \leq i \leq r$, *is a generator*). *Having defined the regular sequences of all basic commutators of weight less than* n, *we define the regular sequence* [c] *for* $W(c) = n$:

$$[c] = [e_1,e_2,\ldots,e_p, c^R] \text{ where } [c^L] = [e_1,e_2,\ldots,e_p].$$

At this point we establish an important method for finding maximal components. It appears as Lemma 2.1 in [17].

Lemma 3.5. *Let* C *and* c *be basic commutators with* $W(C) > 1$, $C > c$ *and* $[C] = [d_1,d_2,\ldots,d_n]$. *Then*

$$[M(C,c)] = [d_1,e_1,e_2,\ldots,e_n] \qquad (3.9)$$

where $e_1,e_2,\ldots,e_n$ *is a rearrangement of* $d_2,\ldots,d_n$, c *such that* $e_1 \leq e_2 \leq \ldots \leq e_n$.

Proof. Let $C^L = a$, $C^R = b$. If $b \leq c$, then (C,c) is also a basic commutator and the conclusion of our lemma clearly holds. It remains to consider the case $c < b$. We start by applying (2.3d) and (2.3b) to $(C,c) = (a,b,c)$ to write it as

$$(a,b,c) \equiv (a,c,b)(a,(b,c)) \bmod F_{W+1}$$

where $W = W(a) + W(b) + W(c)$. When $W(C) = 2$ then $(a,c,b) > (b,c,a)$ and they are both basic commutators by Definition 2.1. Having established the lemma for $W(C) = 2$, we assume it for $2 \leq W(G) \leq k$ and proceed to $W(C) = k+1$. By the induction hypothesis, $[M(a,c)] = [d_1,e_1,\ldots,e_{n-1}]$ where $e_1 \leq \ldots \leq e_{n-1}$ is a rearrangement of $d_2,\ldots,d_{n-1},c$. Thus $M(a,c)^R = e_{n-1} = \max\{d_{n-1},c\}$, but $d_{n-1} \leq d_n = b$ and $c < b$ by hypothesis, so that $e_{n-1} \leq b$ in any case. Now Lemma 3.4 implies that $M(a,c,b) = M(M(a,c),b)$. Since $M(a,c)^R \leq b$, $(M(a,c),b)$ is basic and so $M(a,c,b) = (M(a,c),b)$. Moreover this evidently implies that $[M(a,c,b)]$ has the form (3.9). Thus it will suffice to prove that in this case, i.e. $c < b$, $M(a,b,c) = M(a,c,b)$.

If (a,(b,c)) has weight W in F, we note that by the induction hypothesis, Lemma 3.4 and Definition 2.1

$$M(a,(b,c)) = \begin{cases} (a,M(b,c)) & \text{if } (b,c) < a \\ (M(b,c),a) & \text{if } (b,c) > a \\ (a,d) & \text{if } M(b,c) = a \end{cases} \qquad (3.10)$$

where d is found as follows. The b.c. rep of (b,c) is

$$(b,c) \equiv \prod_{j=1}^{k} c_{i_j}^{\varepsilon_j} \bmod F_{W(a)+1} \qquad (3.11)$$

where $a = M(b,c) = c_{i_k}$ and the basic commutators c_{i_j} have $W(c_{i_j}) = W(a)$ with $c_{i_1} < \ldots < c_{i_k} = a$. By (2.3b)

$$(a,(b,c)) \equiv \prod_{j=1}^{k-1} (a,c_{i_j})^{\varepsilon_j} \bmod F_{2W(a)+1}.$$

Clearly each (a,c_{i_j}) is basic with $W(a,c_{i_j}) = 2W(a)$ for $j = 1,2,\ldots,k-1$. Thus $M(a,(b,c)) = (a,c_{i_{k-1}})$ by Definition 2.1; so that $d = c_{i_{k-1}}$ in (3.11).

Now $a > b > c$ implies that $(a,c) > (b,c)$ by Theorem 3.1 and so $((a,c),b) > (a,(b,c))$ according to Definition 2.1 and (3.10) when $(a,(b,c))$ has weight W in F. This concludes our proof.

3(b) *The Collection Process*

The collection process was first introduced by Philip Hall in [6]. Its main use will be to represent group elements by basic commutators. We shall review the collection process for our applications.

We use the following notation. For elements a, b of the free group F

$$\begin{aligned} &(b,\ a;\ 0) = b, \\ &(b,\ a;\ 1) = (b,a), \text{ and} \\ &(b,\ a;\ n+1) = ((b,\ a;\ n),a) \end{aligned} \tag{3.12}$$

for $n = 1,2,\ldots$. Using this notation, we note that the collection process is based on the identities

$$ba = ab\,(b,a) \tag{3.13a}$$

$$ba^{-1} = a^{-1}b \cdot \prod_{k=1}^{m} (b,a;2k) \cdot ((b,a;2m),a^{-1}) \cdot \left[\prod_{k=1}^{m} (b,a;2k-1)\right]^{-1} \tag{3.13b}$$

$$b^{-1}a = a\,(b,a)^{-1}b^{-1} \tag{3.13c}$$

$$b^{-1}a^{-1} = a^{-1} \cdot \prod_{k=1}^{m} (b,a;2k-1) \cdot ((b,a;2m),a^{-1})^{-1} \cdot \left[\prod_{k=1}^{m} (b,a;2k)\right]^{-1} b^{-1}. \tag{3.13d}$$

We note that (3.13a) and (3.13c) follow from (2.1). But (3.13b) and (3.13d) are easily derived from the well known identities [5]:

$$(ab,c) = (a,c)(a,c,b)(b,c) \tag{3.14a}$$

$$(a,bc) = (a,c)(a,b)(a,b,c). \tag{3.14b}$$

An arbitrary $f \neq 1$ in F is expressible as

$$f = \prod_{j=1}^{J} c_{i_j}^{n_j} \tag{3.15}$$

where $1 \leq i_j \leq r$ for $j = 1,2,\ldots,J$, i.e., c_{i_j} are generators. A typical step in the collection of f takes us from

$$f = \left(\prod_{i=1}^{I} c_i^{\varepsilon_i}\right) f_I\; g_{N+1,I} \tag{3.16}$$

to

$$f = \left(\prod_{i=1}^{I+1} c_i^{\varepsilon_i} \right) f_{I+1} \, g_{N+1,I+1}. \tag{3.17}$$

To describe this precisely, we require:

Definition 3.4. *The generators of* F *and their inverses* $c_1^{\pm 1}, \ldots, c_r^{\pm 1}$ *are the 1-commutators. Suppose that we have defined the* k*-commutators for* $1 \leq k \leq m$. *An* (m+1)*-commutator is any* $c = (u,v)$ *where* u *is a* s*-commutator,* v *is a* t*-commutator and* $s + t = m + 1$. (Note that a k-commutator has weight at least k in F by inequality (2.3a).)

Let N be a given positive integer and let q(N) be the number of basic commutators of weight $\leq N$. Here q(N) is the number of such commutators in F but q(N) will also be used to denote this number in other groups. It should be clear from the context which group the number q(N) refers to. Let $0 \leq I \leq q(N)$. The word

$$\prod_I = \prod_{i=1}^{I} c_i^{\varepsilon_i}$$

in (3.16) is the collected part of f if $I > 0$. (If $I = 0$ then the collected part $\Pi_I = 1$.) If $I = q(N)$ then $f_I = 1$ in (3.16). Moreover if $0 < I < q(N)$ then f_I is a word in basic commutators c_k and is such that (3.16) has the following properties:

(a) $I < k \leq q(N)$

(b) If $W(c_k) > 1$, then $c_k^R \leq c_I$

(c) The $g_{N=1,I}$ in (3.16) is a word in finitely many m-commutators such that each $m > N$.

We continue to show inductively why the above properties (a), (b) and (c) hold.

Having obtained the collected part Π_I, we find Π_{I+1} by the rewriting of f_I in (3.16). (Note that for $I = 0$ the collected part $\Pi_I = 1$, $f_I = f$ and $g_{N+1,I} = 1$ so that properties (a), (b) and (c) hold vacuously and our induction has already begun.) Let us focus on f_I. For $0 \leq I < q(N)$, we first assume that f_I has the form $f_I = c_{k_1} \ldots c_{k_n}$. If c_u is the earliest commutator in f_I then evidently $\Pi_I f_I = \Pi_{u-1} f_{u-1}$. Thus we take $u = I + 1$ without loss of generality. Suppose $c_{k_j} = c_u$ is the left-most c_u in f_I such that $j > 1$ and $k_{j-1} > u$. Then we replace $c_{k_1} \ldots c_{k_{j-1}} c_{k_j} \ldots c_{k_n}$ by $c_{k_1} \ldots c_{k_j} c_{k_{n-1}} (c_{k_{j-1}}, c_{k_j}) \ldots c_{k_n}$ using identity (3.13a). This moves c_{k_j} to the left and also introduces the new commutator $(c_{k_{j-1}}, c_{k_j})$ which is surely later than c_{k_j} by weight. Thus c_{k_j} is still the earliest commutator in the uncollected part if $j > 1$. Moreover $(c_{k_{j-1}}, c_{k_j})$ is a basic commutator because in the replacement we collect c_{k_j} before $c_{k_{j-1}}$ so $k_{j-1} > k_j$ and if $c_{k_{j-1}} = (c_s, c_t)$ then we have already collected c_t before collecting this c_{k_j} thus $k_j = u > t$. After enough steps c_{k_j} will be moved to the front of f_I and to the end of Π_I. Continuing this procedure until every such c_u which occurs in f_I is moved to the end of Π_I we obtain

$$f = \Pi_{I+1} \, \bar{f}_I \, g_{N+1,I}$$

where $\overline{f}_I = c_{i_1} c_{i_2} \cdots c_{i_m}$ and the c_{i_j} are basic commutators such that $i_j > u$ for all $j = 1,2,\ldots,m$. If $W(c_{i_j}) \leq n$ for all j, then we take $f_{I+1} = \overline{f}_I$ and $g_{N+1,I+1} = g_{N+1,I}$ and we have satisfied properties (a), (b) and (c). Otherwise there is a left most c_{i_j}, call it c_{i_t}, in $\overline{f}_I$ such that $W(c_{i_t}) > N$. We then write, for $m - t > 0$,

$$\begin{aligned}\overline{f}_I &= c_{i_1} \cdots c_{i_{t-1}} c_{i_t} c_{i_{t+1}} \cdots c_{i_m} \\ &= c_{i_1} \cdots c_{i_{t-1}} c_{i_{t+1}} \cdots c_{i_m} c_{i_t} (c_{i_t}, c_{i_{t+1}} \cdots c_{i_m})\end{aligned} \tag{3.18}$$

and note that if $m - t = 0$ in (3.18), we just take $f_{I+1} = c_{i_1} c_{i_2} \cdots c_{i_{m-1}}$ and $g_{N+1,I+1} = c_{i_m} g_{N+1,I}$ and again properties (a), (b) and (c) are satisfied. Otherwise we claim that the commutator $(c_{i_t}, c_{i_{t+1}} \cdots c_{i_m})$ for $m - t \geq 1$ which occurs in (3.18) may be expressed as a word in k-commutators with each $k > N$. This follows from an eacy induction on $m - t$ which uses the identity (3.14b), the fact that $W(c_{i_t}) > N$ implies it is a k-commutator with $k > N$ and the obvious identity:

$$(A,b) = A^{-1} \prod_i [a_i(a_i,b)]^{\varepsilon_i} \tag{3.19}$$

where $A = \prod_i a_i^{\varepsilon_i}$. For ease of notation, we suppress the double subscripts and just write our commutator as $(c_t, c_{t+1} \cdots c_m)$. Now if $m - t = 1$ then our claim is trivial because the commutator (c_t, c_{t+1}) has weight $> N$. Let us assume we have proven that for all $m - t$, with $1 \leq m - t < M$, $(c_t, c_{t+1} \cdots c_m)$ can be expressed in terms of k-commutators with $k > N$. Consider $(c_t, c_{t+1} \cdots c_m)$ with $m - t = M$. Then by (3.14b) with $a = c_t$, $b = c_{t+1} \cdots c_{m-1}$, and $c = c_m$, we find that

$$(c_t, c_{t+1} \cdots c_m) = (c_t, c_m)(c_t, c_{t+1} \cdots c_{m-1})((c_t, c_{t+1} \cdots c_{m-1}), c_m).$$

Since $W(c_t) > N$, (c_t, c_m) is a k-commutator with $k > N$ and by our induction hypothesis $(c_t, c_{t+1} \cdots c_{m-1}) = \prod_i a_i^{\varepsilon_i}$ where each a_i is a k-commutator with $k > N$. Thus applying identity (3.19) to $(c_t, c_{t+1} \cdots c_{m-1}, c_m)$ with $A = \prod_i a_i^{\varepsilon_i}$ and $b = c_m$, we find

$$(c_t, c_{t+1} \cdots c_{m-1}, c_m) = A^{-1} \prod_i [a_i(a_i, c_m)]^{\varepsilon_i}$$

and so $(c_t, c_{t+1} \cdots c_m)$ is itself a word in k-commutators with $k > N$ as claimed.

We now continue with (3.18). Applying the same procedure to every c_{i_j} with $W(c_{i_j}) > N$ as was just applied to c_{i_t} moves all such c_{i_j} to the end of $\overline{f}_I$. This finally shows how we proceed from (3.16) to (3.17) so that properties (a), (b) and (c) have been satisfied.

As it stands the collection process has not been defined for all elements f_I. We must take into account the possible presence of inverses. Let us now assume that $f_I = c_{k_1}^{\eta_1} c_{k_2}^{\eta_2} \cdots c_{k_n}^{\eta_n}$ where the $\eta_i = \pm 1$. Again c_u is the earliest basic commutator in f_I and $c_{k_j} = c_u$ is the leftmost occurrence of c_u in f_I where $j > 1$ and $k_{j-1} > u$. Thus we consider collecting c_u or c_u^{-1}

in expressions $c_v c_u^{-1}$, $c_v^{-1} c_u$, and $c_v^{-1} c_u^{-1}$. Now $c_v^{\pm 1} c_u^{\pm 1}$ is easily rewritten in the form $c_u^{\pm 1} X c_v^{\pm 1} Y$ by the identities (3.13) where X and Y may either be 1 or words in commutators as given by (3.13). Thus we replace

$$c_{k_1}^{n_1} \ldots c_{k_{j-1}}^{n_{j-1}} c_{k_j}^{n_j} c_{k_{j+1}}^{n_{j+1}} \ldots c_{k_n}^{n_n} \text{ by } c_{k_1}^{n_1} \ldots c_{k_j}^{n_j} X c_{k_{j-1}}^{n_{j-1}} Y c_{k_{j+1}}^{n_{j+1}} \ldots c_{k_n}^{n_n}.$$

By repeated application of (3.14b) and (3.19) just as was done in (3.18) we find that

$$X c_{k_{j-1}}^{n_{j-1}} Y c_{k_{j+1}}^{n_{j+1}} \ldots c_{k_n}^{n_n} = R c_{k_{j+1}}^{n_{j+1}} \ldots c_{k_n}^{n_n} S$$

where (i) R is a word in basic commutators c_t with $t > u$, $W(c_t) \leq N$ and such that $c_t^R < c_u$ if $W(c_t) > 1$. (Note property (b) of f_I. Note also that just as long as (c_v, c_u) is basic all the iterated commutators $(c_v, c_u; k)$ in (3.13b) and (3.13d) are basic.) (ii) S is a word in finitely many k-commutators so that each $k > N$.

Continuing the above procedure until every $c_u^{\pm 1}$ in f_I is moved to the front of f_I, we obtain (3.17) from (3.16) so that properties (a), (b) and (c) hold. We observe that if $I < q(N)$, then

$$c_{I+1}^{n_{I+1}} f_{I+1} g_{N+1,I+1} = f_I g_{N+1,I} .$$

In our applications of the collection process, we will require the following terminology given in Definition 3.5 and the modification given in Remark 3.1 below.

Definition 3.5. *If* $I = q(N)$ *in* (3.16) *then* (3.16) *will be written as*

$$f \equiv \prod_{i=1}^{q(N)} c_i^{\varepsilon_i} \bmod F_{N+1}$$

and $\prod_{q(N)} = \prod_{i=1}^{q(N)} c_i^{\varepsilon_i}$ *will be called the* N-*composite basic commutator representation of* f *(N-c.b.c. rep. of f).*

Remark 3.1. In our discussion of the collection process we began with a word (3.15) in generators. We may just as well begin with a word in basic commutators c_{k_i}

$$w = \prod_i c_{k_i}^{\varepsilon_i}$$

provided the c_{k_i} which occur in w satisfy the following requirement: If c_u is the smallest basic commutator among the c_{k_i} in w, then $c_{k_i}^R < c_u$, for every c_{k_i} with $W(c_{k_i}) > 1$. Moreover we began by collecting that $c_u = c_{k_j}$ which was the leftmost one occurring such that $j > 1$ and $k_{j-1} > u$. We may, however, also begin collecting c_u wherever it occurs in w to the right of c_{k_j}.

3(c) *The F-simple Basic Commutators.*

We shall obtain results by working with special sets of basic commutators. The simplest of these is the F-simple basic commutators.

Definition 3.6. *A basic commutator* c *is called* F-simple *if either* $W(c) = 1$ *or* $W(c) > 1$ *but* $W(c^R) = 1$.

As a corollary to Theorem 2.1, we immediately obtain:

Corollary 3.2. *Every element of* F *can be represented modulo* F_{n+1} *by* F-*simple basic commutators.*

We will require the well known identity [5, p.150]

$$(a,b,c)(b,c,a)(c,a,b)$$
$$= (b,a)(c,a)a^{-1}(c,b)a(a,b)b^{-1}(a,c)ba^{-1}(b,c)a(a,c)b^{-1}(c,a)b$$

which we will write in the form

$$(a,b,c) = (b,a)(c,a)(c,b)(c,b,a)(a,b)\cdot(a,c)(a,c,b)(b,c). \tag{3.20}$$

Making use of (3.19) and (3.20), we will establish:

Lemma 3.6. *Let* c *and* d *be* F-*simple basic commutators such that* $c = (c_{j_1},c_{j_2},\ldots,c_{j_n})$, $n > 1$, *and* d *is a generator with* $d < c_{j_n}$. *Then the identity*

$$(c,d) = \prod_1 M(c,d) \prod_2 \tag{3.21}$$

holds in F *where* Π_1 *and* Π_2 *are words in* F-*simple basic commutators* v_i *such that* (i) $1 < W(v_i) \leq n+1$, (ii) *if* $v_i = (w_1,w_2,\ldots,w_t)$ *then* $w_1,w_2,\ldots,w_t$ *is a rearrangement of a subsequence of* $c_{j_1},\ldots,c_{j_n}$, d *and* (iii) $v_i < (c,d)$.

Proof. We shall proceed by induction on the place of c in our ordering of basic commutators. For $n = 2$, we note that $c_{j_1} > d$ since necessarily $c_{j_1} > c_{j_2}$; hence $M(c,d) = (c_{j_1},d,c_{j_2})$ by Lemma 3.5. By (3.20), we find

$$(c,d) = (c_{j_1},c_{j_2})^{-1}(c_{j_1},d)^{-1}(c_{j_2},d,c_{j_1})^{-1}(c_{j_2},d)^{-1}$$
$$(c_{j_1},c_{j_2})(c_{j_1},d)(c_{j_1},d,c_{j_2})(c_{j_2},d)$$

which evidently yields our conclusion in the present case. Next suppose that the lemma has been demonstrated for all $c < c_k$ with $W(c_k) > 2$. We proceed to the smallest F-simple basic commutator c such that $c > c_k$. Let $c^L = A$, $c^R = b$. Then by (3.20)

$$(c,d) = c^{-1}(A,d)^{-1}A^{-1}(b,d)^{-1}Ac\cdot(A,d)(A,d,b)(b,d). \tag{3.22}$$

To proceed from (3.22), we consider two cases: (I) $c^R = c_{j_n} = b > d > c_{j_{n-1}}$ and (II) $c^R = c_{j_n} = b > c_{j_{n-1}} > d$. In both cases we will write z for the commutator (A,d) which appears in (3.22).

In case (I), $z = (A,d)$ is F-simple. Hence (3.22) gives our conclusion since $M(c,d) = (A,d,b)$ by Lemma 3.5.

In case (II), $z = (A,d)$ is not a basic commutator. Since $A = c^L$

and so $W(A) < W(c)$, thus $A < c$. So we may apply our induction to A to write

$$z = \prod_{11} M(A,d) \prod_{21} \tag{3.23}$$

where Π_{11} and Π_{21} are words in F-simple basic commutators v_{i1} such that (i') $1 < W(v_i) \leq n$, (ii') if $v_{i1} = (w_{11}, w_{12}, \ldots, w_{t1})$, then $w_{11}, w_{12}, \ldots, w_{t1}$ is a rearrangement of a subsequence of $c_{j_1}, c_{j_2}, \ldots, c_{j_{n-1}}, d$, and (iii') $v_{i1} < z$.

Next, we substitute (3.23) into (3.22) for (A,d). Applying identity (3.19) to compute $(\Pi_{11} M(A,d)\Pi_{21}, b)$ and the induction hypothesis as necessary completes the induction proof of our lemma for the following reasons:

(a) $M(c,d) = (M(A,d),b)$ by Lemma 3.3;

(b) $z = (A,d) < (A,b) = c$ by Theorem 3.1 ($d < b$ by hypothesis);

(c) $(v_{i1},b) < (z,b)$ by Corollary 3.1 and (iii');

(d) by (iii') and (b) above $v_{i1} < c$ and so either (v_{i1},b) is F-simple or can be expressed according to our induction hypothesis (applied to v_{i1}) in terms of F-simple basic commutators. (To verify property (ii) of our conclusion we recall (ii') above.)

4. *THE LOWER CENTRAL SERIES OF* G

4(a) *Free Generators of* G_2

The groups G of interest to us are free products of finitely generated abelian groups. Thus $G = G(1) * G(2) * \ldots * G(s)$ where each G(i) is a finitely generated abelian group. Hence every G(i) is a direct product of a finite number of cyclic groups of infinite or prime power order. We order the generators of G as follows. Suppose that G as a homomorphic image of F has generators $c_1, c_2, \ldots, c_r$. We know there exists integers

$$0 = n_0 < n_1 < n_2 < \ldots < n_s = r \tag{4.1}$$

such that $c_{n_{i-1}+1}, \ldots, c_{n_i}$ generate G(i) where $i = 1,2,\ldots,s$ and any c_k $(0 = n_0 < k \leq n_s = r)$ has either infinite order or order a power of a prime p(k).

The following notation will be used here. The generator c_k has order in G denoted by $O(c_k)$. Thus $O(c_k) = \infty$ or $p(k)^{\beta_k}$. We write $\alpha_k = O(c_k)$ for $O(c_k) < \infty$ but $\alpha_k = 0$ for $O(c_k) = \infty$. In the rest of this paper, we assume that at least one $\alpha_k \neq 0$. We do so because if all $\alpha_k = 0$, then G reduces to a group for which the quotient groups of the lower central series have been completely determined [3].

Remark 4.1. We note that the groups we are interested in, $\overline{G^n} = G/G_{n+1}$, are finitely generated nilpotent groups (see Corollary 2.1). Moreover if $\alpha_k \neq 0$ for all k, $1 \leq k \leq r$, then $\overline{G^n}$ is a finite group. Since every finite nilpotent group is a direct product of prime power groups, i.e., its Sylow subgroups [5], we may consider each of these direct factors one at a time. This reduces the investigation of $\overline{G^n}$ to the special case of just considering each $\alpha_k = p^{\beta_k}$ for some fixed prime p; i.e. we only consider one prime. Moreover since a finitely generated nilpotent group can be embedded into a direct product of a torsion free group and a finite group (see Theorem 2.1 in [1]), we may assume that either $\alpha_k = p^k$ or $\alpha_k = 0$. Accordingly in what follows, we shall investigate only the special case where the generators $c_1,\ldots,c_r$ of G have either infinite order or order a power of a fixed prime, p.

We next note in the notation just introduced and that of Definition 2.2 that G has the presentation

$$G = \langle c_1,c_2,\ldots,c_r;s_1,s_2,\ldots,s_t\rangle \tag{4.2}$$

where the relators s_m $(1 \leq m \leq t)$ are of two classes:

Class A: $s_m = c_m^{\alpha_m}$ for $m = 1,2,\ldots,r$ (Note that $\alpha_m = 0$ if $O(c_m) = \infty$);

Class B: $s_m = (c_k,c_j)$ for $n_{i-1} < j < k \leq n_i$, $i = 1,2,\ldots,s$ and $m = r+1,\ldots,t$. (See (4.1).)

In what follows, we shall also need the groups F, H and J. We define each of these by means of its presentation. In particular F is the free group

$$F = \langle c_1,c_2,\ldots,c_r\rangle. \tag{4.3}$$

Moreover H and J are presented by

$$H = \langle c_1,c_2,\ldots,c_r;s_1,s_2,\ldots,s_r\rangle \tag{4.4}$$

and

$$J = \langle c_1,c_2,\ldots,c_r;s_{r+1},\ldots,s_t\rangle. \tag{4.5}$$

Thus H is a free product of a finite number of cyclic groups and J is a free product of a finite number of finitely generated torsion-free abelian groups. Moreover the orders of the generators in H are precisely the same as their orders in G and the generators which commute in J are precisely the ones which commute in G. Finally we note that imposing only the relators of Class A on F as in (4.3) gives H, imposing only the relators of Class B on F gives J, but imposing relators of both Classes A and B gives G. This gives rise to the various homomorphisms: $F \to H$, $F \to J$, $F \to G$, $H \to G$, $J \to G$.

Remark 4.2. From now on we usually will use the same letter to stand for an element of F as well as its image in H, J or G. It should be clear from the context which group we are considering the element to be in.

As mentioned earlier, we shall obtain results by working with special sets of basic commutators. The basic commutators of weight n are either F-simple or are commutators of F-simple basic commutators (cf. Corollary 3.1); they are mapped into a basis of $\overline{F}_n = F_n/F_{n+1}$ under the homomorphism $F_n \to \overline{F}_n$ according to Theorem 2.1. By analogy, we shall discuss the construction of bases of $\overline{G}_n = G_n/G_{n+1}$. In order to do this we introduce the G-simple basic commutators. We also define H-simple and J-simple basic commutators. Each of these sets of basic commutators plays an analogous role in their corresponding group. In particular the X-simple commutators correspond to the group X.

Definition 4.1. *A basic commutator* c *is* G-*simple if it satisfies four conditions:*

(i) *Either* c *is a generator or if* $W(c) = n > 1$ *then* $c = (c_{j_1}, c_{j_2}, \ldots, c_{j_n})$ *such that* $c_r > c_{j_1} > c_{j_2} > c_1$ *and* $c_{j_2} \leq c_{j_3} \leq \ldots \leq c_{j_n} \leq c_r$; i.e. c *is* F-*simple*.

(ii) *Criterion* 1: *If* $W(c) > 1$ *then* $(c_{j_1}, c_{j_2}) \neq 1$ *in* J.

(iii) *Criterion* 2: *If the generator* c_{j_i} *occurs* k *times in* c *then* $1 \leq k < O(c_{j_i})$.

(iv) *Criterion* 3: *If* $W(c) > 2$ *and* $(c_{j_1}, c_{j_t}) = 1$ *in* J *for* $2 < t \leq n$ *then* $c_{j_t} \leq c_{j_1}$.

A basic commutator which satisfies (i) *and* (iii) *is called* H-*simple while one that satisfies* (i), (ii) *and* (iv) *is called* J-*simple. Finally, a basic commutator* c *is called* X-*basic if either* c *is* X-*simple or else* c *is not* X-*simple but* $c = (c^L, c^R)$ *is such that both* c^L *and* c^R *are* X-*basic, where* X = F, H, J *or* G. (Note that for X = F, the F-basic commutators are just the basic commutators.)

We next define a set of commutators which contains the set of G-simple basic commutators of weight > 1. Moreover, these commutators are free generators of the commutator subgroup G_2 of G. This result, Theorem 4.1 below, is strongly related to a theorem of Gruenberg [4]. As mentioned earlier, we omit its proof.

Definition 4.2. *Let* $n > 1$. *A commutator* $e = (c_{j_1}^{\varepsilon_1}, c_{j_2}^{\varepsilon_2}, \ldots, c_{j_n}^{\varepsilon_n})$ *is quasi-G-simple provided it has the following four properties:*

(a) $c = (c_{j_1}, c_{j_2}, \ldots, c_{j_n})$ *is G-simple;*

(b) $\varepsilon_i = \pm 1$ *for* $i = 1, 2, \ldots, n$;

(c) $\varepsilon_i = -1$ *only if* $O(c_{j_i}) = \infty$;

(d) $j_k = j_i$ *implies* $\varepsilon_k = \varepsilon_i$.

Theorem 4.1 (See Theorem 2.1 [4]). *The quasi-G-simple commutators are free generators of* G_2.

Corollary 4.1. *The* X-*simple commutators are free generators of the subgroup of* X_2 *which they generate, where* X = F, G, H *or* J.

4(b) *Generators of* $\overline{G^n}$

In this subsection, we prove

Lemma 4.1. *The* G-*simple basic commutators of weight* $\leq n$ *in* F *map into generators of* $\overline{G^n} = G/G_{n+1}$ *under the homomorphism* $F \to G \to \overline{G^n}$ *for* $n \geq 1$.

We will prove Lemma 4.1 through a sequence of preliminary lemmas. We first note that since the groups G(i), the free factors of G, were assumed to be abelian, we know that if the F-simple basic commutator c is not mapped into 1 under the homomorphism $F \to G$, then c must satisfy Criterion 1 of Definition 4.1, i.e. the generators c_{j_1} and c_{j_2} do not belong to the same free factor of G.

To prove our preliminary lemmas, we require some consequences of identity (3.14b). Using the notation of (3.12), repeated use of (3.14b) gives

$$(b,a^2) = (b,a;1)^2(b,a;2)$$

and for $n > 2$ (4.6)

$$(b,a^n) = (b,a;1)^2 w(n-1)(b,a;n)$$

where $w(n-1)$ is a word in $(b,a;1),\ldots,(b,a;n-1)$ which is easily computed by induction on n. As a matter of fact if we let $(b,a;0) = 1$ rather than b as in (3.12) and denote $(b,a;n)$ by n for $n \geq 0$, then we find that (4.6) becomes

$$(b,a^2) = 0112,$$

$$(b,a^3) = 01121223,\ldots,$$

$$(b,a^n) = 01121223 \ldots 1223 \ldots (n-1)(n-1)n,$$

where the sequence for (b,a^n) has 2^n terms for every $n \geq 2$. Evidently $(b,a) = (b,a;1)$ occurs exactly n times on the right hand side of (4.6) and $w(n-1)$ does not contain any $(b,a;k)^{-1}$. In particular for a and b generators of the free group $\langle a,b\rangle$ with $b > a$, we may use the collection process (see (3.13a)) to find

$$(b,a^n) = (b,a;1)^n H_n(a,b) \tag{4.7}$$

where $H_n(a,b)$ is a word in basic commutators of weight > 2 in $\langle a,b\rangle$.

We now proceed to our first preliminary lemma.

Lemma 4.2. *Suppose* c *is an* F-*simple basic commutator of weight* $n > 1$ *in* F. *If* c *does not satisfy Criterion 2, then* $c = w$ *in* H *where* w *is a word in* F-*simple basic commutators of weight* $< n$ *all of which satisfy Criterion 2;* i.e. w *is a word in* H-*simple commutators*. (Note this equality, $c = w$, also holds in the group G because G is a homomorphic image of H.)

Proof. We shall proceed by induction on the place of c in the ordering of Definition 2.1. We begin by considering the smallest possible $n = W(c)$ for which the hypothesis of our lemma may be satisfied; i.e. $W(c) = n \geq p + 1$ by Definition 4.1. For $n = p+1$, we must consider two cases: (i) $c = (b,a;p)$, (ii) $c = ((b,a),b;p-1)$. (Note here for (i) $O(a) = p$ but for (ii) $O(b) = p$.) For case (i), we note that $(b,a;p)$ is a word in $(b,a;1),\ldots,(b,a;p-1)$ in H according to (4.6) since $a^p = 1$; this shows the lemma holds in case (i). In case (ii), we consider $((b,a),b;1) = ((a,b)^{-1},b;1),\ldots,((b,a),b;n) = ((a,b)^{-1},b;n)$. Applying identity (3.19) to these, we first find

$$((b,a),b;1) = ((a,b)^{-1},b) = (a,b)((a,b)(a,b,b))^{-1} = (a,b;1)(a,b;2)^{-1}(a,b;1)^{-1}$$

and continuing to apply (3.19) induction on n yields $((b,a),b;n) = S_n(n+1)$ where $S_n(n+1)$ is a word in $(a,b;1),\ldots,(a,b;n+1)$. For $n = p-1$ in particular, we find $((b,a),b;p-1) = S_{p-1}(p)$. But just as in case (i) $(a,b;p)$ is a word itself in $(a,b;1),\ldots,(a,b;p-1)$ in H since $O(b) = p$ in this case. Thus $((b,a),b;p-1)$ is also a word in $(a,b;1)$ $(a,b;1),\ldots,(a,b;p-1)$ in H. Now $(a,b;1) = (a,b) = (b,a)^{-1}$. But applying (3.19) to $(a,b;2) = ((b,a)^{-1},b)$ gives $(a,b;2) = (b,a)((b,a)(b,a,b))^{-1}$. Using the notation of (3.12), this gives that $(a,b;2)$ is a word in $((b,a),b;0)$ and $((b,a),b;1)$. Inductively, we find by repeated use of (3.19) that $(a,b;n)$ is a word in ((b,a),b;0),...,((b,a),b;n-1) in F for $n = 1,2,\ldots$. Thus again for $n = p-1$, ((b,a),b;p-1) is a word in ((b,a),b;0),...,((b,a),b;p-2) in H. Thus we have demonstrated the lemma for $W(c) = n = p+1$.

Next suppose that the lemma has been demonstrated for all $c < c_k$ with $W(c_k) > p+1$. We then proceed to the smallest F-simple basic commutator c which is such that $c > c_k$ and c satisfies the hypothesis of the lemma. By the induction hypothesis it is evidently now sufficient to prove that in H, c is a word in F-simple basic commutators all of which have weight $< n = W(c)$. Here we again consider two cases: (iii) c^L does not satisfy Criterion 2; (iv) c^L satisfies this criterion.

In case (iii) c^L is by the induction hypothesis a word in F-simple basic commutators of weight $< W(c) - 1 = n - 1$. (We may assume by the induction hypothesis that all these are actually H-simple.) Applying identity (3.19) and Lemma 3.6, where necessary, we find that c is a word in F-simple basic commutators of weight $< n = W(c)$. Thus we obtain the conclusion of our lemma in case (iii).

In case (iv), we start with $c = (c_{j_1}, c_{j_2}, \ldots, c_{j_n})$. Evidently c_{j_n} occurs exactly $O(c_{j_n}) = \alpha_{j_n}$ times in c. Let us write $\alpha = \alpha_{j_n}$. If $c_{j_1} \neq c_{j_n}$, then $c = (e, c_{j_n}; \alpha)$ where $e = (c_{j_1}, c_{j_2}, \ldots, c_{j_{n-\alpha}})$. But using identity (4.6), since $\alpha = O(c_{j_n})$, $c = (e, c_{j_n}; \alpha)$ is a word in $(e, c_{j_n}; 1), \ldots, (e, c_{j_n}; \alpha-1)$ in H. It therefore remains to consider the case $c_{j_1} = c_{j_n}$. We note that $c_{j_2} \neq c_{j_n}$ and $c_{j_3} \neq c_{j_n}$ since $W(c) > p+1$. Now $n = W(c) > 3$, so that $d = (c_{j_1}, c_{j_2}, c_{j_4}, \ldots, c_{j_n})$ is also an F-simple basic commutator by Definition 2.1. Since $W(d) = n-1$, $d < c$ and d does not satisfy Criterion 2 since all occurrences of the generator c_{j_n} in c are still present in d. Thus by the induction hypothesis d is a word in F-simple basic commutators of weight $< n-1$ all of which satisfy Criterion 2; i.e., $d = \prod_k c_k^{e_k}$ where each c_k is H-simple. Applying identity (3.19) to (d, c_{j_3}), we must contend with the commutators (c_k, c_{j_3}). If such a commutator is not basic we apply Lemma 3.6 to rewrite it in terms of F-simple basic commutators all of which are $\leq M(c_k, c_{j_3})$. If any among these F-simple basic commutators does not satisfy Criterion 2, we may apply the induction hypothesis to it since it is earlier than c by weight in our ordering of the basic commutators. (Note $W(c_k) \leq n-2$ for all such k.) In this way, we may rewrite (d, c_{j_3}) as a word in F-simple basic commutators of weight $< n$ all of which satisfy Criterion 2, i.e. all of which are H-simple.

Applying Lemma 3.6 to (d, c_{j_3}), we find

$$(d, c_{j_3}) = \prod_1 M(d, c_{j_3}) \prod_2 \tag{4.8}$$

where Π_1 and Π_2 are words in F-simple basic commutators, z_s, all of which are such that $z_s < (d, c_{j_3})$. But Lemma 3.5 implies that $M(d, c_{j_3}) = c$. Applying the induction hypothesis as necessary to any of the z_s and solving equation (4.8) for $c = M(d, c_{j_3})$ writes c as a word in H-simple basic commutators of weight $< n$. This completes our induction proof.

We now proceed to our last two auxiliary lemmas. It is evident that Lemma 4.1 is a trivial consequence of Lemma 4.2 and of:

Lemma 4.3. *Every F-simple basic commutator c of weight $n > 1$ which satisfies Criteria 1 and 2 equals w in G where w is a word in G-simple basic commutators of weight > 1 but $\le n$.*

To prove Lemma 4.3 it is sufficient to consider the special case of:

Lemma 4.4. *Let c be an F-simple basic commutator which satisfies Criterion 1 but not Criterion 3. Let $c = (c_{j_1}, c_{j_2}, \ldots, c_{j_n})$. Let t be the largest integer such that c_{j_t} occurs in c and $(c_{j_t}, c_{j_1}) = 1$ in J.* (The hypothesis that c does not satisfy Criterion 3 guarantees that such a t exists, $t \ge 3$, and that $j_t > j_1$.) *Let the F-simple commutator $d(c)$ be defined by*

$$d(c) = (c_{j_t}, d_2, d_3, \ldots, d_n) \tag{4.9}$$

where $d_2, d_3, \ldots, d_n$ is that rearrangement of $c_{j_1}, \ldots, c_{j_{t-1}}, c_{j_{t+1}}, \ldots, c_{j_n}$ for which $d_2 \le \ldots \le d_n$.

Then the following holds in J

$$c = H_1\ d(c)\ H_2 \tag{4.10}$$

where H_1 and H_2 are words in J-simple basic commutators u_i such that (i) $1 < W(u_i) < n$; (ii) *if* $u_i = (v_1, v_2, \ldots, v_k)$ *then* $v_1, v_2, \ldots, v_k$ *is a rearrangement of a subsequence of* $c_{j_1}, c_{j_2}, \ldots, c_{j_n}$; (iii) $W(H_1 d(c) H_2) \ge n$ *in* F.

Proof of Lemma 4.3 from Lemma 4.4. Either the F-simple basic commutator c satisfies Criterion 3 or not. If it does, then c is G-simple and there is nothing to prove. On the other hand if c does not satisfy Criterion 3 then by Lemma 4.4 $c = w$ in G where w is a word in J-simple basic commutators, u_i, with $1 < W(u_i) \le n = W(c)$. Finally we note by (ii) of the conditions on (4.10) that each of the u_i must also satisfy Criterion 2 since c does. Thus w is a word in G-simple basic commutators u_i such that $1 < W(u_i) \le n$.

Proof of Lemma 4.4. We shall again proceed by induction on the place of c in our natural linear ordering of Definition 2.1. Clearly $W(c) = n \ge 3$. For $n = 3$ we have $c = (c_{j_1}, c_{j_2}, c_{j_3})$ with $(c_{j_1}, c_{j_3}) = 1$ in J and $c_{j_3} > c_{j_1}$; thus (3.20) yields

$$c = (c_{j_1}, c_{j_2})^{-1}(c_{j_3}, c_{j_2}) \cdot (c_{j_3}, c_{j_2}, c_{j_1}) \cdot (c_{j_1}, c_{j_2})(c_{j_3}, c_{j_2})^{-1}.$$

This shows (4.10) has properties (i) and (ii) for $n = 3$. In order to show that (4.10) has property (iii), we will also show by induction that

$$c \equiv d(c) f(c) \bmod J_{n+1} \tag{4.11}$$

where either $f(c) = 1$ in J or $f(c)$ is a word in J-basic commutators c_k ($k > q(3)$) where each c_k has $W(c_k) = W(d(c)) = n$ in F and no c_k is F-simple. This is sufficient for showing (iii) of (4.10) because if one collects the word $H_1 d(c) H_2$ in F only J-basic commutators arise but this collected word must be congruent to $d(c)f(c)$ modulo J_{n+1} and $d(c)f(c)$ is a word in J-basic commutators all of which have weight at least n in F. Thus $W(H_1 d(c) H_2) \ge n$

in F. Applying (2.3d) and (2.3b) to c and using the fact that $(c_{j_1},c_{j_3}) = 1$ in J gives

$$c = (c_{j_1},c_{j_2},c_{j_3}) \equiv (c_{j_3},c_{j_2},c_{j_1}) \bmod J_4.$$

This proves the lemma for $W(c) = 3$. Next suppose that we have already proven our lemma for all $c < c_k$ with $W(c_k) > 3$. We then proceed to the smallest commutator c which satisfies the hypotheses of our lemma and is such that $c > c_k$.

We now define $\delta = n-t$. For $\delta > 0$, let us write $e = c^L$ and we note that, in J, $e = (c_{j_1},\ldots,c_{j_{n-1}})$ is then by the induction hypothesis a word of the form (4.10),

$$e = H_{11}\, d(e)\, H_{12},$$

where H_{1j} $(j = 1,2)$ are words in J-simple commutators z_i which satisfy (i) and (ii) of our conclusion. In particular every z_i has $1 < W(z_i) < n-1$ and also has $z_i^R \leqslant c_{j_{n-1}}$ (recall by hypothesis $c_{j_1} < c_{j_t} \leqslant c_{j_{n-1}}$). Applying identity (3.19) to the computation of $c = (e,c_{j_n}) = (H_{11}\, d(e)\, H_{12},c_{j_n})$, we find that c is a word in the F-simple basic commutators $d(e)$, z_i, $(d(e),c_{j_n})$ and (z_i,c_{j_n}). Now $d(c) = (d(e),c_{j_n})$ and $W(z_i,c_{j_n}) \leqslant n-1$. So that either the F-simple commutators (z_i,c_{j_n}) satisfy all the required properties of our conclusion or we may use the induction hypothesis to write those which do not as words of the form (4.10) in J-simple basic commutators. (We observe that the only property which (z_i,c_{j_n}) may violate is Criterion 3.) In this way, we finally express c itself as a word in J of the form (4.10) which has all the properties required by our conclusion. Moreover by the induction hypothesis $e \equiv d(e)\, f(e) \bmod J_n$ where $f(e) = 1$ in J or

$$f(e) = \prod_i (x_i,y_i)^{\varepsilon_i}$$

and the (x_i,y_i) are all basic commutators such that

(i) $W(x_i) + W(y_i) = n-1$;

(ii) $W(y_i) > 1$;

(iii) both x_i and y_i are J-commutators.

Now if $f(e) = 1$ in J, just as in the above argument $d(c) = (d(e),c_{j_n})$ (since $\delta > 0$), and so by (2.3c), $c = (e,c_{j_n}) \equiv d(c) \bmod J_{n+1}$ with $f(c) = 1$. But if $f(e) \neq 1$ in J, we apply (2.3c) and (2.3b) to $c = (e,c_{j_n})$ to obtain

$$c \equiv d(c) \prod_i (x_i,y_i,c_{j_n})^{\varepsilon_i} \bmod J_{n+1}.$$

Applying (2.3d) and (2.3b) to (x_i,y_i,c_{j_n}) and observing that all these commutators commute modulo J_{n+1} yields

$$c \equiv d(c) \prod_i (x_i,(y_i,c_{j_n}))^{\varepsilon_i}(x_i,c_{j_n},y_i)^{\varepsilon_i} \bmod J_{n+1}. \tag{4.12}$$

We will now focus on $(x_i,(y_i,c_{j_n}))$ but an exactly analogous argument also takes care of $((x_i,c_{j_n}),y_i)$.

According to Theorem 2.1, the commutators (y_i,c_{j_n}) have b.c. rep.'s of the form

$$(y_i, c_{j_n}) \equiv c_{k_1}^{\gamma_1} \dots c_{k_m}^{\gamma_m} \bmod F_{W(y_i)+2}$$

(For (x_i, c_{j_n}, y_i) we would find the b.c. rep.'s of (x_i, c_{j_n}).) Here each basic commutator c_{k_t} $(1 \leq t \leq m)$ has $W(c_{k_t}) = W(y_i, c_{j_n}) = W(y_i) + 1$ which we shall call W. Moreover each c_{k_t} is either itself F-simple or else c_{k_t} is a commutator of finitely many F-simple commutators each of weight > 1. (This follows from an easy induction on the weight of a basic commutator.) If c_{k_t} is F-simple and satisfies Criterion 1 (note otherwise $c_{k_t} = 1$ in J), we may apply the induction hypothesis to it to find $c_{k_t} = d(c_{k_t})f(c_{k_t})R_{k_t}$ in J where $W(d(c_{k_t})) = W(f(c_{k_t})) = W$ if $f(c_{k_t}) \neq 1$ in J and $W(R_{k_t}) > W$. If c_{k_t} is not F-simple, we apply the induction hypothesis to each F-simple commutator which occurs in c_{k_t}. Thus if c_{k_t} is a commutator of the F-simple commutators $e_1, \dots, e_s$ then we write $c_{k_t} = c(e_1, \dots, e_s)$ to denote this. Now applying the induction hypothesis to each e_i $(1 \leq i \leq s)$, repeated application of (2.3c) gives

$$c_{k_t} \equiv c(d(e_1)f(e_1), \dots, d(e_s)f(e_s)) \bmod J_{W+1}.$$

By the induction hypothesis $W(d(e_i)f(e_i)) = W(e_i)$ for every i, so that (2.3a) gives

$$W(c(d(e_1)f(e_1), \dots, d(e_s)f(e_s)) \geq W.$$

Thus in any case if $(y_i, c_{j_n}) \not\equiv 1 \bmod J_{W+1}$ then it is modulo J_{W+1} a word w_i in basic commutators which are either J-simple themselves or are built up from J-simple commutators u_j. Moreover as an element of F, $W(w_i) = W$ and each u_j is such that $1 < W(u_i) \leq W$. Thus we find that

$$(x_i, (y_i, c_{j_n})) \equiv (x_i, w_i) \bmod J_{n+1}.$$

Now if $(x_i, w_i) \in J_{n+1}$, then $W(x_i, w_i) = n$ in F. If we rewrite (x_i, w_i) in F as a word in F-simple commutators, all the F-simple commutators will have weight $< n$. Thus if we apply the collection process in F (modified to begin with a word in F-simple basic commutators as described in Remark 3.1) to find the b.c. rep. of (x_i, w_i), we find that (i) no basic commutator c_k which occurs in it is F-simple and (ii) every such c_k has $W(c_k) = n$ and is a J-commutator. Here (i) comes from the fact that the word (x_i, w_i) has weight n in F but every F-simple commutator which occurs in this word has weight $< n$ and (ii) comes from the fact that the b.c. rep. of (x_i, w_i) is found from collecting a word in J-simple commutators. Thus we finally have established that modulo J_{n+1}, $(x_i, (y_i, c_{j_n}))$ is a word in basic commutators c_k with properties (i) and (ii) above. According to our previous remark concerning (x_i, c_{j_n}, y_i) and of (4.12), this completes the proof of our lemma in the case $\delta > 0$.

It remains to consider the case $\delta = 0$ for which we write $(c^L)^L = A$, $(c^L)^R = c_{j_{n-1}} = g$ and $c^R = c_{j_n} = h$. Then identity (3.20) gives

$$c = (A,g)^{-1} (A,h)^{-1} A^{-1}(h,g) A(A,g)(A,h)(A,h,g)(h,g)^{-1}.$$

Either $(h,g) = 1$ in J, in which case we may neglect it, or all the commutators A, (A,g), (A,h) and (h,g) are F-simple basic commutators of weight > 1 but $< n$ such that (i) each is obtained by commutation of a reordered subset of the set of generators which are involved in c; (ii) each satisfies Criterion 1. Thus by the induction hypothesis each of these commutators is a word in J-simple commutators of weight > 1 but $< n$. Finally we now rewrite $Z = (A,h,g)$ as a word in the desired form.

We have just seen that in J, $(A,h) = b$ is a word of the form (4.10) in $d(b)$ and J-simple basic commutators s_i of weight $< n-1$ with all of the special properties of our conclusion; thus Z has by identity (3.19) the form

$$Z = H_{11}\,(d(b),g)\,H_{21}$$

where H_{j1} are words in $d(b)$ and all of the commutators s_i and (s_i,g). But applying Lemma 3.6 and the induction hypothesis when necessary, we find easily that (s_i,g) itself is a word in J-simple basic commutators of weight > 1 but $< n$ all of which have the required properties of our conclusion. To conclude our induction proof of (4.10) and (4.11), it remains to examine $(d(b),g)$. We note that $d(b) = d_1 = (c_{j_n},e_2,e_3,\ldots,e_{n-1})$ where $e_2 < e_3 < \ldots < e_{n-1}$ is a rearrangement of $c_{j_1},c_{j_2},\ldots,c_{j_{n-2}}$ by (4.5). We proceed in two cases: Case I: $c_{j_1} < g = c_{j_{n-1}}$. Case II: $c_{j_1} > g = c_{j_{n-1}}$. In Case I, $(d_1,g) = d(c)$ and we have obtained the required form (4.10). Moreover applying the induction hypothesis on (4.11) to b gives

$$b \equiv d_1\, f(b) \bmod J_n\,. \tag{4.13}$$

So that (2.3c) and (2.3b) yield

$$(b,g) \equiv (d_1,g)(f(b),g) \bmod J_{n+1} \tag{4.14}$$

Now $(b,g) = (A,h,g)$ so that applying (2.3d) and (2.3b) gives

$$c = (A,g,h) \equiv (b,g)(A,(h,g))^{-1} \bmod J_{n+1}. \tag{4.15}$$

But since we are assuming that $c_{j_1} < g = c_{j_{n-1}} < h = c_{j_n}$ with $(c_{j_1},c_{j_n}) = 1$ in J then $(g,h) = 1$ in J by the way we have ordered the generators of J (see (4.1)). Thus

$$c \equiv (b,g) \equiv d(c)(f(b),g) \bmod J_{n+1},$$

where either $f(b) = 1$ in J or else an argument analogous to that used in the case $\delta > 0$ to handle $(f(e),c_{j_n})$ expresses $(f(b),g)$ in the desired form. Thus we have obtained (4.11) in the Case I.

We now consider Case II. To express (d_1,g) by J-simple basic commutators in this case $(c_{j_1} > g)$, we again make use of identity (3.20). We find that

$$(d_1,g) = d_1^{-1}\,(d_1^L,g)^{-1}\,(d_1^L)^{-1}\,(d_1^R,g)^{-1}\,(d_1^L)\cdot d_1\,(d_1^L,g)\,(d_1^L,g,d_1^R)\,(d_1^R,g).$$

Since $c_{j_1} > g$, Definition 2.1 implies that $d_1 = (c_{j_n},c_{j_2},\ldots,c_{j_{n-2}},c_{j_1})$ and thus $d_1^R = c_{j_1}$. Again if $(d_1^R,g) = 1$ in J we may neglect it, or all the commutators $d_1^L, d_1, (d_1^L, g)$ and (d_1^R, g) are J-simple basic commutators of weight > 1 but $< n$ with the required properties of our conclusion for (4.10). Also it is evident that $(d_1^L, g, d_1^R) = d(c)$.

Applying the induction hypothesis just as in the case of $c_{j_1} < g$ gives (4.13) where now $d(b) = d_1 = (c_{j_n}, c_{j_2}, \ldots, c_{j_{n-2}}, c_{j_1})$ as above. We again get (4.14), but since $g < c_{j_1} < h = c_{j_n}$ it is possible that $(h,g) \neq 1$ in J. If $(h,g) \neq 1$ in J, then $(A,(h,g))^{-1}$ will be absorbed into f(c) because it is a J-basic commutator as A satisfies Criterion 3 in this case. Applying (2.3d) and (2.3b) to $(d_1, g) = (d_1^L, d_1^R, g)$ gives

$$(d_1, g) \equiv (d_1^L, g, d_1^R)(d_1^L, (d_1^R, g)) \bmod J_{n+1}.$$

Thus substituting this into (4.14) via (2.3c) and using (4.15) yields

$$c \equiv d(c)\ (d_1^L, (d_1^R, g))\ (f(b), g)\ (A, (h,g))^{-1} \bmod J_{n+1}.$$

Here $(f(b),g)$ is treated as done previously; whereas $(d_1^L, (d_1^R, g))$ and $(A,(h,g))$ being J-basic commutators of the appropriate weight are absorbed into f(c).

We have thus proven that the F-simple basic commutator c of our lemma can under all circumstances be rewritten in J in the manner of both (4.10) and (4.11).

4(c) *The F-simple Commutators Expressed by Auxiliary Simple Commutators*

To continue our investigation, it will be necessary to introduce the auxiliary group $\mathcal{F}$ defined below. In particular $\mathcal{F}$ will be a free group of larger rank than the free group F. We will use $\mathcal{F}$ together with a homomorphism $\Phi: \mathcal{F} \to F$ to introduce special sets of commutators in F. These commutators will be images of basic commutators in $\mathcal{F}$.

More precisely we assume that H as given by (4.4) has ρ generators of finite order (i.e. the number of those exponents α_i in (4.4) such that $\alpha_i \neq 0$ which by assumption is positive). Then $\mathcal{F}$ has the presentation

$$\mathcal{F} = \langle a_1, a_2, \ldots, a_{r+\rho} \rangle. \tag{4.16}$$

Here the homomorphism

$$\Phi: \mathcal{F} \to F \tag{4.17}$$

may be constructed inductively as follows:

(i) For the generator c_1 we introduce only a_1 with $\Phi(a_1) = c_1$ if $O(c_1) = \infty$ $(\alpha_1 = 0)$, but we introduce the generators a_1 and a_2 with $\Phi(a_1) = c_1$ and $\Phi(a_2) = c_1^{\alpha_1}$ if $\alpha_1 = O(c_1) < \infty$.

(ii) Suppose that μ among the generators $c_1, c_2, \ldots, c_j$ $(1 \leq j \leq r)$ have finite order in H and that we have introduced for these j generators of F the $j+\mu$ generators $a_1, a_2, \ldots, a_{j+\mu}$ of $\mathcal{F}$ together with their images in F. For the generator c_{j+1} we introduce one generator $a_{j+\mu+1}$ with $\Phi(a_{j+\mu+1}) = c_{j+1}$ if $O(c_{j+1}) = \infty$, but we introduce the generators $a_{j+\mu+1}$ and $a_{j+\mu+2}$ with $\Phi(a_{j+\mu+1}) = c_{j+1}$ and $\Phi(a_{j+\mu+2}) = c_{j+1}^{\alpha_{j+1}}$ if $\alpha_{j+1} = O(c_{j+1}) < \infty$. (We note that $\rho > 0$ since we are assuming in this paper that $\sum_i \alpha_i > 0$. Also by analogy with

G, we are using the notation $O(c_j)$ = order of c_j in H = order of c_j in G.)

We now single out for special attention certain sets of basic commutators in $\mathcal{F}$. (Here we are applying Definition 3.6 to $\mathcal{F}$ rather than F.)

Definition 4.3. *Let* a_i *be an* $\mathcal{F}$*-simple basic commutator of weight* > 1 ($i > r + \rho$), i.e.

$$a_i = (a_{i_1}, a_{i_2}, \ldots, a_{i_n}) \tag{4.16}$$

where $r + \rho \geq i_1 > i_2 \leq \ldots \leq i_n \leq r + \rho$.

(I) *Then* a_i *is a relation simple commutator provided the following happens. Replacement of the generators* a_{i_j} *of* $\mathcal{F}$ *in* (4.16) *by their images* $\Phi(a_{i_j})$ *in* F *yields* $(c_{k_1}^{\varepsilon_1}, c_{k_2}^{\varepsilon_2}, \ldots, c_{k_n}^{\varepsilon_n})$ *so that* (i) $(c_{k_1}, c_{k_2}) \neq 1$ *in* F, (ii) (n-1) *of the exponents* $\varepsilon_1, \ldots, \varepsilon_n$ *are* 1, (iii) *one exponent* ε_t ($1 \leq t \leq n$) *is* $\alpha_{k_t} \neq 0$ (here c_{k_t} must have $\alpha_{k_t} = O(c_{k_t}) < \infty$), *and finally* (iv) *if* $t < n$ *then for* $t < f \leq n$, $k_f \neq k_t$ *and* k_f *occurs fewer than* $O(c_{k_f})$ *times among* $k_1, k_2, \ldots, k_n$.

(II) *The* $\mathcal{F}$*-simple basic commutator* a_i *of* (4.16) *is a simple fundamental commutator provided that the above replacements of* a_{i_j} *by their images* $\Phi(a_{i_j})$ *yield*

$$\Phi(a_i) = c = (c_{k_1}, c_{k_2}, \ldots, c_{k_n})$$

So that c *is* H*-simple.* (See Definition 4.1).

(III) *A* $\mathcal{F}$*-simple basic commutator* a_i, *as in* (4.16) *is auxiliary simple if it is either relation simple or simple fundamental.*

We also need to consider basic commutators in $\mathcal{F}$ which are themselves commutators of the simple commutators introduced in Definition 4.3.

Definition 4.4. *A basic commutator* $a = (a^L, a^R)$ *in* $\mathcal{F}$ *is an* x*-commutator if either it is* x*-simple or else it is built up from* x*-simple commutators such that both* a^L *and* a^R *are* x*-commutators where* x = *fundamental or auxiliary. A relation commutator is a basic commutator* $a = (a^L, a^R)$ *in* $\mathcal{F}$ *such that either* a *is relation simple or else at least one among* a^L *and* a^R *is a relation commutator.*

Remark 4.3. We shall speak of the image of an x-commutator in $\mathcal{F}$ under Φ of (4.15) as an x-commutator in F, i.e., the commutator obtained after the replacement of the generators of $\mathcal{F}$ by their corresponding images in F under Φ. (Here x = relation simple, relation, simple fundamental, fundamental, auxiliary simple or auxiliary.) We shall also assume that the x-commutators in F are ordered by the ordering for basic commutators in $\mathcal{F}$.

To illustrate some of the commutators introduced by Definition 4.3, let us take the special case of

$$H = \langle c_1, c_2; c_1^2 = c_2^2 = 1 \rangle$$

for which $F = \langle c_1, c_2 \rangle$ and $\mathcal{F} = \langle a_1, a_2, a_3, a_4 \rangle$ with $\Phi(a_1) = c_1$, $\Phi(a_2) = c_1^2$, $\Phi(a_3) = c_2$, and $\Phi(a_4) = c_2^2$. The basic commutators in F of weight ≤ 3 are all F-simple and are

$$c_1 < c_2 < c_3 = (c_2,c_1) < c_4 = (c_2,c_1,c_1) < c_5 = (c_2,c_1,c_2).$$

Here c_4 and c_5 are not H-simple because the generators c_1 and c_2, respectively, are repeated too many times in them. In $\mathcal{F}$, there are six basic commutators of weight two and twenty basic commutators of weight three. Again, of course, all the basic commutators of weight ≤ 3 in $\mathcal{F}$ are $\mathcal{F}$-simple and they are $a_1 < a_2 < a_3 < a_4 < a_5 = (a_2,a_1) < \ldots < a_{10} = (a_4,a_3) < a_{11} = (a_2,a_1,a_1) < \ldots < a_{30} = (a_4,a_3,a_4)$.

The relation simple commutators of weight ≥ 2 but ≤ 3 in $\mathcal{F}$ are $a_7 = (a_3,a_2)$ with $\Phi(a_7) = (c_2,c_1^2)$, $a_8 = (a_4,a_1)$ with $\Phi(a_8) = (c_2^2,c_1)$, $a_{16} = (a_3,a_1,a_2)$ with $\Phi(a_{16}) = (c_2,c_1,c_1^2)$, $a_{18} = (a_3,a_1,a_4)$ with $\Phi(a_{18}) = (c_2,c_1,c_2^2)$. We note, for example, that $a_{20} = (a_3,a_2,a_3)$ with $\Phi(a_{20}) = (c_2,c_1^2,c_2)$ is not relation simple because it violates Definition 4.3(I)(iv). The only simple fundamental commutator of weight ≤ 3 in $\mathcal{F}$ is $a_6 = (a_3,a_1)$ with $\Phi(a_6) = (c_2,c_1)$. Thus the auxiliary simple commutators of weight ≤ 3 in $\mathcal{F}$ are a_6 with $\Phi(a_6) = (c_2,c_1)$, a_7 with $\Phi(a_7) = (c_2,c_1^2)$, a_8 with $\Phi(a_8) = (c_2^2,c_1)$, a_{16} with $\Phi(a_{16}) = (c_2,c_1,c_1^2)$, and a_{18} with $\Phi(a_{18}) = (c_2,c_1,c_2^2)$.

The usefulness of the auxiliary commutators as given by Definitions 4.3 and 4.4 rests on the following three properties:

(I) Those auxiliary commutators in F which are not basic commutators are equal to 1 in H, i.e. they are relation commutators. Moreover any non-trivial element of F_2 which maps into 1 in H under the homomorphism $F \to H$ will be represented in terms of relation commutators.

(II) The H-simple commutators are among the free generators of H_2 according to Theorem 4.1 applied to H.

(III) The truth of the following lemma.

Lemma 4.5. *The subgroup of* F *generated by the* F-*simple commutators of weight* > 1 *in* F *is also generated by the auxiliary simple commutators.*

To establish Lemma 4.5, we must first introduce a correspondence between F-simple commutators and auxiliary simple commutators. Let us suppose that $c = (c_{k_1},c_{k_2},\ldots,c_{k_f})$ is a F-simple commutator which is not H-simple. According to Definition 4.1 Criterion 2, c contains a unique generator c_{k_μ} with the following properties: (We note that in H Criterion 3 is automatically satisfied because it is a free product of cyclic groups.)

(i) c_{k_μ} occurs at least $\alpha_{k_\mu} = O(c_{k_\mu})$ times in c. We will write $\alpha = \alpha_{k_\mu}$.

(ii) If $\mu < \nu < f$, then c_{k_ν} occurs in c fewer than $O(c_{k_\nu})$ times. Then the auxiliary simple commutator $\tilde{c}$ in F which corresponds to c is given by

$$\begin{aligned}
&(c_{k_1},c_{k_2},\ldots,c_{k_{\mu-\alpha}},c^{\alpha}_{k_{\mu-\alpha+1}}) \text{ if } \mu = f \text{ and } k_{\mu-\alpha+1} = k_\mu;\\
&(c^{\alpha}_{k_1},c_{k_2},\ldots,c_{k_{\mu-\alpha+1}}) \text{ if } \mu = f \text{ and } k_{\mu-\alpha+1} \neq k_\mu;\\
&(c_{k_1},c_{k_2},\ldots,c_{k_{\mu-\alpha}},c^{\alpha}_{k_{\mu-\alpha+1}},c_{k_{\mu+1}},c_{k_{\mu+2}},\ldots,c_{k_f}) \text{ if } \mu < f \text{ and } k_{\mu-\alpha+1} = k_\mu;\\
&(c^{\alpha}_{k_1},c_{k_2},\ldots,c_{k_{\mu-\alpha+1}},c_{k_{\mu+1}},c_{k_{\mu+2}},\ldots,c_{k_f}) \text{ if } \mu < f \text{ and } k_{\mu-\alpha+1} = k_\mu.
\end{aligned} \tag{4.17}$$

When c is H-simple however then $\tilde{c} = c$.

Having introduced our correspondence $c \to \tilde{c}$, we observe by mathematical induction that Lemma 4.5 is a consequence of

Lemma 4.6. *For an* F-*simple commutator* c *we have* $\tilde{c} = u\,c\,v$ *in* F, *where* u *and* v *are either* 1 *or are words in* F-*simple basic commutators* u_i *and* v_j, *respectively, such that* (i) $u_i < c$ *and* $v_j < c$; (ii) $W(u_i) > 1$ *and* $W(v_j) > 1$.

Proof. Evidently we only need to prove the lemma for c not H-simple. We shall proceed by induction on the place of c in the ordering of Definition 2.1. To start our induction, we observe by Definition 4.1 that the smallest F-simple commutator which is not H-simple belongs to one of the special cases (a) or (b) given below in the notation of (3.12):

(a) $c = (A, c_{k_f}; \mu)$ where $\mu = O(c_{k_f})$;

(b) $c = ((c_{k_1}, c_{k_2}), c_{k_1}; \nu-1)$ where $\nu = O(c_{k_1})$.

But

$$\tilde{c} = (A, c_{k_f}^{\mu}) \text{ in case (a)}; \quad \tilde{c} = (c_{k_1}^{\nu}, c_{k_2}) \text{ in case (b)}$$

according to (4.17).

We now recall the proof of Lemma 4.2 and apply it to our present cases (a) and (b). For case (a), we apply (4.2) with $b = A$, $a = c_{k_f}$, and $n = \mu$, to find that $(A, c_{k_f}^{\mu}) = (A, c_{k_f}; 1)^2 \cdot w(\mu-1) \cdot (A, c_{k_f}; \mu)$ where $w(\mu-1)$ is a word in $(A, c_{k_f}; 1), \ldots, (A, c_{k_f}; \mu-1)$. Thus taking $u = (A, c_{k_f}; 1)^2 w(\mu-1)$, $c = (A, c_{k_f}; \mu-1)$, and $v = 1$ shows that the lemma holds in case (a).

In case (b), we consider $((c_{k_1}, c_{k_2}), c_{k_1}; 1) = ((c_{k_2}, c_{k_1})^{-1}, c_{k_1}; 1), \ldots,$ $((c_{k_1}, c_{k_2}), c_{k_1}; \nu-1) = ((c_{k_2}, c_{k_1})^{-1}, c_{k_1}; \nu-1)$. Applying identity (3.19) to these just as in the proof of Lemma 4.2, we find that $((c_{k_1}, c_{k_2}), c_{k_1}; \nu-1) = S_{\nu-1}(\nu)$ where $S_{\nu-1}(\nu)$ is a word in $(c_{k_2}, c_{k_1}; 1), \ldots, (c_{k_2}, c_{k_1}; \nu)$. More specifically, we note that

$$S_{\nu-1}(\nu) = T(\nu-1)(c_{k_2}, c_{k_1}; \nu)^{-1}\,U(\nu-1) \tag{4.18}$$

where $T(\nu-1)$ and $U(\nu-1)$ are words in $(c_{k_2}, c_{k_1}; 1), \ldots, (c_{k_2}, c_{k_1}; \nu-1)$. This is easily seen by repeated application of identity (3.19). But just as in case (a), we find from (4.2) that $(c_{k_2}, c_{k_1}^{\nu}) = (c_{k_2}, c_{k_1}; 1)^2 w(\nu-1) \cdot (c_{k_2}, c_{k_1}; \nu)$ where $w(\nu-1)$ is a word in $(c_{k_2}, c_{k_1}; 1), \ldots, (c_{k_2}, c_{k_1}; \nu-1)$. Thus $(c_{k_2}, c_{k_1}; \nu)^{-1}$ $= (c_{k_1}^{\nu}, c_{k_2})(c_{k_2}, c_{k_1}; 1)^2 \cdot w(\nu-1)$. Substituting this into (4.18) gives

$$S_{\nu-1}(\nu) = T(\nu-1)(c_{k_1}^{\nu}, c_{k_2})W(\nu-1) \tag{4.19}$$

where $W(\nu-1)$ and $T(\nu-1)$ are again words in $(c_{k_2}, c_{k_1}; 1), \ldots, (c_{k_2}, c_{k_1}; \nu-1)$.

Continuing just as in Lemma 4.2 by applying identity (3.19), we find that each $(c_{k_2}, c_{k_1}; n)$ is a word in $((c_{k_1}, c_{k_2}), c_{k_1}; 0), \ldots, ((c_{k_1}, c_{k_2}), c_{k_1}; n-1)$ for $1 \leq n \leq \nu-1$. Using this to rewrite each of $T(\nu-1)$ and $W(\nu-1)$ in (4.19) and recalling that $\tilde{c} = (c_{k_1}^{\nu}, c_{k_2})$, we find that

$$c = ((c_{k_1}, c_{k_2}), c_{k_1}; \nu-1) = T(\nu-1)\tilde{c}\,W(\nu-1)$$

where $T(\nu-1)$ and $W(\nu-1)$ are words in $((c_{k_1},c_{k_2}),c_{k_1};0),\ldots,((c_{k_1},c_{k_2}),c_{k_1};\nu-2)$. Clearly, this implies our lemma holds in case (b).

Next we suppose that the lemma has been demonstrated for all $c < c_k$. We then proceed to the smallest F-simple basic commutator, $c = (c_{k_1},c_{k_2},\ldots,c_{k_f})$ which is such that (i) $c > c_k$, (ii) c is not H-simple, (iii) c does not belong to a special case (a) or (b), since we have already proven our lemma for such c's.

To continue our induction, we again consider two cases: case (i) c^L is not H-simple and c^R occurs in c fewer than $O(c^R)$ times; case (ii) c^L is H-simple and c^R occurs in c at least $O(c^R)$ times. In case (i), the induction hypothesis implies that $\tilde{c}^L = u_1\, c^L v_1$ where $u_1 = \prod_i u_i^{\varepsilon_i}$, $v_1 = \prod_j v_j^{\gamma_j}$, u_i and v_j are F-simple basic commutators with $u_i < c^L$, $v_j < c^L$, and $W(u_i) > 1$, $W(v_j) > 1$. We note that in this case $\tilde{c} = (\tilde{c}^L,c^R)$ by (4.17). Applying identity (3.19) to $\tilde{c} = (u,c^L v_1,c^R)$ yields

$$\tilde{c} = (u_1 c^L v_1)^{-1} \prod_i [u_i(u_i,c^R)]^{\varepsilon_i}\, c^L \cdot c \cdot \prod_j [v_j(v_j,c^R)]^{\gamma_j}.$$

Since each of the basic commutators u_i and v_j is less than c^L, we obtain $(u_i,c^R) < c$ and $(v_j,c^R) < c$ from the ordering theorem (Theorem 3.1). Finally it is possible that (u_i,c^R) or (v_j,c^R) are not basic commutators. We may then apply Lemma 3.6 to write those commutators which are not basic in terms of F-simple basic commutators all of which are $\leq M(u_i,c^R)$ or $M(v_j,c^R)$ but are of weight > 1. This establishes our lemma in case (i).

In case (ii), we assume that c^L is H-simple and c^R occurs in c at least $O(c^R)$ times. Moreover since we are also assuming that c does not belong to one of the special cases (a) or (b), c must have the form

$$c = ((c_{k_1},c_{k_2},\ldots,c_{k_{f-\nu+1}}),c_{k_1};\nu-1)$$

where $\nu = O(c_{k_1})$ and $f-\nu > 1$. (Note that if $f-\nu = 1$, then c is in special case (b).) It then makes sense to consider $c' = ((c_{k_1},c_{k_2},\ldots,c_{k_{f-\nu}}),c_{k_1};\nu-1)$. By the induction hypothesis, c' satisfies the relation $\tilde{c}' = u'c'v'$ where u' and v' are words in F-simple basic commutators all of which are $< c'$ and have weight > 1 in F. Now $\tilde{c} = (\tilde{c}',c_{k_{f-\nu+1}})$ by (4.17). Just as in case (i) applying identity (3.19) and Lemma 3.6, as necessary, to the computation of $\tilde{c} = (u'c'v',c_{k_{f-\nu+1}})$ gives us $c = u''(c',c_{k_{f-\nu+1}})v''$ where u'' and v'' are words in F-simple basic commutators all of which are less than $M(c',c_{k_{f-\nu+1}})$ by the ordering theorem. Moreover $(c',c_{k_{f-\nu+1}}) = \Pi_1\, M(c',c_{k_{f-\nu+1}})\, \Pi_2$ according to Lemma 3.6 where Π_1 and Π_2 are words in F-simple basic commutators all of which are smaller than $M(c',c_{k_{f-\nu+1}})$. Lemma 3.5, however, implies that $M(c',c_{k-\nu+1}) = c$, so that we finally obtain the conclusion of our lemma also for case (ii). This concludes the proof of Lemma 4.6.

Using the 1-1 correspondence (4.17), Lemma 4.6, Corollary 2.13.1 [9] we immediately find:

Theorem 4.2. *The subgroup of* F_2 *generated by the* F-*simple commutators* c_{i_k} *of weight* > 1 *but* $\leq n$ *has as free generators the auxiliary commutators* $\tilde{c}_{i_k}$

which correspond to the above F-simple commutators according to (4.17).

5. *THE REPRESENTATION ALGORITHM*

5(a) *Formulation of the problem*

Let G, H, F and $\mathcal{F}$ have the meanings given to them previously. The groups of interest to us, $\overline{G^n} = G/G_{n+1}$, consist of the cosets of G_{n+1} in G. According to Corollary 2.1, these cosets have the words

$$\prod_{i=1}^{q(n)} c_i^{\varepsilon_i}$$

as their representatives, where the c_i $(1 \leq i \leq q(n))$ are the basic commutators of weight $\leq n$ in F.

In order to investigate the nilpotent group $\overline{G^n}$, we analyze the groups $\overline{G_m} = G_m/G_{m+1}$ for $1 \leq m \leq n$. For this purpose, let us consider the words

$$\prod_m = \prod_{i=q(m-1)+1}^{q(m)} c_i^{\varepsilon_i} \tag{5.1}$$

in the basic commutators of weight m in F. (Note that q(0) = 0.) Let Π_m be a representative of a coset of $\overline{G}_m$. Corollary 2.1 then implies that representatives of all the cosets of $\overline{G^n}$ are those words

$$\prod_1 \prod_2 \cdots \prod_n \tag{5.2}$$

which are such that every Π_m is in a complete set of coset representatives of $\overline{G}_m$.

Therefore to determine $\overline{G^n}$ we proceed as follows. First we obtain a rule for choosing the representatives Π_m in (5.1) of the cosets of $\overline{G}_m$. Then we compute a multiplication table for the group of coset representatives (5.2) found by our rule.

Accordingly we will start the task of choosing the coset representatives Π_m in (5.1) by finding those Π_m which are in G_{m+1}. In order to do this, we will first consider the special case of the group H given by (4.4).

5(b) *The relators of* $\overline{H}_m$ *expressed by relation commutators* (m > 1)

In this subsection we just consider the group H. By analogy with G, we will write $\overline{H}_m$ for H_m/H_{m+1}. We begin with two preliminary definitions.

Definition 5.1. *A word* $w \in F$ *is said to be a relator in* H *if* w *is mapped into the identity under the homomorphism* $F \to H$. *In particular, a relator* $c_i^{\alpha_i}$ $(\alpha_i \neq 0)$ *which occurs in* (4.4) *is said to be a defining relator.*

Definition 5.2. *Let* $w \in F_2$ *and suppose that*

$$w \equiv \prod_i d_i^{n_i} \bmod F_{m+1} \tag{5.3}$$

where the product is taken over all relation commutators of weight $\le$ m *in* $\mathcal{F}$. *By Remark 4.3, however, we are assuming that the* d_i *in* (5.3) *are elements of* F. *Moreover we shall assume the relation commutators are ordered according to Remark 4.3. Then the right hand side of* (5.3) *is said to be an* m-*composite relation commutator representation of* w (m - *c.r.c. rep. of* w).

Having given your preliminary definitions, we are ready to state the important lemma:

Lemma 5.1. *Let* $m > 1$. *Let* Π_m *be a word of the form* (5.1). *Then* $\Pi_m \in H_{m+1}$ *if and only if it has an* m-*c.r.c. rep. as an element of* F.

It is evident from Definition 5.2 that $\Pi_m \in H_{m+1}$ if Π_m has an m-c.r.c. rep. because each relation commutator is a relator in H. Thus we only need to prove:

Lemma 5.2. *If* $\Pi_m \in H_{m+1}$, *then* Π_m *has an* m-*c.r.c. rep.*

Since the homomorphism $F_{m+1} \to H_{m+1}$ induced by the presentation (4.4) is onto, there exists an $f_{m+1} \in F_{m+1}$ such that $\Pi_{m+1} f_{m+1}$ is a relator in H in the sense of Definition 5.1. Thus it is sufficient to prove:

Lemma 5.3. *Let* $w \in F_2$. *Suppose that* $W(w) = m$ *and* w *is a relator in* H. *Then* w *has an* m-*c.r.c. rep.*

In the following we will establish Lemma 5.3 through the use of the collection process. We will start by dividing the k-commutators in $\mathcal{F}$ (here we are applying Definition 3.4 to $\mathcal{F}$) into two categories.

Definition 5.3 (Refer to (4.14) and (4.15)). *The* 1-*commutator* $a_i^{\pm 1}$ *in* $\mathcal{F}$ *is in category* I *if* $\Phi(a_i)$ *is not a defining relator;* $a_i^{\pm 1}$ *is in category* II *if* $\Phi(a_i)$ *is a defining relator. Let* d *be a* k-*commutator in* $\mathcal{F}$, *where* $k > 1$. *Then* d *is in category* II *if at least one among* d^L *and* d^R *is in category* II; *otherwise* d *is in category* I, *if it is not in category* II.

Let

$$\widetilde{F} = \langle a_1, a_2, \ldots, a_{r+\rho}; a_t = 1 \text{ for certain } t \rangle \tag{5.4}$$

where the t's in (5.4) are such that a_t ranges over all generators of $\mathcal{F}$ of category II. We note that $\widetilde{F}$ is a free group of the same rank as F, thus $\widetilde{F}$ is isomorphic to F.

We are now ready to apply the collection process to the relators w of Lemma 5.3. It is well known that w is a product in F of conjugates of defining relators [9]. Let $\overline{w} \in \mathcal{F}$ be such that $\Phi(\overline{w}) = w$. Such a $\overline{w}$ is given by

$$\overline{w} = \prod_j (w_j^{-1} a_{i_j} w_j)^{\varepsilon_j} \tag{5.5}$$

in $\mathcal{F}$, where (i) the $\varepsilon_j = \pm 1$, (ii) the a_{i_j} are generators of $\mathcal{F}$ of category II, (iii) the w_j are words in generators of $\mathcal{F}$ of category I. We note that the generators of $\mathcal{F}$ of category I have as their images in F precisely all of the generators of F. Thus (iii) does allow the images of the w_j, $\Phi(w_j)$, to range over all elements of F. Applying the collection process to $\bar{w}$ as given by (5.5) we find:

Lemma 5.4. *Let* $a_1, a_2, \ldots, a_{q(m)}$ *be the basic commutators of weight* $\leq m$ *in* $\mathcal{F}$. *Let*

$$\bar{g} = \prod_{j=1}^{q(m)} a_j^{\eta_j}$$

be the m-*c.b.c. rep. of* $\bar{w}$ *as given by* (5.5) *in* $\mathcal{F}$ (here Definition 3.5 is being applied to $\mathcal{F}$). *Then* $\bar{g}$ *has the property that* $\eta_j = 0$, *if* a_j *is in category* I. *Let* $\bar{h} = \bar{g}^{-1}\bar{w}$, $h = \Phi(\bar{h})$ *and* $g = \Phi(\bar{g})$. *Then* $h = g^{-1}w$ *is a relator in* H *and since* $\bar{w} \equiv \bar{g} \bmod \mathcal{F}_{m+1}$, $\bar{h} \in \mathcal{F}_{m+1}$ *and so* $h \in F_{m+1}$.

Proof. The m-c.b.c. rep. of $\bar{w}$ in $\mathcal{F}$ is computed by means of the collection process applied in $\mathcal{F}$ as discussed in subsection 3(b). Thus we have

$$\bar{w} = \prod_{j=1}^{q(m)} a_j^{\eta_j}\, \bar{w}_{m+1} \tag{5.6}$$

where $\bar{w}_{m+1} \in \mathcal{F}_{m+1}$. We observe that in (5.6)

$$\eta_j = 0 \text{ for } 1 \leq j \leq q(1) = r+\rho \tag{5.6a}$$

since $w = \Phi(\bar{w}) \in F_2$. For if any $\eta_j \neq 0$, then $w = \Phi(\bar{w}) \notin F_2$ by definition of Φ.

We note that $\bar{w}$ as given by (5.5) is mapped into 1 in the free group $\tilde{F}$ under the homomorphism $\mathcal{F} \to \tilde{F}$ induced by (5.4). Let $\tilde{w}_{m+1}$ be the image of $\bar{w}_{m+1}$ under this homomorphism. Clearly $\tilde{w}_{m+1} \in \tilde{F}_{m+1}$ by definition of the lower central series. We also observe that under this homomorphism, $\mathcal{F} \to \tilde{F}$, the image of the basic commutator a_j, which we will denote by $\tilde{a}_j$, is such that $\tilde{a}_j = 1$ if a_j is in category II but $\tilde{a}_j = a_j$ is a basic commutator in $\tilde{F}$ if a_j is in category I.

Consider now

$$g_1 = \prod_{j=1}^{q(m)} \tilde{a}_j^{\eta_{j1}}$$

in $\tilde{F}$ where

$$\eta_{j1} = \begin{cases} \eta_j \text{ if } a_j \text{ is in category I} \\ 0 \text{ if } a_j \text{ is in category II.} \end{cases} \tag{5.7}$$

Applying the homomorphism $\mathcal{F} \to \tilde{F}$ to both sides of (5.6) gives $1 = g_1 \tilde{w}_{m+1}$. But g_1 is then the m-c.b.c. rep. of 1 in $\tilde{F}$. Hence all $\eta_{j1} = 0$ by Theorem 2.1 applied to $\tilde{F}$. We then obtain our conclusion according to (5.7).

To establish Lemma 5.3 as a consequence of Lemma 5.4, we require:

Lemma 5.5. *Let* u *be a collected word in basic commutators of weight* $\leq m$ *but* > 1 *in* $\mathcal{F}$, i.e.

$$u = \prod_{i=q(1)+1}^{q(m)} a_i^{\varepsilon_i}.$$

Then $\phi(u)$ *is a word in* F*-simple basic commutators of weight* > 1. *Hence* $\phi(u)$ *is also a word in auxiliary simple commutators by Lemma* 4.5.

To prove Lemma 5.5, we require an auxiliary lemma, Lemma 5.6 below. To state it we first need:

Definition 5.4. *A* n*-commutator* ($n \geq 2$) *of the form*

$$c = (c_{i_1}^{\varepsilon_1}, c_{i_2}^{\varepsilon_2}, \ldots, c_{i_n}^{\varepsilon_n}) \tag{5.8}$$

in F *will be called left normed if* $i_1 \neq i_2$ *and* $c_{i_1}, \ldots, c_{i_n}$ *are among the generators of* F. *Here the* ε_i *are nonzero integers. Moreover if all the* $\varepsilon_i > 0$ *in* (5.8), *then* c *will be called a positive left normed commutator.*

Lemma 5.6. *Let* c *be a positive left normed commutator. Then* $c = w$ *in* F *where* w *is a word in* F*-simple basic commutators of weight* > 1.

Proof. We will proceed by induction on n in (5.8). The collection process will be modified here so that after c has been written as a word in generators we will always only be collecting positive powers of generators. In particular we will repeatedly apply only identities (3.4a) and (3.4c) as necessary. Let us repeat these identities here for the sake of convenience

$$ba = ab(b,a); \quad b^{-1}a = a(b,a)^{-1}b^{-1}.$$

We start with $n = 2$ in which case

$$c = (c_{i_1}^{\varepsilon_1}, c_{i_2}^{\varepsilon_2}) = c_{i_1}^{-\varepsilon_1} c_{i_2}^{-\varepsilon_2} c_{i_1}^{\varepsilon_1} c_{i_2}^{\varepsilon_2}. \tag{5.9}$$

Our process is begun by moving the smaller of $c_{i_1}^{\varepsilon_1}$ and $c_{i_2}^{\varepsilon_2}$ in (5.9) to the left. In particular if $i_1 > i_2$, we repeatedly apply (3.4c) to moving $a = c_{i_1}$ in $c_{i_1}^{\varepsilon_1}$ to the left in (5.9). Hence after (ε_2)-applications of (3.4c) in (5.9), we find that

$$c = c_{i_1}^{-\varepsilon_1+1} (c_{i_2}, c_{i_1})^{-1} c_{i_2}^{-1} (c_{i_2}, c_{i_1})^{-1} c_{i_2}^{-1} \ldots c_{i_2}^{-1} c_{i_1}^{\varepsilon_1 - 1} c_{i_2}^{\varepsilon_2} \tag{5.10}$$

where (ε_2) - $c_{i_2}^{-1}$'s still appear but (ε_2)-commutators, $(c_{i_2}, c_{i_1})^{-1}$, have been introduced. Whereas if $i_2 < i_1$, we repeatedly apply (3.4a) to moving $a = c_{i_2}$ in $c_{i_2}^{\varepsilon_2}$ to the left in (5.9). Hence after (ε_1)-applications, we find that

$$c = c_{i_1}^{-\varepsilon_1} c_{i_2}^{-\varepsilon_2+1} c_{i_1} (c_{i_1}, c_{i_2}) c_{i_1} (c_{i_1}, c_{i_2}) \ldots (c_{i_1}, c_{i_2}) c_{i_2}^{\varepsilon_2 - 1} \tag{5.10'}$$

where (ε_1) - c_{i_1}'s still appear but (ε_1)-commutators (c_{i_1}, c_{i_2}) have been introduced.

Continuing to apply (3.4c) to moving $a = c_{i_1}$ in $c_{i_1}^{\varepsilon_1 - 1}$ in (5.10) to the left or (3.4a) to moving $a = c_{i_2}$ in $c_{i_2}^{\varepsilon_2 - 1}$ in (5.10') to the left until all the $(\varepsilon_1 - 1)$ - c_{i_1}'s or $(\varepsilon_2 - 1)$ - c_{i_2}'s have been collected to the left, we get

$$c = w_1 c_{i_2}^{\varepsilon_2} \tag{5.11}$$

where w_1 is a word in which only the following inverses appear: $c_{i_2}^{-1}$, $(c_{i_2},c_{i_1};1)^{-1},\ldots,(c_{i_2},c_{i_1};\varepsilon_1)^{-1}$. This is true because only (3.4c) has been applied in (5.9). Alternatively, we get

$$c = c_{i_1}^{-\varepsilon_1} w_2 \tag{5.11'}$$

where w_2 is a word in c_{i_1}, $(c_{i_1},c_{i_2};1),\ldots,(c_{i_1},c_{i_2};\varepsilon_2)$. Here no inverses appear since only (3.4a) has been applied in (5.9).

We then continue to apply (3.4c) to moving $a = c_{i_2}$ in $c_{i_2}^{\varepsilon_2}$ in (5.11) to the left or (3.4a) to moving $a = c_{i_1}$ in w_2 in (5.11') wherever it occurs. After all the c_{i_2}'s or c_{i_1}'s have been moved to the left, c will be written as a word in F-simple basic commutators of weight > 1. This completes the induction for $n = 2$.

Now suppose that we have proven the lemma for all positive left normed n-commutators with $2 \leq n \leq N$. We then proceed to $n = N+1$ and consider $c = (c_{i_1}^{\varepsilon_1},\ldots,c_{i_{n-1}}^{\varepsilon_{n-1}},c_{i_n}^{\varepsilon_n}) = (c^L,c_{i_n}^{\varepsilon_n})$. By the induction hypothesis, $c^L = \prod_j a_j^{\gamma_j}$ where the a_j are F-simple commutators with $W(a_j) > 1$. Applying identity (3.19), we find that

$$c = \left(\prod_j a_j^{\gamma_j}\right)^{-1} \cdot \prod_j [a_j(a_j,c_{i_n}^{\varepsilon_n})]^{\gamma_j}.$$

Furthermore (4.6) applied to each $(a_j,c_{i_n}^{\varepsilon_n})$ gives

$$(a_j,c_{i_n}^{\varepsilon_n}) = (a_j,c_{i_n};1)^2 W(\varepsilon_n - 1)(a_j,c_{i_n};\varepsilon_n)$$

where $w(\varepsilon_n - 1)$ is a word in $(a_j,c_{i_n};1),\ldots,(a_j,c_{i_n};\varepsilon_n-1)$. If every (a_j,c_{i_n}) is basic, we are done since all the $(a_j,c_{i_n};k)$ are then also basic.

On the other hand, if any (a_j,c_{i_n}) is not basic, we apply Lemma 3.6 to write it as a word, call it w, in F-simple commutators v_k with $W(v_k) > 1$. We then apply identity (3.19) to compute $(a_j,c_{i_n};2) = (w,c_{i_n})$. Then if any (v_k,c_{i_n}), which occurs in this computation, is not basic we apply Lemma 3.6 to rewrite it in terms of F-simple commutators. In this way repeated applications of identity (3.19) and Lemma 3.6, where necessary, express all of the commutators $(a_j,c_{i_n};k)$, $1 \leq k \leq \varepsilon_n$, in terms of F-simple basic commutators of weight at least 2. This completes the proof of Lemma 5.6.

We observe that since the homomorphism $\phi : \mathcal{F} \to F$ of (4.15) replaces the generator a_i by c_t or $c_t^{\alpha_t}$, then every $\phi(a_i)$, $q(1) + 1 \leq i \leq q(m)$ is either a positive left normed commutator or is a word in positive left normed commutators. Thus by Lemma 5.6 each such $\phi(a_i)$ is a word in F-simple basic commutators of weight > 1. So that, finally, $\phi(u)$ in Lemma 5.5 is also such a word and we see that Lemma 5.6 yields Lemma 5.5 as an easy consequence.

We now apply Lemma 5.4 and 5.5 to establish Lemma 5.3.

Proof of Lemma 5.3. If w is a relator in H and $w \in F_2$, then there is a $\overline{w} \in \mathcal{F}$ given by (5.5) such that $\phi(\overline{w}) = w$ and $\overline{w} \in \mathcal{F}_2$ (see (5.6a)). Let us write the m=c.b.c. rep. of w in $\mathcal{F}$ and call it $\overline{g}$ as in Lemma 5.4 where

$$\bar{w} \equiv \bar{g} \bmod \mathcal{F}_{m+1} \quad \text{and by (5.6a)} \quad \bar{g} = \prod_{j=q(1)+1}^{q(m)} a_j^{\eta_j}.$$

Then according to Lemma 5.4, $\eta_j = 0$ if a_j is in category I. So that if we write $g = \Phi(\bar{g})$ again as in Lemma 5.4, then g is a relator in H. Moreover Lemma 5.5 now implies that $g = \Phi(\bar{g})$ is a word in auxiliary simple commutators

$$g = \prod_i d_{j_i}^{\gamma_i} \tag{5.12}$$

in F.

Let us next consider a preimage under Φ of the right hand side of (5.12). Such a preimage may be given by $\prod_i a_{j_i}^{\gamma_i}$ where the a_{j_i} as elements of $\mathcal{F}$ are $\mathcal{F}$-simple basic commutators of weight > 1 which are auxiliary simple and $\Phi\left(\prod_i a_{j_i}^{\gamma_i}\right) = g$. We now apply the collection process in $\mathcal{F}$ to $\prod_i a_{j_i}^{\gamma_i}$ modified as in Remark 3.1 to begin with a word in $\mathcal{F}$-simple basic commutators. We then find

$$\prod_i a_{j_i}^{\gamma_i} = \prod_{i=q(1)+1}^{q(m)} a_j^{\delta_j} \bar{g}_{m+1} \tag{5.13}$$

in $\mathcal{F}$ where (i) the a_j are basic commutators in $\mathcal{F}$, (ii) $2 \leq W(a_j) \leq m$, (iii) $\delta_j = 0$ if a_j is not an auxiliary commutator, and (iv) $\bar{g}_{m+1}$ is a word in finitely many k-commutators u_z with each $k > m$. We observe that (iii) follows from the fact that (5.13) is the result of collecting a word, $\prod_i a_{j_i}^{\gamma_i}$, in which only auxiliary simple commutators occur. Furthermore this fact also implies that each k-commutator u_z is itself a commutator of auxiliary simple commutators.

Let us next consider

$$\prod_{j=q(1)+1}^{q(m)} a_j^{\delta_{j1}} \bar{g}_{m+1,1} \tag{5.14}$$

where

$$\delta_{j1} = \begin{cases} \delta_j \text{ if } a_j \text{ is not a relation commutator} \\ 0 \text{ if } a_j \text{ is a relation commutator} \end{cases} \tag{5.15}$$

and $\bar{g}_{m+1,1}$ is obtained from $\bar{g}_{m+1}$ by replacing those k-commutators, u_z, the images, $\Phi(u_z)$, of which are relators in H by 1. Now since $g = \Phi\left(\prod_{j=q(1)+1}^{q(m)} a_j^{\delta_j} \bar{g}_{m+1}\right)$ is a relator in H and all the terms which have been deleted from the right hand side of (5.13) are relators in H according to (5.15) and the definition of $\bar{g}_{m+1,1}$, then the image of (5.14) under Φ is also a relator in H. Furthermore Definition 4.4 implies that those basic commutators a_i which occur in (5.14) with nonzero exponents must be fundamental commutators. Also the commutators u_z which occur in $\bar{g}_{m+1,1}$ are words in fundamental commutators. Thus when we map (5.14) into F under Φ, we find that its image, which we shall call g_1, is a word in H-simple commutators in F.

Finally we observe that the H-simple commutators are, according to Corollary 4.1, free generators of subgroups of F as well as of H. Since g_1 is a relator in H and is a word in these free generators, $g_1 = 1$ identically in both F and H. Thus from (5.14)

$$1 \equiv \prod_{j=q(1)+1}^{q(m)} \Phi(a_j)^{\delta_{j1}} \bmod F_{m+1}.$$

Hence since the $\Phi(a_j)$ are basic commutators in F, all $\delta_{j1} = 0$ by Theorem 2.1. So only relation commutators appear in equation (5.13) among the a_j $(q(1)+1 \leq j \leq q(m))$ with nonzero exponents according to (5.15). We shall write the result of applying Φ to both sides of (5.13) as

$$w \equiv \prod_j d_j^{\delta_j} \bmod F_{m+1} \tag{5.16}$$

where the product is taken over all the relation commutators of weight $\leq m$ in $\mathcal{F}$ and the relation commutators, $d_j \in F$, are ordered according to Remark 4.3. Hence (5.16) gives our m-c.r.c. rep. of w.

We have now established Lemmas 5.1, 5.2 and 5.3. By the discussion of subsection 5(a) and by Corollary 2.1, we easily obtain:

Theorem 5.1. *Let* $m > 1$. *Also let* c_i $(q(m-1)+1 \leq i \leq q(m))$ *be the basic commutators of weight* m *in* F. *Then* $\overline{H}_m = H_m/H_{m+1}$ *is the abelian group generated by these basic commutators subject to the additional relations*

$$\prod_m = \prod_{i=q(m-1)+1}^{q(m)} c_i^{\varepsilon_i} = 1$$

in $\overline{H}_m$ *if and only if* Π_m *has an* m-c.r.c. *rep. as an element of* F.

5(c) *The Relators of* $\overline{G}_m$ *expressed by Relation Commutators* $(m > 1)$

In this subsection, we shall apply the results previously established to find "relators" for $\overline{G}_m = G_m/G_{m+1}$. In analogy to Definition 5.1, any word $w \in F$ which is mapped into 1 in G, under the homomorphism $F \to G$ induced by (5.17) below, will be called a relator in G. Similarly a relator in G_m is any word in F_m which is mapped into 1 in G_m under the homomorphisms $F_m \to \overline{G}_m \to G_m$.

To proceed from Theorem 5.1 and establish results for our groups $\overline{G}_m$ from the groups $\overline{H}_m$, we require the group J defined in (4.5).

We next observe by Lemma 4.4 that every F-simple basic commutator $c \neq 1$ in J is such that $c = w$ in J where w is a word in J-simple commutators. This is the content of equation (4.10) in Lemma 4.4. Moreover Theorem 2.1 implies that every $g \in F_n$ is expressible uniquely in F as

$$g = \prod_{i=q(n-1)+1}^{q(n)} c_i^{\varepsilon_i} g_{n+1} \tag{5.17}$$

where $g_{n+1} \in F_{n+1}$ and the c_i's are the basic commutators of weight n in F. Analogously, we claim that every $h \in J_n$ is expressible uniquely in J as

$$h = \prod_{i=q(n-1)+1}^{q(n)} c_i^{\epsilon_{i1}} h_{n+1} \tag{5.18}$$

where $h_{n+1} \in J_{n+1}$, the c_i's are basic commutators of weight n in F and $\epsilon_{i1} = 0$ if c_i is not a J-basic commutator. Clearly any $h \in J_n$ is the image of a $g \in F_n$ under the homomorphism $F_n \to J_n$ induced by the presentation (4.5). Now it is also evident that every c_i in (5.17) is a unique word in F-simple basic commutators c_{i_k} of weight > 1. But then every such F-simple commutator c_{i_k} is in J a unique word $w_k = H_1\, d(c_{i_k}) H_2$ in J-simple basic commutators by (4.10). Replacing every c_{i_k} which arises from the c_i in (5.17) by the w_k and collecting in F, we then find (5.18). (Here by collecting in F it must be understood that we begin the collection process with a word in J-simple commutators in accordance with Remark 3.1.)

It is obvious that when we collect a word in J-simple basic commutators only J-basic commutators will arise. Moreover since $W(w_k) \geq W(c_{i_k})$ in F by (iii) on (4.10), this collection only yields J-basic commutators of weight n in F. Thus we find from Lemma 4.4 that the J-simple commutators play the same role in J as the F-simple commutators do in F. Thus the J-basic commutators of weight n in F are mapped into a basis of the free abelian group $\overline{J}_n = J_n/J_{n+1}$ under the homomorphisms $F_n \to J_n \to \overline{J}_n$.

To proceed from J to G, we now apply the Second Isomorphism Theorem [13]. In order to do this we define the following groups:

K_A = the kernel of the homomorphism $\overline{F}_n \to \overline{H}_n$ i.e. the group of relators in $\overline{F}_n = F_n/F_{n+1}$ induced by the relators of Class A.

K_B = the kernel of the homomorphism $\overline{F}_n \to \overline{J}_n$; i.e. the group of relators in $\overline{F}_n$ induced by the relators of Class B.

$K = K_A \cdot K_B$ (product of groups).

The second Isomorphism Theorem implies that $\overline{G}_n \cong \overline{F}_n/K \cong (\overline{F}_n/K_B)/(K/K_B) \cong \overline{J}_n/(K/K_B)$. Thus to find $\overline{G}_n$ we must impose on $\overline{J}_n$ the relators of Class A and we will just work with the J-basic commutators in F. Hence Theorem 5.1 becomes:

Corollary 5.1. *Let* $m > 1$. *Also let*

$$c_{i_1} < c_{i_2} < \ldots < c_{i_t} \tag{5.19}$$

be the J-commutators of weight m *in* F. *Then* $\overline{G}_m = G_m/G_{m+1}$ *is the abelian group generated by the basic commutators in* (5.19) *subject to the additional relations that*

$$\overline{\prod_m} = \prod_{k=1}^{t} c_{i_k}^{\epsilon_k} = 1 \tag{5.20}$$

in $\overline{G}_m$ *if and only if* $\overline{\Pi}_m$ *has an* m*-c.r.c. rep. as an element of* F.

5(d) $\overline{G}_m$ *determined by Ideal Theory for* $m > 1$

We shall obtain a rule for choosing the representative of the cosets of $\overline{G}_m = G_m/G_{m+1}$ $(m > 1)$ by means of elementary ideal theory [13]. We start this process by dividing the relators which occur in equation (5.22) into t relator classes $C_1, C_2, \ldots, C_t$ corresponding to the J-basic commutators of weight m in F of (5.21).

Definition 5.5. *The relator class* $C_j(m) = C_j$ $(1 \leq j \leq t)$ *consists of collected words* w *in J-basic commutators of weight* m *in* F

$$w = \prod_{k=1}^{t} c_{i_k}^{\varepsilon_k} . \tag{5.21}$$

Here $w \in C_j$ *provided* (i) $\varepsilon_1 = \varepsilon_2 = \ldots = \varepsilon_{j-1} = 0$ *if* $j > 1$, (ii) *if* $\varepsilon_j = 0$, *then* $w = 1$ *in* F (and so $\varepsilon_k = 0$ for all k in (5.23)), *and* (iii) w *is a relator in* G_m. (Note that the word w in (5.23) is in collected order because of inequalities (5.21).) *Moreover if* $w \in C_j$ *then the exponent* ε_j *in* (5.23), *which we shall denote by* $E_j(w)$, *is said to be the minimal exponent of* w. *Finally, we shall refer to the elements of* C_j *as relators of* C_j.

It is evident that every relator of G_n from (5.22) is in a unique C_j. A class C_j may, however, consist only of 1. The class C_j is said to be trivial if it contains only 1 and non-trivial otherwise.

We next observe in the following lemma, the proof of which is self-evident, that the set of minimal exponents $E_j(w)$ of relators of C_j constitutes an ideal in the ring of integers Z. We shall denote this set by I_j.

Lemma 5.7. *Suppose that* C_j $(1 \leq j \leq t)$ *contains the relators* $w_1 = \prod_{k \geq j} c_{i_k}^{\varepsilon_{k1}}$ *and* $w_2 = \prod_{k \geq j} c_{i_k}^{\varepsilon_{k2}}$ *such that* $\varepsilon_{k1} + \varepsilon_{k2} \neq 0$. *Then* $w_3 = \prod_{k \geq j} c_{i_k}^{\alpha\varepsilon_{k1}} \equiv w_1^{\alpha} \bmod F_{m+1}$ *and* $w_4 = \prod_{k \geq j} c_{i_k}^{\varepsilon_{k1}+\varepsilon_{k2}} \equiv w_1 w_2 \bmod F_{m+1}$ *are also elements of* C_j *for any* $\alpha \in Z$. *Hence the set* I_j *is an ideal in* Z.

But since Z is a principal ideal domain, we easily find:

Lemma 5.8. *Suppose that the class* C_j *is non-trivial. The ideal* I_j *is then generated by a positive integer* A_{jj}. *Hence there exists a relator*

$$R_j = \prod_{k=j}^{t} c_{i_k}^{A_{jk}} \tag{5.22}$$

of class C_j *such that* $E_j(R_j) = A_{jj}$.

Definition 5.6. *If* C_j *is non-trivial then the* R_j *of* (5.22) *is said to be the representative of the relator class* C_j. *A trivial relator class* C_j *has representative* $R_j = 1$ *for which we take all exponents, in* (5.22), $A_{jk} = 0$.

Making use of Corollary 5.1, Definitions 5.5 and 5.6, and Lemma 5.8, let us now characterize the relators in (5.20) of $\overline{G}_m$ and also determine the elements of $\overline{G}_m$.

Lemma 5.9. $\overline{\prod}_m = \prod_{k=1}^{t} c_{i_k}^{\varepsilon_k}$ *is a relator in* $\overline{G}_m$ *if and only if there exist integers* $\gamma_1,\gamma_2,\ldots,\gamma_t$ *such that* $\overline{\prod_m} \equiv \prod_{k=1}^{t} R_k^{\gamma_k} \bmod F_{m+1}$ *where the* R_j *are as in* (5.22) *or such that* $\varepsilon_k = \sum_{j=1}^{k} \gamma_j A_{jk}$ *for* $1 \leq k \leq t$ *where the* A_{jk} *are as in* (5.22).

Theorem 5.2. *A complete set of representatives of the cosets of* $\overline{G}_m$ *consists of those words in J-basic commutators of weight* m *in* $F, \prod_k c_{i_k}^{\varepsilon_k}$, *which have the following property:* $0 \leq \varepsilon_k < A_{kk}$ *if* C_k *is non-trivial.*

We have completed our investigation of $\overline{G}_m$. We are thus ready to determine $\overline{G^n} = G/G_{n+1}$ where n is a given positive integer.

5(e) *The group* $\overline{G^n} = G/G_{n+1}$

The representatives of the cosets of $\overline{G^n}$ were discussed in subsection 5(a). To determine them we must still examine $\overline{G_1} = G/G_2$. It is evident from Corollary 2.1 that $\overline{G}_1$ is an abelian group generated by $c_1,c_2,\ldots,c_r$ subject to the additional relations that $c_1^{\alpha_1} = \ldots = c_r^{\alpha_r} = 1$ where the α_i are given in the relations of Class A of presentation (4.2). Applying Theorem 5.2, we immediately obtain Theorem 5.3 below. To state it, however, we require the auxiliary:

Definition 5.7. *Let* $1 \leq j \leq r = q(1)$. *The relator class* C_j *is trivial and consists only of* 1 *if* $\alpha_j = 0$. *The relator class* C_j *is non-trivial and consists of all powers* $c_j^{\alpha_j}$ *if* $\alpha_j \neq 0$. *If* C_j *is non-trivial then* $A_{jj} = \alpha_j$.

Theorem 5.3. *A complete set of representatives of the cosets of* $\overline{G^n} = G/G_{n+1}$ *consists of those collected words in basic commutators of weight* $\leq$ n, $\prod_{i=1}^{q(n)} c_i^{\varepsilon_i}$, *which have the following properties:*

(i) $\varepsilon_i = 0$ *for* $i > r = q(1)$ *if* c_i *is not a J-basic commutator;*

(ii) $0 \leq \varepsilon_i < A_{ii}$ *for* $1 \leq i \leq q(n)$ *if* C_i *is non-trivial.*

Having found the elements of $\overline{G^n}$ it remains to compute a multiplication table for this group. We will see that this can be done by the "representation algorithm" given below. This algorithm finds the representative given by Theorem 5.3 of that coset which contains a specified freely reduced word in the generators $c_1,\ldots,c_r$.

To descrive this algorithm we must assign to relator classes C_j not only their representatives R_j (given by Definition 5.6) but we must also assign to them elements $\widetilde{R}_j$ of F which are relators in G. We do this in:

Definition 5.8. *If* C_j *is trivial then* $\widetilde{R}_j = R_j = 1$. *If* C_j *is non-trivial and* $1 \leq j \leq r$ *then* $\widetilde{R}_j = R_j = c_j^{\alpha_j}$. *Now suppose* $r = q(1) < j \leq q(n)$. *If* c_j *is not a J-basic commutator we take* $\widetilde{R}_j = R_j = 1$; *otherwise if* C_j *is non-trivial then* $\widetilde{R}_j$ *is a* $W(c_j)$*-c.r.c. rep. of* R_j *in* F. (Note that $\widetilde{R}_j$ always exists by Corollary 5.1. However, the choice of $\widetilde{R}_j$ need not be unique. That the $\widetilde{R}_j$

are relators in G is evident from Definitions 4.3 and 5.2.)

We are now ready to consider a freely reduced word $w = \prod_i c_{j_i}^{n_i} \neq 1$ in F where the c_{j_i} are among the generators $c_1, \ldots, c_r$. This word, when thought of as an element of G, is in a coset of $\overline{G^n}$. This coset has a representative of the form $\prod_{i=1}^{q(n)} c_i^{\varepsilon_i}$ as given in Theorem 5.3. In the "representation algorithm" we will compute the exponents $\varepsilon_1, \varepsilon_2, \ldots, \varepsilon_{q(n)}$ in order of increasing subscripts.

To find ε_j $(1 \leq j \leq q(n))$ we proceed as follows. By means of the algorithm to be presented, we express w in the form

$$w = u_{j-1} v_{j-1} f_{j-1} \tag{5.23}$$

valid in the group J such that (5.23) has the three properties

(I)

$$u_{j-1} = \begin{cases} \prod_{i=1}^{j-1} c_i^{\varepsilon_i} & \text{if } 1 < j \leq q(n); \\ 1 & \text{if } j = 1. \end{cases}$$

(II) $v_0 = w$. In general either $v_{j-1} = 1$ in J or $v_{j-1} > c_j$. (See Definition 3.2.)

(III) $f_{j-1} \in G_{n+1}$ when f_{j-1} is considered as an element of G.

We then obtain the exponent ε_j from the expression (5.23) as follows:

(A) If $j > r = q(1)$ and c_j is not a J-basic commutator, then we take $\varepsilon_j = 0$.

(B) Making use of property (II), we rewrite v_{j-1} considered as a word in F, in the form

$$v_{j-1} = \left(\prod_{i=j}^{q(n)} c_i^{\varepsilon_{ij}} \right) h_j$$

through the collection process, where $\varepsilon_{ij} \neq 0$ only if c_i is a J-basic commutator for $j > q(1)$ and $h_j \in F_{n+1}$. (See subsection 3(b).)

(C) We take $\varepsilon_j = \varepsilon_{jj}$ and $\rho_j = 0$ when the relator class C_j is trivial. When C_j is non-trivial, however, the ε_j and ρ_j are the unique solutions of

$$\varepsilon_{jj} = \rho_j A_{jj} + \varepsilon_j; \quad 0 \leq \varepsilon_j < A_{jj}.$$

(See Definitions 5.5 and 5.7 and Lemma 5.8.)

Having completed steps (A), (B) and (C), we express the word w in the form

$$w = u_j v_j f_j \tag{5.24}$$

valid in J such that

$$u_j = u_{j-1} c_j^{\varepsilon_j}, \quad v_j = \tilde{R}_j^{-\rho_j} v_{j1}$$

$$v_{j1} = c_j^{-\varepsilon_j} v_{j-1} h_j^{-1}, \quad f_j = v_{j1}^{-1} \tilde{R}_j^{\rho_j} v_{j1} h_j f_{j-1}.$$

We first observe that applying the collection process to w in F rewrites w in terms of basic commutators. However when we take the image of w in J (also denoted by w), Lemma 4.4 guarantees that all we require for weight > 1 are the basic commutators which are also J-basic commutators. Thus condition (A) may be imposed and this also explains why (5.23) is valid in J. Continuing the collection and noting that $v_{j-1} \geqslant c_j$ by hypothesis, we find first that

$$v_{j1} = c_j^{\varepsilon_{jj}-\varepsilon_j} \left(\prod_{i>j} c_i^{\varepsilon_{ij}} \right) = c_j^{\rho_j A_{jj}} \prod_{i>j} c_i^{\varepsilon_{ij}}$$

in J. But $\tilde{R}_j \equiv R_j \bmod F_{W(c_j)+1}$ according to Definition 5.2 where $R_j = \prod_{i \geqslant j} c_i^{A_{ji}}$ by Lemma 5.7 and Definition 5.6. (Here it must be understood that $A_{ji} = 0$ if c_i is not a J-basic commutator for $i > q(1)$. Also we note that $\tilde{R}_j \equiv R_j \bmod J_{W(c_j)+1}$ must hold.) Thus

$$v_j = \tilde{R}_j^{-\rho_j} v_{j1} \equiv \prod_{i=j+1}^{q(n)} c_i^{\varepsilon_{ij}} \bmod J_{W(c_j)+1} \tag{5.25}$$

and so $v_j \geqslant c_{j+1}$, i.e. v_j also has property (II). Moreover since $\tilde{R}_j$ is a relator in G, evidently $f_j \in G_{n+1}$. Thus having expressed w in the form (5.24) for $1 \leqslant j < q(n)$ we are ready to compute ε_{j+1} by the above steps (A), (B) and (C).

We refer to our computation of the q(n) exponents ε_j by the successive steps (A), (B) and (C) as the "representation algorithm". We note that the last exponent to be computed is $\varepsilon_{q(n)}$. If $c_{q(n)}$ is not a J-basic commutator then $\varepsilon_{q(n)} = 0$ by (A). Otherwise from (5.25), we obtain that $v_{q(n)} = \tilde{R}_{q(n)}^{-\rho_{q(n)}} v_{q(n),1} \equiv 1 \bmod J_{n+1}$. So that $v_{q(n)} \equiv 1 \bmod J_{n+1}$ in any case. Thus we conclude that this algorithm yields a coset representative as given in Theorem 5.3.

Our algorithm evidently provides a means for obtaining a multiplication table of $\overline{G^{\pi}}$. In particular multiplication is carried out by finding the coset representatives of products of freely reduced words. In this way we have completed both of the tasks stated in subsection 5(a).

REFERENCES

1. Gilbert Baumslag, *Lecture notes on nilpotent groups*, American Mathematical Society, Providence, R. I. (1971).
2. Rex S. Dark, *On nilpotent products of groups of prime order*, Ph.D. Thesis, Cambridge University, England (1969).
3. Anthony M Gaglione, Factor groups of the lower central series for special free products, *J. Algebra*, 37 (1975), 172-185.
4. Anthony M. Gaglione, On free products of finitely generated abelian groups, *Trans. Amer. Math. Soc.* 195 (1974), 421-430.
5. Marshall Hall, Jr., *The Theory of Groups*, Macmillan, New York (1959).
6. Philip Hall, A contribution to the theory of groups of prime power order, *Proc. London Math. Soc.* 36 (1934), 29-95.

7. G. Higman, On finite groups of exponent 5, *Proc. Cambridge Phil. Soc.* 52 (1956), 381-390.
8. A. I. Kostrikin, The Burnside Problem (Russian), *Izv. Akad. Nauk. USSR-Math. Ser.* 23 (1959), 3-34; (English), *Amer. Math. Soc. Transl.* 36 (1964), 63-99.
9. Wilhelm Magnus, Abraham Karass & Donald Solitar, *Combinatorial Group Theory,* Interscience, New York (1966).
10. J. Rhodes & E. Shamir, Complexity of grammars by group theoretic methods, *J. Combin. Theory* 4 (1968), 222-239.
11. Ruth R. Struik, On nilpotent products of cyclic groups, *Canad. J. Math.* 12 (1960), 447-462.
12. Ruth R. Struik, On nilpotent products of cyclic groups II, *Canad. J. Math.* 13 (1961), 557-568.
13. B. L. Van der Waerden, *Algebra,* 7th ed., Springer-Verlag, Berlin (1966).
14. Hermann V. Waldinger, A natural linear ordering of basic commutators, *Proc. Amer. Math. Soc.* 12 (1961), 140-147.
15. Hermann V. Waldinger, Addendum to the lower central series of groups of a special class, *J. Algebra* 25 (1973), 172-175.
16. Hermann V. Waldinger, The lower central series of groups of a special class, *J. Algebra* 14 (1970), 229-244.
17. Hermann V. Waldinger & Anthony M. Gaglione, On nilpotent products of cyclic groups - reexamined by the commutator calculus, *Canad. J. Math.* 27 (1975), 1185-1210.

LATTICE-ORDERED GROUPS - A VERY BIASED SURVEY

A.M.W. Glass
Bowling Green State University, Bowling Green, Ohio 43403, USA

Many theorems of combinatorial group theory have analogues which hold for lattice-ordered groups despite the miserable failure of the amalgamation property. Since the proofs rely heavily on groups of order-preserving permutations of linearly ordered sets, this survey should motivate a renewed study of permutation groups as a tool as well as provide propaganda for their power.

Recall that a *lattice* is a partially ordered set in which any two elements a and b have a least upper bound (denoted $a \vee b$) and a greatest lower bound (denoted $a \wedge b$). A lattice that is also a group in which the group operation distributes over the lattice operations is called a *lattice-ordered group;* so $a(b \vee c)d = abd \vee acd$ and $a(b \wedge c)d = abd \wedge acd$ for all a,b,c,d.

The integers $\mathbb{Z}$, the rationals $\mathbb{Q}$ and the reals $\mathbb{R}$ are all lattice-ordered groups under their usual group operations and order. There are many ways to make $\mathbb{Z} \oplus \mathbb{Z}$ a lattice-ordered group. Three favourites are (i) *lexicographic* $\mathbb{Z} \overset{\leftarrow}{\oplus} \mathbb{Z}$: $(m_1,n_1) > (m_2,n_2)$ if ($m_1 > m_2$ or [$m_1 = m_2$ and $n_1 > n_2$]) (ii) $\mathbb{Z} \boxplus \mathbb{Z}$, the *cardinal product:* $(m_1,n_1) > (m_2,n_2)$ if ($m_1 > m_2$ and $n_1 > n_2$) (iii) choose your favourite irrational ξ: $(m_1,n_1) > (m_2,n_2)$ if $m_1 + \xi n_1 > m_2 + \xi n_2$ as real numbers. $C(X,\mathbb{R})$, the group of all continuous functions from a topological space X into the reals under the pointwise ordering ($f < g$ if and only if $xf < xg$ in $\mathbb{R}$ for all $x \in X$) is a lattice-ordered group. For a non-abelian example, let $(\Omega,<)$ be any linearly ordered set. Then $\mathrm{Aut}(\Omega,<)$ the group (under composition) of all order-preserving permutations of Ω is a lattice-ordered group under the pointwise ordering. Where no ambiguity on the linear ordering on Ω is likely, $A(\Omega)$ will be used instead of $\mathrm{Aut}(\Omega,<)$. So, for example, $A(\mathbb{R})$ denotes the lattice-ordered group of all order-preserving permutations of the real line.

One expects some restrictions on the class of groups that can be lattice ordered, and there are some obvious ones:

(i) lattice-ordered groups are torsion-free.

This is easy to see since a simple induction (using the fact that multiplication distributes over the lattice operations) shows that $(x \vee e)^n = x^n \vee x^{n-1} \vee \ldots \vee e$ for all $0 < n \in \mathbb{Z}$, where e denotes the group identity. Hence if $x^n = e$, $x^n \vee e = e$ and the right hand side is just $(x \vee e)^{n-1}$. Thus $x \vee e = e$; so $x < e$. Similarly $x > e$, proving that $x = e$.

Although lattice-ordered groups certainly fail to have unique extraction of roots, the following is true:

(ii) $x^n = y^n \Rightarrow x = z^{-1}yz$ for some z.

(A routine check shows that $y^{n-1} \vee y^{n-2}x \vee y^{n-3}x^2 \vee \ldots \vee x^{n-1}$ is such a z.)

At this stage one may hope for a nice classification of the class of groups which can be made into lattice-ordered groups. However, this is not possible as the following example due to Vinogradov [20] shows:

Let A be the splitting extension of $\mathbb{Q}$ by $\mathbb{Z}$, where if a is the generator of $\mathbb{Z}$, then $\frac{m}{n}$ conjugated by a is just $-\frac{m}{n}$. Now $\mathbb{Q} \oplus \mathbb{Z}$ and $\mathbb{Z}$ satisfy the same group-theoretic sentences and hence so do the metabelian groups $A \oplus \mathbb{Q} \oplus \mathbb{Z}$ and $A \oplus \mathbb{Z}$. The former is easily seen to be lattice orderable since $A \oplus \mathbb{Q}$ and $\mathbb{Z}$ both are, but $A \oplus \mathbb{Z}$ is not. (If it were, $a^{-1}\left(\frac{1}{1} \vee \frac{-1}{1}\right)a = \frac{-1}{1} \vee \frac{1}{1}$, so $\frac{1}{1} \vee \frac{-1}{1}$ has no non-zero "$\mathbb{Q}$ component". But $k\left(\frac{1}{k} \vee \frac{-1}{k}\right) = \frac{1}{1} \vee \frac{-1}{1}$; a contradiction follows by divisibility.)

Hence no set of first order axioms in the language of groups can select precisely those groups that can be made into lattice-ordered groups.

However, one famous theorem from group theory does have an analogue for lattice-ordered groups and provides an essential tool - indeed the only tool - in the study.

The Cayley-Holland Theorem [2, Appendix I]. *Every lattice-ordered group* G *can be embedded in* $A(\Omega)$ *for some linearly ordered set* Ω *with* $|\Omega| \leq |G|$.

Here "embedding" means with respect to both the group and lattice operations.

Indeed, $(\Omega,\leq)$ can be chosen so that it is *doubly homogeneous* [2, Corollary 2.L]: if $\alpha_i < \beta_i$ $(i = 1,2)$, there is $g \in A(\Omega)$ such that $\alpha_1 g = \alpha_2$ and $\beta_1 g = \beta_2$. $\mathbb{Q}$ and $\mathbb{R}$ are examples of such linearly ordered sets; $\mathbb{Z}$ is not.

Once this is believed, it is easy to show further that G can be embedded in the normal subgroup $B(\Omega)$ of $A(\Omega)$; $B(\Omega) = \{g \in A(\Omega) : \exists\alpha,\beta \in \Omega$ $\alpha < supp(g) < \beta\}$, where $supp(g) = \{\gamma \in \Omega : \gamma g \neq \gamma\}$.

But W.C. Holland has shown (see [2, Theorem 2E]) that if $(\Omega,\leq)$ is doubly homogeneous, $A(\Omega)$ and $B(\Omega)$ are divisible and G. Higman has shown (see [2, Theorem 2G]) that such $B(\Omega)$ are simple (as *groups*). Hence

Corollary 1. *Every lattice-ordered group can be embedded in a divisible simple lattice-ordered group.*

An easy adaption of a proof due to B.H. and Hanna Neumann [19] of the group theoretic analogue using wreath products shows that:

Theorem 2 [2, Theorem 10A]. *Every countable lattice-ordered group can be embedded in a two generator lattice-ordered group.*

Note that any conjugate of an element strictly exceeding e must strictly exceed e. The following, due to K.R. Pierce, is therefore the natural analogue of the theorem in group theory that every group can be embedded in one in which any two elements of the same order are conjugate.

Theorem 3 [2, Theorem 10B]. *Every lattice-ordered group can be embedded in one in which any two elements strictly exceeding* e *are conjugate.*

Although the proofs use groups of order-preserving permutations, the actual results certainly have a strong odour, if not of sanctity, at least of combinatorial group theory. The obvious question arises: do free products of lattice-ordered groups with amalgamated sublattice subgroup exist as in groups?

Consider the weaker question of the amalgamation property. A class $\mathbf{C}$ of algebras is said to satisfy the *amalgamation property* if given $G, H_1, H_2 \in \mathbf{C}$ and embeddings $\sigma_i : G \to H_i$ $(i = 1,2)$, there exist $L \in \mathbf{C}$ and embeddings $\tau_i : H_i \to L$ $(i = 1,2)$ making the following diagram commute:

$$\begin{array}{ccc} G & \xrightarrow{\sigma_1} & H_1 \\ \sigma_2 \downarrow & & \downarrow \tau_1 \\ H_2 & \xrightarrow[\tau_2]{} & L \end{array}$$

If $\mathbf{C}$ is the class of groups this holds with L the free product of H_1 and H_2 with G amalgamated. Indeed, $h_1\tau_1 = h_2\tau_2$ $(h_i \in H_i)$ implies there is $g \in G$ such that $h_i = g\sigma_i$ $(i = 1,2)$; i.e., the only elements of H_1 and H_2 that get identified in L are those emanating from the same element of G.

Unfortunately, as K.R. Pierce has proved:

Theorem 4 [2, Theorem 10C]. *The class of all lattice-ordered groups fails to enjoy the amalgamation property.*

More recently, it has been shown that the amalgamation property fails for the class of subdirect products of linearly ordered groups [10]. An easy example is provided by Powell & Tsinakis. Let $\theta \in A(\mathbb{R})$ fix no real number and θ_1, θ_2 be distinct square roots of θ in $A(\mathbb{R})$. Make $A = \sum_{r \in \mathbb{R}} \mathbb{Z}$ into a linearly ordered group in any way. Let $G = A \overleftarrow{\rtimes} \langle\theta\rangle$ and $H_i = A \overleftarrow{\rtimes} \langle\theta_i\rangle$ $(i = 1,2)$. Embed G in H_i in the natural way $(i = 1,2)$. Since linearly ordered groups have unique extraction of roots and there is $a \in A$ with $a^{\theta_1} \neq a^{\theta_2}$, no amalgam of $\{G, H_1, H_2\}$ is possible even in the class of groups with unique extraction of square roots ($\theta_1^2 = \theta = \theta_2^2$ but $\theta_1 \neq \theta_2$).

It follows that any form of the HNN property fails for both these classes [1].

Why might one expect the amalgamation property to fail for lattice-ordered groups? *I* don't know! The only hint comes from the lack of uniqueness of a normal form for words in free lattice-ordered groups.

By the Cayley-Holland theorem, the lattice of a lattice-ordered group is distributive; i.e. $a \vee (b \wedge c) = (a \vee b) \wedge (a \vee c)$ and $a \wedge (b \vee c) = (a \wedge b) \vee (a \wedge c)$. This together with the axioms of lattice-ordered groups show that if a lattice-ordered group G is generated by $x_1, \ldots, x_m$ and $g \in G$, then g is equal to G to an expression $\bigvee_i \bigwedge_j w_{ij}$, where w_{ij} are group words in $x_1, \ldots, x_m$ and i and j run over finite index sets. Such expressions are said to be in *normal form*. In any lattice-ordered group, $|a| = a \vee a^{-1} \geq e$; moreover, $|a| = e$ if and only if $a = e$. Hence $(a \vee a^{-1}) \wedge e = e$ for any a, and therefore $(a \wedge e) \vee (a^{-1} \wedge e) = e$ for any a. Thus the normal form is not unique and there is no known canonical such form. A more complicated example to which we will return later is due to Holland [12]:

$$[[(g \vee e)^x, g^{-1} \vee e], [(g^{-1} \vee e)^x, g \vee e]] = e \qquad (*)$$

for any $g \in G$, where $[a,b]$ is the commutator of a & b and $a^b = b^{-1}ab$.

This example suggests that it will be no easy matter to determine whether or not two arbitrary elements of even a free lattice-ordered group are equal; it is certainly a more difficult problem than that for free groups. To begin with, the free lattice-ordered group on one generator is not $\mathbb{Z}$ but $\mathbb{Z} \boxplus \mathbb{Z}$ which is generated by $(1,-1)$ since $(1,-1) \vee (0,0) = (1,0)$. (The same idea can be used to show that $\mathbb{Z} \boxplus \ldots \boxplus \mathbb{Z}$ is generated by two elements and indeed if a lattice-ordered group G is generated by m elements, $G \boxplus \ldots \boxplus G$ is generated by 2m elements for any finite cardinal sum of copies of G [9].) The free abelian lattice-ordered group on two generators is, as a group, the free abelian group on $\aleph_0$ generators.

Despite these problems, Holland and McCleary [13] have used the Cayley-Holland theorem to give an explicit algorithm to determine whether or not an arbitrary element of a free lattice-ordered group is the identity:

Theorem 5. *Free lattice-ordered groups have solvable word problem.*

What properties do free lattice-ordered groups on more than one generator have? For example, do they have trivial centre and are they cardinally indecomposable? The answer is "of course" and the most perspicuous proof employs a representation theorem. The drawback of this approach is that it takes a sledge hammer to crack a peanut - something I was brought up to believe was too inelegant to be satisfactory. If, in the definition of doubly homogeneous, the element g is restricted to a sublattice subgroup G of $A(\Omega)$, then (G,Ω) or simply G is said to be *doubly transitive*. Such lattice-ordered groups are m transitive for all positive integers m [2, Lemma 1.10.1]; they are immediately seen to have trivial centre and to be directly indecomposable (even as groups). Hence the questions are answered by:

Theorem 6. *Any free lattice-ordered group on more than one generator is isomorphic to a doubly transitive sublattice subgroup of some* $A(\Omega)$.

The proof is a cheap logical trick for an infinite number of generators [2, Theorem 6.7] but the "representation" in $A(\mathbb{Q})$ for a finite number of generators is more subtle [14] or [16]. Actually, Kopytov's representation gives that free lattice-ordered groups do have unique extraction of roots ($x^n = y^n$ implies $x = y$). Just in case the reader thinks that free lattice-ordered groups are understood, I should point out that the conjugacy problem for them remains open - the representation being, so far, inadequate for such existential questions:

Question. *Does there exist an algorithm which when given two arbitrary elements of the free lattice-ordered group on, say, two generators, determines whether or not they are conjugate?*

I next wish to consider free products in the class of lattice-ordered groups. If G is any lattice-ordered group which is not linearly ordered, let $g \in G$ with $g \vee e \neq e \neq g^{-1} \vee e$. Then (*) shows that if $\mathbb{Z} = \langle x \rangle$, the subgroup of the free product (in the class of lattice-ordered groups) of G and $\mathbb{Z}$ is not the free product (in the class of groups) of G and $\mathbb{Z}$. There is no known workable representation theorem for free products of lattice-ordered groups. For example, does the free product of two arbitrary non-singleton lattice-ordered groups have trivial centre? The only results known are generalisations of Theorem 6 - again the "take a sledge hammer to crack a peanut" approach:

Theorem 7 [5]. *If* G *is an arbitrary lattice-ordered group and* F *is a free lattice-ordered group on at least* $|G|$ *generators, then the free product of* G *and* F *is isomorphic to a doubly transitive sublattice subgroup of some* $A(\Omega)$. *The same is true if* G *is countable and* F *is any free lattice-ordered group on at least two generators.*

(Added in proof: The free product of any two non-trivial countable lattice-ordered groups is isomorphic to a doubly transitive sublattice subgroup of $A(\mathbb{Q})$ and hence has trivial centre and contains no non-trivial normal subgroup with non-trivial centraliser. (A.M.W. Glass, Free Products of Lattice-ordered Groups, submitted).)

A *finitely presented lattice-ordered group* is a quotient of a free lattice-ordered group F on a finite number of generators by an "ideal" of F generated by a finite number of elements. Symbolically, $G = (x_1,\ldots,x_m;\ r_1(\underset{\sim}{x}) = e,\ldots,r_n(\underset{\sim}{x}) = e)$ if $G = F/R$, where F is the free lattice-ordered group on $\{x_1,\ldots,x_m\}$ and R is the ideal generated by $\{r_1(x),\ldots,r_n(\underset{\sim}{x})\} \subseteq F$. (Ideals are normal sublattice subgroups that are convex ($a,b \in I$ and $a < c < b$ implies $c \in I$); they are precisely the kernels of homomorphisms.) Since $|g| \geq e$ with $|g| = e$ if and only if $g = e$, $r_1(\underset{\sim}{x}) = e$ & ... & $r_n(\underset{\sim}{x}) = e$ if and only if $|r_1(\underset{\sim}{x})| \vee \ldots \vee |r_n(\underset{\sim}{x})| = e$. Hence any finitely presented lattice-ordered group is a finitely generated one relator lattice-ordered group. Moreover, the free product of $G = (x_1,\ldots,x_m;\ r(\underset{\sim}{x}) = e)$ and $H = (y_1,\ldots,y_k;\ s(\underset{\sim}{y}) = e)$ is just $(x_1,\ldots,x_m,\ y_1,\ldots,y_k;\ |r(\underset{\sim}{x})| \vee |s(\underset{\sim}{y})| = e)$. Caution: finite generation as a lattice-ordered group does *not* imply finite generation as a group.

The last part of the Theorem 7 establishes the following result whose group theoretic analogue is trivial:

Corollary 8. *Every finitely presented lattice-ordered group can be embedded in one which has trivial centre (and is directly indecomposable).*

Given the lack of uniqueness of normal forms, how can analogues of combinatorial group theoretic results be established for lattice-ordered groups? The answer is to use sublattice subgroups of $A(\mathbb{R})$ to show that certain words do not collapse to the identity, although of course other words may. The reason for choosing $A(\mathbb{R})$ is that, by the strengthening of the Cayley-Holland Theorem, every countable lattice-ordered group can be embedded in $A(\mathbb{Q})$ and hence $A(\mathbb{R})$. To avoid any further vagueness, let me provide an example.

Consider the old chestnut $\langle x,y;\ y^{-1}x^2y = x^3\rangle$ in the class of groups (due to G. Baumslag and D. Solitar). This finitely presented group is isomorphic to a proper quotient of itself [15, Theorem IV 4.9]. Now consider the presentation instead in the class of lattice-ordered groups; i.e., $(x,y;\ y^{-1}x^2y = x^3)$. Again the map $x \to x^2$, $y \to y$ determines a homomorphism θ of this lattice-ordered group onto itself. Note that $[x,y^{-1}xy]$ belongs to the kernel of θ. To see $[x,y^{-1}xy]$ is not the identity in $(x,y;\ y^{-1}x^2y = x^3)$, let

$e < f \in B(\mathbb{R})$ have supp(f) a single interval, and $\alpha \in$ supp(f). It is easy to find $g \in A(\mathbb{R})$ such that $g^{-1}f^2g = f^3$ with $\alpha g = \alpha$, $\alpha fg = \alpha f$ and $\alpha f^2 g = \alpha f^3$ (see [3] for the details). Now $\alpha[f,g^{-1}fg] \neq \alpha$ so $[f,g^{-1}fg] \neq e$. If H is the sublattice subgroup of $A(\mathbb{R})$ generated by f and g, then as $g^{-1}f^2g = f^3$, the map $x \to f$, $y \to g$ is a homomorphism of $(x,y;\ y^{-1}x^2y = x^3)$ onto H. Since $[f,g^{-1}fg] \neq e$, $[x,y^{-1}xy] \neq e$. Consequently, the kernel of θ is non-trivial:

Theorem 9 [3]. *The lattice-ordered group* $(x,y;\ y^{-1}x^2y = x^3)$ *is non-Hopfian.*

There is another proof of the group theoretic analogue of Theorem 2 that every countable group can be embedded in a two-generator group. It uses free products with amalgamation and establishes that every countable group G can be embedded in a two generator group whose defining relations are finite, recursive or recursively enumerable if those of G are. But, as Trevor Evans has kindly pointed out, the proof of Theorem 2 achieves this already since the generators x_i of G are mapped to $[b^{2i-1}ab^{-(2i-1)},a]$ $(i \in I \subseteq \mathbb{Z}^+)$, where a and b are the generators of the lattice-ordered group into which we embed G. Hence G can be embedded in the lattice-ordered group with two generators a and b and defining relations obtained from those of G by replacing each x_i by $[b^{2i-1}ab^{-(2i-1)},a]$:

Theorem 10. *Every countable lattice-ordered group* G *can be embedded in a two generator lattice-ordered group that is finitely (recursively) presented if* G *is.*

Another proof of Theorem 10 (with three instead of two) illustrates an idea which is central in the proof of Theorem 11. One needs to obtain from a finite number of defining relations that some specified element a is disjoint from all its conjugates by powers of an element b; i.e., $a \wedge b^{-n}ab^n = e$ for all $0 \neq n \in \mathbb{Z}$. Unfortunately, this is a recursive set of relations instead of a finite one. However, if $a_0, b_0 \in A(\mathbb{R})$ satisfy this recursive set of relations, let $c_0 \in A(\mathbb{R})$ be (well-) defined by:

$$\alpha c_0 = \begin{cases} \alpha b_0^{-n}a_0b_0^n & \text{if } \alpha \in \mathrm{supp}(b_0^{-n}a_0b_0^n) \text{ for some } 0 < n \in \mathbb{Z} \\ \alpha & \text{otherwise.} \end{cases}$$

Then $a_0 \wedge c_0 = e$, $b_0^{-1}a_0b_0 \vee c_0 = c_0$ and $b_0^{-1}c_0b_0 \vee c_0 = c_0$. Conversely, in any lattice-ordered group, $f \wedge h = e$, $g^{-1}fg \vee h = h$ and $g^{-1}hg \vee h = h$, imply by induction $g^{-(n+1)}fg^{n+1} \leq g^{-n}hg^n \leq h$; since $f \wedge h = e$, $f \wedge g^{-n}fg^n = e$ for all $0 < n \in \mathbb{Z}$. Hence $(a,b;\ b^{-n}ab^n \wedge a = e)_{0 \neq n \in \mathbb{Z}}$ is embeddable in $(a,b,c;\ a \wedge c = e,\ b^{-1}ab \vee c = c,\ b^{-1}cb \vee c = c)$.

Next since $x = (x \vee e)(x^{-1} \vee e)^{-1}$, we may always assume that the set of generators of a lattice-ordered group exceed e (replace each x by $x \vee e$ and $x^{-1} \vee e$). Moreover, $G = (x_1,\ldots,x_m;\ r(\underline{x}) = e)$ can be embedded in $A((0,1))$ and and hence in the sublattice subgroup L of $A(\mathbb{R})$ generated by $x_1,\ldots,x_m$, b_0 and c_0 where b_0 is translation by +1 and c_0 is obtained from b_0 and $a_0 = \bigvee_{i=1}^{m} x_i$ as above. The proof of Theorem 3 gives the existence of a lattice-ordered group K containing L as a sublattice subgroup and $d_0 \in K$ such that $d_0^{-1}b_0^{-i}x_ib_0^id_0 = b_0^{-i}a_0b_0^i$ $(1 \leq i \leq m)$. Hence G can be embedded in $H = (x_1,\ldots,x_m,\ a,b,c,d;$

$r(\underset{\sim}{x}) = e,\ a = \bigvee_{i=1}^{m} x_i,\ a \wedge c = e,\ b^{-1}ab \vee c = c,\ b^{-1}cb \vee c = c,\ d^{-1}b^{-i}x_ib^id = b^{-i}ab^i)_{1 \leq i \leq m}$. But H is generated by a,b,c,d since $x_i = b^idb^{-i}ab^id^{-1}b^{-i}$ $(1 \leq i \leq m)$. Thus, by an easy expansion of the proof, we obtain Theorem 10 with four in place of two.

Actually, as David Feldman has pointed out, four can be reduced to three [5].

Given any recursively enumerable set X of the natural numbers, it is possible to define a finitely presented lattice-ordered group G_X with elements $a,b,c_X \in G_X$ such that $c_X^{-1}b^{-m}ab^mc_X = b^{-m}ab^m$ if $m \in X$. Moreover, if the defining relations of G_X are chosen with sufficient care, it is possible to construct elements $a_0,b_0,c_{0,X},\ldots$ of $A(\mathbb{R})$ satisfying them but with $c_{0,X}^{-1}b_0^{-m}a_0b_0^mc_{0,X} \neq b_0^{-m}a_0b_0^m$ if $m \notin X$. Hence $c_X^{-1}b^{-m}ab^mc_X = b^{-m}ab^m$ if and only if $m \in X$. Consequently, if X is not recursive:

Theorem 11 [7]. *There exists a finitely presented lattice-ordered group with insoluble group word problem.*

The Turing degree of the word problem for G_X is at least that of X. I do not know if equality holds.

By Theorem 10 - and in sharp contrast with Magnus' result for groups [15].

Corollary 12. *There exists a two generator one relator lattice-ordered group with insoluble group word problem.*

I do not know if in Theorem 11 such a lattice-ordered group can be defined just by group relations or whether any finitely presented group with insoluble word problem can be lattice-ordered. Also, I do not know of any finitely presented lattice-ordered group with soluble word problem and insoluble conjugacy problem. The technique in [17, §IIIA] runs aground because of "convex" in the definition of ideal among other things.

The isomorphism problem is also insoluble. Indeed, a property $\mathfrak{P}$ of finitely presented lattice-ordered groups is said to be *Markov* if

(i) $G \cong H$ and H enjoys $\mathfrak{P}$ imply G enjoys $\mathfrak{P}$,

(ii) some finitely presented G_1 enjoys $\mathfrak{P}$, and

(iii) some finitely presented G_2 cannot be embedded in any finitely presented lattice-ordered group enjoying $\mathfrak{P}$.

For example, $\mathfrak{P}$ might be the property of having just one element, being abelian, or being free (the last is not hereditary; it is Markov by Theorems 5 and 11).

Given $G = (x_1,\ldots,x_m;\ r(\underset{\sim}{x}) = e)$ and $w(\underset{\sim}{x}) \in F$, where F is the free lattice-ordered group on $\{x_1,\ldots,x_m\}$, it is possible to construct a lattice-ordered group $G(w) = (x_1,\ldots,x_m,\ a_0,a_1,a_2,a_3,a_4;\ r(\underset{\sim}{x}) = e,\ s(\underset{\sim}{x},\underset{\sim}{a}) = e)$ such that (a) $G(w) \cong \{e\}$ if $w(\underset{\sim}{x}) = e$ in G and (b) G can be embedded in G(w) otherwise. The construction of G(w) is uniform and effective (indeed explicit) in G and w, but quite messy and uses groups of order-preserving permutations (no surprise!) - see [4] for the proof. Explicitly G(w) =
$(x_1,\ldots,x_m,a_0,a_1,a_2,a_3,a_4;\ r(\underset{\sim}{x}) = e,\ a_0^{-1}|w|a_0 = \bigvee\{|x_i| : 1 \leq i \leq m\}$,
$|wa_4^{-1}| \wedge |a_4| = e,\ a_2^{-1}|a_4|a_2 = |a_0|,\ |a_2| \wedge a_1^6 = |a_2|,\ |a_3| \wedge a_1^6 = |a_3|$,

$(|a_0| \vee a_1^{-1}|a_0|a_1 \vee a_1^{-3}|a_0|a_1^3 \vee a_1^{-4}|a_0|a_1^4)^4 = a_3^{-1}a_1^3a_3)$.

Now if G_0 is given by Theorem 11 and G_1 and G_2 by the definition of a Markov property, it is easily seen that if w is any element of the free lattice-ordered group on the generators of G_0, then the free product of G_1 and H(w) enjoys $\mathfrak{P}$ if and only if $w = e$ in G_0, where H is the free product of G_0 and G_2. Hence we obtain the analogue of [15, Theorem IV.4.1]:

Theorem 13 [4]. *Given any Markov property $\mathfrak{P}$ for finitely presented lattice-ordered groups, there is no algorithm to determine whether or not an arbitrary finitely presented lattice-ordered group enjoys $\mathfrak{P}$; in particular, whether or not an arbitrary finitely presented lattice-ordered group has only one element!*

It follows that there is no algorithm to determine whether or not two arbitrary finitely presented lattice-ordered groups are isomorphic. Indeed,

Theorem 14 [8]. *In any non-trivial variety of lattice-ordered groups, the isomorphism problem for* 10 *generator* 1 *relator finitely presented members is insoluble.*

Theorem 14 applies to finitely presented abelian lattice-ordered groups which have uniformly soluble word problem. Its proof does not use permutation groups!

The existence of G(w) mentioned above is useful in other ways.

Theorem 15 [3]. *Every countable lattice-ordered group can be embedded in a seven generator simple one.*

This follows from the properties of G(w) and a general argument due to C.F. Miller III. The theorem is the natural analogue of a theorem of Philip Hall [11].

A similar proof to that of the existence of G(w) establishes:

Theorem 16 [5]. *Every finitely presented lattice-ordered group can be embedded in an eight generator one relator perfect one.*

Here *perfect* means that the convex sublattice subgroup of G generated by the commutator subgroup [G,G] is all of G.

Groups of order-preserving permutations can also be used to show that the product varieties of lattice-ordered groups $\mathfrak{A}^2, \mathfrak{A}^3, \mathfrak{A}^4, \ldots$ ($\mathfrak{A}$ the variety of all abelian lattice-ordered groups) are each generated by a single finitely presented (in the variety of all) lattice-ordered group(s) - see [6].

One result I would dearly like to prove (using permutation groups - how else!) is the analogue of Graham Higman's great theorem for groups [15, §IV7]:

The finitely generated sublattice subgroups of finitely presented groups are precisely those defined by recursively enumerable sets of relations.

Such a connection between lattice-ordered groups and recursion theory would be most welcome.

The above survey should be adequate to persuade the reader of the power of permutation groups. Since Philip Hall has given a proof of the existence of free products of groups with amalgamated subgroup using permutation groups [18], this power is not altogether surprising. In the absence of the amalgamation property it is most propitious though it certainly lacks the grace of combinatorial group theory and its relation to any sophisticated geometry and topology - rather like going by boat from London to St Andrews, and clockwise at that, instead of taking the train!

Acknowledgement. Research supported in part by NSF grant no. 8401745.

REFERENCES

1. A.M.W. Glass, Results in partially ordered groups, *Comm. in Algebra* 3 (1975), 749-761.
2. A.M.W. Glass, *Ordered Permutation Groups*, London Math. Soc. Lecture Notes Series no. 55 (University Press, Cambridge, 1981).
3. A.M.W. Glass, Countable lattice-ordered groups, *Math. Proc. Cambridge Phil. Soc.* 94 (1983), 29-33.
4. A.M.W. Glass, The isomorphism problem and undecidable properties for finitely presented lattice-ordered groups, in *Orders: description and roles*, Annals of Discrete Math. 23, N. Holland Math. Studies 99 (ed. M. Pouzet and D. Richard, N. Holland, Amsterdam, 1984), pp. 157-170.
5. A.M.W. Glass, Effective embeddings of countable lattice-ordered groups, in *Proc. 1st International Conference on Ordered Algebras*, Luminy 1984, ed. M. Jambu-Giraudet and S. Wolfenstein.
6. A.M.W. Glass, Generating varieties of lattice-ordered groups: approximating wreath products, Illinois J. Math. 30 (1986).
7. A.M.W. Glass & Y. Gurevich, The word problem for lattice-ordered groups, *Trans. Amer. Math. Soc.* 280 (1983), 127-138.
8. A.M.W. Glass & J.J. Madden, The word problem versus the isomorphism problem, *J. London Math. Soc.* 30 (1984), 53-61.
9. A.M.W. Glass & H.H.J. Riedel, Growth sequences - a counterexample, *Algebra Universalis*, to appear.
10. A.M.W. Glass, D. Saracino & C. Wood, Non-amalgamation of ordered groups, *Math. Proc. Cambridge Phil. Soc.* 95 (1984), 191-195.
11. P. Hall, Embedding a group in a join of given groups, *J. Australian Math. Soc.* 17 (1974), 434-495.
12. W.C. Holland, Group equations which hold in lattice-ordered groups, *Symposia Math.* 21 (1977), 365-378.
13. W.C. Holland & S.H. McCleary, Solvability of the word problem in free lattice-ordered groups, *Houston J. Math.* 5 (1979), 99-105.
14. V.M. Kopytov, Free lattice-ordered groups, *Siberian Math. J.* 24 (1983), 98-101.
15. R.C. Lyndon & P.E. Schupp, *Combinatorial Group Theory*, Ergebnisse Math. Grenzgeb. 89 (Springer-Verlag, Heidelberg, 1977).
16. S.H. McCleary, Free lattice-ordered groups represented as 0-2 transitive ℓ-permutation groups, *Trans. American Math. Soc.* 290 (1985), 69-79.
17. C.F. Miller III, *On group-theoretic decision problems and their classification*, Annals of Math. Studies, 68 (University Press, Princeton, 1971).
18. B.H. Neumann, An essay on free products of groups with amalgamations, *Philos. Trans. Royal Soc. London Ser. A* 246 (1954), 503-554.
19. B.H. Neumann & H. Neumann, Embedding theorems for groups, *J. London Math. Soc.* 34 (1959), 465-479.
20. A.A. Vinogradov, Non-axiomatizability of lattice orderable groups, *Siberian Math. J.* 13 (1971), 331-332.

TOTALLY ORTHOGONAL FINITE GROUPS

L.C. Grove
University of Arizona, Tucson, Arizona 85721, U.S.A.

K.S. Wang
University of Arizona, Tucson, Arizona 85721, U.S.A.

1. *ORTHOGONALITY*

A complex representation of a finite group is called *orthogonal* if it is equivalent with a real representation. A group is called *totally orthogonal* if all its complex representations are orthogonal. We present some results here on totally orthogonal groups and, more generally, on the problem of determining the number of inequivalent absolutely irreducible orthogonal representations of a group. Full details will appear elsewhere.

Denote by Irr(G) the set of all absolutely irreducible characters of a group G, and by $\mathfrak{L}(G)$ its subset of linear (degree 1) characters. If $\chi \in \mathrm{Irr}(G)$ then its *Frobenius-Schur indicator* is $\nu(\chi) = |G|^{-1}\Sigma\{\chi(x^2):x \in G\}$. The Frobenius-Schur Theorem asserts that if χ is orthogonal (type 1) then $\nu(\chi) = 1$, if χ is real-valued but not orthogonal (type 2) then $\nu(\chi) = -1$, and if χ is not real-valued (type 3) then $\nu(\chi) = 0$. Define $\nu^+(G)$, $\nu^-(G)$, and $\nu^0(G)$ to be the numbers of $\chi \in \mathrm{Irr}(G)$ that are of types 1, 2, and 3, respectively. Thus G is totally orthogonal if and only if $\nu^+(G) = |\mathrm{Irr}(G)|$.

An element $x \in G$ is called *real* if it is conjugate in G to x^{-1}. If every $x \in G$ is real then G is called *ambivalent*. We will write $\mathfrak{A}$ for the class of ambivalent groups and $\mathfrak{TO}$ for the class of totally orthogonal groups. It is well known, and easy to prove, that $G \in \mathfrak{A}$ if and only if all complex characters of G are real-valued, so $\mathfrak{TO} \subseteq \mathfrak{A}$.

For any element $x \in G$ we will write $J_G(x) = \{y \in G:y^2 = x\}$.

Proposition 1.1. *If* $|J_G(1)| > |G|/2$ *then* $G \in \mathfrak{TO}$.

Proposition 1.2. *If* A *is abelian,* $\sigma \in \mathrm{Aut}(A)$ *has order* 2, *and* ${}^{\sigma}a=a^{-1}$ *for all* $a \in A\setminus J_A(1)$, *then* $G = A \rtimes \langle\sigma\rangle \in \mathfrak{TO}$.

Proposition 1.3. *If* $G \in \mathfrak{A}$ *then* $\mathfrak{L}(G) \cong G/G'$ *and* Z(G) *are both elementary abelian 2-groups.*

Theorem 1.4. *If* $G \in \mathfrak{TO}$ *then* G *is generated by involutions.*

Corollary 1.5. *If* $G \in \mathfrak{TO}$ *and* $[G:N] = 2$ *then* $G = N \rtimes \langle\sigma\rangle$ *for some* $\sigma \in J_G(1)$.

Proposition 1.6. *If* $G \in \mathfrak{TO}$ *and* $[G:N] = 2$ *then* $\nu^+(N)$ *is equal to the number of real conjugacy classes in* N.

The roles of G and N in Proposition 1.6 can be reversed, as follows.

Proposition 1.7. *If* $[G:N] = 2$ *and* $N \in \mathfrak{TO}$ *then* $\nu^+(G)$ *is equal to the number of real conjugacy classes in* G.

Theorem 1.8. *If* G *is a 2-group then* $G \in \mathfrak{TO}$ *if and only if*

(1) G *is ambivalent,*

(2) G *is generated by involutions, and*

(3) *if* $[G:N] = 2$ *then* $\nu^-(N) = 0$.

If E is an elementary abelian 2-group and G is a semidirect product $E \rtimes H$ then H acts on $\mathfrak{L}(E)$ as follows: $\phi^h(v) = \phi({}^h v)$ for all $\phi \in \mathfrak{L}(E)$, $h \in H$, and $v \in E$, where ${}^h v = hvh^{-1}$. In that setting we have the following result.

Proposition 1.9. *If* $G = E \rtimes H$, *with* E *an elementary abelian 2-group, then* $G \in \mathfrak{TO}$ *if and only if* $\mathrm{Stab}_H(\phi) \in \mathfrak{TO}$ *for all* ϕ *in* $\mathfrak{L}(E)$.

A group G is called k-*uniform* if $\{1,k\}$ is the set of degrees of characters in Irr(G). For example, if G is nonabelian and has an abelian normal subgroup of prime index p, then G is p-uniform by Ito's Theorem.

Proposition 1.10. *If* G *is ambivalent and* k-*uniform then either* G *is a 2-group or else* $k = 1$ *or* 2. *In particular* k *is a power of* 2.

Theorem 1.11. *A nonabelian group* G *is totally orthogonal and 2-uniform if and only if* $G = A \rtimes \langle\sigma\rangle$, *where* A *is abelian,* $|\sigma| = 2$, *and* ${}^\sigma a = a^{-1}$ *for all* $a \in A \setminus J_A(1)$.

2. *EXAMPLES*

All finite groups generated by reflections in Euclidean spaces are known to be totally orthogonal. That fact, together with Proposition 1.6, yields the next result.

Proposition 2.1. *If* H *is the rotation subgroup of a finite reflection group then* $\nu^+(H)$ *is equal to the number of real conjugacy classes in* H.

As a special case we obtain a combinatorial expression for $\nu^+(\mathrm{Alt}(n))$. Write t_n for the number of Sym(n)-conjugacy classes in Alt(n) that split into 2 Alt(n)-conjugacy classes, and r_n for the number of them that split into 2 real classes (both conditions translate into easily verifiable arithmetical restrictions on cycle types).

Proposition 2.2. *If* $P(n)$ *denotes the partition function then* $\nu^+(\mathrm{Alt}(n)) = (P(n) - t_n + 4r_n)/2$.

Proposition 2.3. *If* T *is a 2-Sylow subgroup in* Alt(n) *then* $T \in \mathfrak{TO}$.

Theorem 2.4. *If* T *is a 2-Sylow subgroup in a finite reflection group then* $T \in \mathfrak{TO}$.

Theorem 2.4 has been proved by an exhaustive (and exhausting) case-by-case analysis, as often happens with general results about reflection groups. A unified conceptual proof would be of considerable interest.

ONE-RELATOR PRODUCTS OF GROUPS

James Howie
University of Glasgow, Glasgow, G12 8QW, Scotland.

1. *INTRODUCTION*

A *one-relator product* of a family $(A_\lambda)_{\lambda \in \Lambda}$ of groups is the quotient $G = A/N(r)$ of their free product $A = *_\lambda A_\lambda$ by the normal closure $N(r)$ of a single element r, called the *relator*, which is always assumed to be a cyclically reduced word of length at least 2. In the case where the groups in A_λ are all free, the resulting group G is just a one-relator group, and indeed all infinite one-relator groups arise in this way. It is thus natural to try to generalize results about one-relator groups to more general one-relator products, and in fact one can do so under suitable restrictions. This was the main theme of my survey article [5]. This note is intended as a sequel to [5]. The purpose is to report on recent progress on one-relator groups in general, and on some of the problems raised in [5] in particular. Proofs of the results announced here will be published elsewhere.

Possible restrictions on G are of two kinds: on the groups A_λ, and on the relator r. In §2 I will concentrate on the case where the A_λ are locally indicable, and in §3 on the case where r is a proper power, $r = s^m$, $m \geq 2$.

2. *LOCALLY INDICABLE GROUPS*

A group is said to be *locally indicable* if each of its nontrivial, finitely generated subgroups has an infinite cyclic homomorphic image. Interesting examples of locally indicable groups include: free groups, torsion-free locally nilpotent groups, torsion-free one-relator groups, and many 3-manifold groups including all knot and link groups. Suppose $G = A/N(r)$ is a one-relator product of a family of locally indicable groups $(A_\lambda)_{\lambda \in \Lambda}$. Then many results of one-relator group theory, such as the Freiheitssatz, the solution to the word problem, Lyndon's Identity Theorem and the resulting computation of cohomology, the Newman-Gurevich-Pride-Schupp Spelling Theorems, etc., have natural analogues for G. See [5], and the references cited there, for more details.

The following result is an analogue of the Cohen-Lyndon Theorem [1] for one-relator groups, and answers Question 6.9 of [5].

Theorem 2.1 (Edjvet and Howie [3]). *Let* C *be the centralizer of* r *in* $A = *_\lambda A_\lambda$. *Then there exists a left transversal* U *for* $N(r).C$ *in* A *such that* $N(r)$ *is free with basis* $\{uru^{-1};\ u \in U\}$.

One-relator groups have few non-obvious abelian subgroups [7], and the same holds for one-relator products of locally indicable groups.

Theorem 2.2. *Let* S *be an abelian subgroup of* G *such that* $S \not\subset gA_\lambda g^{-1}$ *for all* $g \in G$ *and for all* $\lambda \in \Lambda$.

(a) *If* r *is a proper power, then* S *is cyclic, and* $S \cap gA_\lambda g^{-1} = 1$ *for all* g, λ.

(b) *If* r *is not a proper power, then* S *is either free abelian of rank one or two, or locally cyclic, and* $S \cap gA_\lambda g^{-1}$ *is cyclic for all* g, λ.

In particular, G has trivial centre, with a few exceptions.

Theorem 2.3. *Suppose the centre* C *of* G *is nontrivial. Then*

(a) r *is not a proper power;*

(b) $\Lambda = \{1,2\}$ *is a 2-element set;*

(c) *for each* $\lambda \in \Lambda$, *either* $A_\lambda \cong \mathbb{Z}$ *or* $C \subset A_\lambda$.

Furthermore, if C *is not cyclic, then* $A_1 \cong A_2 \cong \mathbb{Z}$ *and* $C = G \cong \mathbb{Z} \times \mathbb{Z}$.

3. *HIGH-POWERED RELATORS*

Suppose now that the groups A_λ are arbitrary, but that r is a proper power in A, that is $r = s^m$ for some $m \geq 2$. Without loss of generality, we may suppose that m is maximal, in other words that s is not itself a proper power. This uniquely determines s, which is then known as the root of r. The philosophy behind the present section is that, the greater m is, the better one-relator theory works.

If m is very high ($m \geq 6$), then the symmetrized closure of (r) satisfies a small cancellation condition, and so life is easy. For example, the Freiheitssatz holds [2,4]. There is also an analogue of the identity Theorem [2], provided that r is not conjugate in A to its own inverse (that is s is not conjugate to a word of the form $xuyu^{-1}$ for some word u and elements x, y of order 2).

In fact, a slightly more detailed analysis of small cancellation diagrams handles the case $m = 5$ (see [4] for the Freiheitssatz, for example), and a much deeper analysis handles the case $m = 4$.

Theorem 3.1 (Freiheitssatz). *If* $m \geq 4$, *then each* A_λ *embeds in* G *via the natural map.*

Theorem 3.1 partially answers Question 6.2 of [5].

Theorem 3.2. *If* $m \geq 4$, *then the word problem for* G *reduces to that for* $A = *_\lambda A_\lambda$.

Theorem 3.3 (Identity Theorem). *If* $m \geq 4$, *then* $N(r)/[N(r),N(r)]$ *is isomorphic, as a* $\mathbb{Z}G$*-module, to* $\mathbb{Z}G/(1-s)\mathbb{Z}G$, *except in the case*

(*): s *is conjugate to a word of the form* $xuyu^{-1}$ *for some word* u *and elements* x, y *of orders* p, q *respectively, where* $1/m + 1/p + 1/q < 1$.

The classical identity Theorem of Lyndon [6], as well as that of Collins and Perraud [2], is used to calculate the cohomology of G. The analogous result holds here also.

Corollary 3.4. *Suppose* $m \geq 4$ *and* s *does not have the form* (*). *Then there exist natural isomorphisms for each* $k \geq 3$, *and a natural epimorphism for* $k = 2$:

$$H^k(G;-) \to (\Pi_\lambda H^k(A_\lambda;-)) \times H^k(\mathbb{Z}_m;-).$$

This helps to illustrate why the condition in Theorem 3.3 is necessary. Suppose the collection $\{A_\lambda\}$ consists of only two finite cyclic groups, generated by x, y respectively, and that u is the empty word. Then G is the finite, noncyclic group

$$F(p,q,m) = \langle x,y \mid x^p = x^q = (xy)^m = 1 \rangle,$$

whereas Corollary 3.4 suggests that it should ultimately have periodic cohomology of period 2, and thus be cyclic. In fact, it turns out that this counterexample is generic, in the sense that we have a pushout of groups

$$\begin{array}{ccc} \mathbb{Z}_p * \mathbb{Z}_q & \to & F(p,q,m) \\ \downarrow & & \downarrow \\ A & \to & G . \end{array} \qquad (\dagger)$$

Furthermore, we can partially rescue the cohomology calculation of 3.4 as follows.

Theorem 3.5. *Suppose* $m \geq 4$ *and* s *has the form* (*). *Then the pushout square* ($\dagger$) *induces a Mayer-Vietoris sequence on cohomology:*

$$\to H^k(G;-) \to H^k(A;-) \times H^k(F(p,q,m);-) \to H^k(\mathbb{Z}_p * \mathbb{Z}_q;-) \to H^{k+1}(G;-) \to .$$

The proofs of the above results, as I indicated, involve a careful scrutiny of small cancellation diagrams, or "pictures". This becomes progressively more difficult as m decreases, so it is not clear to what extent such methods might be successful for $m = 3$ or $m = 2$. Nevertheless, the consistency of the results for high m make it tempting to formulate a conjecture.

Conjecture 3.6. *The condition* $m \geq 4$ *in* 3.1 - 3.5 *can be weakened to* $m \geq 2$.

Acknowledgement. This work was supported by an SERC Advanced Fellowship.

REFERENCES

1. D. E. Cohen & R. C. Lyndon, Free bases for normal subgroups of free groups, *Trans. Amer. Math. Soc.* 108 (1963), 528-537.
2. D. J. Collins & J. Perraud, Cohomology and finite subgroups of small cancellation quotients of free products, *Math. Proc. Camb. Phil. Soc.* 97 (1985), 243-259.
3. M. Edjvet & J. Howie, A Cohen-Lyndon Theorem for free products of locally indicable groups, *J. Pure Appl. Alg.*, to appear.
4. F. Gonzalez-Acuña & H. Short, Knot surgery and finiteness, *Math. Proc. Camb. Phil. Soc.* 99 (1986), 89-102.

5. J. Howie, How to generalize one-relator group theory, *Proceedings of the Alta Conference on Combinatorial Group Theory* (eds. S. M. Gersten and J. R. Stallings), Annals of Mathematics Studies, Princeton University Press, to appear.
6. R. C. Lyndon, Cohomology theory of groups with a single defining relation, *Ann. of Math.* 52 (1950), 650-665.
7. B. B. Newman, Some results on one-relator groups, *Bull. Amer. Math. Soc.* 74 (1968), 568-571.

THE CAVICCHIOLI GROUPS ARE PAIRWISE NON-ISOMORPHIC

David L. Johnson
University of Nottingham, Nottingham, NG7 2RD, England
Richard M. Thomas
St Mary's College, Twickenham, TW1 4SX, England

In [1], coloured graphs are used to construct a set of pairwise non-homeomorphic prime homology spheres with Heegaard genus two. There is one such for each positive integer n coprime to 6, whose fundamental group G(n) has presentation (see [1]):

$$\langle a_0, a_1, a_2 \mid a_0a_1a_2 = 1,\ a_0^{-1}a_{-2}^{-1}a_{-4}^{-1} \ldots a_{-n+3}^{-1}a_{-n+2}a_{-n+4} \ldots a_{-3}a_{-1}a_1 = 1,$$
$$a_1^{-1}a_{-1}^{-1}a_{-3}^{-1} \ldots a_{-n+4}^{-1}a_{-n+3}a_{-n+5} \ldots a_{-2}a_0a_2 = 1,$$
$$a_2^{-1}a_0^{-1}a_{-2}^{-1} \ldots a_{-n+5}^{-1}a_{-n+4}a_{-n+6} \ldots a_{-1}a_1a_3 = 1 \rangle,$$

where all subscripts are modulo 3.

We give an independent proof that these groups are all different by using Tietze transformations to change generators (see [2, page 35]). Replacing each a_i by a_0, a_1 or a_2, according as $i \equiv 0$, 1 or 2 (mod 3), we get:

$$\langle a_0, a_1, a_2 \mid a_0a_1a_2 = 1,\ \ldots a_0a_2a_1a_0a_2a_1 = \ldots a_2a_1a_0a_2a_1a_0,$$
$$\ldots a_1a_0a_2a_1a_0a_2 = \ldots a_0a_2a_1a_0a_2a_1,\ \ldots a_2a_1a_0a_2a_1a_0 = \ldots a_1a_0a_2a_1a_0a_2 \rangle,$$

where the last three relations contain $(n+1)/2$ terms on the left and $(n-1)/2$ terms on the right. We split the proof into two cases:

Case 1. $n = 6k+1$. In this case we have:

$$\langle a_0, a_1, a_2 \mid a_0a_1a_2 = 1,\ a_1(a_0a_2a_1)^k = (a_2a_1a_0)^k,$$
$$(a_2a_1a_0)^k a_2 = (a_0a_2a_1)^k,\ (a_0a_2a_1)^k a_0 = (a_1a_0a_2)^k \rangle.$$

The first three relations give:

$$(a_1a_0a_2)^k = a_1(a_0a_2a_1)^k a_1^{-1} = (a_2a_1a_0)^k a_1^{-1} = (a_0a_2a_1)^k a_2^{-1}a_1^{-1} = (a_0a_2a_1)^k a_0,$$

so that the last relation is redundant. We introduce a new generator $b = a_2a_1$ to get:

$$\langle a_0, a_1, a_2, b \mid a_0a_1a_2 = 1,\ b = a_2a_1,\ a_1(a_0b)^k = (ba_0)^k,\ (ba_0)^k a_2 = (a_0b)^k \rangle,$$

and we then delete $a_1 = (ba_0)^k(a_0b)^{-k}$, $a_2 = (ba_0)^{-k}(a_0b)^k$ to get:

$$\langle a,b \mid a(ba)^k(ab)^{-k}(ba)^{-k}(ab)^k = 1,\ b = (ba)^{-k}(ab)^k(ba)^k(ab)^{-k} \rangle,$$

writing a_0 as a. We now introduce $r = (ab)^k$, $s = (ba)^k$ to get:

$$\langle a,b,r,s \mid r = (ab)^k,\ s = (ba)^k,\ a = r^{-1}srs^{-1},\ b = s^{-1}rsr^{-1} \rangle,$$

which, on deleting a and b becomes:

$$\langle r,s \mid r = (srs^{-2}.rsr^{-2})^k,\ s = (rsr^{-2}.srs^{-2})^k \rangle.$$

The second relation is equivalent, via the first, to $rsr^{-2}.r = s.rsr^{-2}$, and hence to $rsr = srs$.

We now introduce new generators $x = rs$, $y = rsr$:

$$\langle r,s,x,y \mid x = rs,\ y = rsr,\ r = x^{-1}y,\ s = y^{-1}x^2,$$
$$x^{-1}y = (y^{-1}xyx^{-2}yx^{-2}yxy^{-1}xy^{-1}x)^k,\ y^2 = x^3 \rangle.$$

We now delete r and s and introduce $z = x^3 = y^2$, which is central in the group. Noting that $yx^{-2} = y^{-1}x$ and that $y = y^{-1}z$, we get:

$$\langle x,y,z \mid x^3 = y^2 = z,\ x^{-1}y = (y^{-1}x)^{6k}z^k \rangle,$$

in other words:

$$\langle x,y,z \mid x^3 = y^2 = z,\ (x^{-1}y)^n = z^k \rangle.$$

Case 2. $n = 6k - 1$. This case is very similar to case 1, and we shall just give an outline of what happens. Here G(n) has presentation:

$$\langle a_0,a_1,a_2 \mid a_0a_1a_2 = 1,\ (a_0a_2a_1)^k = a_2^{-1}(a_2a_1a_0)^k,$$
$$(a_1a_0a_2)^k = (a_2a_1a_0)^k a_0^{-1},\ (a_2a_1a_0)^k = (a_0a_2a_1)^k a_1^{-1} \rangle.$$

Here we may omit the third relation, and we introduce a, b as before to get:

$$\langle a,b \mid a(ba)^{-k}(ab)^k(ba)^k(ab)^{-k} = 1,\ b = (ba)^k(ab)^{-k}(ba)^{-k}(ab)^k \rangle,$$

and then r, s to get:

$$\langle r,s \mid r = (s^{-1}r^{-1}s^2.r^{-1}s^{-1}r^2)^k,\ s = (r^{-1}s^{-1}r^2.s^{-1}r^{-1}s^2)^k \rangle,$$

which, after transforming the second relation to $rsr = srs$ and introducing x and y as above, becomes:

$$\langle x,y,z \mid x^3 = y^2 = z,\ x^{-1}y = (x^{-1}y)^{6k}z^{-k} \rangle,$$

and hence finally:

$$\langle x,y,z \mid x^3 = y^2 = z,\ (x^{-1}y)^n = z^k \rangle.$$

So, in both cases 1 and 2, we have:

$$\langle x,y,z \mid x^3 = y^2 = z,\ (x^{-1}y)^n = z^k \rangle$$

as a presentation for G(n). If we factor out the central subgroup $\langle z \rangle$, we get the group D(3,2,n) with presentation:

$$\langle x,y \mid x^3 = y^2 = (xy)^n = 1 \rangle.$$

These groups are F-groups (in the sense of [3, Chapter III]). It follows that every element of finite order in D(3,2,n) is conjugate to a power of x, y or xy ([3, III (6.2)]) and that D(3,2,n) has trivial centre ([3, III 7.10]), so that G(n) has centre $\langle z \rangle$, and G(n)/Z(G(n)) is not isomorphic to G(m)/Z(G(m)) for n not equal to m. Hence the groups G(n) are all distinct.

Acknowledgements. The authors are grateful to Roger Lyndon for originally telling us about the problem and for showing us his calculations on G(5), and to him and Jim Howie for confirming the above properties of the $D(\ell,m,n)$; the second author would also like to thank Hilary Craig for all her help and encouragement.

REFERENCES

1. A. Cavicchioli, A countable class of non-homeomorphic homology spheres with Heegaard genus two, to appear.
2. D. L. Johnson, *Topics in the theory of group presentations*, LMS Lecture Notes, Vol. 42, Cambridge University Press (1980).
3. R. C. Lyndon & P. E. Schupp, *Combinatorial group theory*, Springer-Verlag (1977).

CONGRUENCE AND NON-CONGRUENCE SUBGROUPS OF THE MODULAR GROUP: A SURVEY

Gareth A. Jones
University of Southampton, Southampton, SO9 5NH, England.

1. *INTRODUCTION*

My aim in this note is to give a brief survey of one particular aspect of the modular group $\Gamma = PSL_2(\mathbb{Z})$, namely the balance (or rather the lack of it) between its congruence and non-congruence subgroups. Among the arithmetic subgroups (those of finite index) in Γ, the congruence subgroups have proved to be the most important and the most widely studied. Nevertheless, it has been known for some time that, in a certain sense,

"most of the arithmetic subgroups of Γ are non-congruence subgroups"; (1.1)

I shall describe several recent lines of investigation which, in different ways, add substance to this rather tenuous statement.

The modular group (together with the closely related groups $SL_2(\mathbb{Z})$, $GL_2(\mathbb{Z})$ and $PGL_2(\mathbb{Z})$) is like an octopus, with tentacles reaching out into many branches of pure mathematics; a complete survey is out of the question here, but I hope that, at the very least, the bibliography will enable the reader to discover more about this fascinating group and its applications. For background reading, the classic reference is still Klein and Fricke [38]; more modern treatments of various aspects of Γ can be found in [1], [20], [27], [35], [41], [49], [65], [70], [75], [76], [80], [83]. Fine [19] surveys the similarities and differences between Γ and the Picard group $PSL_2(\mathbb{Z}[i])$, while other classes of groups closely related to Γ are considered in [5], [30], [84], [95].

2. *ARITHMETIC SUBGROUPS*

Let $\mathbb{A}$ denote the set of *arithmetic* subgroups of Γ, that is, those of finite index. There is one obvious family of such subgroups: for each integer $\ell \geq 2$, the reduction $\mathrm{mod}(\ell) : \mathbb{Z} \to \mathbb{Z}/\ell\mathbb{Z}$ induces, in the natural way, an epimorphism $\Gamma \to PSL_2(\ell) := PSL_2(\mathbb{Z}/\ell\mathbb{Z})$; the kernel $\Gamma(\ell)$ is the *principal congruence subgroup* of level ℓ, a normal subgroup of index

$$|PSL_2(\ell)| = \begin{cases} \frac{1}{2}\ell^3 \prod_{p|\ell} (1 - p^{-2}) & (\ell > 2), \\ 6 & (\ell = 2), \end{cases} \tag{2.1}$$

where p ranges over the distinct primes dividing ℓ. We have $\Gamma(m) \leq \Gamma(\ell)$ if and only if $\ell | m$, and $\bigcap_{\ell} \Gamma(\ell) = 1$ whenever ℓ ranges over an unbounded set.

Any subgroup of Γ containing some $\Gamma(\ell)$ is a *congruence subgroup*, and the least such ℓ is its *level*; we let $\mathbb{C}$ denote the set of congruence subgroups, while $\mathbb{N} = \mathbb{A} \setminus \mathbb{C}$ consists of the remaining arithmetic subgroups, the *non-congruence subgroups*.

Subgroups $\Lambda \geq \Gamma(\ell)$ are in one-to-one correspondence with subgroups of $PSL_2(\ell)$, a group which has been intensively studied, from the time of Galois [22] for ℓ prime, and since Klein (see [38] for example) for general ℓ, so that its structure is now quite well understood. For instance McQuillan [46], completing work by Newman [60], has classified its normal subgroups. Because

$$PSL_2(\ell) \cong SL_2(\ell)/\{\pm I\} \tag{2.2}$$

and

$$SL_2(\ell) \cong \prod_q SL_2(q), \tag{2.3}$$

with q ranging over the prime powers in the prime power decomposition of ℓ, it is generally sufficient to consider $PSL_2(\ell)$ where ℓ is a prime power; McQuillan [47], following work of Gierster [24], [25], has given its conjugacy classes, its automorphism group and a set of defining relations for odd prime powers ℓ, and Dennin [15] has done the same for powers of 2.

One therefore tends to regard congruence subgroups as, in some vague sense, "known", and it would make the study of Γ a great deal simpler if these were the only arithmetic subgroups. This pleasant state of affairs actually occurs if one replaces Γ by its analogue $PSL_d(\mathbb{Z})$ for $d > 2$, with congruence subgroups defined in the obvious way: this is the positive solution to the Congruence Subgroup Problem, due to Bass, Lazard, Serre [3] and Mennicke [53], which has generated a major industry in which PSL_2 (or SL_2) is replaced by some other algebraic group, and $\mathbb{Z}$ by some other ring of coefficients. (See Mennicke [54], [55] and Beyl [6] for Sp_{2d} and $SL_2\left(\mathbb{Z}\left[\frac{1}{p}\right]\right)$, Bass, Milnor and Serre [4] for SL_d over rings of integers in algebraic number fields, Matsumoto [51] for Chevalley groups, Serre [82] for SL_2, and Raghunathan [73] for algebraic groups; a good introduction is Humphreys' account of how Matsumoto's methods can be applied to $SL_d(\mathbb{Z})$, $d \geq 3$, in [30].)

Unfortunately, as is often the case, low-dimensional behaviour is exceptional, and Γ does indeed possess non-congruence subgroups. Klein [37] first pointed this out at a meeting of the Munich Academy on 6th December 1879, and in 1887 Fricke [21] and Pick [71] independently and simultaneously published the first proofs of this; see also the detailed treatment of their subgroups in [38, Vol. I, III, 5, §2-3]. Little more seems to have been done in this direction until Reiner [77] generalised their construction in 1958; since then many authors have produced classes of non-congruence subgroups (for instance [92], [62], [63], [64], [74], [50], [23], [32], [78], [9], [33]) using techniques which range from linear algebra to Dedekind sums, triangulated surfaces, and Chebyshev polynomials.

A very attractive demonstration of the existence of non-congruence subgroups is that presented by Magnus [49, III.2], which we will summarise here. First we need two facts:

Lemma 1. *The only non-abelian composition factors of* $PSL_2(\ell)$ *have the form* $PSL_2(p)$ *for some prime* p *dividing* ℓ.

Lemma 2. $\Gamma \cong C_2 * C_3$.

Lemma 1 follows easily from (2.2) and (2.3), together with the simplicity of $PSL_2(p)$ for primes $p \geq 5$, and the fact that the kernel of the natural epimorphism $SL_2(p^{i+1}) \to SL_2(p^i)$ is a p-group. To prove Lemma 2, we use the presentation

$$\Gamma = \langle X, Y \mid X^2 = Y^3 = 1 \rangle, \tag{2.4}$$

where X and Y are the images in Γ of the matrices $\begin{pmatrix} 0 & -1 \\ 1 & 0 \end{pmatrix}$ and $\begin{pmatrix} 0 & 1 \\ -1 & 1 \end{pmatrix}$. One can obtain this presentation by considering the action of Γ, by Möbius transformations, on the upper half-plane $\mathbb{U}$ [35] or on the rational projective line $\hat{\mathbb{Q}} = \mathbb{Q} \cup \{\infty\}$ [28], or by matrix theory [65], or by letting Γ act on quadratic forms [49] (a representation of Γ which goes back to Gauss).

Lemma 2 implies that Γ must have many epimorphic images: any group G generated by elements x, y satisfying $x^2 = y^3 = 1$. Now it is easy to find such a group G which is finite, and which has a non-abelian composition factor not isomorphic to any group $PSL_2(p)$: Magnus takes $G = A_{11}$, and indeed any alternating group A_n with $n \geq 9$ will suffice here [56], [17]. Then the kernel Λ of the epimorphism $X \to x$, $Y \to y$, is an arithmetic subgroup which cannot be a congruence subgroup since, by Lemma 1, $PSL_2(\ell)$ cannot be mapped onto G. For a generalisation of this argument to other rings of coefficients, see [7].

This argument not only shows that non-congruence subgroups exist, it also suggests that, because Γ has so few defining relations, there will be many such epimorphic images G, and consequently the non-congruence subgroups will tend to outweigh the congruence subgroups (Garbe [23] and Cohen [9] have shown that all but a few of the finite simple groups of type PSL_2, PSL_3 or PSU_3 are images of Γ). This lends support to statement (1.1), but we cannot substantiate it merely by comparing the cardinalities $|\mathbb{C}|$ and $|\mathbb{N}|$: since Γ is finitely-generated, $\mathbb{A}$ is countable, and we have just seen that both $\mathbb{C}$ and $\mathbb{N}$ are infinite, so $|\mathbb{C}| = |\mathbb{N}| = \aleph_0$. Nevertheless, as we shall see in the rest of this paper, several "reasonable" methods of sampling arithmetic subgroups show that elements of $\mathbb{N}$ occur far more frequently than those of $\mathbb{C}$.

3. *COUNTING SUBGROUPS OF A GIVEN INDEX*

The most obvious method of sampling arithmetic subgroups is to estimate, for each integer $n \geq 1$, the numbers a_n and c_n of arithmetic and congruence subgroups of index n in Γ. First we consider a_n.

Given any permutation representation $\Gamma \to S_n$, the stabiliser $\Lambda = \Gamma_1$ of 1 is a subgroup of index $r \leq n$, where r is the length of the orbit containing 1. Conversely, if Λ is any subgroup of index $r \leq n$ in Γ, we get all permutation representations $\Gamma \to S_n$, with $\Lambda = \Gamma_1$, by letting Γ act by right-multiplication on the r cosets of Λ (labelled with r of the symbols 1,...,n, with Λ labelled 1), and any action on the other symbols. Given Λ, there are $(n-1)!/(n-r)!$ ways of labelling its non-trivial cosets with r-1 symbols from 2,...,n, so the number of permutation representations $\Gamma \to S_n$ corresponding to Λ in this way is $(n-1)!h_{n-r}/(n-r)!$, where h_{n-r} is the number of representations of Γ on the remaining n-r symbols, that is, the number of homomorphisms $\Gamma \to S_{n-r}$. Defining $h_0 = 1$, and summing over all Λ of index $r \leq n$, we obtain the recurrence relation

$$h_n = (n-1)! \sum_{r=1}^{n} \frac{h_{n-r}a_r}{(n-r)!} . \tag{3.1}$$

This result (valid for any finitely-generated group, not just Γ) is essentially due to Dey [16]; the above proof is due to Wohlfahrt [94].

Now for $\Gamma \cong C_2 * C_3$, we have $h_n = \tau_2(n)\tau_3(n)$, where $\tau_p(n)$ is the number of elements of S_n of order dividing p. Moser and Wyman [59] determined the asymptotic behaviour of $\tau_p(n)$, as $n \to \infty$, for any prime p; by combining this with (3.1), Newman [66] showed that

$$a_n \sim (12\pi e^{\frac{1}{2}})^{-\frac{1}{2}} \exp\left(\frac{n \log n}{6} - \frac{n}{6} + n^{\frac{1}{2}} + n^{\frac{1}{3}} + \frac{\log n}{2}\right). \tag{3.2}$$

Stirling's formula reduces this to the slightly more attractive expression

$$6^{n/6}\left(\frac{n}{6}\right)! \exp (n^{\frac{1}{2}} + n^{\frac{1}{3}})/2\pi e^{\frac{1}{4}}, \tag{3.3}$$

given by Stothers in [88], but however one writes it, it is clear that a_n grows rapidly as $n \to \infty$; see, for example, the values of a_n, for $n \leq 255$, computed by Newman in [67].

This line of thought has been explored further by Stothers [87], Wohlfahrt [94], Imrich [31], and by Godsil, Imrich and Razen [26], for instance to count free subgroups of a given index in Γ, or to count subgroups of $SL_2(\mathbb{Z})$.

In estimating c_n, the important step is to bound the level ℓ of a congruence subgroup in terms of its index n, and to do this we need an alternative, purely group-theoretic definition of level. If Λ is any subgroup of index n in Γ, then Γ acts transitively on the coset space Γ/Λ. Let the element $Z = (XY)^{-1} = \pm\begin{pmatrix}1 & 1\\ 0 & 1\end{pmatrix}$ have t cycles, of lengths $n_1,\ldots,n_t$, in this representation. Then t is the *parabolic class number* of Λ, equal to the number of Λ-conjugacy classes of maximal parabolic subgroups in Λ, and also to the number of orbits of Λ on $\hat{\mathbb{Q}}$, while $n_1,\ldots,n_t$ are the *cusp-amplitudes* of Λ (they determine the shape of the cusps on the boundary of a fundamental region for Λ on $\mathbb{U}$). The partition

$$n = n_1 + \ldots + n_t \tag{3.4}$$

of n is called the *cusp-split* of Λ. Now the order of the permutation induced by Z on Γ/Λ is the least common multiple of the cusp-amplitudes n_i, and Wohlfahrt [93] has shown that, if Λ is a congruence subgroup, then this is equal to the level ℓ of Λ (actually Wohlfahrt credits Fricke with this idea, but it has become traditional, and not unfair, to call this Wohlfahrt's Theorem since it was he who first stated and proved it clearly). We can therefore define the *level* ℓ of *any* subgroup of finite index to be the least common multiple of its cusp-amplitudes.

Recently Larcher [43] and Stothers [88], by different methods, have shown that a congruence subgroup of level ℓ must have a cusp-amplitude $n_i = \ell$, so that (3.4) implies

$$n \geq \ell, \tag{3.5}$$

a bound due to Galois [22] in the case where ℓ is prime. Stothers [88] uses this to estimate c_n, the number of congruence subgroups of index n. These correspond to subgroups of $PSL_2(\ell)$ for $\ell \leq n$; now (2.1) gives $|PSL_2(\ell)| < \ell^3$, and it is not hard to show that a group of order N has at most $N^{\log_2 N}$ subgroups, so Stothers deduces that

$$c_n < \sum_{\ell=1}^{n} \ell^{9\log_2 \ell} \leq n^{1 + 9\log_2 n}. \tag{3.6}$$

This bound grows faster than any polynomial (indeed it follows from the remark at the top of p.121 in [88], and the Prime Number Theorem, that there is an infinite sequence of integers n for which c_n is not bounded by any polynomial); nevertheless, comparison of (3.3) and (3.6) confirms statement (1.1), and also the assertion in [2] that, among all subgroups of index at most n, the ratio of congruence to non-congruence subgroups approaches 0 as $n \to \infty$.

4. *CUSP-SPLITS*

The cusp-split (3.4) of a congruence subgroup is, in fact, even more severely restricted than we indicated in the previous section: for instance Larcher [43], [44] has shown that if n_i and n_j are cusp-amplitudes for a congruence subgroup Λ, then so are their least common multiple and their highest common factor.

While it seems unlikely that one can give an effective characterisation of the cusp-splits of *all* arithmetic subgroups, it is clear that there are many more possibilities than there are for congruence subgroups. For example Stothers [85], [86] has shown that if $0 < \varepsilon \leq 1$ there is a constant $N(\varepsilon)$ such that, provided $n \geq (6+\varepsilon)t + N(\varepsilon)$, every partition of n into t parts is the cusp-split for some subgroup of index n. His approach is a topological version, using coset-diagrams, of Millington's use of permutations [58] to construct subgroups of Γ with certain specified parameters. Similarly, Jones and Singerman [34], [32] have developed a correspondence between orientable maps on surfaces and subgroups of triangle groups, which can be used to construct subgroups of Γ from triangular maps; for instance, (3.4) is the cusp-split of an arithmetic free subgroup of Γ if and only if some compact orientable surface has a triangulation with vertices of valencies $n_1,\ldots,n_t$. This connection, which has been extended by Bryant and Singerman [8] to $PGL_2(\mathbb{Z})$, will be outlined in the next section where we introduce the concept of the genus of a subgroup of Γ.

Millington [57], Petersson [69] and Rosenberger [79] have given constructions for *cycloidal* subgroups of Γ (arithmetic subgroups with $t = 1$); there are infinitely many of them, but only finitely many are congruence subgroups [69].

5. *COUNTING SUBGROUPS OF A GIVEN GENUS*

Γ acts, by Möbius transformations, on $\overline{\mathbb{U}} = \mathbb{U} \cup \hat{\mathbb{Q}}$, and the quotient space $\overline{\mathbb{U}}/\Gamma$ inherits the structure of a Riemann surface; the elliptic modular function $j : \mathbb{U} \to \mathbb{C}$ induces a conformal equivalence between $\overline{\mathbb{U}}/\Gamma$ and the Riemann sphere. If Λ is any subgroup of finite index n in Γ, then $\overline{\mathbb{U}}/\Lambda$ is an n-sheeted branched covering of $\overline{\mathbb{U}}/\Gamma$; it is a compact, connected, orientable surface, and its genus is defined to be the *genus* γ of Λ. We now show how γ may be calculated.

For each $g \in \Gamma$, let n_g be the number of cycles of g, acting by right-multiplication on Γ/Λ; for example, $n_1 = n$, the index of Λ, while $n_Z = t$, the parabolic class number. Taking $g = X$ (an involution), there will be $e_2 := 2n_X - n$ fixed-points and $\frac{1}{2}(n - e_2) = n - n_X$ cycles of length 2; similarly Y (of order 3) will have $e_3 := \frac{1}{2}(3n_Y - n)$ fixed-points and $\frac{1}{2}(n - n_Y)$ cycles of length 3, while the cycle-lengths for Z are the cusp-amplitudes n_i. The algebraic significance of e_2 and e_3 is that they are the numbers of conjugacy classes of (elliptic) subgroups of orders 2 and 3 in Λ, while t plays the same role with respect to the maximal parabolic subgroups. The topological analogue of this is that these numbers determine the local branching of the covering $\overline{\mathbb{U}}/\Lambda \to \overline{\mathbb{U}}/\Gamma$. This occurs at the points $z^\Gamma \in \overline{\mathbb{U}}/\Gamma$, where $z = i, -\bar{\omega}, \infty$ is the unique fixed-point of the element $g = X, Y, Z$ in $\overline{\mathbb{U}}$: there are n_g points in $\overline{\mathbb{U}}/\Lambda$ lying above z^Γ, the n sheets of $\overline{\mathbb{U}}/\Lambda$ coming together in n_g cycles around z^Γ corresponding to the cycles of g on Γ/Λ. Using this, we can calculate the Euler characteristic $\chi = 2 - 2\gamma$ of $\overline{\mathbb{U}}/\Lambda$ by lifting a triangulation from the sphere $\overline{\mathbb{U}}/\Gamma$ to $\overline{\mathbb{U}}/\Lambda$; this gives the Riemann-Hurwitz formula

$$\chi = n_X + n_Y + n_Z - n_1, \tag{5.1}$$

or equivalently

$$\gamma = 1 + \frac{n}{12} - \frac{t}{2} - \frac{e_2}{4} - \frac{e_3}{3}. \tag{5.2}$$

Finally, we note that Λ can be generated by e_2 elliptic elements X_i, e_3 elliptic elements Y_j, t parabolic elements Z_k, and 2γ hyperbolic elements A_ℓ, B_ℓ, with defining relations:

$$X_i^2 = Y_j^3 = \prod_i X_i \cdot \prod_j Y_j \cdot \prod_k Z_k \cdot \prod_\ell [A_\ell, B_\ell] = 1. \tag{5.3}$$

We now consider the first examples of non-congruence subgroups, given by Fricke [21] and Pick [71], and also re-examined in [38, Vol. I, III, 5, §§2-3] and [62].

The principal congruence subgroup $\Gamma(2)$ has $n = 6$, $e_2 = e_3 = 0$, and $t = 3$ (corresponding to the cusp-split $6 = 2 + 2 + 2$), so that by (5.2) it has genus 0, and by (5.3) it is a free group of rank 2, in fact generated by

$Z^2 = \pm\begin{pmatrix}1 & 2\\0 & 1\end{pmatrix}$ and $XZ^2X = \pm\begin{pmatrix}1 & 0\\2 & 1\end{pmatrix}$. The normal closure Λ of Z^{2m} and XZ^2X in $\Gamma(2)$ is, for each integer $m \geq 1$, a subgroup of index $6m$ in Γ, with $e_2 = e_3 = 0$ and $t = 2 + m$ (corresponding to the cusp-split $6m = 2m + 2m + 2 + \ldots + 2$), so by (5.2) it has genus 0. Since it has level $2m$, Λ will, by Wohlfahrt's Theorem, be a congruence subgroup if and only if it contains $\Gamma(2m)$; this can be shown to happen if and only if m divides 8, so we obtain infinitely many non-congruence subgroups of genus 0.

This argument is due to Newman [62]; in [61] he classified the normal subgroups of genus 1 in Γ (they all lie in the derived group Γ' and contain Γ''), and in [63] he showed that only four of these are congruence subgroups, so that there are infinitely many non-congruence subgroups of genus 1. In [32], Jones generalised these results to show that for each $\gamma \geq 0$ there are infinitely many non-congruence subgroups of genus γ. First we form triangulations of a surface of genus γ, with bounded valencies (in fact dividing 24) but with unbounded numbers E of edges; next we truncate each of these triangulations, "blowing up" each vertex of valency v into a small v-gon; the resulting graph, with the old edges labelled X, and the new edges labelled Z and directed according to the orientation of the surface, is the coset diagram for a subgroup Λ of Γ. The Riemann-Hurwitz formula (5.2), with $e_2 = e_3 = 0$, shows that Λ has genus γ, and since the cusp-amplitudes are the valencies in the original triangulation, Λ has level dividing 24. Thus Λ will be a congruence subgroup if and only if it contains $\Gamma(24)$. However the index $n = 2E$ of Λ is unbounded, so all but finitely many of the subgroups Λ obtained in this way are non-congruence subgroups.

Rademacher conjectured that there are only finitely many congruence subgroups of genus 0 in Γ; see also Newman's comments at the end of [62]. Special cases of this were proved by Knopp and Newman [39], McQuillan [48] and Dennin [12], [13], [14], and a much stronger form of the conjecture was eventually proved by Thompson [91] and by Cox and Parry [11].

Thompson's work was motivated by problems involving finite simple groups (there is an intriguing connection between modular functions and characters of the Fischer-Griess Monster group; see [10], [68], [72], [89], [90] for instance). Thompson actually considers subgroups of $PSL_2(\mathbb{R})$ commensurable with Γ, but restricting our attention to Γ itself, what he proves is that for congruence subgroups,

$$\frac{\gamma}{n} \longrightarrow \frac{1}{12} \quad \text{as} \quad n \longrightarrow \infty, \tag{5.4}$$

in the sense that, given any $\varepsilon > 0$, all but finitely many congruence subgroups satisfy

$$\left|\frac{\gamma}{n} - \frac{1}{12}\right| < \varepsilon.$$

The proof is deep, but the basic idea is to show that if Λ is a congruence subgroup then the average cycle-length n/n_g of the element $g = X, Y, Z$ on Γ/Λ approaches the order $2, 3, \infty$ of g as $n \to \infty$; the Riemann-Hurwitz formula

(5.1) then gives $\chi/n \to -1/6$ and so $\gamma/n \to 1/12$. It follows that if γ is fixed then n is bounded, so there are only finitely many congruence subgroups of any given genus γ in Γ. As we have seen, there are infinitely many non-congruence subgroups of genus γ, so this confirms statement (1.1).

This result was obtained independently, and by different methods, by Cox and Parry [11]. They generalise the result in a different direction by considering congruence subgroups in the group of units of reduced norm 1 in some maximal order in an indefinite $\mathbb{Q}$-quaternion algebra $\mathfrak{M}$; when $\mathfrak{M}$ is the 2×2 matrix algebra $M_2(\mathbb{Q})$, the group of units obtained is $SL_2(\mathbb{Z})$, so they deduce information about congruence subgroups of Γ. On their way towards (5.4), they obtain a wealth of numerical information about congruence subgroups of Γ, such as:

(i) there are 133 conjugacy classes of congruence subgroups of genus 0, all having level $\ell \leq 48$ and index $n \leq 72$;
(ii) the level of a congruence subgroup of genus γ is divisible by no prime $p > 12\gamma + 13$ (and this is best-possible);
(iii) a congruence subgroup of genus γ has level $\ell \leq 12\gamma + \frac{1}{2}(13\sqrt{(48\gamma + 121)} + 145)$.

6. *THE CONGRUENCE KERNEL*

Another topological method for comparing arithmetic and congruence subgroups is to consider the profinite completions they induce; here it is traditional, and a little more convenient, to work with the unimodular group $SL_2(\mathbb{Z})$, which we shall denote by Σ.

The congruence subgroups of Σ are those containing, for some ℓ, the kernel $\Sigma(\ell)$ of the reduction mod(ℓ): $\Sigma \to SL_2(\mathbb{Z}/\ell\mathbb{Z})$. By using these as a fundamental system of open neighbourhoods of the identity we obtain a topology on Σ, the *congruence topology* $\mathfrak{T}_c$, with respect to which Σ is a topological group. The *profinite completion* $\overline{\Sigma}$ of Σ, with respect to $\mathfrak{T}_c$, is the inverse limit $\varprojlim \Sigma/N$, where N ranges over all normal congruence subgroups, and the morphisms are the natural projections $\Sigma/N \to \Sigma/M$ where $N \leq M$. Thus $\overline{\Sigma}$ is the subgroup of $\prod_N \Sigma/N$ consisting of all elements $\prod g_N N$ such that $g_N \equiv g_M \bmod M$ whenever $N \leq M$. If $\mathbb{Z}_p$ denotes the ring $\varprojlim \mathbb{Z}/p^i\mathbb{Z}$ of p-adic integers, where p is prime, then we can use (2.3) to identify $\overline{\Sigma}$ with $\prod_p SL_2(\mathbb{Z}_p)$. Since $\bigcap_\ell \Sigma(\ell) = 1$, the function $\phi : g \to \prod_N gN$ imbeds Σ in $\overline{\Sigma}$. If we put the discrete topology on each Σ/N, then $\overline{\Sigma}$ inherits from $\prod \Sigma/N$ the structure of a totally disconnected compact topological group, with $\phi : \Sigma \to \overline{\Sigma}$ a continuous imbedding of Σ as a dense subgroup. Similarly, if we allow N to range over all normal *arithmetic* subgroups of Σ, we obtain the *arithmetic topology* $\mathfrak{T}_a$ and its corresponding completion $\hat{\Sigma}$, again containing Σ as a dense subgroup.

Since $\mathfrak{T}_a$ is a refinement of $\mathfrak{T}_c$, the identity map $(\Sigma, \mathfrak{T}_a) \to (\Sigma, \mathfrak{T}_c)$ is continuous, and it induces a continuous epimorphism $\hat{\Sigma} \to \overline{\Sigma}$; the kernel of this is the *congruence kernel* C of Σ, which measures how much finer $\mathfrak{T}_a$ is than $\mathfrak{T}_c$, and hence (by mapping Σ onto Γ) how much larger $\mathbf{A}$ is than $\mathbf{C}$.

This group C, together with its analogues for other algebraic groups and rings of coefficients, plays a crucial role in the study of the

Congruence Subgroup Problem. Serre [82] has shown that in our case C has cardinality $c = 2^{\aleph_0}$, and Mel'nikov [52], answering a question of Shafarevich [40, Problem 3.40] has shown that C is a free profinite group of countable rank, that is, $C = \varprojlim F/N$, where F is the free group on a countably infinite basis $\mathbf{X}$, and N ranges over the normal subgroups of finite index in F with $\mathbf{X}\backslash N$ finite. In this sense, C is "large", so the congruence subgroup property fails badly for Σ, and thus $\mathbf{C}$ is a rather small subset of $\mathbf{A}$. (For a generalisation of Mel'nikov's result to other rings, see Lubotzky [45].)

7. *THE EFFECT OF AUTOMORPHISMS*

Clearly the set $\mathbf{A}$ of arithmetic subgroups of Γ is invariant under automorphisms, but is the same true for $\mathbf{C}$? To put it another way, are congruence subgroups a property of the *abstract* group $C_2 * C_3$, or must we impose some additional structure (such as a particular isomorphism with $PSL_2(\mathbb{Z})$, or a particular set of generators) in order to distinguish them from the non-congruence subgroups? Thus we are looking for a conceptual, rather than numerical, response to (1.1), and this time the answer is a little surprising.

It is easy to show that Aut Γ can be identified with $\Pi := PGL_2(\mathbb{Z})$ acting by conjugation on its normal subgroup Γ; for instance Schreier [81] has determined the automorphisms of Γ. Now the principal congruence subgroups $\Gamma(\ell)$ are clearly normal in Π, so they are characteristic subgroups of Γ, and hence $\mathbf{C}$ is invariant under Aut Γ. Thus congruence subgroups *can* be defined group-theoretically, though it is not easy to do this: for example Langer [42] and Jones [33], extending a result of Newman [64], have shown that for each prime $p \geqslant 13$ there are (about $\frac{1}{2}p$) non-congruence subgroups $N \triangleleft \Gamma$ with $\Gamma/N \cong PSL_2(p)$; how do we distinguish them group-theoretically from $\Gamma(p)$?

The situation is similar with $SL_2(\mathbb{Z})$ and $GL_2(\mathbb{Z})$, since in both cases all automorphisms are induced by $GL_2(\mathbb{Z})$. However, when we consider $\Pi = PGL_2(\mathbb{Z})$, there is an outer automorphism α of order 2 which does not leave Γ invariant (see Dyer [18], correcting the error in [29]). Jones and Thornton [36] have shown that all but finitely many congruence subgroups of Π are mapped by α to non-congruence subgroups, so in this case, congruence subgroups are *not* group-theoretically defined.

REFERENCES

1. T.M. Apostol, *Modular functions and Dirichlet series in number theory*, Springer, New York (1976).
2. A.O.L. Atkin & H.P.F. Swinnerton-Dyer, Modular forms on non-congruence subgroups, *Proc. Sympos. Pure Math.* 19, Combinatorics (1971), 1-26, ed. by T.S. Motzkin, Amer. Math. Soc., Providence, R.I.
3. H. Bass, M. Lazard & J-P. Serre, Sous-groupes d'indices finis dans SL(n,Z), *Bull. Amer. Math. Soc.* 70 (1964), 385-392.
4. H. Bass, J. Milnor & J-P. Serre, Solution of the congruence subgroup problem for SL_n ($n \geqslant 3$) and Sp_{2n} ($n \geqslant 2$), *Inst. Hautes Études Sci. Publ. Math.* 33 (1967), 59-137.
5. A.F Beardon, *The geometry of discrete groups*, Springer, New York (1983).
6. F.R. Beyl, The Schur multiplicator of SL(2,Z/mZ) and the congruence subgroup property, *Math. Z.* 191 (1986), 23-42.

7. J. Britto, On the construction of non-congruence subgroups, *Acta Arith.* 33 (1977), 261-267.
8. R.P. Bryant & D. Singerman, Foundations of the theory of maps on surfaces with boundary, *Quart. J. Math. Oxford* (2) 36 (1985), 17-41.
9. J.M. Cohen, On Hurwitz groups and non-congruence subgroups of the modular group, *Glasgow Math. J.* 22 (1981), 1-7.
10. J.H. Conway & S.P. Norton, Monstrous moonshine, *Bull. London Math. Soc.* 11 (1979), 308-339.
11. D.A. Cox & W.R. Parry, Genera of congruence subgroups in Q-quaternion algebras, *J. Reine Angew. Math.* 351 (1984), 66-112.
12. J.B. Dennin, Jr., Fields of modular functions of genus 0, *Illinois J. Math.* 15 (1971), 442-455.
13. J.B. Dennin, Jr., Subfields of $K(2^n)$ of genus 0, *Illinois J. Math.* 16 (1972), 502-518.
14. J.B. Dennin, Jr., The genus of subfields of $K(p^m)$, *Illinois J. Math.* 18 (1974), 246-264.
15. J.B. Dennin, Jr., The automorphisms and conjugacy classes of $LF(2,2^n)$, *Illinois J. Math.* 19 (1975), 542-552.
16. I.M.S. Dey, Schreier systems in free products, *Proc. Glasgow Math. Assoc.* 7 (1965), 61-79.
17. I.M.S. Dey & J. Wiegold, Generators for alternating and symmetric groups, *J. Austral. Math. Soc.* 12 (1971), 63-68.
18. J.L. Dyer, Automorphism sequences of integer unimodular groups, *Illinois J. Math.* 22 (1978), 1-30.
19. B. Fine, The Picard group and the modular group, these Proceedings.
20. L.R. Ford, *Automorphic functions*, 2nd ed., Chelsea, New York (1951).
21. R. Fricke, Ueber die Substitutionsgruppen, welche zu den aus dem Legendre'schen Integralmodul $k^2(\omega)$ gezogenen Wurzeln gehören, *Math. Ann.* 28 (1887), 99-118.
22. É. Galois, Lettre à Auguste Chevalier, in *Revue encyclopédique*, Sept. 1832; also in *Oeuvres mathématiques d'Évariste Galois*, Gauthiers-Villars, Paris (1897).
23. D. Garbe, Über eine Klasse von arithmetisch definierbaren Normalteilern der Modulgruppe, *Math. Ann.* 235 (1978), 195-215.
24. G. Gierster, Die Untergruppen der Galois'schen Gruppe der Modulargleichungen für den Fall eines primzahligen Tranformationsgrades, *Math. Ann.* 18 (1881), 319-365.
25. G. Gierster, Über die Galois'sche Gruppe der Modulargleichungen wenn der Transformationsgrad die Potenz einer Primzahl > 2 ist, *Math. Ann.* 26 (1886), 309-368.
26. C. Godsil, W. Imrich & R. Razen, On the number of subgroups of given index in the modular group, *Monatsh. Math.* 87 (1979), 273-280.
27. R.C. Gunning, *Lectures on modular forms*, Princeton University Press, Princeton, New Jersey (1962).
28. G. Higman & Q. Mushtaq, Coset diagrams and relations for PSL(2,Z), *Arab Gulf J. Sci. Res.* 1 (1983), 159-164.
29. L-K. Hua & I. Reiner, Automorphisms of the unimodular group, *Trans. Amer. Math. Soc.* 72 (1952), 467-473.
30. J.E. Humphreys, *Arithmetic groups*, Lecture Notes in Mathematics, Vol. 789, Springer, Berlin (1980).
31. W. Imrich, On the number of subgroups of given index in $SL_2(Z)$, *Arch. Math.* 31 (1978), 224-231.
32. G.A. Jones, Triangular maps and non-congruence subgroups of the modular group, *Bull. London Math. Soc.* 11 (1979), 117-123.
33. G.A. Jones, Some maximal normal subgroups of the modular group, *Proc. Edinburgh Math. Soc.*, to appear.
34. G.A. Jones & D. Singerman, Theory of maps on orientable surfaces, *Proc. London Math. Soc.* (3) 37 (1978), 273-307.
35. G.A. Jones & D. Singerman, *Complex functions: an algebraic and geometric approach*, Cambridge University Press, to appear.
36. G.A. Jones & J.S. Thornton, Automorphisms and congruence subgroups of the extended modular group, *J. London Math. Soc.*, to appear.
37. F. Klein, Zur Theorie der elliptischen Modulfunktionen, *Math. Ann.* 17 (1880), 62-70.
38. F. Klein & R. Fricke, *Vorlesungen über die Theorie der elliptischen Modulfunktionen*, 2 vols., Teubner, Leipzig (1890, 1892); reprinted by Johnson Reprint, New York (1965).
39. M.I. Knopp & M. Newman, Congruence subgroups of positive genus in the modular group, *Illinois J. Math.* 9 (1965), 577-583.
40. *The Kourovka notebook of unsolved problems in group theory*, 7th ed., Novosibirsk (1980); English transl. by D.J. Johnson and L.J.

Leifman, Amer. Math. Soc. Translations, Ser. 2, vol. 121, Amer. Math. Soc., Providence, R.I. (1983).
41. S. Lang, *Introduction to modular forms*, Springer, Berlin (1976).
42. U. Langer, Doctoral dissertation, Hamburg University (1977).
43. H. Larcher, The cusp amplitudes of congruence subgroups of the classical modular group, *Illinois J. Math.* 26 (1982), 164-172.
44. H. Larcher, The cusp amplitudes of the congruence subgroups of the classical modular group (II), *Illinois J. Math.* 28 (1984), 312-338.
45. A. Lubotzky, Free quotients and the congruence kernel of SL_2, *J. Algebra* 77 (1982), 411-418.
46. D.L. McQuillan, Classification of normal congruence subgroups of the modular group, *Amer. J. Math.* 87 (1965), 285-296.
47. D.L. McQuillan, Some results on the linear fractional groups, *Illinois J. Math.* 10 (1966), 24-38.
48. D.L. McQuillan, On the genus of fields of elliptic modular functions, *Illinois J. Math.* 10 (1966), 479-487.
49. W. Magnus, *Noneuclidean tesselations and their groups*, Academic Press, New York and London (1974).
50. A.W. Mason, Lattice subgroups of free congruence subgroups, *Glasgow Math. J.* 10 (1969), 106-115.
51. H. Matsumoto, Sur les sous-groupes arithmétiques des groupes semi-simples déployés, *Ann. Sci. École Norm. Sup.* 2 (1969), 1-62.
52. O.V. Mel'nikov, Congruence kernel of the group $SL_2(Z)$ (Russian), *Dokl. Akad. Nauk SSSR* 228 (1976), 1034-1036; *Soviet Math. Dokl.* 17 (1976), 867-870 (English transl.).
53. J. Mennicke, Finite factor groups of the unimodular group, *Ann. of Math.* 81 (1965), 31-37.
54. J. Mennicke, Zur Theorie der Siegelschen Modulgruppe, *Math. Ann.* 159 (1965), 115-129.
55. J. Mennicke, On Ihara's modular group, *Invent. Math.* 4 (1967), 202-228.
56. G.A. Miller, On the groups generated by two operators, *Bull. Amer. Math. Soc.* 7 (1901), 424-426.
57. M.H. Millington, On cycloidal subgroups of the modular group, *Proc. London Math. Soc.* (3) 19 (1969), 164-176.
58. M.H. Millington, Subgroups of the classical modular group, *J. London Math. Soc.* (2) 1 (1970), 351-357.
59. L. Moser & M. Wyman, On solutions of $x^d = 1$ in symmetric groups, *Canad. J. Math.* 7 (1955), 159-168.
60. M. Newman, Normal congruence subgroups of the modular group, *Amer. J. Math.* 85 (1963), 419-427; errata ibid 85 (1963), 753; addendum ibid 86 (1964), 465.
61. M. Newman, Complete description of normal subgroups of genus 1 of the modular group, *Amer. J. Math.* 86 (1964), 17-24.
62. M. Newman, On a problem of G. Sansone, *Ann. Mat. Pura Appl.* (4) 65 (1964), 27-33.
63. M. Newman, Normal subgroups of the modular group which are not congruence subgroups, *Proc. Amer. Math. Soc.* 16 (1965), 831-832.
64. M. Newman, Maximal normal subgroups of the modular group, *Proc. Amer. Math. Soc.* 19 (1968), 1138-1144.
65. M. Newman, *Integral matrices*, Academic Press, New York (1972).
66. M. Newman, Asymptotic formulas related to free products of cyclic groups, *Math. Comp.* 30 (1976), 838-846.
67. M. Newman, *The number of subgroups of the classical modular group of index N (tables)*, Nat. Bureau of Standards, Washington (1976).
68. A.P. Ogg, Modular functions, *Proc. Sympos. Pure Math.* 37 (Santa Cruz Conf. on Finite Groups), 521-532, Amer. Math. Soc., Providence, R.I. (1980).
69. H. Petersson, Über die Konstruktion zykloider Kongruenzgruppen in der rationalen Modulgruppe, *J. Reine Angew. Math.* 250 (1971), 182-212.
70. H. Petersson, *Modulfunktionen und quadratische Formen*, Springer, Berlin (1982).
71. G. Pick, Ueber gewisse ganzzahlige lineare Substitutionen, welche sich nicht durch algebraische Congruenzen erklären lassen, *Math. Ann.* 28 (1887), 119-124.
72. L. Queen, Modular functions and finite simple groups, *Proc. Sympos. Pure Math.* 37 (Santa Cruz Conf. on Finite Groups), 561-566, Amer. Math. Soc., Providence, R.I. (1980).
73. M.S. Raghunathan, On the congruence subgroup problem, *Inst. Hautes Études Sci. Publ. Math.* 46 (1976), 107-161.
74. R.A. Rankin, Lattice subgroups of free congruence subgroups, *Invent. Math.* 2 (1967), 215-221.

75. R.A. Rankin, *The modular group and its subgroups*, Ramanujan Institute, Madras (1969).
76. R.A. Rankin, *Modular forms and functions*, Cambridge University Press, Cambridge (1977).
77. I. Reiner, Normal subgroups of the unimodular group, *Illinois J. Math.* 2 (1958), 142-144.
78. G. Rosenberger, Über Tschebyscheff-Polynome, Nicht-Kongruenzuntergruppen der Modulgruppe und Fibonacci-Zahlen, *Math. Ann.*, 246 (1980), 193-203.
79. G. Rosenberger, Über Darstellungen von Elementen und Untergruppen in freien Produkten, *Groups - Korea* 1983 (ed. A.C. Kim and B.H. Neumann), 142-160, Lecture Notes in Mathematics vol. 1098, Springer, Berlin-Heidelberg-New York-Tokyo (1984).
80. B. Schoeneberg, *Elliptic modular functions*, Springer, Berlin (1974).
81. O. Schreier, Über die Gruppen $A^aB^b = 1$, *Abh. Math. Sem. Hamburgischen Univ.* 3 (1924), 167-169.
82. J-P. Serre, Le problème des groupes de congruence pour SL_2, *Ann. of Math.* 92 (1970), 489-527.
83. J-P. Serre, *A course in arithmetic*, Springer, Berlin (1973).
84. J-P. Serre, *Trees*, Springer, Berlin (1980).
85. W.W. Stothers, Impossible specifications for the modular group, *Manuscripta Math.* 13 (1974), 415-428.
86. W.W. Stothers, Subgroups of the modular group, *Proc. Cambridge Phil. Soc.* 75 (1974), 139-153.
87. W.W. Stothers, The number of subgroups of given index in the modular group, *Proc. Roy. Soc. Edinburgh* 78A (1977), 105-112.
88. W.W. Stothers, Level and index in the modular group, *Proc. Roy. Soc. Edinburgh*, 99A (1984), 115-126.
89. J.G. Thompson, Finite groups and modular functions, *Bull. London Math. Soc.* 11 (1979), 347-351.
90. J.G. Thompson, Some numerology between the Fischer-Griess monster and the elliptic modular function, *Bull. London Math. Soc.* 11 (1979), 352-353.
91. J.G. Thompson, A finiteness theorem for subgroups of PSL(2,R) which are commensurable with PSL(2,Z), *Proc. Sympos. Pure Math.* 37 (Santa Cruz Conf. on Finite Groups), 533-555, Amer. Math. Soc., Providence, R.I. (1980).
92. K. Wohlfahrt, Über Dedekindsche Summen und Untergruppen der Modulgruppe, *Abh. Math. Sem. Univ. Hamburg* 23 (1959), 5-10.
93. K. Wohlfahrt, An extension of F. Klein's level concept, *Illinois J. Math.* 8 (1964), 529-535.
94. K. Wohlfahrt, Über einem Satz von Dey und die Modulgruppe, *Arch. Math.* 29 (1977), 455-457.
95. H. Zieschang, E. Vogt & H-D. Coldewey, *Surfaces and planar discontinuous groups*, Lecture Notes in Mathematics vol. 835, Springer, Berlin-Heidelberg-New York (1980).

SMALL CANCELLATION THEORY WITH NON-HOMOGENEOUS GEOMETRICAL CONDITIONS AND APPLICATION TO CERTAIN ARTIN GROUPS

Arye Juhasz
The Weizmann Institute of Science, Rehovot, Israel

INTRODUCTION

The subject of this article is small cancellation techniques which when applied to certain Artin groups solves the word problem, the conjugacy problem and proves a conjecture of J. Tits for these groups. Our method is based on a new nonhomogeneous geometrical condition which is more flexible than the usual ones. This condition is a common generalization of the conditions C(4) and T(4) and the condition C(6). Recall that if a group G has a presentation which satisfies the condition C(6) then the corresponding van Kampen diagrams have the property that every inner region of them has at least 6 neighbours. Similarly, the condition C(4) and T(4) implies that every inner region has at least 4 neighbours and no inner vertex has valency 3. The condition C(6) corresponds in an obvious way to the regular tessellation of the plane by hexagons and the conditions C(4) and T(4) corresponds to the regular tessellation of the plane by squares (see [6]). Our condition which we call condition W(6) applies to maps which in some places look like the hexagonal tessellation (see Fig. 1) and in some other places like the tessellation of the plane by squares (see Fig. 2) and in some more places look like the following two tessellations by pentagons (see Figs. 3a, 3b).

More precisely, given a map M, we say that M satisfies the condition W(6) if for every inner region D one of the following alternatives holds:

(a) D *has* 4 *neighbours and no vertices with valency* 3;
(b) D *has* 5 *neighbours and at most* 3 *vertices with valency* 3;
(c) D *has at least* 6 *neighbours.*

Fig. 1 Fig. 2

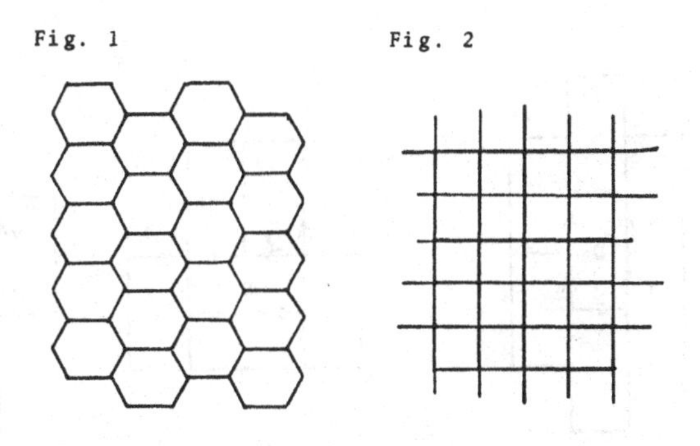

The following group exemplifies a situation where W(6) is applicable while neither condition C(4) and T(4) nor the condition C(6) holds. Let

$$G = \langle x, y \mid x^{-2}y^{-1}x^{2}y,\ y^{-2}x^{-1}y^{2}x \rangle .$$

Let D be an inner region having $y^{-2}x^{-1}y^{2}x$ as a label for a boundary cycle. Then the cases in Fig. 4 occur. The same situation holds for a region with $x^{-2}y^{-1}x^{2}y$ as a boundary cycle.

We develop a small cancellation theory for W(6) maps. In particular we show that groups having a presentation which gives rise to W(6) diagrams, have solvable word and conjugacy problems.

We apply this theory to certain Artin Groups. Before stating the results let us recall some definitions from [2] and [3] in connection with Artin Groups. A Coxeter matrix $\mathbb{M}$ over $\{1,\ldots,n\}$ is a symmetric matrix with entries $m_{i,j} \in \{1,\ldots,n\} \cup \{\infty\}$, where $m_{i,i} = 1$ for $i \in \{1,\ldots,n\}$ and $m_{i,j} \geq 2$ for $i \neq j$. The Artin group G defined by $\mathbb{M}$ is the group with generating set $\{x_i \mid i = 1,\ldots,n\}$ and for each pair $i \neq j$ with $m_{i,j} < \infty$ a defining relation of the form

$$x_i x_j x_i \ldots = x_j x_i x_j \ldots \qquad (*)$$

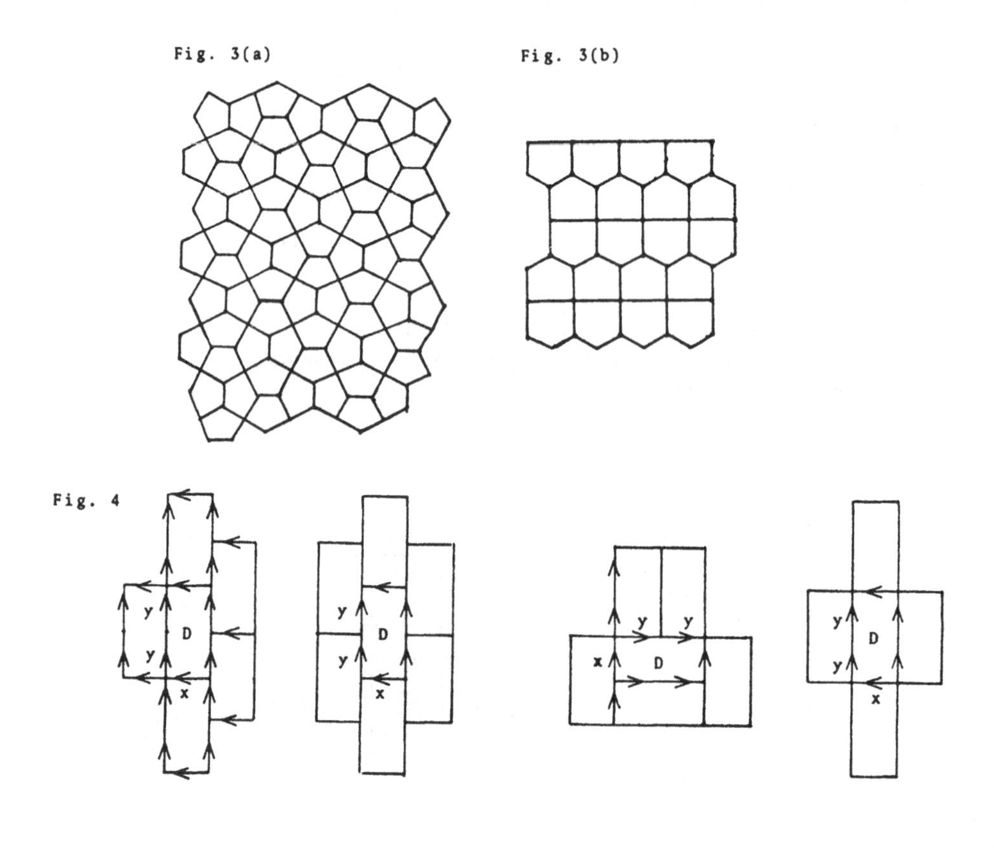

Fig. 3(a)

Fig. 3(b)

Fig. 4

saying that the alternating string of x_i's and x_j's of length $m_{i,j}$ beginning with x_i is equal to the alternating string of length $m_{i,j}$ beginning with x_j. We shall call a relator (*) an Artin relator of length $m_{i,j}$. If $m_{i,j} = 2$ we shall call it a commutation relator.

Theorem A. *Let* $\mathcal{F}$ *be the free group on* $x_1,\ldots,x_4$ *and let* $\mathcal{N}$ *be the normal closure of some Artin relators* $\mathcal{R}$ *on* $x_1,\ldots,x_4$. *Let* $G = \mathcal{F}/\mathcal{N}$. *Then the following hold.*

(a) G *has solvable word and conjugacy problems.*

(b) *(A generalization of Tits conjecture.)* *Let* $W = x_{i_1}^{n_1} \ldots x_{i_r}^{n_r}$, $|n_i| > 1$, $i_1,\ldots,i_r \in \{1,\ldots,n\}$. *If* W *represents* 1 *in* G *then* W *is a consequence of the commutation relators only.*

(c) *(Freiheitssatz).* *Let* $W \in \mathcal{N}$. *Denote by* $\mathrm{Supp}(W)$ *the set of the generators which occur in* W. *Then* W *has a representation* $W = \prod_{i=1}^{r} Q_i^{-1}R_iQ_i$, $Q_i \in F$, $R_i \in \mathcal{R}$ *such that* $\mathrm{Supp}(Q_i) \leqslant \mathrm{Supp}(W)$ *and* $\mathrm{Supp}(R_i) \leqslant \mathrm{Supp}(W)$ *for* $i = 1,\ldots,r$.

1. *THE THEORY OF W(6) DIAGRAMS*

As in the classical case, the basic result is a version of Greendlinger's Lemma and we begin with it.

1.1 *Greendlinger's Lemma and the Area Theorem for W(6) maps*

We recall some definitions from the fundamental work [8] of E. Rips in a version most convenient for us.

Definitions. (1) Let M be a connected and simply connected map and let S be a submap of M. We say that S is *strongly connected* if S is simply connected and the interior of S is connected.

(2) Let M be a connected map and let S be a strongly connected submap of M. The *star neighbourhood* $\mathrm{St}_M(S)$ of S in M is the union of all the regions in M the boundary of which has a common edge with the boundary ∂S of S.

(3) Let S be a strongly connected submap of M and assume that $\mathrm{St}_M(S)$ is also strongly connected. Then define $\mathrm{St}_M^2(S) = \mathrm{St}_M(\mathrm{St}_M(S))$. Similarly, if $\mathrm{St}_M^k(S)$ is defined and is strongly connected we define $\mathrm{St}_M^{k+1}(S) = \mathrm{St}_M(\mathrm{St}_M^k(S))$.

(4) Let S be a strongly connected submap of M and assume that $\mathrm{St}_M(S)$ is also strongly connected. We say that $\mathrm{St}_M(S)$ is *absolutely simply connected*

Fig. 5

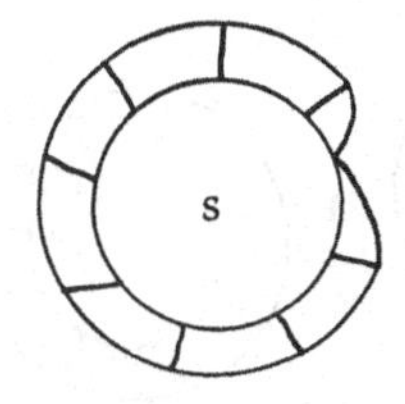

absolutely simple connected

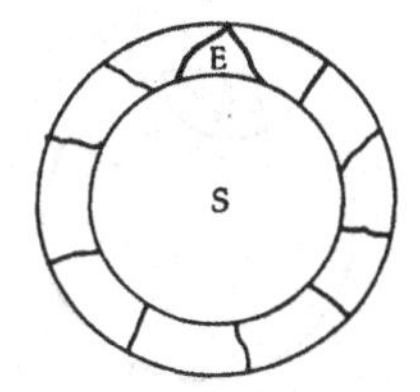

not absolutely simply connected

if every connected submap of $St_M(S)$ which contains S is simply connected. See Fig. 5.

In these terms our basic result is the following:

Theorem 1. *Let* M *be a strongly connected map which satisfies the condition* W(6) *and let* D *be a region in* M. *Then*

(i) $St_M^k(D)$ *is defined for every* k *(hence is strongly connected);*

(ii) $St_M(St_M^{k-1}(D)) = (St_M^k(D))$ *is absolutely simply connected for every* k.

The importance of this theorem stems from the fact that the increasing sequence of the star neighbourhoods $St_M^k(D)$, $k = 0,1,2,\ldots$ defines a "filtration" on M which imposes an "order" on the regions of M. This order serves as a "coordinate system" with D in its origin.

We come now to the Lemma of Greendlinger. But first we need some more definitions.

Definitions. Let D be a boundary region of M and denote by i(D) the number of inner edges of D in M. Call D a *corner region of degree k* for $k = 1,2$ if $i(D) = k$. Call D a *corner region of degree 3* if $i(D) = 3$ and every inner vertex of D has valency 3.

Let D and E be consecutive boundary regions of M such that $i(D) = i(E) = 3$. Let v be the inner vertex common to ∂E and ∂D and let u and w be the adjacent inner vertices on ∂E and ∂D respectively. (See Fig. 6.) Call {D,E} a *corner twin* if the following hold:

(i) v has valency 3,

(ii) u and w have valency $\geqslant 4$.

Denote by b(k) the number of corner regions of ∂M having degree k for $k = 1,2$ and denote by b(3) the number of corner regions of degree 3 plus the number of corner twins.

In these terms we have the following version of Greendlinger's Lemma:

Theorem 2. *Let* M *be a strongly connected map which satisfies the condition* W(6). *Then one of the following holds:*

Fig. 6

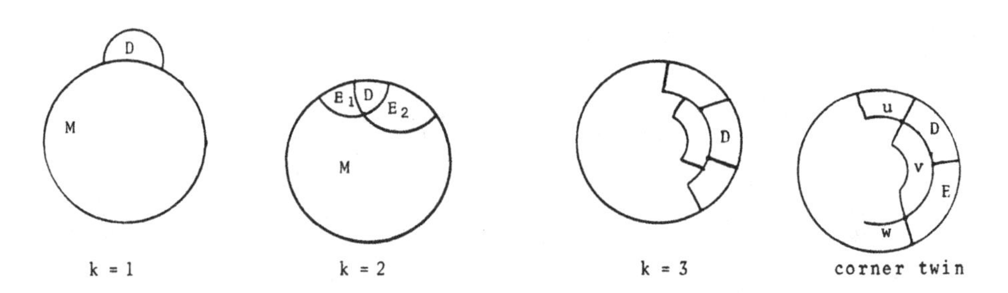

(a) $b(3) \geq 6$,
(b) $b(3) \geq 4$ *and* $b(2) \geq 1$,
(c) $b(3) \geq 3$ *and* $b(1) \geq 1$,
(d) $b(3) \geq 2$ *and* $b(2) \geq 2$,
(e) $b(1) \geq 1$, $b(2) \geq 1$ *and* $b(3) \geq 1$,
(f) $b(2) \geq 3$,
(g) $b(1) \geq 2$.

(With more elaborate work a much stronger version of Greendlinger's Lemma can be proved.)

The proof of the Area Theorem is based on the following key lemma.

Lemma. (The growth condition.) *Let* M *be a strongly connected map which satisfies the condition* W(6) *and let* D *be a region of* M. *Let* μ *be a boundary path of* $St_M^i(D)$ *for some* i. *Denote by* $\lambda(\mu)$ *the number of regions* E *in* $St_M^{i+1}(D) \setminus St_M^i(D)$ *such that* $\partial E \cap \mu$ *contains an edge. Denote by* $\rho(\mu)$ *the number of regions* F *in* $St_M^i(D)$ *such that* $\partial F \cap \mu$ *contains an edge. If* μ *does not intersect the boundary of* M *then* $\rho(\mu) \leq \lambda(\mu)$. *Moreover if* $St_M^{i+1}(D)$ *contains no boundary regions, then there are boundary paths* μ *on* $\partial St_M^{i+1}(D)$ *such that* $\rho(\mu) < \lambda(\mu)$ (see Fig. 7).

Fig. 7

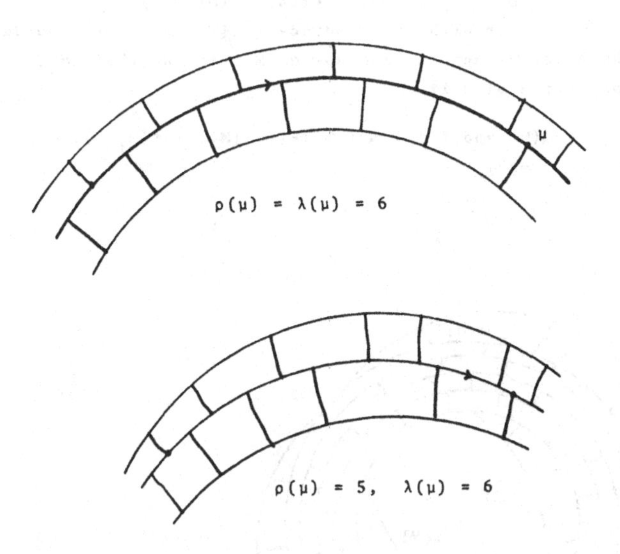

The Area Theorem. *Let* M *be a strongly connected map which satisfies* W(6) *and let* n *be the number of boundary regions of* M *which intersect the boundary of* M *nontrivially. Then the number* $|M|$ *of the regions in* M *is bounded by* n^2. *(Here "nontrivially" means "contains an edge".)*

This of course solves the word problem for groups with W(6) van Kampen diagrams. However in certain groups the Lemma of Greendlinger can be used in a more efficient way to solve the word problem.

1.3 *The conjugacy problem*

Let $\mathcal{F}$ be a finitely generated free group and $\mathcal{R}$ a finite symmetrized set of elements of F. Let $\mathcal{N}$ be the normal subgroup of $\mathcal{F}$ generated by $\mathcal{R}$ and let $G = \mathcal{F}/\mathcal{N}$. Recall that if $u, v \in \mathcal{F}$ are reduced and cyclically reduced then u and v are conjugate mod $\mathcal{N}$ iff there exists an annular $\mathcal{R}$-diagram M with outer boundary $\omega(M)$ and inner boundary $\tau(M)$ such that the label of $\omega(M)$ is u and the label of $\tau(M)$ is v. Consequently in this subsection we shall concentrate on annular maps with condition W(6). To formulate the solution of the conjugacy problem we need some definitions.

Definitions. (1) Let M be a map and D a region in M. The *augmented star neighbourhood* $ASt_M(D)$ of D in M is the union of all the regions of M, the boundary of which has a nonempty intersection with the boundary of D.
(2) Let M be an annular map. We say that M is an *elementary annulus*, if we can arrange the regions $D_1,\dots,D_t$ of M such that for i, $2 \leq i \leq t-1$, $ASt_M(D_i) \subseteq \{D_{i-1},D_i,D_{i+1}\}$ holds, $ASt_M(D\ ) \subseteq \{D_t,D_1,D_2\}$ and $ASt_M(D_t) \subseteq \{D_{t-1},D_t,D_1\}$. Thus an elementary annulus is a one layer map. For an elementary annulus M we denote by $|M|$ the number of regions in M.
(3) Let M be an annular map with outer boundary $\omega(M)$ and inner boundary $\tau(M)$. We denote by $\|\omega(M)\|$ the number of regions of M the boundary of which intersects non-trivially $\omega(M)$. We define $\|\tau(M)\|$ similarly.
(4) Let M be an annular map with outer boundary $\omega(M)$ and inner boundary $\tau(M)$. Let $\{M_1,\dots,M_r\}$ be a set of annular submaps of M. We say that $\{M_1,\dots,M_r\}$ is a *concentric partition* of M if

$$\tau(M_1) = \tau(M),\ \omega(M_r) = \omega(M) \text{ and for } 1 \leq i \leq r-1,\ \omega(M_i) = \tau(M_{i+1}). \qquad (*)$$

Fig. 8

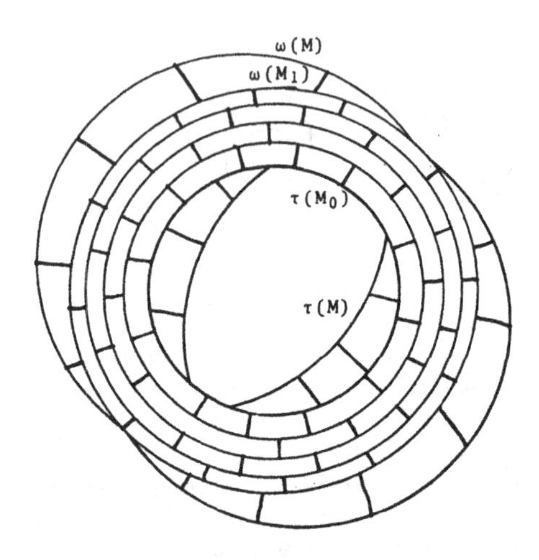

(5) A concentric partition $\{M_1,...,M_r\}$ of M is *elementary* if every M_i is an elementary annulus.

Let $\{M_1,...,M_r\}$ be an elementary concentric partition of M and let $|M_i| = m_i$. The growth type of M is the vector $(m_1,...,m_r)$.

The following theorem solves the conjugacy problem for presentations giving rise to W(6) diagrams, provided that we know the length of the longest relator.

Theorem. *Let* M *be an annular map. If* M *satisfies the condition* W(6) *then* M *contains an annular submap* M_0 *with an elementary concentric partition* $\{M_1,...,M_r\}$ *with growth vector* $(m_1,...,m_r)$ *satisfying*

(i) *there exist* $1 \leq s \leq t \leq r$ *such that* $m_1 > m_2 > ... > m_s = m_{s+1} = ... = m_t < m_{t+1} ... < m_r$;

(ii) $\omega(M_0) \cap \omega(M) \neq \emptyset$ *and* $\tau(M_0) \cap \tau(M) \neq \emptyset$;

(iii) $|\omega(M_0)| \leq |\omega(M)|$ *and* $|\tau(M_0)| \leq |\tau(M)|$.

Acknowledgements. I would like to express my most cordial thanks to Dr E. Rips for the numerous discussions and for his useful suggestions which improved the paper substantially.

This work was written while I was a Feinberg Fellow in the Weizmann Institute, Rehovot. My thanks to the Foundation for the grant and to the Pure Mathematics Department for its generous hospitality. I am especially grateful to Professor A. Joseph for his constant attention and encouragement.

REFERENCES

1. K. Appel, On Artin groups and Coxeter groups of large type, *Contemp. Math.* 33 (1984), 50-78.
2. P. Schupp & K. Appel, Artin groups and infinite Coxeter groups, *Invent. Math.* 72 (1983), 201-220.
3. D.L. Johnson, Analogues of the braid groups, in *Groups-Korea 1983*, Lecture Notes in Mathematics 1098, Springer-Verlag, Berlin-Heidelberg-New York (1984), 63-68.
4. A. Juhàsz, A weak small cancellation condition, in preparation.
5. A. Juhàsz, Some problems in the theory of Artin groups, in preparation.
6. R.C. Lyndon, On Dehn's Algorithm, *Math. Ann.* 166 (1966), 208-228.
7. R.C. Lyndon & P.E. Schupp, *Combinatorial Group Theory*, Springer-Verlag, Berlin-Heidelberg-New York (1977).
8. E. Rips, Generalized small cancellation theory and applications. I. The word problem, *Israel J. Math.* 41 (1982), 1-146.

THE LIE ALGEBRA ASSOCIATED TO THE LOWER CENTRAL SERIES OF A GROUP

John P. Labute
McGill University, Montreal, Quebec H3A 2K6, Canada

The lower central series of a group G is the sequence of subgroups G_n ($n \geq 1$) of G defined inductively by (i) $G_1 = G$, (ii) $G_{n+1} = [G,G_n]$, where, for any two subgroups H,K of G, [H,K] denotes the subgroup of G generated by the commutators $[h,k] = h^{-1}k^{-1}hk$ with $h \in H$, $k \in K$. The subgroups G_n also satisfy (iii) $G_{n+1} \subseteq G_n$, (iv) $[G_m,G_n] \subseteq G_{m+n}$.

Let $L_n(G) = G_n/G_{n+1}$, with the operation in this abelian group denoted additively, and let $i_n : G_n \to L_n = L_n(G)$ be the canonical surjection. Then the graded abelian group $L(G) = \oplus_{n \geq 1} L_n(G)$ has a natural Lie algebra structure over $\mathbb{Z}$ where the bracket of two homogeneous elements $\xi = i_m(x)$, $\eta = i_n(y)$ is defined by $[\xi,\eta] = i_{m+n}([x,y])$ (cf. [8]).

If $x \in G_n$, $x \notin G_{n+1}$, then $n = \omega(x)$ is called the *weight* of x and $\xi = i_n(x)$ is called the *initial form* of x. If $x \in G_n$ for $n \geq 1$, then the initial form of x is defined to be zero.

Problem 1. *Given the group* G, *determine the Lie algebra* L(G).

This is a more general problem than that of just determining the sequence of abelian groups $L_n(G)$ (cf. [1]) and has been solved for relatively few groups. If F is a free group then Magnus [7] and Witt [9] have shown that $L = L(F)$ is a free Lie algebra over $\mathbb{Z}$. Moreover, if $(x_i)_{i \in I}$ freely generate F and $\xi_i = i_1(x_i)$ is the image of x_i in $L_1(F) = F/[F,F]$, then $(\xi_i)_{i \in I}$ freely generate $L = L(F)$. If $G = F/F_n$, then $L(G) = \oplus_{1 \leq k < n} L_k = L/C_n(L)$, where $C_n(L)$ is the n-th term of the lower central series of L. More generally, if $G = F/R$, then $L(G) = L/\mathfrak{R}$, where $\mathfrak{R} = \oplus_{n \geq 1} (R \cap F_n)F_{n+1}/F_{n+1}$ is the ideal of L consisting of the initial forms of the relators $r \in R$.

Problem 2. *Given a presentation* $G = F/R$ *of* G, *determine the ideal* $\mathfrak{R}$ *of initial forms of the relators* $r \in R$.

If R is the normal subgroup of the free group F generated by the elements r_j ($j \in J$) and ρ_j is the initial form of r_j, then the ideal $\mathfrak{r}$ of L generated by the elements ρ_j ($j \in J$) is contained in $\mathfrak{R}$. However $\mathfrak{r} \neq \mathfrak{R}$ in general. For example, let F be the free group on x_1, x_2 and let R be the normal subgroup of F generated by x_1^2. Here $\mathfrak{r}$ is generated by $2\xi_1$. Now $r = [x_2,x_1^2] = [x_2,x_1]^2[[x_2,x_1],x_1]$ and $[[x_2,x_1],r] = [[x_2,x_1],[[x_2,x_1],x_1]]$ are relators in R whose initial forms are respectively $2[\xi_2,\xi_1]$ and $[[\xi_2,\xi_1],[[\xi_2,\xi_1],\xi_1]]$; the second of these is not in $\mathfrak{r}$.

Magnus [8] calls a relator $r \in F$ a *power relator* if the initial form of r is of the form $n\sigma$ with $n > 1$. All other relators $r \in F$ are called *commutator relators*. A power relator r in R is called *trivial* if the initial form of r is of the form $n\sigma$ with $n > 1$ and $\sigma \in \mathfrak{R}$. Non-trivial power relators exist in R if and only if L(F/R) has torsion elements. In the above example R is generated (as a normal subgroup of F) by the single power relator x_1^2 and $[[x_2,x_1],[[x_2,x_1],x_1]]$ is a commutator relator in R. In [8] Magnus posed the following problem.

Problem 3. *If* R *is generated as a normal subgroup of* F *by a single power relator* r, *find the initial forms of all commutator relators in* R.

In [2] we answered this question in the case F is the free group on x_1,x_2 and $r = x_1^p$, p a prime. The following is a more general result (cf. [6]).

Theorem 1. *Let* F *be the free group on* $x_1,\ldots,x_N$ *and let* R *be the normal subgroup of* F *generated by* $r = x_1^p$, *p a prime. Then* $\mathfrak{R}$ *is generated, as an ideal of* L(F) *by the elements*

$$p\xi_1,\quad [ad(\lambda\xi_1)\xi_i, ad(\lambda\xi_1^{n(p-1)+p})\xi_i],\quad [ad(\lambda\xi_1)\xi_j, ad(\mu\xi_1^{n(p-1)+p})\xi_1] +$$

$$[ad(\mu\xi_1)\xi_i, ad(\lambda\xi_1^{n(p-1)+p})\xi_j] \quad (\lambda,\mu \in V,\ n \geq 0,\ 2 \leq i,\ j \leq N),$$

where V *is the enveloping algebra of* L(F) *and* ad *is the adjoint representation of* V.

If p is not prime or if $r = u^p$ with u a commutator relator of weight greater than 1, then $\mathfrak{R}$ is not known. However, we do have the following result (cf. [6]).

Theorem 2. *Let* R *be the normal subgroup of the free group* F *generated by the single relator* $r = u^p v$, *where p is a prime and u,v are commutator relators with v in the normal subgroup of* F *generated by* u *and* $\omega(v) > \omega(u)$. *Let* σ *and* $\tau = ad(\mu)\sigma$ *be respectively the initial forms of* u *and* v. *If* $p > \omega(v)/\omega(u)$, *then* $\mathfrak{R}$ *is the ideal of* L(F) *generated by the elements* $p\sigma$ *and* $[ad(\lambda)\sigma, ad(\lambda\mu^n)\tau]$ $(\lambda \in V,\ n \geq 1)$, *where* V *is the enveloping algebra of* L(F).

Problem 4. *If* R *is the normal subgroup of the free group* F *generated by the relators* $r_1,\ldots,r_T$, *give conditions on the initial forms* $\rho_1,\ldots,\rho_T$ *of these relators so that the ideal* $\mathfrak{r}$ *of* L(F) *they generate is equal to the ideal* $\mathfrak{R}$ *of initial forms of the relators in* R.

In [3] we solve this problem in the case $T = 1$ by showing that, in this case, $\mathfrak{r} = \mathfrak{R}$ if $r = r_1$ is a commutator relator. As a by-product of the proof we also show that if r is a commutator relator, then R contains no non-trivial power relators, i.e. L(F/R) is torsion free, thus answering a question of Magnus [8]. The case $T > 1$ is more delicate as the following example shows.

Example. Let F be the free group on $x_1,\ldots,x_4$ and let $r_1 = [x_1,x_2]$, $r_2 = [x_2,x_3]$, $r_3 = [x_3,x_1][x_2,[x_2,x_4]]^2$. Then $\rho_1 = [\xi_1,\xi_2]$, $\rho_2 = [\xi_2,\xi_3]$, $\rho_3 = [\xi_3,\xi_1]$ and we see that ρ_1, ρ_2, ρ_3 are part of a basis for $L_2(F)$. Since $r = [x_3,r_1][x_1,r_2][x_2,r_3] = [x_2,[x_2,[x_2,x_4]]]^2u$, where u is an element of F_4 whose initial form is in the ideal I of L(F) generated by ξ_1 and ξ_2, it follows that r is an element of R whose initial form is not in I. Since $\mathfrak{r} \subseteq I$, it follows that $\mathfrak{r} \neq \mathfrak{R}$. Moreover r is a non-trivial power relator in R.

For any prime p and any abelian group N we let $N(p) = N/pN$. If, in addition, M is a subgroup of N we let $N_M(p)$ be the image of M(p) in N(p). Let $L = L(F)$ and let U be the enveloping algebra of $L/\mathfrak{r}$. Then $\mathfrak{r}/[\mathfrak{r},\mathfrak{r}]$ is a U-module via the adjoint representation and is generated by the images of $\rho_1,\ldots,\rho_T$. In [5] we proved that $\mathfrak{r} = \mathfrak{R}$ if the following condition was satisfied:

(*) *For any prime* p, $\mathfrak{r}_L(p)/[\mathfrak{r}_L(p),\mathfrak{r}_L(p)]$ *is a free* U(p)*-module on the images of* $\rho_1,\ldots,\rho_T$.

This condition implies that $L/\mathfrak{r}$ is a torsion-free Z-module and that $\mathfrak{r}/[\mathfrak{r},\mathfrak{r}]$ is a free U-module; if g_n is the rank of $L_n/\mathfrak{r}_n$ and d_i is the degree of ρ_i, then

$$1 - Nt + t^{d_1} + t^{d_2} + \ldots + t^{d_T} = \prod_{n \geq 1} (1 - t^n)^{g_n}.$$

If $T = 1$, then the condition holds if and only if r_1 is a commutator relator. If F is free on $x_1,\ldots,x_N$, then it can be shown that condition (*) holds in the following cases (cf. [5])

(i) $T = N-1$, $\rho_1 = [\xi_1,\xi_2]$, $\rho_2 = [\xi_1,\xi_3],\ldots,\rho_{N-1} = [\xi_1,\xi_N]$;

(ii) $T = N-1$, $\rho_1 = [\xi_1,\xi_2]$, $\rho_2 = [\xi_2,\xi_3],\ldots,\rho_{N-1} = [\xi_{N-1},\xi_N]$;

(iii) $N = 3$, $T = 2$, $\rho_1 = [\rho_3,[\xi_1,\xi_2]]$, $\rho_2 = [\xi_2,[\xi_1,\xi_3]]$;

(iv) $N = 3$, $T = 2$, $\rho_1 = [\xi_1,\xi_2]$, $\rho_2 = [[\xi_1,\xi_3],\xi_2]$.

These results have been used by Hain [2] to determine L(G) for certain link groups G.

Problem 5. *If* $\rho_k = \Sigma_{i<j}\, a_{ijk}[\xi_i,\xi_j]$ $(1 \leq k \leq T)$, *find conditions on* a_{ijk} *so that condition* (*) *holds.*

The condition (*) is not necessary for $\mathfrak{r} = \mathfrak{R}$ to hold. For example, if F is free on x_1,x_2,x_3 and $\rho_1 = [\xi_1,\xi_2]$, $\rho_2 = [\xi_2,\xi_3]$, $\rho_3 = [\xi_3,\xi_1]$, then condition (*) fails to be true because of the Jacobi identity, while $\mathfrak{r} = \mathfrak{R}$.

Problem 6. *If* $\rho_k = \Sigma_{i<j}\, a_{ijk}[\xi_i,\xi_j]$ $(1 \leq k \leq T)$, *find conditions on* a_{ijk} *so that* $\mathfrak{r} = \mathfrak{R}$.

Acknowledgement. This work was supported by a grant from the National Science and Engineering Council of Canada and a Quebec FCAR grant.

BIBLIOGRAPHY

1. K.T. Chen, R.H. Fox & R.C. Lyndon, Free differential calculus, IV. The quotient groups of the lower central series, *Ann. of Math.* 68 (1958), 81-95.
2. R.M. Hain, Iterated integrals, intersection theory and link groups, *Topology* 24 (1985), 45-66.
3. J.P. Labute, On the descending central series of groups with a single defining relation, *J. Algebra* 14 (1970), 16-23.
4. J.P. Labute, The lower central series of the group $\langle x,y : x^p = 1 \rangle$, *Proc. Amer. Math. Soc.* 66 (1977), 197-201.
5. J.P. Labute, The determination of the Lie algebra associated to the lower central series of a group, *Trans. Amer. Math. Soc.* 288 (1985), 51-57.
6. J.P. Labute, The Lie algebra associated to the lower central series of certain groups defined by a single power relator, to appear.
7. W. Magnus, Uber Beziehungen zwischen höheren Kommutatoren, *J. Reine Angew. Math.* 177 (1937), 105-115.
8. W, Magnus, A. Karrass & D. Solitar, *Combinatorial group theory: presentations of groups in terms of generators and relations*, Pure and Appl. Math., Vol. 13, Interscience, New York (1966).
9. E. Witt, Treue Darstellung Liescher Ringe, *J. Reine Angew. Math.* 177 (1937), 152-160.

ALGEBRAICALLY CLOSED LOCALLY FINITE GROUPS

Felix Leinen
Johannes Gutenberg University, 6500-Mainz, West Germany

Let $\mathfrak{X}$ be a class of groups. Then $G \in \mathfrak{X}$ is said to be *algebraically closed* (a.c.) in $\mathfrak{X}$, if every system of finitely many equations with coefficients from G, which has a solution in some $\mathfrak{X}$-supergroup of G, can already be solved in G. If the above condition holds for systems of finitely many equations and inequations, then G is called *existentially closed* (e.c.) in $\mathfrak{X}$. The existence of e.c. (and a.c.) $\mathfrak{X}$-groups is guaranteed by [3], Proposition I.1.3, whenever $\mathfrak{X}$ is inductive, i.e. closed under forming unions of ascending chains. Both notions may be seen as a pendant to the (usual) notion of an a.c. field (in the class of all fields).

In general it is much easier to study e.c. $\mathfrak{X}$-groups rather than a.c. $\mathfrak{X}$-groups. For example, the group table of every finite group can be described by finitely many equations and inequations; therefore, if G is e.c. in a class $\mathfrak{X}$ of locally finite ($L\mathfrak{F}$) groups, then for every finite group F, which is contained in an $\mathfrak{X}$-supergroup of G, the identity $id:F \cap G \to G$ can be extended to an embedding $\phi:F \to G$. This kind of injectivity need not hold for a.c. $\mathfrak{X}$-groups, as is demonstrated drastically by the fact, that the trivial group is a.c. in $L\mathfrak{F}$.

However, B.H. Neumann [8] has already shown in 1952, that (except for the trivial case) both notions coincide for the class of all groups. In order to prove that non-trivial a.c. groups are e.c., he replaced inequations by certain systems of finitely many equations; and, roughly speaking, he therefore had to use the simplicity of e.c. groups. R.E. Phillips observed that this idea can be transferred to a.c. $L\mathfrak{F}$-groups, since the e.c. $L\mathfrak{F}$-groups are simple too.

Theorem A. *The non-trivial a.c. $L\mathfrak{F}$-groups are exactly the e.c. $L\mathfrak{F}$-groups.*

Note that the e.c. $L\mathfrak{F}$-groups are the universal $L\mathfrak{F}$-groups introduced by P. Hall [1]. In particular, there exists up to isomorphism a unique countable, e.c. $L\mathfrak{F}$-group.

Few years ago the study of e.c. groups in the class $L\mathfrak{F}_p$ of all locally finite p-groups was initiated by B. Maier [7]. Since minimal normal subgroups of $L\mathfrak{F}_p$-groups are central and cyclic of order p, an e.c. $L\mathfrak{F}_p$-group cannot be simple. B. Maier showed that there exists a unique countable, e.c. $L\mathfrak{F}_p$-group E_p, and that E_p is characteristically simple. But the conjecture, that every e.c. $L\mathfrak{F}_p$-group be char. simple, was refuted by S. Thomas [9].

In the meanwhile, the structure of the normal subgroup lattice of e.c. $L\mathfrak{F}_p$-groups was determined by F. Leinen [4]: In every e.c. $L\mathfrak{F}_p$-group G the normal subgroups form a chain; within this chain the chief factors M/N (formed from pairs of normal subgroups $M,N \triangleleft G$ such that M/N is minimal normal in G/N) are located densely without a maximal or minimal member. In particular, G has no minimal normal subgroup. That the latter need not hold for a.c. $L\mathcal{F}_p$-groups G, is immediately clear: Without formulating inequations there is no way to force the centre of G to become trivial.

Furthermore, every group G, which is e.c. in the class $(L\mathcal{F}_p)_{C_p}$ of all $L\mathcal{F}_p$-groups with centre containing C_p, is a.c. in $L\mathcal{F}_p$: It is known from [5], that $Z(G) = C_p$ is a monolith in G; therefore, if S is a system of finitely many equations with coefficients from G and solution in some $L\mathcal{F}_p$-supergroup H of G, and if K/L is a chief factor in H with $L \cap G = 1 < Z(G) = K \cap G$, then G can be identified via $G/G \cap L \cong GL/L$ with the subgroup GL/L of H/L; now the $(L\mathcal{F}_p)_{C_p}$-supergroup H/L of the e.c. $(L\mathcal{F}_p)_{C_p}$-group G still provides a solution for S; hence S must be solvable in G.

Modifying the idea of B.H. Neumann and using some knowledge of the normal subgroup structure of a.c. $L\mathcal{F}_p$-groups it is possible to prove

Theorem B. *The non-trivial a.c.* $L\mathcal{F}_p$*-groups are exactly the e.c. groups in one of the classes* $L\mathcal{F}_p$ *and* $(L\mathcal{F}_p)_{C_p}$.

Note that for every countable, abelian p-group A there exists a unique countable, e.c. group E_A in the class $(L\mathcal{F}_p)_A$ of all $L\mathcal{F}_p$-groups with centre containing A. Moreover, the Schur-multipliers of these groups are trivial. Therefore, $E_A/A \cong E_B/B$ if and only if $A \cong B$, and the factor groups E_A/A yield $2^{\aleph_0}$ isomorphism types of countable, characteristically simple $L\mathcal{F}_p$-groups with totally ordered normal subgroup lattice. These results are proved in F. Leinen and R.E. Phillips [5]; the distinguishing of isomorphism types by studying central extensions goes back to K. Hickin [2].

Theorem B can be generalized to countable, a.c. groups in the class $L(\mathcal{F}_\pi \cap \mathcal{S})$ of all locally finite-soluble π-groups (π a set of primes). Here $(L\mathcal{F}_p)_{C_p}$ has to be replaced by the classes $(L(\mathcal{F}_\pi \cap \mathcal{S}))_p$, $p \in \pi$, of all $L(\mathcal{F}_\pi \cap \mathcal{S})$-groups with a minimal normal subgroup containing C_p.

Theorem C. *The non-trivial, countable, a.c.* $L(\mathcal{F}_\pi \cap \mathcal{S})$*-groups are exactly the countable, e.c. groups in one of the classes* $L(\mathcal{F}_\pi \cap \mathcal{S})$ *and* $(L(\mathcal{F}_\pi \cap \mathcal{S}))_p$, $p \in \pi$.

In particular, since every a.c. $L(\mathcal{F}_\pi \cap \mathcal{S})$-group has a totally ordered normal subgroup lattice too, there exist at least $|\pi| + 1$ isomorphism types of countable, a.c. $L(\mathcal{F}_\pi \cap \mathcal{S})$-groups $\neq 1$. However, we don't yet have such nice uniqueness statements as for the classes $L\mathcal{F}_p$ and $(L\mathcal{F}_p)_{C_p}$.

The countable, a.c. $L(\mathcal{F}_\pi \cap \mathcal{S})$-groups appear quite naturally as factors of the countable, e.c. $L(\mathcal{F}_\pi \cap \mathcal{S})$-groups.

Theorem D. *The group* G *is a countable, a.c.* $L(\mathcal{F}_\pi \cap \mathcal{S})$*-group if and only if there exists a countable, e.c.* $L(\mathcal{F}_\pi \cap \mathcal{S})$*-group* H *and a normal subgroup* $N \lhd H$ *with* $N \neq \langle h^H \rangle$ *for all* $h \in H \setminus 1$ *such that* $G \cong H/N$.

In the situation of Theorem D it is known from [4] that H splits over N. The proofs of the above theorems can be found in F. Leinen and R.E. Phillips [6].

REFERENCES

1. P. Hall, Some constructions for locally finite groups, *J. London Math. Soc.* 34 (1959), 305-319.
2. K. Hickin, Universal locally finite central extensions of groups, *Proc. London Math. Soc.* (3), to appear.
3. J. Hirschfeld & W.H. Wheeler, *Forcing, arithmetic, division rings,* Berlin - Heidelberg - New York (1975).
4. F. Leinen, Existentially closed groups in locally finite group classes, *Comm. Algebra* 13 (1985), 1991-2024.
5. F. Leinen & R.E. Phillips, Existentially closed central extensions of locally finite p-groups, *Math. Proc. Camb. Phil. Soc.*, to appear.
6. F. Leinen & R.E. Phillips, Algebraically closed groups in locally finite group classes, in preparation.
7. B. Maier, Existenziell abgeschlossene lokal endliche p-Gruppen, *Arch. Math.* 37 (1981), 113-128.
8. B.H. Neumann, A note on algebraically closed groups, *J. London Math. Soc.* 27 (1952), 247-249.
9. S. Thomas, Complete existentially closed locally finite groups, *Arch. Math.* 44 (1985), 97-109.

ON POWER-COMMUTATIVE AND COMMUTATION-TRANSITIVE GROUPS

Frank Levin
The Ruhr University, 463 Bochum-Querenburg, West Germany

Gerhard Rosenberger
University of Dortmund, Dortmund, West Germany

1. *INTRODUCTION*

A group G is a PC-group (power commutative) if for any elements a,b in G of infinite orders, proper powers of a,b permute only if a,b permute, i.e. $[a^m,b^n] = 1$, $m,n \neq 0$, implies $[a,b] = 1$. G is a CT-group (commutation-transitive) if for any a,b,c in G, b noncentral, $[a,b] = [b,c] = 1$ implies $[a,c] = 1$. In [1], B. Fine notes that free groups and Fuchsian groups are both PC and CT groups and poses the problem of finding further classes of PC and of CT groups (the latter problem is attributed to L. Greenberg). In this note we construct various examples of PC and CT groups and consider generalized free products and HNN-extensions of such groups.

2. GL(2,K)

The example of a free nilpotent group of class at least 4 shows that not all PC-groups are CT-groups. Below (Theorem 2) we give an example of a CT-group which is not a PC-group. On the other hand, a group with presentation $\langle a,b,t : tat^{-1} = a^{-1}\rangle$, which is an HNN-extension of a free group is neither a CT-group nor a PC-group.

In this section we consider examples arising from linear groups. Let K be any field and GL(2,K) the general linear group of 2×2 matrices over K. If K contains elements of infinite orders GL(2,K) contains elements of infinite order which are non-central but have finite powers which are central, and, hence, GL(2,K) will not be a PC-group. However, the normal subgroup Λ given by

$$\Lambda = \{A \in GL(2,K) : (\det A)^m = 1 \text{ for some } m \neq 0\},$$

is a PC-group. To show this we use the following observations.

Lemma 1. *For any* $A \in GL(2,K)$, $n > 1$, A^n *is a linear combination of* A *and* I, *the identity matrix. Specifically, if* $a = \det A$ *and* $x = \operatorname{tr} A$, *then*

$$A^n = T_n A - aT_{n-1} I, \tag{1}$$

where $T_n = T_n(A)$ *is a scalar defined inductively by*

$$T_0 = 0,\ T_1 = 1,\ T_n = xT_{n-1} - aT_{n-2}. \tag{2}$$

Proof. The first remark follows directly from the fact that $A^2 = xA - aI$. The proof of (1) is by induction. For $n = 1$ there is nothing to prove. Thus, if $A^n = T_nA - aT_{n-1}I$, $n > 1$, then $A^{n+1} = T_nA^2 - aT_{n-1}A = T_n(xA-aI) - aT_{n-1}A = T_{n+1}A - aT_nI$, by (2).

Lemma 2. *If* $T_n(A) = 0$ *for* $A \in \Lambda$ *and some* $n \geq 1$, *then* A *has finite order.*

Proof. By (1), $A^n = dI$, $d \in K$. Hence, $d^2 = (\det A)^n$, so if $(\det A)^m = 1$, then $d^{2m} = 1$ and $A^{2mn} = I$.

Theorem 1. Λ *is a* PC-*group.*

Proof. Suppose $[A^m, B^n] = I$. By (1), $A^m B^n - B^n A^m = T_m(A) T_n(B)(AB - BA) = 0$. Hence, if A and B have infinite orders, then by Lemma 2, $AB = BA$.

The following corollary follows from the observation that PC-groups are subgroup closed.

Corollary 1. *The following are* PC-*groups: Fuchsian groups,* NEC-*groups, Kleinian groups, free products of cyclic groups, free groups,* $P\Lambda$ *(the image of* Λ *under the canonical map* P *of* GL(2,K) *to* PGL(2,K)).

Theorem 2. GL(2,K) *is a* CT-*group.*

Proof. Suppose $[A,B] = [B,C] = I$, B not central. To show that $[A,C] = I$ we may further assume A and C not central and K algebraically closed. Since $AB = BA$ and $BC = CB$ it follows that A,B,C have the same eigenvectors. Hence, a similarity transformation which diagonalizes B or reduces B to upper triangular with equal diagonal elements will do the same to A and C, and, hence, $[A,C] = I$.

In general PGL(2,K) and PSL(2,K) are not CT-groups but it is easy to see that PSL(2,$\mathbb{R}$) is a CT-group. Therefore Fuchsian groups, free product of cyclic groups and free groups are CT-groups.

3. HNN-*EXTENSIONS AND GENERALIZED FREE PRODUCTS*

Theorem 3. *For any* $r \neq 1,0$, $G = \langle t,x : txt^{-1} = x^r \rangle$ *is a* PC-*group and a* CT-*group.*

Proof. If $r = 1$, G is free abelian and trivially in both classes. Thus, suppose $|r| \geq 2$ and $a,b \in G$ of infinite orders with $[a^m, b^n] = 1$. Conjugating, if necessary, by a sufficiently high power of t^{-1} we may assume that $a = x^{\alpha} t^{-\beta}$ and $b = x^{\gamma} t^{-\delta}$ with $\beta, \delta \geq 0$ and one, say $\beta \neq 0$. Thus, $[a,b] = t^{\beta} x^{-\alpha} t^{\delta} x^{-\gamma} x^{\alpha} t^{-\beta} x^{\gamma} t^{-\delta} = x^k$, where $k = \alpha(r^{\beta+\delta} - r^{\beta}) - \gamma(r^{\beta+\delta} - r^{\delta})$. A similar expansion, observing that $a^m = t^{-m\beta} x^{\alpha'}$, $b^n = t^{-n\delta} x^{\gamma'}$, where $\alpha' = \alpha(r^{m\beta} + \ldots + r^{2\beta} + r^{\beta})$ and $\gamma' = \gamma(r^{n\delta} + \ldots + r^{2\delta} + r^{\delta})$, gives $[a^m, b^n] = x^p$, where $p = \alpha'(r^{n\delta} - 1) - \gamma'(r^{m\beta} - 1)$, and it follows that $k = 0$ if and only if $p = 0$ and, hence, $[a,b] = 1$. Finally, suppose $[a,b] = [b,c] = 1$ for $a,b,c \in G$, $b \neq 1$. As above, we may assume that after conjugation, $a = x^{\alpha} t^{-\beta}$, $b = x^{\gamma} t^{-\delta}$, $c = x^{\varepsilon} t^{-\tau}$ with $\beta, \delta, \tau \geq 0$. Thus, $[a,b] = x^k$, $[b,c] = x^{k'}$, $[a,c] = x^{k''}$, where k is as above with analogous expressions for k' and k'', and a direct comparison shows that $k = k' = 0$ implies that $k'' = 0$. Thus, $[a,c] = 1$ and G is a CT-group.

A free product of PC-groups is clearly a PC-group, and a free product of CT-groups is again a CT-group if the factors are either abelian or CT-groups without centers. This gives the following corollary.

Corollary 2. *For any* $r \neq -1,0$, $G = \langle a_i, i \in I, x, t : txt^{-1} = x^r \rangle$, *an* HNN-*extension of the free group* $\langle x, a_i, i \in I : \emptyset \rangle$, *is a* PC-*group and a* CT-*group.*

Before stating the next example we recall a definition: A subgroup H of a group G is *malnormal* in G if for any $g \notin H$, $g^{-1}Hg \cap H = 1$. The following lemma incorporates results from S. Pride [6] and N. Pecynski and W. Reiwer [5], cf. Rosenberger [8].

Lemma 3. *Let* $B_1 = \langle B, t : \text{rel } B, t^{-1}K_1t = K_{-1} \rangle$ *be an* HNN-*extension with basis* B, *stable letter* t *and conjugated subgroups* K_1, K_{-1} *of* B, *both malnormal in* B. *Let* $\langle a,b \rangle$ *be a two-generator subgroup of* B_1 *which is not cyclic. Then* $\langle a,b \rangle$ *is conjugate to* $\langle a',b' \rangle$ *satisfying one of the following conditions:* (i) $\langle a',b' \rangle$ *is a free product of cyclic groups;* (ii) $\langle a',b' \rangle$ *is in* B; (iii) a' *has the form* $th_1 \ldots th_r$, $r > 1$, $b' \in K_{-1}$ *and* $a'^{-1}b'a'$ *is in* B.

Theorem 4. *Let* B *be a* PC-*group and* K *be a malnormal subgroup of* B. *Then the* HNN-*extension* $B_1 = \langle B, t : \text{rel } B, txt^{-1} = x$ for all x in $K \rangle$ *is a* PC-*group.*

Proof. Suppose $a,b \in B_1$ have infinite orders and $[a^m, b^n] = 1$. To show that $[a,b] = 1$ we may assume that $\langle a,b \rangle$ is not cyclic nor is conjugate to a subgroup of a PC-group. By the lemma this implies that $\langle a,b \rangle$ is conjugate to $\langle a',b' \rangle$ satisfying (iii). However, if $a' = th_1 \ldots th_r$ and b' is in K, it follows by an easy induction that $a'^{-1}b'a' \in B$ implies $h_i \in K$ for $i < r$, that is, $a' = t^r h$, for some h in B. Using the fact that K is malnormal in B and t acts trivially on K, it follows that a', b' generate a free group if $h \notin K$. However, $\langle a,b \rangle$ is clearly not free, so $h \in K$ and a is conjugate to $t^p h_1$, b to $t^q h_2$, say, with $h_i \in K$. Since t centralizes K, $[a^m, b^n] = 1$ implies $[h_1^m, h_2^n] = 1$ and, hence, $[a,b] = [h_1, h_2] = 1$ since B is a PC-group.

The analogue of Theorem 4 for CT-groups will be false if K is nonabelian and nontrivial in B. Before stating the analogue for K abelian we first observe the following lemma, which follows directly from the definitions.

Lemma 4. *Let* $B_1 = \langle B, t : \text{rel } B, txt^{-1} = x$ for all x in $K \rangle$, K *malnormal in* B. *Then* $\langle t, K \rangle$ *is malnormal in* B_1.

Theorem 5. *Let* B *be a* CT-*group and* K *be a nontrivial abelian malnormal subgroup of* B. *Then the group* $B_1 = \langle B, t : \text{rel } B, txt^{-1} = x$ for all x in $K \rangle$ *is a* CT-*group.*

Proof. Since K is malnormal in B, if B has nontrivial center, then $K = B$ and B_1 is abelian. Hence, assume B has trivial center. Suppose $a,b,c \in B_1$, $b \neq 1$, and $[a,b] = [b,c] = 1$. If either $\langle a,b \rangle$ or $\langle b,c \rangle$, say $\langle a,b \rangle$, is not cyclic, then by Lemma 3, $\langle a',b' \rangle$ will satisfy (ii) or (iii), since $\langle a,b \rangle$ is abelian. If $\langle a',b' \rangle$ satisfies (iii), by the above argument, $\langle a',b' \rangle$ is in $\langle t,K \rangle$. Conjugating $\langle b,c \rangle$ by the same element which produced $\langle a',b' \rangle$ will produce an abelian group with one element in $\langle t,K \rangle$, so, by Lemma 4, the whole subgroup will be in $\langle t,K \rangle$. Thus, $[a,c] = 1$, since $\langle t,K \rangle$ is abelian. If, on the other hand, (ii) holds, that is, $\langle a',b' \rangle$ is an abelian subgroup of B, then again conjugating $\langle b,c \rangle$ by the same element produces an abelian group with one element, the conjugate of b, in B. However, since we are essentially assuming that this conjugate is not in K (otherwise we have (iii)), the whole subgroup must be in B, so $[a,c] = 1$ since B is a CT-group. Thus, we may assume that both $\langle a,b \rangle$ and $\langle b,c \rangle$ are cyclic, that is, $\langle a,b \rangle = \langle d^p \rangle$ and $\langle b,c \rangle = \langle f^q \rangle$ for some $d,f \in K$. If $\langle d,f \rangle$ is not abelian, then by Lemma 3, $\langle d,f \rangle$ must be conjugate to a subgroup of B since $\langle d,f \rangle$ is not free. Hence, $\langle a,b,c \rangle$ is conjugate to a subgroup of B, and it follows that $[a,c] = 1$ since B is a CT-group.

Next we consider free products with malnormal subgroups amalgamated. In this case, results of Karrass and Solitar [2], Rosenberger [7], Zieschang [9] yield properties of two-generator subgroups analogous to those of Lemma 3 (cf., Rosenberger [8]).

Theorem 6. *Let* $G = \langle *H_1;A \rangle$ *be a free product of* PC-*groups with* A, *malnormal in each* H_i, *amalgamated. Then,* G *is a* PC-*group.*
Proof. Let $a,b \in G$ have infinite orders and $[a^m,b^n] = 1$. If $\langle a,b \rangle$ is not cyclic it follows from results in the papers cited above that $\langle a,b \rangle$ is a free product of cyclic groups or lies in a conjugate of some H_i, but in either case this implies that $[a,b] = 1$.

Theorem 7. *Let* $G = \langle *H_i;A \rangle$ *be a free product of* CT-*groups with the non-trivial group* A, *malnormal and proper in each* H_i, *amalgamated. Then* G *is a* CT-*group.*
Proof. By hypothesis, each H_i has trivial center (otherwise, A is not proper). Thus, suppose $a,b,c \in G$, $b \neq 1$, and $[a,b] = [b,c] = 1$. If one of $\langle a,b \rangle$ or $\langle b,c \rangle$ is noncyclic, it follows as in the proof of Theorem 6 that $\langle a,b \rangle$ and $\langle b,c \rangle$ are conjugate to subgroups of the factors H_i. If b is not conjugate to an element of A, then these conjugates must be equal, so $[a,c] = 1$ since each H_i is a CT-group. If b is conjugate to an element of A, it follows from the malnormality of A that both a and c are in A, and $[a,c] = 1$ since H_i is a CT-group. Finally, if both $\langle a,b \rangle$ and $\langle b,c \rangle$ are cyclic, then $\langle a,b \rangle = \langle d^p \rangle$, $\langle b,c \rangle = \langle f^q \rangle$ for integers p,q, but now a similar analysis shows that either $\langle d,f \rangle$ is cyclic, so $[a,b] = 1$, or $\langle d,f \rangle$ is in a factor, in which case all a,b,c are in a conjugate of the factor, so, again, $[a,c] = 1$.

The malnormal conditions imposed in Theorems 4 through 7 cannot be dispensed with, in general. For instance, the groups $\langle a,b,t : ta^2t^{-1} = a^2 \rangle$ and, for any $p,q > 1$, the free product $\langle a \rangle \underset{K}{*} \langle b,c \rangle$ with $K = \langle a^p \rangle = \langle b^q \rangle$ amalgamated are neither PC nor CT.

Finally, we mention another large class of groups which are both PC and CT-groups, the one-relator groups with torsion. The proof of this fact comes directly from results of Pride [6] Ree and Mendelsohn [3] and B.B. Newman [4].

REFERENCES

1. B. Fine, On power conjugacy and SQ-universality for Fuchsian groups and Kleinian groups, preprint.
2. A. Karrass & D. Solitar, The free product of two groups with a malnormal amalgamated subgroup, *Canad. J. Math.* 23 (1971), 933-959.

3. N.S. Mendelsohn & R. Ree, Free subgroups of groups with a single defining relation, *Arch. Math. (Basel)* 19 (1968), 577-580.
4. B.B. Newman, Some results on one-relator groups, *Bull. Amer. Math. Soc.* 74 (1968), 568-571.
5. N. Peczynski & W. Reiwer, On cancellations in HNN-groups, *Math. Z.* 158 (1978), 79-86.
6. S. Pride, The two-generator subgroups of one-relator groups with torsion, *Trans. Amer. Math. Soc.* 234 (1977), 483-496.
7. G. Rosenberger, Gleichungen in freien Produkten mit Amalgam, *Math. Z.* 173 (1980), 1-12. Korrektur: *Math. Z.* 178 (1981), 178.
8. G. Rosenberger, Applications of Nielsen's reduction method to the solution ofcombinatorial problems in group theory: a survey, *Homological Group Theory*, LMS Lecture Note Series No. 36 (1979), 339-357.
9. H. Zieschang, Über die Nielsensche Kürzungsmethode in freien Produkten mit Amalgam, *Invent. Math.* 10 (1970), 4-37.

DIMENSION FUNCTION FOR DISCRETE GROUPS

Alexander Lubotzky
Hebrew University, Givat-Ram, Jerusalem 91904, Israel

In this note we will describe briefly some recent work on p-adic analytic groups ([LM]) and strong approximation results for linear groups ([W], [N], [MVW]). These works suggest a notion of "dimension" for discrete groups. This notion will be presented, we will indicate some properties of "groups of finite dimension" and present some examples.

As proofs will only be sketched, the reader is referred for unexplained definitions and arguments to [Bo], [La] and [S1] for notions related to p-adic analytic groups, to [B] and [H] on algebraic groups, [We] for linear groups, [R] for pro-finite groups, [S] poly-cyclic groups and to [R1] about lattices and discrete subgroups of Lie groups.

Notation. p is always a prime. $\mathbb{Z}, \mathbb{Q}, \mathbb{R}, \mathbb{C}, \hat{\mathbb{Z}}_p$ denote the integers, rationals, reals, complex numbers and p-adic integers, respectively.

1. *THE DIMENSION OF PRO-p GROUPS AND OF DISCRETE GROUPS*

Let G be either a discrete group or a pro-finite group, $\mathcal{S} = \mathcal{S}(G)$ the set of (closed) finite index subgroups of G and $\mathcal{NS} = \mathcal{NS}(G)$ the subset of $\mathcal{S}(G)$ of normal subgroups. $\mathcal{S}$ is partially ordered by the order opposite to the inclusion (i.e. $A, B \in \mathcal{S}$ then $A > B$ if $A \subseteq B$). If $f : \mathcal{S} \to \mathbb{R}$ is a function then $\{f(A) \mid A \in \mathcal{S}\}$ forms a net (in the sense of [Ke, p.65]) and $\{f(A) \mid A \in \mathcal{NS}\}$ is a subnet. The limits of f on $\mathcal{S}$ and $\mathcal{NS}$ do not necessarily exist, we can still look at $\overline{L}_f(G) = \limsup\{f(A) \mid A \in \mathcal{S}\}$, $\overline{NL}_f(G) = \limsup\{f(A) \mid A \in \mathcal{NS}\}$ and $\underline{L}_f(G)$ and $\underline{NL}_f(G)$ the corresponding lim inf.

The function we look at first is d where d(A) = minimal number of generators of A. If A is a pro-finite group we mean, of course, the number of topological generators.

For a general group G (either discrete or pro-finite) the four limits of G can be quite different from each other. It is, therefore, quite surprising that in the case where G is a pro-p group, three of the four limits are equal:

Theorem 1.1 [LM, Thm. 3.2 and Prop. 4.1]. *Let* G *be a pro*-p *group. Then:*

(1) $\overline{L}_d(G) = \overline{NL}_d(G) = \underline{NL}_d(G)$, *call this number* r.

(2) $r < \infty$ *if and only if* G *is a* p-*adic Lie group in which case* $r = \dim G$ *where* $\dim G$ *denotes the dimension of* G *as a* p-*adic manifold.*

(3) *If* G *is a* p-*adic Lie group then* $\underline{L}_d(G)$ *is equal to the number of generators of the Lie algebra of* G.

So, for p-adic Lie groups the dimension is determined by the asymptotic number of generators. This might suggest that for a discrete group Γ one would take this number as a definition of dimension of Γ. But here come some problems:

I. Unlike pro-p groups, for discrete groups the three limits $\overline{L}_d(\Gamma)$, $\overline{NL}_d(\Gamma)$ and $\underline{NL}_d(\Gamma)$ are not necessarily equal.

II. We formulated Theorem 1.1 with the function d, but in fact we could formulate it with at least two other functions δ and d_p where $\delta(A) = d(A/[A,A])$ and $d_p(A) = d(A/[A,A]A^p)$ (where $[A,A]$ denotes the commutator subgroup of A and A^p the subgroup generated by the p powers). The reason is that for a pro-p group A, $d(A) = \delta(A) = d_p(A)$. So, which function should one take for discrete groups: d, δ or d_p and which one of the limits?

The answer to that is not very clear. Let us first eliminate some of the possibilities: Here we will use some recent work of Weisfeiler [W] (see also [N], [MWV]) who proved strong approximation results for finitely generated linear groups:

Theorem 1.2 [W, Thm. 9.3.1]. *Let Γ be a finitely generated subgroup of $GL_n(k)$ where k is an algebraically closed field of characteristic $\neq 2,3$. Then there exists a subgroup Γ' of finite index in Γ, a finitely generated ring R, a simply connected semi-simple scheme G over R and a homomorphism* $f : \Gamma' \to G(R)/Z(G(R))$ s.t.

(i) $\operatorname{Ker} f$ *is solvable.*

(ii) $\operatorname{Im} f$ *is dense in* $G(\hat{R})/Z(G(\hat{R}))$ *where* $\hat{R}$ *is the pro-finite completion of the ring* R.

This theorem gives a method to find a lot of finite quotients of Γ, and can be used, therefore, to give the following:

Proposition 1.3. *If Γ is as in the theorem but not solvable by finite then $\overline{L}_{d_p}(\Gamma) = \infty$ for every prime p and in particular $\overline{L}_\delta(\Gamma) = \overline{L}_d(\Gamma) = \infty$.*

Another instructive example is the following:

Proposition 1.4. *Let* $\Gamma = \Gamma_n = SL_n(\mathbb{Z})$, $n \geq 3$. *Then:*

(i) $\overline{L}_d(\Gamma) = \overline{L}_\delta(\Gamma) = \overline{L}_{d_p}(\Gamma) = \infty$ *for every prime* p.

(ii) $\overline{NL}_{\delta_p}(\Gamma_n) = \begin{cases} \infty & \text{if } p \mid n \\ n^2-1 & \text{if } p \nmid n \end{cases}$ *and thus* $\overline{NL}_d(\Gamma_n) = \overline{NL}_\delta(\Gamma_n) = \infty$.

(iii) $\underline{NL}_\delta(\Gamma_n) = n^2-1$.

(iv) $\underline{L}_d(\Gamma_n) \leq n^2-n$.

We do not know what is $\underline{NL}_d(\Gamma_n)$.

The proof of Propositions 1.3 and 1.4 will be sketched in §2. Meanwhile we observe that these results rule out the lim sup of any of the functions d, δ or d_p from serving as a good dimension function. Moreover, what one would really like is that the dimension of $SL_n(\mathbb{Z})$ will be n^2-1, just like the dimension of $SL_n(\mathbb{R})$ as a real Lie group and that of $SL_n(\mathbb{Q}_p)$ as a p-adic Lie group. The underlying idea is that a discrete group of dimension k should be thought of as a "Lie group" over $\mathbb{Z}$ of dimension k (see §3 for more details). We thus make the following definition:

Definition. *Let* Γ *be a finitely generated group. The dimension of* Γ*, denoted by* $\mathrm{Dim}(\Gamma)$ *is defined to be* $\underline{NL}_\delta(\Gamma)$ *where* δ *is the function* $\delta(A)$ = *the number of generators of the commutator quotient of* A.

2. *SOME METHODS OF COMPUTING* $\mathrm{Dim}(\Gamma)$ *AND EXAMPLES*

In this section we will show how one can use the pro-finite completion or the pro-p completion (for a prime p) of the group Γ as a tool to compute $\mathrm{Dim}(\Gamma)$ (and the other limits mentioned in §1). In particular we will compute them for poly-cyclic groups and for $SL_n(\mathbb{Z})$.

Let Γ be a dense subgroup of a pro-finite group G. Then we have a map of directed systems $\sigma : \mathcal{S}(\Gamma) \to \mathcal{S}(G)$ given by $\sigma(A) = \bar{A}$ where $\bar{A}$ denotes the closure of A in G. This is a surjective map as for every K open in G, $\overline{K \cap \Gamma} = K$. Moreover $\sigma(\mathcal{NS}(\Gamma)) = \mathcal{NS}(G)$ and if f is one of the functions d, δ or d_p then $f(\sigma(A)) \leq f(A)$. We therefore deduce the following:

Lemma 2.1. *Let* Γ *be a dense subgroup of a pro-finite group* G *and let* f *be one of the functions* d, δ *or* d_p. *Then:*

(i) $\underline{L}_f(G) \leq \underline{L}_f(\Gamma)$ *and* $\underline{NL}_f(G) \leq \underline{NL}_f(\Gamma)$;

(ii) $\overline{L}_f(G) \leq \overline{L}_f(\Gamma)$ *and* $\overline{NL}_f(G) \leq \overline{NL}_f(\Gamma)$.

If G is the pro-finite completion of Γ, then the map σ is an isomorphism of the directed systems $\mathcal{S}(\Gamma)$ and $\mathcal{S}(G)$ and if f is either δ or d_p then $f(\sigma(A)) = f(A)$ for every $A \in \mathcal{S}(\Gamma)$. The next lemma follows:

Lemma 2.2. *Let* Γ *be a finitely generated discrete group,* $G = \hat{\Gamma}$ *the pro-finite completion of* Γ *and* f *is either* δ *or* d_p. *Then*

(i) $\overline{L}_f(\Gamma) = \overline{L}_f(G)$ *and* $\overline{NL}_f(\Gamma) = \overline{NL}_f(G)$;

(ii) $\underline{L}_f(\Gamma) = \underline{L}_f(G)$ *and* $\underline{NL}_f(\Gamma) = \underline{NL}_f(G)$.

The following proposition gives some methods to compute the limits for pro-finite groups. In the light of the previous lemmas this can be useful for discrete groups as well.

Proposition 2.3. *Let* G *be a pro-finite group,* K *a closed subgroup of* G *and* f *one of the functions* d, δ *or* d_p. *Then:*

(i) $\overline{L}_f(K) \leq \overline{L}_f(G)$.

(ii) *If* K *is normal in* G *and* $f(K) = \infty$ *then* $\overline{NL}_f(G) = \infty$.

(iii) *If* $K = G_p$ *the* p*-sylow subgroup of* G *then* $\overline{L}_{d_p}(G) = \overline{L}_{d_p}(K)$.

Proof. All three parts of the proposition follow from the following fundamental property of pro-finite groups: If A is an open subgroup of K then G has an open normal subgroup N such that $N \cap K \subseteq A$.

We are now ready to make some computations for discrete groups.

Proposition 2.4. *Let* Γ *be a poly-cyclic group and* f *one of the functions* d, δ *or* d_p. *Then:* $\underline{NL}_f(\Gamma) = \overline{NL}_f(\Gamma) = \overline{L}_f(\Gamma) = h(\Gamma)$ *where* $h(\Gamma)$ *is the Hirsch rank of* Γ. *In particular* $\mathrm{Dim}(\Gamma) = h(\Gamma)$.

Proof. The proposition claims that nine limits are equal to $h(\Gamma)$ among which $\underline{NL}_{d_p}(\Gamma)$ is the minimal and $\overline{L}_d(\Gamma)$ is the maximal. It is therefore sufficient to prove that $\overline{L}_d(\Gamma) \leq h(\Gamma)$ and that $\underline{NL}_{d_p}(\Gamma) \geq h(\Gamma)$. The first inequality follows from the well known fact that Γ contains a subgroup of finite index which is poly-$\mathbb{Z}$. The second inequality follows from the less known fact that Γ contains a finite index subgroup Δ (depending on p) whose pro-p completion $\Delta_{\hat{p}}$ is a p-adic Lie group of dimension $h(\Gamma)$. Thus $\underline{NL}_{d_p}(\Gamma) \underset{(1)}{\geq} \underline{NL}_{d_p}(\Delta) \underset{(2)}{\geq} \underline{NL}_{d_p}(\Delta_{\hat{p}}) \underset{(3)}{=} h(\Gamma)$. Inequality (1) is a general fact, (2) follows from Lemma 2.1 and (3) follows from Theorem 1.1.

The last proposition implies that Γ has a finite index subgroup Δ s.t. if $N < \Gamma$ and $N \subset \Delta$ then $d_p(\Gamma) = h(\Gamma)$. This is a stronger version of the main result in [Mc]. But, as was pointed out to us by B.A.F. Wehrfritz, this stronger version can also be proved directly.

We are now ready to prove Propositions 1.3 and 1.4.

Proof of 1.3. In [L2], it is shown how Weisfeiler's Theorem 1.2 implies that for every prime p the p-sylow subgroup of $\hat{\Gamma}$ is infinitely generated. This fact with Proposition 2.3(iii) implies the desired result.

Proof of 1.4. By the affirmative solution to the congruence subgroup problem for $SL_n(\mathbb{Z})$ (cf [BMS]) one knows that $\hat{\Gamma}_n = \prod_{\ell} SL_n(\hat{\mathbb{Z}}_\ell)$ where $\hat{\mathbb{Z}}_\ell$ is the ring of ℓ-adic integers. Part (i) follows directly from that (or from 1.3). For part (ii); take $K = Z(\hat{\Gamma}_n) = \prod_{\ell} Z(SL_n(\hat{\mathbb{Z}}_\ell))$ where $Z(G)$ denotes the center of G. $Z(SL_n(\hat{\mathbb{Z}}_\ell))$ is the group of scalar matrices of the invertible elements of $\hat{\mathbb{Z}}_\ell$ of order n and so it is finite of order g.c.d.$(\ell-1,n)$. Hence if $p \mid n$, $d_p(K) = \infty$ and so by Proposition 2.3 (ii) and Lemma 2.2, $\overline{L}_{d_p}(\Gamma_n) = \infty$. On the other hand if $p \nmid n$ and N is a normal subgroup of $\hat{\Gamma}_n$ then $NZ(\hat{\Gamma}_n)$ is of the form $\prod_{\ell \notin S} SL_n(\hat{\mathbb{Z}}_\ell) \times U_{p_1} \times \dots \times U_{p_k}$ where $S = \{p_1,\dots,p_k\}$ is a finite set of primes and U_{p_i} is proper normal in $SL_n(\hat{\mathbb{Z}}_{p_i})$. $d_p(N)$ is greater equal zero iff $p = p_i$ for some i, in which case $d_p(N) = d_p(U_p)$. It follows now from Theorem 1.1 that $\overline{NL}_{d_p}(SL_n(\hat{\mathbb{Z}}_p)) = n^2-1$ so the same is true for $\hat{\Gamma}_n$ and Γ_n. To prove part (iii), note first that if $\Gamma_n(m) = \mathrm{Ker}(SL_n(\mathbb{Z}) \to SL_n(\mathbb{Z}/m\mathbb{Z}))$ is a principal congruence subgroup then $\delta(\Gamma_n(m)) = d(\Gamma_n(m)/\Gamma_n(m^2)) = n^2-1$ and so $\underline{NL}_\delta(\Gamma_n) \leq n^2-1$. To see that $\underline{NL}_\delta(\Gamma_n)$ is not less than n^2-1, note that if this were the case then the pro-p completion of any finite index subgroup of Γ_n would be analytic of dim $\lneq n^2-1$ (see Proposition 3.1), but the pro-p completion of $\Gamma_n(p)$ is a p-adic analytic group of dimension n^2-1. Finally part (iv) follows from a result of Tits [T].

Proposition 1.4 should be contrasted with the case $\Gamma_2 = SL_2(\mathbb{Z})$ which fails to have the congruence subgroup property. In fact Γ_2 contains a non-abelian free subgroup of finite index and one can easily deduce:

Proposition 2.5. *Let* f *be either* d, δ *or* d_p *and* $\Gamma = SL_2(\mathbb{Z})$. *Then* $\underline{L}_f(\Gamma) = \underline{NL}_f(\Gamma) = \overline{NL}_f(\Gamma) = \overline{L}_f(\Gamma) = \infty$ *and in particular* $\mathrm{Dim}(SL_2(\mathbb{Z})) = \infty$.

We will see later that this is the case for every discrete subgroup of finite co-volume in $SL_2(\mathbb{R})$ or $SL_2(\mathbb{C})$ (in fact in $SL_2(K)$ for any locally compact field K).

3. *ON GROUPS OF FINITE DIMENSION*

In this section, we will show that the finite dimensionality of a discrete group Γ has an impact on the completions of Γ, its representation theory, its cohomology and even its presentations by generators and relations.

Throughout this section assume that Γ *is a finitely generated group of dimension* $\ell < \infty$.

Proposition 3.1. *For every prime* p *and for every finite index subgroup* Δ *of* Γ, *the pro-*p *completion* $\Delta_{\hat{p}}$ *of* Δ *is a* p*-adic Lie group of dimension* $\leq \ell$.

Proof. $\underline{NL}_\delta(\Delta_{\hat{p}}) \leq \underline{NL}_\delta(\Delta) \leq \underline{NL}_\delta(\Gamma) = \ell$ so by Theorem 1.1, $\Delta_{\hat{p}}$ is analytic of dimension $\leq \ell$.

This last property of the pro-p completions has a lot of applications to the representation theory of Γ as the following propositions will show.

Proposition 3.2. *Let* ρ *be a finite dimensional representation of* Γ *over* $\mathbb{C}$. *Let* G *be the connected component of the Zariski closure of* $\rho(\Gamma)$. *Then* $\dim([G,G]U(G)) \leq \ell$ *where* $[G,G]$ *is the commutator subgroup of* G *and* U(G) *its unipotent radical. Moreover if* ρ *is irreducible then* $\dim G \leq \ell$.

Proof. As Γ is finitely generated then $\rho(\Gamma) \subseteq GL_n(A) \subseteq GL_n(\mathbb{C})$ where A is a finitely generated ring. Moreover we can assume A is a subring of the ring of integers of some p-adic field, or after restriction of scalars even that $\rho(\Gamma) \subseteq GL_n(\mathbb{Z}_{\hat{p}})$.

Claim. If Γ is a subgroup of $GL_n(\hat{\mathbb{Z}}_p)$, H the p-adic closure of Γ and G its Zariski closure, then $\dim H \geq \dim([G,G]U(G))$.

To prove the claim let $\underset{\sim}{G}$ and $\underset{\sim}{H}$ be the Lie algebras of G and H respectively. Then $\underset{\sim}{G}$ normalizes $\underset{\sim}{H}$, $\mathrm{ad}(\Gamma)$ acts trivially on $\underset{\sim}{G}/\underset{\sim}{H}$ which implies that $[\underset{\sim}{G},\underset{\sim}{G}] \subseteq \underset{\sim}{H}$. This reduces the claim to the abelian case in which it is clear.

Applying this to our $\rho(\Gamma)$ we deduce that $\dim H \geq \dim([G,G]U(G))$ but $\dim H \leq \ell$ as H is a compact p-adic Lie group and so $\dim H = \underline{NL}_\delta(H) \leq \underline{NL}_\delta(\Gamma) = \ell$.

The proof of the second part is similar.

Corollary 3.3. *Let* $A(\Gamma)$ *be the pro-algebraic hull of* Γ (i.e. *the Hochschild-Mostow group* cf. [HM], [M], [LMg, §4]). *Then* $\dim([A^0(\Gamma),A^0(\Gamma)]U(A(\Gamma))) \leq \ell$, *in particular* $\dim U(A(\Gamma)) \leq \ell$.

The last propsotion in fact asserts that Γ is a "super-rigidity" group i.e. there exists a finite dimensional algebraic group G and a map

$i : \Gamma \to G$ such that for every irreducible representation $\rho : \Gamma \to GL_n(\mathbb{C})$ there exists $\tilde{\rho} : G \to GL_n(\mathbb{C})$ such that $\tilde{\rho} \circ i$ agrees with ρ on a finite index subgroup of Γ. This is called "Γ is intimately embedded in G" in [Ba, p.49]. In fact in that paper Bass defined a group to be of $\mathbb{Q}$-representation type if for every finite dimensional irreducible representation ρ over $\mathbb{C}$ and for every $\gamma \in \Gamma$, $\mathrm{Tr}\,\rho(\gamma)$ is algebraic. Of course every f.g. group of $\mathbb{Q}$-representation type satisfies the condition:

(FAb). For every subgroup Γ_1 of Γ of finite index the commutator quotient Γ_1^{ab} is finite.

Bass also asked there whether every group of $\mathbb{Q}$-representation type is a "super-rigidity" group. We will show now that if Γ is a group of finite dimension satisfying (FAb) then Γ is of $\mathbb{Q}$-representation type (in fact we will get a stronger result) and by the previous results such a group is indeed a super-rigidity group.

Proposition 3.4. *Assume now that* Γ *also satisfies* (FAb). *Then for every irreducible representation* ρ *there exists a number field* k s.t. $(k:\mathbb{Q}) < \ell$ *and* $\mathrm{tr}\,\rho(\gamma) \in k$ *for every* γ *in some finite index subgroup* Γ_1 *of* Γ. *In particular* $\mathrm{tr}\,\rho(\gamma)$ *is algebraic for every* $\gamma \in \Gamma$.

Proof. Let G be the connected component of the Zariski closure of $\rho(\Gamma)$ and $\Gamma_1 = G \cap \Gamma$. Let A be the (finitely generated) ring generated by $\{\mathrm{tr}(\gamma) | \gamma \in \Gamma_1\}$. By Theorem 1.1 of [W] there exists $b \in A$ and a group scheme $\underline{G}$ defined over A such that Γ_1 is dense in $\underline{G}(\hat{A}_b)$. This immediately implies that A is algebraic as otherwise an open subgroup of $G(\hat{A}_b)$ would be mapped onto a group of the form $\underline{G}(\mathbb{F}_{\hat{q}}[x])$ and so a finite index subgroup of Γ would be mapped densely into the latter group; this is impossible by Proposition 3.1. So A is algebraic i.e. A is in some algebraic number field k. Weisfeiler and 3.1 imply that k cannot be of dimension $> \ell$ (otherwise for most of the primes p in k, $\underline{G}(\hat{A}_p)$ would be of dimension $> \ell$ and Γ_1 is dense there).

Corollary 3.5. *Assume again that* Γ *satisfies* (FAb), *then for every* n, $\dim(S_n(\Gamma)) = 0$ *where* $S_n(\Gamma)$ *is the variety of equivalent classes of irreducible representations of dimension* n *of* Γ. *Namely, for every* n *there are only finitely many equivalent classes of* n-*dimensional irreducible representations.*

Proof. This follows from [Ba, p.33], or directly as the traces generate the affine algebra of $S_n(\Gamma)$. Thus, all the points of $S_n(\Gamma)$ are algebraic. This implies $S_n(\Gamma)$ is of dimension 0.

In fact, even if Γ does not satisfy (FAb), $S_n(\Gamma)$ is "bounded". It was shown in [LMg, (4.8) and (4.11)] that $\dim(S_n(\Gamma)) \leq \dim(U(A(\Gamma)^{ab})$ and $\dim(U(A(\Gamma))^{ab}) = \Sigma\, (\dim H^1(\Gamma,V))\cdot(\dim V)$ where the sum runs over those finite dimensional irreducible modules V for which $H^1(\Gamma,V) \neq 0$. Corollary 3.3 therefore implies:

Corollary 3.6. (i) $\dim S_n(\Gamma) \leq \ell$ *for every* n.

(ii) $\sum_V (\dim H^1(\Gamma,V)) \cdot (\dim V) \leq \ell$, *in particular there are only finitely many finite dimensional irreducible modules* V *for which* $H^1(\Gamma,V) \neq 0$.

Finally we mention an application to presentations of Γ by generators and relations. Recall that $\mathrm{def}(\Gamma) = \max\{|X| - |R| \mid \langle X;R\rangle$ is a presentation of $\Gamma\}$.

Proposition 3.7. (i) $\mathrm{def}(\Gamma) \leq \frac{-\delta(\Gamma)^2}{4} + \delta(\Gamma) + 1$.

(ii) *If* Γ *is linear* (i.e. *a subgroup of* $GL_n(\mathbb{C})$) *but not solvable by finite then* $\underline{L}_{def}(\Gamma) = -\infty$.

Proof. (i) follows from [L1, Theorem 2.3] and Proposition 3.1. (ii) follows from (i) applied to all finite index subgroups of Γ and from Proposition 1.3.

4. *SOME MORE EXAMPLES, REMARKS AND PROBLEMS*

The results mentioned in §3 have some applications and raise some questions. Some of these we will indicate here.

Propositions 4.1. *Let* Γ *be a finitely generated subgroup of* $GL_n(\mathbb{C})$ *of dimension* ≤ 2. *Then* Γ *is meta-abelian by finite.*

Proof. Γ has a finite index subgroup Δ which is a residually finite p-group for some prime p [We, Chap. 4]. Thus, Δ is a subgroup of $\Delta_{\hat{p}}$. But $\underline{NL}_\delta(\Delta_{\hat{p}}) \leq \underline{NL}_\delta(\Delta) \leq 2$. So $\Delta_{\hat{p}}$ is a p-adic Lie group of dimension ≤ 2. As every Lie algebra of dimension two is meta-abelian (or abelian) $\Delta_{\hat{p}}$ has an open (i.e. finite index) subgroup which is meta-abelian and so does Δ.

Just to illustrate that the above proposition has a non trivial content, let us reformulate it. We do not know any proof for the following which does not go through p-adic Lie groups.

Corollary 4.2. *Let* Γ *be a finitely generated linear group which is not meta-abelian by finite. Then* Γ *has a finite index subgroup* Δ *such that for every normal finite index subgroup* N *contained in* Δ, $\delta(N) \geq 3$ *(and in particular* $d(N) \geq 3$*).*

Problem I. *Can one omit the condition "normal"? Does* $SL_n(\mathbb{Z})$ $(n \geq 3)$ *have arbitrarily small 2-generator finite index subgroups?*

Contrary to Proposition 4.1, there are linear groups of dimension 3 which are not solvable by finite. An example is $SL_2(\mathbb{Z}[1/p])$. This group has the congruence subgroup property (cf [S2]) and the proof is the same as the proof of Proposition 1.4(iii). In fact one can see that if an arithmetic group has the congruence subgroup property then it has finite dimension. It is therefore compatible with Serre's conjecture [S2, p.489, footnote] to make the following conjecture:

Conjecture II. *If* Γ *is an irreducible lattice in a semi-simple real Lie group of* $\mathbb{R}$*-rank* ≥ 2 *then* Γ *has finite dimension.*

By Margulis [M] every such lattice is arithmetic and many of them are known to have the congruence subgroup property [R2]. On the other hand for lattices in rank-1 semi-simple Lie groups one should expect the contrary.

Problem III. *Do all lattices in* $\mathbb{R}$*-rank* 1 *simple Lie groups have infinite dimension?*

Some elaboration is in order here: For the most important rank-1 group $PSL_2(\mathbb{R}) = SO(2,1)$ this is clearly the case as every lattice there either contains a surface group of finite index (if it is co-compact) or contains a free non-abelian group of finite index (if it is not co-compact). In either case the lattice has a finite index subgroup which is mapped onto a non-abelian free group. This clearly implies that it has infinite dimension. It is less trivial but also true that lattices in $SO(3,1) = PSL_2(\mathbb{C})$ have infinite dimension: in [L1, Remark 5.3] it is shown that such a lattice has a finite index subgroup whose pro-2 completion is not a p-adic Lie group. This implies by Proposition 3.1 that Γ has infinite dimension. It should be remarked here that if K is a p-adic field then lattices in $SL_2(K)$ are virtually free [S3, p.82] and so again of infinite dimension. For higher dimensional rank-1 groups the situation is less clear: One can see that the family of arithmetic lattices in $SO(n,1)$ studied by Millson in [Mi] all have infinite dimension as it is proved there (implicitly) that Γ has a finite index subgroup which is mapped onto a non-abelian free group. But for the other lattices in $SO(n,1)$ or the lattices in $SU(n,1)$ we do not know the dimension. A question of special interest will be what is the dimension of lattices in $Sp(n,1)$ and $\mathbb{F}_4$. These groups (and therefore their lattices) have the Kazhdan property (T) ([Ka]) and it is not clear whether they are "similar" to the other rank-1 groups or to higher rank groups.

Finally we mention some more questions concerned directly with the limits defined in §1.

Problem IV. *Let* Γ *be a finitely generated group. Is* $\underline{NL}_\delta(\Gamma) = \sup_p \underline{NL}_{d_p}(\Gamma)$?

We just mention that this is not the case for pro-finite groups: Consider the example given in [LV, Example 2.6(D)] - this is a pro-finite group G for which $\underline{NL}_\delta(G) = \infty$ but $\underline{NL}_{d_p}(G) = 0$ for every p.

Conjecture V. *Let* Γ *be a finitely generated linear group then either* $\underline{NL}_{d_p}(\Gamma) = \infty$ *for almost all primes* p *or* $\underline{NL}_{d_p}(\Gamma)$ *is bounded for almost all* p.

Problem VI. *What is the relation between* $\underline{NL}_d(\Gamma)$ *and* $\underline{NL}_d(\hat{\Gamma})$ *where* $\hat{\Gamma}$ *is the pro-finite completion of* Γ?

Compare with Lemma 2.2, but notice that according to [No] (see also [LMW]) $d(\hat{\Gamma})$ might be less than $d(\Gamma)$. Still it might be that the asymptotic behaviour is similar (especially for linear groups). One might ask the same question for the other limits (i.e. $\overline{L}_d$ etc.) and a solution might have various applications.

Acknowledgements. The author would like to thank Avinoam Mann, Madhav Nori and Boris Weisfeiler for stimulating discussions which helped the author to understand their works.

This work was partially done while the author was visiting the University of Michigan and the University of Chicago. Their hospitality and support is gratefully acknowledged.

REFERENCES

[Ba] H. Bass, Groups of integral representation type, *Pacific J. Math.* 86 (1980), 15-51.
[BMS] H. Bass, J. Milnor & J.P. Serre, Solution to the congruence subgroup problem for $SL_n (n \geq 3)$ and $Sp(2n) (n \geq 2)$, *Inst. Hautes Etudes Sci. Publ. Math.* 33 (1967), 59-137.
[B] A. Borel, *Linear Algebraic Groups*, Benjamin, New York (1969).
[Bo] N. Bourbaki, *Groupes et Algebres de Lie*, Chapitres 2 et 3, Herman, Paris (1972).
[H] J. Humphreys, *Linear Algebraic groups*, Springer, New York (1975).
[HM] G.P. Hochschild & G.D. Mostow, Representations and representative functions of Lie groups, *Ann. of Math.* 66 (1957), 495-542.
[Ka] D.A. Kazhdan, On the connection of the dual space of a group with the structure of its closed subgroups, *Functional Anal. Appl.* 1 (1967), 63-65.
[Ke] J.L. Kelley, *General-Topology*, Van Nostrand, New York (1955).
[La] M. Lazard, Groupes analytique p-adiques, *Inst. Hautes Etudes Sci. Publ. Math.* 26 (1965), 389-603.
[LMW] P.A. Linnel, A.J. McIsaac & P.J. Webe, Bounding the number of generators of a meta-abelian group, *Arch. Math.* 38 (1982), 501-505.
[L1] A. Lubotzky, Group presentation, p-adic analytic groups and lattices in $SL_2(\mathbb{C})$, *Ann. of Math.* 118 (1983), 115-130.
[L2] A. Lubotzky, On finite index subgroups of linear groups, in preparation.
[LM] A. Lubotzky & A. Mann, Powerful p-groups II. p-adic analytic groups, *J. Algebra*, to appear.
[LMg] A. Lubotzky & A. Magid, Varieties of representations of finitely generated groups, *Mem. Amer. Math. Soc.* 336, Providence (1985).
[LV] A. Lubotzky & L. van-der Dries, Subgroups of free pro-finite groups and large subfields of $\tilde{Q}$, *Israel J. Math.* 39 (1981), 25-45.
[M] G.A. Margulis, Arithmetically of irreducible lattices in the semi-simple groups of rank greater than 1, *Invent. Math.* 76 (1984), 93-120.
[Mc] J.J. McCutcheon, On certain polycyclic groups, *Bull. London Math. Soc.* 1 (1969), 179-186.
[MVW] C.R. Matthews, L.N. Vaserstein & B. Weisfeiler, Congruence properties of Zariski-dense subgroups, *Proc. London Math. Soc.* 48 (1984), 514-532.
[Mi] J.J. Millson, On the first Betti number of a constant negatively curved manifold, *Ann. of Math.* 104 (1976), 235-247.
[M] G.D. Mostow, Representative functions on discrete groups, *Amer. J. Math.* 92 (1970), 1-32.
[N] M.V. Nori, Subgroups of $SL_n(\mathbb{Z})$ and $SL_n(\mathbb{F}_p)$, *Invent. Math.*, to appear.
[No] G.A. Noskov, Number of generators of a group, *Math. Notes* 33 (1983), 249-255.
[R1] M.S. Raghunathan, *Discrete Subgroups of Lie Groups*, Berlin, Springer-Verlag (1972).
[R2] M.S. Raghunathan, On the congruence subgroup problem, *Inst. Hautes Etudes Sci. Publ. Math.* 46 (1976), 107-161.
[R] L. Ribes, *Introduction to Pro-finite Groups and Galois Cohomology*, Queen's papers in Math. 24, Queen's University, Kingston (1970).
[S] D. Segal, *Poly-Cyclic Groups*, Cambridge University Press, Cambridge (1983).
[S1] J.P. Serre, *Lie Algebras and Lie Groups*, Benjamin, New York (1965).
[S2] J.P. Serre, Le probleme de groupes de congruence pour SL_2, *Ann. of Math.* 92 (1970), 489-527.
[S3] J.P. Serre, *Trees*, Springer-Verlag, Berlin-Heidelberg-New York (1980).
[T] J. Tits, Systèmes gènèrateurs de groupes de congruence, *C.R. Acad. Sci. Paris* 289 (1976), 693-698.
[W] B. Weisfeiler, Strong approximation for Zariski-dense subgroups of semi-simple algebraic groups, *Ann. of Math.* 120 (1984), 271-315.
[We] B.A.F. Wehrfritz, *Infinite Linear Groups*, Springer-Verlag, Berlin-Heidelberg-New York (1973).

COSET GRAPHS

Roger C Lyndon
University of Michigan, Ann Arbor, Michigan 48109, U.S.A.

1. This is a report on a series of joint papers with Joel Brenner [BL 1-6]. Our work was prompted by a 1933 paper of Bernhard Neumann [N1] arising, I am told, from a problem in the foundations of geometry, or, more immediately, by Carol Tretkoff's 1975 sequel [T1] to Neumann's paper.

Neumann was led to study maximal nonparabolic subgroups of the modular group. The *modular group* M has a presentation

$$M = \langle a,b : a^2 = b^3 = 1 \rangle.$$

It is well known to be isomorphic to the group PSL(2,$\mathbb{Z}$) and to have a representation on the extended complex plane $\mathbb{C}^*$ given by

$$a : z \to -1/z, \qquad b : z \to (z-1)/z.$$

The element $c = ab : z \to z+1$ is *parabolic* in the sense of having only a single fixed point in $\mathbb{C}^*$, and it is easily shown that the parabolic elements of M are exactly the conjugates of nontrivial powers of c.

A subgroup P of M is a *parabolic group* if all its nontrivial elements are parabolic, and a subgroup S is *nonparabolic* if it contains no parabolic element. The maximal parabolic sugroups are exactly the conjugates of the infinite cyclic group C generated by c. Neumann observed that if P is a maximal parabolic subgroup and if S is a complement to P in the sense that $S \cap P = 1$ and $SP = M$, then S is a maximal nonparabolic subgroup. Neumann and Tretkoff enumerated many such *Neumann subgroups* S and Tretkoff gave substantial evidence that every such group S is a free product of infinitely many cyclic groups, of orders 2, 3 or infinite and that, if the number of infinite factors is finite, then it is even.

W. Wilson Stothers [Sto 1,2] anticipated us slightly in proving Tretkoff's theorem, by essentially the same methods.

Digression. If D is a fundamental region for a Fuchsian group G, it is easy to see that every nontrivial maximal elliptic subgroup (maximal finite subgroup) is conjugate to the stabilizer in G of a vertex of D. The analogous result gives the nontrivial maximal parabolic subgroups as conjugate to the stabilizers of cusps. For the modular group M this result is easily obtained by arithmetic considerations but, for general G, I know of no simpler proof than the not entirely trivial one given in Alan Beardon's book [Be1].

2. Our method consists in associating with any subgroup S of M a *graph* Γ. The *vertices* of Γ are the cosets Sg, g in M. For each vertex Sg there

is a (directed) *edge* from Sg to Sga with *label* a, and an edge from Sg to Sgb with label b. Each orbit of a, b, c on the set of vertices is then represented naturally by a finite closed path or an infinite path in Γ. An a-orbit has length 1 or 2, a b-orbit length 1 or 3, while a c-orbit can have any finite length or be infinite. Note the following

(I) S *is nonparabolic if and only if* Γ *has no finite* c*-orbit.*
(II) S *is a Neumann subgroup if and only if* Γ *is infinite and has exactly one* c*-orbit, that is,* c *is transitive on vertices.*

Nothing important is lost if we simplify Γ as follows. We delete all a-edges joining a vertex to itself while if $Sga \neq Sg$ we unite the two a-edges, from Sg to Sga and from Sga to Sg, into a single undirected edge. Finally, we pass to the quotient graph $\Gamma^* = \Gamma/b$ obtained by contracting all b-orbits to points. The graph Γ^* is now *cuboid* in the sense that there are at most three edges at each vertex. For Γ^* the conditions (I) and (II) now take the following form.

(I*) Γ^* *possesses an* *Eulerian system* Σ *of paths, all infinite and without U-turns except at dead ends, such that each edge of* Γ^* *is traversed exactly once in each direction by some path in* Σ
(II*) Γ^* *possesses an Eulerian system consisting of a single, infinite, Eulerian path.*

Conversely, apart from some minor ambiguity about fixed points, a connected cuboid graph Γ^* with Eulerian system Σ satisfying (I*) determines a parabolic subgroup S, while one satisfying (II*) determines a Neumann subgroup. A straightforward analysis of cuboid graphs with Eulerian path now yields a proof of Tretkoff's theorem.

Digression. We embarked on a catalog of finite cuboid graphs with Eulerian path, observing that in all our examples the graph had an automorphism group of order 1, 2, 3 or 6. G. Bianchi and R. Cori [BC1], who had encountered such graphs in computer science, proved that this is always the case and, indeed, with 2 and 3 replaced by any two distinct primes.

Digression. Tretkoff gave a definition of a group G associated with a planar graph Γ (or indeed a graph embedded in any orientable manifold), which we paraphrase as follows. The generators of G are the directed edges e. The relations are of two kinds, first, relations $ee^{-1} = 1$ where e^{-1} is the edge e with orientation reversed and, second, relations $e_1 \ldots e_t = 1$ where $e_1, \ldots, e_t$ are the edges at a vertex in cyclic order. These groups have quadratic presentations as in the papers of A.H.M. Hoare, A. Karrass, and D. Solitar [HKS 1,2] on subgroups of Fuchsian groups, and also arise in the work of S. Stahl [Sta1] in connection with automorphisms of Riemann surfaces and with the genus of a graph.

3. It is easy to paste together finite graphs to obtain infinite graphs with an Eulerian path or with an Eulerian system consisting of more than one infinite path. In this way one could answer a question of Wilhelm

Magnus [Ma1] by exhibiting a maximal nonparabolic subgroup of M that is not a Neumann subgroup. To obtain a more natural example, we started with a group Q of symmetries of the regular hexagonal tessellation of the Euclidean plane that had been studied by H.S.M. Coxeter [C1] and A. Sinkov [Si1]. This group has a presentation

$$Q = \langle a,b : a^2 = b^3 = (abab^{-1})^3 = 1 \rangle.$$

If S is the kernel of the obvious map from M onto Q, the graph Γ^* associated with S is easily described geometrically, and has a triply infinite Eulerian system of infinite paths. If S_1 is any nonparabolic subgroup containing S, then Γ_1^* is a quotient of the graph Γ^* and, in fact, $\Gamma_1^* = \Gamma^*/R$ for some group R of rigid motions of the Euclidean plane. Inspection now shows that no such Γ_1^* has a single infinite Eulerian path, whence no maximal nonparabolic subgroup containing S is a Neumann subgroup.

4. The rest is digression. We next used our graphical methods to study generalizations of the conditions that $a^2 = 1$, $b^3 = 1$ and $c = ab$ be transitive. A typical result is the following.

If b *is any permutation of a countable set that has at least one infinite orbit, then there exists a permutation* a *such that* $a^2 = 1$ *and* $c = ab$ *is transitive.*

In fact, K. Luoto showed that the associated graph Γ can be embedded in the plane with all edges straight segments of the same length. Another generalization lay in sharpening a theorem proved in 1900 by G.A. Miller [Mi1], and later by R.H. Fox.

Given integers $2 < \alpha < \beta < \gamma$, *there exist permutations* a, b, $c = ab$ *of some finite set* Ω *such that* a, b *and* c *have orders* α, β *and* γ.

We showed that c can be taken with only one or, in cases where the parity clearly demands it, two nontrivial orbits. Miller [Mi2] showed that typically a and b can be chosen with the group they generate transitive on Ω of unbounded finite cardinality, whence it followed that the triangle group with exponents α, β, γ has infinitely many nonisomorphic finite homomorphic images. We conjectured that Ω could be taken with the smallest cardinality d compatible with G containing elements of orders α, β, γ, or cardinality $d+1$ or $d+2$ when dictated by parity. We found an infinite class of exceptions for $\alpha = 2$, but none for $\alpha = 3, 4, 5, 6$ or for other large special cases, supporting the modified conjecture that there are no further exceptions.

5. Digressing in another direction, we asked what familiar infinite graphs have an Eulerian path. A somewhat complicated argument shows that the 1-skeleton of the regular hexagonal tessellation of the Euclidean plane has an Eulerian path. Further argument extends this result to all tessellations of the plane that satisfy a 'small cancellation condition', that each face has at least p sides and each vertex lies on at least q edges, where (p,q) is one of (3,6), (4,4) and (6,3). This includes all regular tessellations of

the plane in either the Euclidean or the hyperbolic metric.

We know of only one type of infinite nonplanar graph with an Eulerian path. A rather simple argument shows that the 1-skeleton of the regular tessellation of n-dimensional Euclidean space by n-cubes has an Eulerian path.

Digression. Small cancellation theory entered the argument mentioned above in showing that the tessellations under consideration were 'concentric' (with an arbitrary face as 'center') in the following sense:

Let B_1 *be the subcomplex of* T *consisting of any single face and define* B_{n+1} *inductively to consist of* B_n *together with all faces having a side on* B_n. *Then all the* B_n *are simply connected.*

This follows, in fact, from a lemma that I had proved [L1] in 1967 for some forgotten reason:

For T *as above, let* D *be a finite subcomplex of* T *bounded by a simple closed path, and let* z_0 *be any vertex of* T. *Then the maximum value of the path distance* $d(z_0,z)$ *for vertices* z *of* D *is not attained at any vertex interior to* D.

I believe it was from H. Zieschang I first learned that the simple connectedness of the B_n gives an easy solution of the word problem for Fuchsian groups. It does not seem to be known whether an analogous result holds for discontinuous groups in higher dimensional hyperbolic space. This question seems also to have some relevance to the structure of the growth function of a presentation of a discontinuous group.

REFERENCES

[Be1] A.F. Beardon, *The Geometry of Discrete Groups*, Springer (1983).

[BC1] G. Bianchi & R. Cori, Colorings of hypermaps and a conjecture of Brenner and Lyndon, *Pacific J. Math.* 110 (1984), 41-48.

[BL1] J.L. Brenner & R.C. Lyndon, Nonparabolic subgroups of the modular group, *J. Algebra* 77 (1982), 311-322.

[BL2] J.L. Brenner & R.C. Lyndon, Permutations and cubic graphs, *Pacific J. Math.* 104 (1983), 285-315.

[BL3] J.L. Brenner & R.C. Lyndon, Maximal nonparabolic subgroups of the modular group, *Math. Ann.* 263 (1983), 1-11.

[BL4] J.L. Brenner & R.C. Lyndon, The orbits of the product of two permutations, *European J. Combin.* 4 (1983), 279-293.

[BL5] J.L. Brenner & R.C. Lyndon, Doubly Eulerian trails on rectangular grids, *J. Graph Theory* 8 (1984), 379-385.

[BL6] J.L. Brenner & R.C. Lyndon, A theorem of G.A. Miller on the order of the product of two permutations, I, *Jñānābha* 14 (1984), 1-16; II, *Indian J. Math.* 26 (1984), 105-133; III, *Pure Appl. Math. Sci.* 20 (1984), 37-51.

[C1] H.S.M. Coxeter, The groups determined by the relations $S^\ell = T^m = (S^{-1}T^{-1}ST)^p = 1$, *Duke Math. J.* 2 (1936), 61-73.

[HKS1] A.H.M. Hoare, A. Karrass & D. Solitar, Subgroups of finite index of Fuchsian groups, *Math. Z.* 120 (1971), 289-298.

[HKS2] A.H.M. Hoare, A. Karrass & D. Solitar, Subgroups of infinite index in Fuchsian groups, *Math. Z.* 125 (1972), 59-68.

[L1] R.C. Lyndon, A maximum principle for graphs, *J. Combin. Theory* 3 (1967), 34-37.

[Ma1] W. Magnus, *Noneuclidean Tessellations and Their Groups*, Academic Press (1974).

[Mi1] G.A. Miller, On the product of two substitutions, *Amer. J. Math.* 22 (1900), 185-190.

[Mi2] G.A. Miller, Groups defined by the order of two generators and the order of their product, *Amer. J. Math.* 24 (1902), 96-100.

[N1] B.H. Neumann, Über ein gruppentheoretisch-arithmetiches Problem, *Preuss. Akad. Wiss. Phys. Math. Kl.* 10 (1933).

[Si1] A. Sinkov, The groups determined by the relations $S^{\ell} = T^{m} = (S^{-1}T^{-1}ST)^{p} = 1$, *Duke Math. J.* 2 (1936), 74-83.

[Sta1] S. Stahl, Permutation-pairs. I. A combinatorial generalization of graph embeddings; II. Bounds on the genus of the amalgamation of graphs, *Trans. Amer. Math. Soc.* 259 (1980), 129-145; 271 (1982), 175-182.

[Sto1] W.W. Stothers, Diagrams associated with subgroups of Fuchsian groups, *Glasgow Math. J.* 20 (1979), 103-114.

[Sto2] W.W. Stothers, Subgroups of infinite index in the modular group, I, II, III, *Glasgow Math. J.* 19 (1978), 33-43; 22 (1981), 101-118; 22 (1981), 119-131.

[T1] C. Tretkoff, Non-parabolic subgroups of the modular group, *Glasgow Math. J.* 16 (1975), 91-102.

NILPOTENT QUOTIENT ALGORITHMS

I.D. Macdonald
University of South Carolina at Aiken, Aiken, SC 29801, U.S.A.

Four variants of the original nilpotent quotient algorithm for groups are described. All have significant advantages over the versions currently in use.

1. *INTRODUCTION*

The coset enumeration and nilpotent quotient algorithms are almost the only group-theoretical algorithms which lend themselves to machine computation. In this note we consider only the NQA. The first paper written on the NQA was [2], and this opened up an entirely new field of study. The success of the NQA may be gauged from the number of papers featuring it which have subsequently appeared. They are too many to list.

Computer technology has greatly advanced in the 15 years or so since the original NQA program was developed. In particular fast access to random entries in very large arrays is not the problem it once was. This suggests that a search for new algorithms, or at least for variants of old ones, designed to take advantage of modern facilities, might be a good idea.

We shall describe four new, closely related variants of the NQA. The last three have been successfully programmed in PASCAL. It seems best to follow the order of actual development in their description, for ease of exposition; in addition to giving practical improvements in NQA implementations, they offer insight into how and why the NQA works.

All four new variants possess the following advantages.

1. *Only basic commutators are collected.*
2. *The Jacobi-Witt formula does not explicitly appear in the calculations.*

Point 2 is especially advantageous. We work at all times in groups and not sometimes (as in the original NQA) in a larger structure which reduces to a group after repeated Jacobi-Witt calculations. We also note that the nilpotent quotients are not necessarily finite p-groups.

2. *NOTATION, IDENTITIES, LEMMA, RULES*

Commutators etc. are defined as follows:

$$[x,y] = x^{-1}y^{-1}xy, \tag{1}$$

$$x^y = x[x,y], \tag{2}$$

$$[x_1,\ldots,x_{n-1},x_n] = [[x_1,\ldots,x_{n-1}],x_n] \text{ for } n > 2. \tag{3}$$

We recall the identities:

$$[x,yz] = [x,z][x,y]^z, \tag{4}$$

$$[x,y^{-1}] = [x,y,y^{-1}]^{-1}[x,y]^{-1} \tag{5}$$

and their respective generalizations:

$$[x,y_1\dots y_{n-1}y_n] = [x,y_n][x,y_{n-1}]^{y_n}\dots[x,y_1]^{y_2\dots y_n}, \tag{6}$$

$$[x,y^{-1}] = [x,2y][x,4y]\dots[x,ny,y^{-1}]^{(-1)^n}\dots[x,3y]^{-1}[x,y]^{-1} \tag{7}$$

for each $n \geq 1$; $[x,ny]$ means $[x,y_1,\dots,y_n]$ with $y = y_1 = \dots = y_n$.

Lemma. *Let* G *be a nilpotent group generated by* $\{a_1,a_2,\dots,a_m\}$ *such that every element of* G *is expressible as* $a_1^{\alpha_1}a_2^{\alpha_2}\dots a_m^{\alpha_m}$ *for suitable* $\alpha_i \in \mathbb{Z}$. *Let the normal subgroup* N *of* G *be the normal closure of* $\{b_1,b_2,\dots,b_n\}$. *Then* N *is generated by the set of all elements of the form*

$$[b_i,c_1,\dots,c_j] \tag{8}$$

for $1 \leq i \leq n$, $j \geq 0$, *and each* $c_p \in \{a_1,a_2,\dots,a_m\}$; *and if* $1 \leq p < q \leq j$ *and* $c_p = a_{f(p)}$, $c_q = a_{f(q)}$ *then* $f(p) \leq f(q)$.

Proof. Clearly $N = \langle b_i,[b_i,g] : 1 \leq i \leq m,\ g \in G\rangle$. Let $g = a_1^{\alpha_1}a_2^{\alpha_2}\dots a_m^{\alpha_m}$. If every $\alpha_i > 0$ then (6) along with an easy induction argument shows that $[b_i,g]$ is the product of elements of the form (8). If some $\alpha_i < 0$ then we use (7) as well; here the nilpotence of G is required.

Finally we state the commutator rules used in collection:

$$ba = ab[b,a]; \tag{9}$$

$$ba^{-1} = a^{-1}b[b,2a][b,4a]\dots[b,na,a^{-1}]^{(-1)^n}\dots[b,3a]^{-1}[b,a]^{-1}; \tag{10}$$

$$b^{-1}a = a[b,a]^{-1}b^{-1}; \tag{11}$$

$$b^{-1}a^{-1} = [b,a]a^{-1}b^{-1}; \tag{12}$$

see [1], page 167.

3. *BASIC COMMUTATORS AND COLLECTION*

Let $\mathcal{P}$ be a group presentation consisting of generators and relators. If $\mathcal{P}$ defines the group G then an NQA constructs $G/\gamma_{w+1}(G)$ for $w = 1,2,\dots$; it is understood $G/\gamma_{w+1}(G)$ is presented in some straightforward standard way. For instance, in [2] we presented $G/\gamma_{w+1}(G)$ by means of a "commutator table" and a "power table".

But now we want to use only basic commutators. In order merely to simplify the exposition, we shall assume that if $\mathcal{P}$ has d generators then $G/\gamma_2(G)$ cannot be presented with fewer than d generators and is finite. So let $c_1,c_2,\dots,c_n$ be a full set of basic commutators in the generators, ordered in some fixed way, up to and including weight w. We aim to present $G/\gamma_{w+1}(G)$ by a set of relators of the form

$$c_1^{\alpha_{i1}}c_2^{\alpha_{i2}}\dots c_n^{\alpha_{in}} \tag{13}$$

for $1 \leq i \leq n$; here $\alpha_{ij} \in \mathbb{Z}$. (We have in addition, of course, relations among $c_1, c_2, \ldots, c_n$.) Once this is done we might as well write (13) as an $n \times n$ matrix,

$$M = [\alpha_{ij}], \tag{14}$$

which we describe as the BC-matrix. If the weight w is important we write M_w for M.

An important step in our algorithms consists in reducing M to upper triangular form. This is perfectly feasible provided we modify the usual matrix row operations. (For example, to add row 2 to row 1 we have to form

$$c_1^{\alpha_{11}} c_2^{\alpha_{12}} \ldots c_n^{\alpha_{1n}} c_1^{\alpha_{21}} c_2^{\alpha_{22}} \ldots c_n^{\alpha_{2n}}$$

and perform a commutator collection.) We can then read off from the diagonal such information as the order of $\gamma_i(G)/\gamma_{i+1}(G)$ for $1 \leq i \leq w$ and in particular we see whether or not $\gamma_i(G) = \gamma_{i+1}(G)$.

We now come to our key result, the theorem which states that the only commutators we need are the basic commutators (which we call BC's). It is trivial that, in any group G, $\gamma_n(G)/\gamma_{n+1}(G)$ is generated by BC's in a set of generators of G. Our result is much stronger and is crucial to what follows.

Theorem. *Let $G = \langle c_1, c_2, \ldots, c_d \rangle$ and let $\mathfrak{B}$ be a full set of basic commutators, in some ordering, on $\{c_1, c_2, \ldots, c_d\}$. If $\gamma_i(G)/\gamma_{i+1}(G)$ for $1 \leq i < w$ is generated by the images of a given subset $\mathfrak{C}_i$ of commutators in $\mathfrak{B}$ with weight i then $\gamma_w(G)/\gamma_{w+1}(G)$ is generated by the image of*

$$\{[x,y] \in \mathfrak{B} : x \in \mathfrak{C}_i,\ y \in \mathfrak{C}_j,\ i + j = w\}.$$

Proof. Clearly $\gamma_w(G)/\gamma_{w+1}(G)$ is generated by the image of

$$\mathfrak{D} = \{[x,y] : x \in \mathfrak{C}_i,\ y \in \mathfrak{C}_j,\ x > y,\ i + j = w\}.$$

We show that each element in $\mathfrak{D}$ is the product of elements in $\mathfrak{B} \cap \mathfrak{D}$ and inverses. Take $[x,y] \in \mathfrak{D}$. If $[x,y] \notin \mathfrak{B}$ then $x = [x_1, x_2]$ with $x_1 \in \mathfrak{B}$, $x_2 \in \mathfrak{B}$, and $x_1 > x_2 > y$. The Jacobi-Witt identity gives

$$[x,y] = [x_2,y,x_1]^{-1}[x_1,y,x_2]$$

modulo $\gamma_{w+1}(G)$. Use induction on the place of y in the ordering in $\mathfrak{B}$. This allows us to assume that both $[x_2,y,x_1]$ and $[x_1,y,x_2]$ are products of elements in $\mathfrak{B}$ and inverses. Therefore the same is true of $[x,y]$. The last two statements are of course modulo $\gamma_{w+1}(G)$.

Finally we have to deal with the fact that the commutator of two BC's need not be a BC or the inverse of one. Suppose we want to collect c_i to the left in $\ldots c_j c_i \ldots$ with $i < j$. If $[c_j, c_i]$ is not a BC then write c_j as $c_r^{-1} c_s^{-1} c_r c_s$ with $r > s > i$. Thus we want to collect c_i in

$\ldots c_r^{-1}c_s^{-1}c_r c_s c_i \ldots$. If $[c_s,c_i]$ is not a BC then repeat the process. Eventually we arrive at some BC $[c_t,c_i]$ and collection resumes. There is no additional difficulty in discussing $\ldots c_j^{\pm 1}c_i^{\pm 1}\ldots$.

This process takes the place of the Jacobi-Witt calculation of [2]. It therefore plays a fundamental role in our algorithms. Perhaps it can be called "uncollection".

4. *ALGORITHM #1*

Definition. *An ordered set of basic commutators* $a_1,a_2,\ldots,a_m$ *is said to be a basic sequence IFF* $a_1 \leq a_2 \leq \ldots \leq a_m$. *The sum of the weights of the* a_i *is said to be the weight of the sequence.*

To construct the BC-matrix M we use the lemma with the a's basic commutators and the b's relators from $\mathfrak{P}$. We have to define all basic sequences of weight $< w$; clearly a recursive treatment is appropriate. To each pair consisting of a relator and a basic sequence there corresponds a row of M.

The row operations used in the reduction of M to row echelon form are recorded in a sequence of matrices $L_1,L_2,\ldots$; the first w-1 of these are used inductively at step w. Given M_1, an L_1 for which L_1M_1 is reduced is found in the familiar linear-algebraic way. Suppose that M_w $(w > 1)$ is known and that $L_1,\ldots,L_{w-1}$ are such that $L_{w-1}(L_{w-2}(\ldots L_1)M_w)$ is reduced apart from the columns referring to BC's of weight w. (Note that the dimensions of the L's have to be adjusted in an obvious way.) We then find L_w for which $L_w(L_{w-1}(\ldots L_1)M_w)$ is reduced.

The number of basic sequences of weight w in d generators is d^w ([1], Theorem 11.2.1). The number of rows of M_w will therefore be nearly d^{w+1}. Since in general

$$L_w(L_{w-1}(\ldots L_1)M_w) \neq (L_wL_{w-1}\ldots L_1)M_w,$$

each of the L's has to be stored individually. These facts suggest that Algorithm #1 may not be particularly efficient (and it has not in fact been programmed).

5. *ALGORITHM #2*

We now abandon L and concentrate on M. As soon as a new row vector $\underset{\sim}{v} = (\alpha_1,\alpha_2,\ldots,\alpha_n)$ is found it is placed in M as follows. Let α_i be the first non-zero entry in $\underset{\sim}{v}$. If entry m(i,i) of M is 0 then $\underset{\sim}{v}$ becomes row i of M, and row i of M becomes $\underset{\sim}{v}$. If not, then row i of M is replaced by a new row beginning with the highest common factor of m(i,i) and α_i, and $\underset{\sim}{v}$ is replaced by a new vector with zero i-th component. Proceed recursively.

At this point basic sequences may also be abandoned with no loss of efficiency. In replacing M_w with M_{w+1} we have to calculate row vectors for M_{w+1} by

(i) collection of relators in $\mathfrak{P}$ up to weight w+1; and
(ii) commutation of each row of M_w with each generator in $\mathfrak{P}$.

Suppose some $m(i,i)$ becomes 1 (and keep all $m(i,i) > 0$). Then we can clearly make every entry of column i of M (apart from $m(i,i)$) equal to 0. This leads to the idea of a "forbidden" BC; c_i is "forbidden" IFF $m(i,i) = 1$ at some weight w (and therefore at all greater weights). We could in principle get rid of these "forbidden" BC's; their number will in general be much greater than the other BC's which we call "allowed".

6. *ALGORITHM #3*

This is a simple modification of #2. A commutator table (called a "C-table") is now constructed and recorded; it specifies $[c_i, c_j]$ with c_i and c_j basic and $c_i > c_j$, at each weight. Of course we might as well store $[c_i^{\pm 1}, c_j^{\pm 1}]$ while we are at it.

The collection routine necessary to calculate the BC-table must be modified, and "uncollection" is abandoned except in so far as it is used to calculate the C-table. No longer is it the case that every commutation is reduced to a sequence of calculations each resulting in a BC. (In practice the old collection routine was used to construct the C-table and a new routine was written to construct the BC-table.)

7. *ALGORITHM #4*

We now have a choice. One option is the NQA of [2]. When the weight increases from w to w+1, we know each entry in the C-table modulo $\gamma_{w+1}(G)$ and we can introduce "indeterminates" of weight w+1, eventually to be determined as products of BC's of weight w+1 and inverses. The C-table becomes the commutator table of [2], and the BC-table becomes the power table; "allowed" BC's correspond to "definitions" of commutators.

The alternative is to get rid of the "forbidden" BC's as completely and as early as possible. As soon as we have $m(i,i) = 1$ in M_{w+1} we can eliminate c_i from the C-table, and from the rest of column i in the BC-table. Once we have performed commutations of row i of M, c_i will play no further role until, of course, the weight w increases.

REFERENCES

1. Marshall Hall Jr., *The Theory of Groups*, Macmillan, New York (1959).
2. I.D. Macdonald, A computer application to finite p-groups, *J. Austral. Math. Soc.* 17 (1974), 102-112.

GENERATORS OF p-GROUPS

Avinoam Mann
Hebrew University of Jerusalem, 91904 Jerusalem, Israel

Since I started to work in the field of finite p-groups I have encountered, sometimes to my surprise, results concerned with the number of generators, and gradually I became convinced that this is an area worth investigating in its own right, with applications to other areas (such as p-adic Lie groups or Schur multipliers). In the present paper I try to collect some of the results about generators that seem to me to be the most interesting, admitting a natural bias towards my own work. This being a survey article, proofs are not usually given, except when not available elsewhere or as an illustration.

Some notation. The word "group" usually means a finite p-group, cl G and exp G are the class and exponent of G, d(G) and r(G) are the minimal numbers of generators and relations of G, G', G_i, $\Pi_i(G)$, $\Phi(G)$, Z(G), M(G) are the commutator subgroup, the i-th term of the lower central series, the subgroup generated by p^i-th powers, the Frattini subgroup, the centre and the Schur multiplier, C_n is a cyclic group of order n, wr stands for wreath product. Also, $\langle x \rangle$ is the smallest integer not less than the real number x, and logarithms are always to the base 2.

1. *NUMBER OF GENERATORS*

The simplest restriction is, of course, just to assume that d(G) is given (or bounded). This is a very weak assumption. Indeed, any p-group can be embedded in a 2-generator one [NN]. Still, we mention two deep results. The first is Kostrikin's, stating that there are only finitely many p-groups of exponent p with a given number of generators. To be sure, the main restriction here is the one on exp(G), not d(G). I have applied this result (for d(G) = 2) on at least two very different occasions [M1, M2, also M4], and I am sure that group-theorists wish, with me, that there were a reasonable exposition of this result, see Note. While on the subject, let me also mention my namesake's results on the orders of groups of exponent 4 [M].

The second result I have referred to, also originally related to the Burnside problem, is the Golod-Shafarevitch inequality $r(G) > \frac{1}{4}d(G)^2$, where r(G) denotes the minimal number of relations needed to define G. Actually, we can replace r(G) by $r_p(G)$, the minimal number of relations needed if we know already that G is a p-group. (To define $r_p(G)$ formally we use the category of pro-p groups - i.e. inverse limits of p-groups - as this category has free objects. Write then G = F/R, where F is a free pro-p group. Let $d_F(R)$ be the minimal number of generators of R as a normal closed subgroup of F (as an inverse limit, F has a natural topology, with all subgroups of finite index forming a basis for the neighbourhoods of the identity). The minimum of $d_F(R)$, over all such presentations of G, is $r_p(G)$). In this form, the inequality is equivalent to a statement about M(G), the Schur multiplier of G, as $d(M(G)) = r_p(G) - d(G)$. The Golod-Shaferevitch inequality has been shown to be asymptotically best possible [Ws].

2. *RANK*

Definition. *The* *rank* *of* G, rk(G), *is the maximum of* d(H), *for all* $H \subseteq G$.

Blackburn [B4] determined all p-groups of rank 2. They are either metacyclic, or 3-groups of maximal class, or the non-abelian group of order p^3 and exponent p, p odd, or a certain group of order p^4 (p odd, of class 3 and exponent p^2). We recall that a group G of order p^n is of *maximal class*, if cl G = n-1. An example is the wreath product $C_p \operatorname{wr} C_p$, the Sylow p-subgroup of S_{p^2}. Excepting this example, all groups of maximal class have rank p-1 at most (p odd), so, for p = 3, they have rank 2. The 2-groups of maximal class also have rank 2, but actually these are only the dihedral, quaternion and quasi-dihedral groups, so they are metacyclic. The 3-groups of maximal class are not metacyclic, but have a metacyclic maximal subgroup. (Blackburn's result is actually stronger than stated, assuming only that the normal subgroups of one given order are 2-generated - the only additional groups thus allowed have a cyclic subgroup of index p^2. For groups of maximal class, see [B3] or [Hu], also [LMK] and [M5]).

No similar result is possible for higher rank, since P. Hall has shown that there are enough p-groups of rank 3 to generate the variety of all groups. Thus such groups cannot satisfy any non-trivial law ([Ha] or [Hu]).

There are some papers discussing groups of a given rank which need not be p-groups. E.g. it is shown in [HM] that if G is soluble of rank r and *torsion-free*, the derived length $\ell(G)$ of G is bounded in terms of r, in sharp contrast to what we have just stated about p-groups. Indeed, [HM] allows also some torsion, bounding $\ell(G)$ in terms of r and the maximum t of the orders of torsion elements (t is assumed to exist). The bound is $3.58 \log(r+1) + 1.36$ for torsion-free groups, and $5.58 \log(r+1) + \log\log t + 5.36$ in general (recall that logarithms are to the base 2). This implies, in particular, a bound for $\ell(G)$ for a p-group G in terms of $r = \operatorname{rk} G$ and $t = \exp G = p^e$ (say), but actually we have reversed the true order here. Humphreys and McCutcheon first establish the bound $2\log r + \log\log t + 4$ for p-groups, and use it to get the general result. We shall see later that the last given bound can be slightly improved to $\log r + \log e + 2$, which improves the general bound to $4.58\log(r+1) + \log\log t + 3.36$ (the existence of bounds for $\ell(G)$ in terms of rk G and exp G was proved again in [LM], where we have forgotten the precedence of [HM]).

Returning to p-groups, it is obvious that if we want to restrict only rk G, we have to look for results of a different type. Indeed, this is where my interest in this problem arose first, on noting relations between the rank and the "power structure" of p-groups. Thus, it is shown in [M2] that if $\operatorname{rk} G_{\frac{1}{2}(p+1)} < \frac{1}{2}(p-1)$ (p odd) then G is a regular p-group (for regular p-groups, see [Hu]), and the following stronger results can be shown:

If $d(G_{\frac{1}{2}(p+1)}) \leq \frac{1}{2}(p-1)$, *or* $d(G_i) \leq i$ *for* $i \leq \frac{1}{2}(p-1)$ *then* G *is regular.* (1)

If we assume that $\operatorname{rk} G \leq \frac{1}{2}(p-1)$, we get that G is *totally regular* [M3]. That means that whenever $G \triangleleft H$, for a p-group H, and $a \in H$, then $\langle G,a \rangle$ is regular. The property of total regularity is not inherited by sections, but of course the restriction on the rank does go over to sections, so if $\operatorname{rk} G \leq \frac{1}{2}(p-1)$, then all sections of G are totally regular. Conversely, if all sections of G are totally regular, then $\operatorname{rk} G \leq p-1$. This last inequality is, however, a very simple result, merely stating than an elementary abelian group of order p^p is not totally regular (e.g. because it is a maximal subgroup of the irregular group $C_p \operatorname{wr} C_p$). A result similar to (1) states that, if $d(\Pi_1(G)G_{\frac{1}{2}(p+1)}) \leq \frac{1}{2}(p+1)$, then G is *power-closed* [M4], i.e. in G, and also in its sections, products of p-th powers are p-th powers (so $\Pi_1(G)$ is the set of all p-th powers).

Finally in this section, we mention that if $\operatorname{rk} G \leq p$, then the subgroups $\Pi_i(G)$ are "well-behaved". We have $\exp \Pi_i(G) = p^{e-i}$ (where $p^e = \exp G$), and $\Pi_i(\Pi_j(G)) = \Pi_{i+j}(G)$, $[\Pi_i(G),\Pi_j(G)] \subseteq \Pi_{i+j}(G)$, etc.

3. *POWERFUL GROUPS*

Definition. *A* p*-group* G *is* ***powerful****, if either* P *is odd and* $\Pi_1(G) \supseteq G'$ *(so that* $\Phi(G) = \Pi_1(G)$*) or* $p = 2$ *and* $\Pi_2(G) \supseteq G'$.

For simplicity we shall assume now that $p > 2$. The results for $p = 2$ are sometimes different. Most proofs can be found in [LM].

Examples of such groups are readily found. Thus, if G is regular, then $[\Pi_1(G),\Pi_1(G)] = \Pi_2(G')$, so $\Pi_1(G)$ is powerful. This holds also if G is only power-closed, since we still have $[\Pi_1(G),\Pi_1(G)] \subseteq \Pi_2(G)$ $(\subseteq \Pi_1(\Pi_1(G)))$. King [K1] gives other examples. Hobby [Ho1] investigated powerful groups without giving them a specific name. He proved, for instance, that if G is powerful, then so is G′, and $d(G') \leq d(G)$. According to [K2], King has developed a theory of powerful groups, but has not published it. It is noted in [K2], that a product of pairwise permutable cyclic groups is powerful. A special class of powerful groups was considered by McCutcheon [MC]. He assumes that $G_{i+1} = \Pi_1(G_i)$, for each i, and obtains that $d(H) \leq d(G)$ for each $H \subseteq G$ (I have quoted, of course, only what seems to me the most interesting results).

In [LM] we have introduced also the following concept: a subgroup $N < G$ is *powerfully embedded,* if $\Pi_1(N) \supseteq [N,G]$. Then: if M and N are powerfully embedded in G, so are MN, $[M,N]$, $\Pi_1(N)$ and $[N,G]$. It follows that if G is powerful, so powerfully embedded in itself, then G_i, $\Pi_i(G)$, $G^{(i)}$, $\Phi(G)$ etc. are powerfully embedded in G. Thus a powerful group is rich in powerful subgroups. It also follows that $\operatorname{cl} G \leq e$, where $\exp G = p^e$. Powerful groups can be characterized in terms of generators as follows: G is powerful if, and only if, whenever $G = \langle a_1,\dots,a_d \rangle$, with $d = d(G)$, then G is the product of the cyclic groups $\langle a_i \rangle$ (they need not be pairwise permutable). An important result is the following: *if* G *is powerful, and* $H \subseteq G$, *then* $d(H) \leq d(G)$.

Thus, for powerful groups, $\operatorname{rk} G = d(G)$. A sort of converse is given by: let $\operatorname{rk} G = r$. *Then* G *contains a normal powerful subgroup whose index is bounded by a function of* r *(and* p*).*

Proof. Let N be a maximal normal powerful subgroup of G. Let $s = d(N) \leq r$. Then $\Phi(N) = \Omega_1(N) \lhd G$. Let $C = C_G(N/\Omega_1(N))$, so $N \subseteq C \lhd G$. If $C \neq N$, let $D \lhd G$, $N \subseteq D \subseteq C$, $|D:N| = p$. Then $D' = [D,N] \subseteq \Omega_1(N) \subseteq \Omega_1(D)$, so D is powerful, contradicting the maximality of N. Thus C = N and G/N is isomorphic to a subgroup of $\mathrm{Aut}(N/\Omega_1(N)) \cong GL(s,p)$, so $|G:N| \leq p^{\binom{s}{2}} \leq p^{\binom{r}{2}}$. It also follows that G/N, like the Sylow p-subgroup S(r,p) of GL(r,p), has a normal series of length $\langle \log r \rangle$ with elementary abelian factors. Since these factors have at most r generators, we get also $|G:N| \leq p^{r \cdot \langle \log r \rangle}$, which is better than the bound $p^{\binom{r}{2}}$ for $r \geq 7$.

One corollary of this is that, if $\mathrm{rk}\, G = r$ and $\exp G = p^e$, then $\ell(G) \leq \ell(G/N) + \ell(N) \leq \langle \log r \rangle + \log \mathrm{cl}\, N \leq \log r + \log e + 2$, as was claimed in the previous section.

In [LM], the existence of N above is established in a different way. Let V be any variety that contains the group S(r,p). Then it is proved, that if G is a p-group, $N \lhd G$, $N \subseteq V(G)$ and $d(N) \leq r$, then N is powerfully embedded in V(G). This generalizes previous results of Hobby and King. If $\mathrm{rk}\, G = r$, let V_r be the variety generated by S(r,p), then we obtain $|G:V_r(G)| \leq p^{r \langle \log r \rangle}$ as above, and $V_r(G)$ is powerful. The previous proof now shows that $V_r(G)$ is contained in all maximal normal powerful subgroups of G. As an example, assume that $d(G_i) \leq i$ for some i. Since $\mathrm{cl}(S(r,p)) = r-1$, we get that G_i is powerful, and hence $|G_i:\Omega_1(G_i)| \leq p^i$. If $i \leq \frac{1}{2}(p-1)$, then $i \leq p-i$, so a well-known criterion of P. Hall shows that G is regular, and proves (1) of the previous section. The statement following that, about totally regular groups, is an immediate corollary. The final claims of Section 2, regarding groups of rank $\leq p$, are also proved by noting that $\Omega_1(G)$ is powerful in such groups, because $\exp S(p,p) = p$. As another example, $V_2(G) = \Phi(G)$, so a 2-generator normal subgroup of G contained in $\Phi(G)$ satisfies $|G:\Omega_1(G)| = p^2$, hence is metacyclic, generalizing results of [Bl,Ho2].

Next we apply these ideas to solve a problem posed by J. Wiegold in the previous St Andrews Groups Conference. Wiegold [Wi] asked whether there exists a bound for d(M(G)) in terms of rk G. We first consider a powerful G. Let $H/Z(H) = G$ be powerful. Then it can be shown that H' is powerfully embedded in H. If $d(G) = d$, then H' is the normal closure of the $\binom{d}{2}$ commutators of the generators of G. But for a powerfully embedded subgroup of H, any set of normal generators is already a set of ordinary generators, as follows immediately from the definition. Thus $d(H') \leq \binom{d}{2}$. Since H' is powerful, also $d(H' \cap Z(H)) \leq \binom{d}{2}$. Taking H as a covering group of G, we have that $M(G) \subseteq H' \cap Z(H)$ and $d(M(G)) \leq \binom{d}{2}$.

Now let G be any p-group of rank r. Then G contains a powerful subgroup N with $|G:N| \leq p^{r \langle \log r \rangle}$. By [Jo], $d(M(G)) \leq d(M(N)) + r^2 \langle \log r \rangle \leq \binom{r}{2} + r^2 \langle \log r \rangle$.

As mentioned earlier, this result can be interpreted as one about $r_p(G)$. It can also be proved via this interpretation. First, we show that a powerful group with d generators can be defined, as a p-group, by $\binom{d+1}{2}$

relations. Of these, $\binom{d}{2}$ express the commutators of the generators in terms of their p-th powers, while the remaining are relations between these p-th, or higher, powers. If G is not necessarily powerful, we again find N as above and go up from N to G by steps of index p, each time adding at most r+1 relations.

One can certainly do no better than a quadratic bound (in r) for d(M(G)), but we do not know if the presence of the logarithm is necessary. For a given r, this is a finite problem. Thus, using previous notation, suppose that G/N is defined by certain r generators $\{x_i\}$ and s relations $\{w_j(x)\}$ between them, while N is generated by $\{y_1,\ldots,y_r\}$, with at most $\binom{r+1}{2}$ relations between them. Then a presentation for G is obtained by adding to the one for N the generators $\{x_i\}$, r^2 relations describing $\{y_j^{x_i}\}$, and s relations expressing each $w_j(x)$ as a word in the y's. Thus the logarithm, if present, must come from s. So we are interested in the number of relations needed to define a rank r subgroup of S(n,p) as a p-group.

I hope that I have given enough examples to convince you that powerful groups are interesting and important. They can also be applied outside the field of p-groups. Thus, Lazard [Lz] has investigated p-adic Lie groups. Such groups are locally pro-p. Now Lazard proved a criterion that can be formulated as follows: *a finitely generated pro-p-group is a p-adic Lie group if and only if it has a powerful subgroup of finite index* (a pro-p-group G is powerful, of $\Pi_1(G) \supseteq G'$, where we refer to the *closed* subgroups generated by the p-th powers and the commutators). Our results then show that a pro-p-group is a p-adic Lie group if, and only if, it has finite rank.

A pro-p-group being infinite, it makes sense to look, not only at the maximum, but also at limiting values of generators of subgroups. Such values can also be used to characterize p-adic Lie groups, but they can also be defined and applied for any group. For details, see [LM] and Lubotzky's talk in this Conference [L2].

We end this section by mentioning possible variations. First, one can consider conditions of the type $\Pi_i(G_j) \supseteq G_k$. Some such results are given in [Ar]. Next, Thompson has proved a "dual" result: if all elements of order p of G are central then $d(G) \leq d(Z(G))$. Thus rk G = d(Z(G)) ([Hu]; for p = 2 we have to assume that all elements of order 2 or 4 are central [M6]). Conversely, if G is any group of rank r, then $V_r(G)$ is not only powerful, but also satisfies this dual property of having its elements of order p central, and of course we have our previous bound for $|G:V_r(G)|$. Another way to get such a subgroup is to let E be a maximal elementary abelian normal subgroup of G, and let $C = C_G(E)$. Then E contains all elements of order p in C, so all such elements are central (see [Hu]). Here $|G:C| \leq p^{\binom{r}{2}}$.

As an illustration, we mention the paper [Ho3]. Hobby considers there a p-group G with the property that all normal abelian subgroups contained in G_n can be generated by n elements (for some n; p is still odd!). Let E be maximal among the normal elementary abelian subgroups contained in G_n. Then again E contains all elements of order p in G_n. Since $d(E) \leq n$, G_n centralizes E, so again $d(G_n) \leq d(Z(G_n)) = d(E) \leq n$, which is Hobby's result. Of course we have now that G_n is powerful.

4. *INEQUALITIES*

In this section we consider relations between the number of generators of different subgroups. We always have Schreier's formula, that if $H \subseteq G$ then $d(H) - 1 \leq |G:H| (d(G) - 1)$. Sometimes this can be improved. In [B2] it is shown that if $d(G) = 2$ and G/G' has type (p^n, p^m), then $d(G') \leq (p^n-1)(p^m-1)$. It has been remarked by P.M. Weichsel, that this can be greatly improved for regular groups, because if G above is regular, we have $G_p \subseteq \Phi(G')$, so counting basic commutators generating G_i/G_{i+1}, $i < p$, gives $d(G') \leq \frac{1}{2}(p-1)(p-2)$. A similar argument shows, that if G is regular, $d(G) = 2$ and H is maximal in G, then $d(H) \leq p$. Another result of Blackburn is, that if $d(G) = 2$, $\exp G/G' = p^e$, and $|G_i:G_{i+1}| = p$ for $i \geq 2$, then $d(G) < p^e$ [B3]. This result was useful in recent work on groups of large class [LGM].

Trying to systematize such relations, I first considered the following class of groups: a p-group G is called *monotone* if $d(H) \leq d(K)$ whenever $H \subseteq K \subseteq G$. We are going to discuss such groups.

Let, then, G be monotone. If H is a 2-generator subgroup of G, it follows that $\operatorname{rk} H = 2$, so H is given by Blackburn's result quoted in the beginning of §2. Now the p-groups, for $p > 2$, whose 2-generator subgroups are all metacyclic are known. These are precisely the so-called *modular* p-groups, i.e. whose lattice of subgroups is modular, and are abelian-by-cyclic. In our case, though the 2-generator subgroups are not necessarily metacyclic, they are so restricted that, for $p > 3$, the two possible non-metacyclic H can usually be eliminated, yielding the result that a monotone group of exponent at least p^3, is modular for $p > 3$. For $p = 3$ we find, at least, a modular maximal subgroup. For exponents p and p^2 the groups are also severely limited. Thus monotone p-groups are essentially completely known for odd p. For $p = 2$ the problem is more difficult, because metacyclic 2-groups are more complicated than odd-order ones. Power closed 2-groups, by the way, are monotone, and they have been determined (the proofs of those and the following results are in [M7]).

Since monotone groups turn out to be so special, it seems that we should consider a weaker condition. Bearing in mind Schreier's formula, I make the following definition.

Definition. *Let* α *be a real number,* $1 < \alpha \leq p$. *The class* $\mathfrak{M}_\alpha$ *consists of those* p-*groups* G *satisfying:*

$$\textit{If } H \subseteq K \subseteq G,\ |K:H| = p \textit{ and } d(K) > 1, \textit{ then } d(H) - 1 < \alpha(d(K) - 1). \quad (2)$$

(We could have allowed also $\alpha \leq 1$, but this is too restrictive, and uninteresting; for $\alpha > p$, all p-groups satisfy (2)).

It seems that we have here a continuum of classes, but this is not the case, as it turns out that we can have $\mathfrak{M}_\alpha = \mathfrak{M}_\beta$ for $\alpha \neq \beta$. To describe when this happens, we divide the interval $(1,p)$ as follows. For any two integers k, ℓ, with $2 \leq k \leq p-1$, $\ell > 0$, let $I(k,\ell) = \left\{\alpha \middle| k - \frac{1}{\ell} < \alpha \leq k - \frac{1}{\ell+1}\right\}$. We call each of these half-open intervals a "special interval". We also consider each integer point $2,\ldots,p-1$ as a special interval, as well as the interval $\{p-1 < \alpha \leq p\}$.

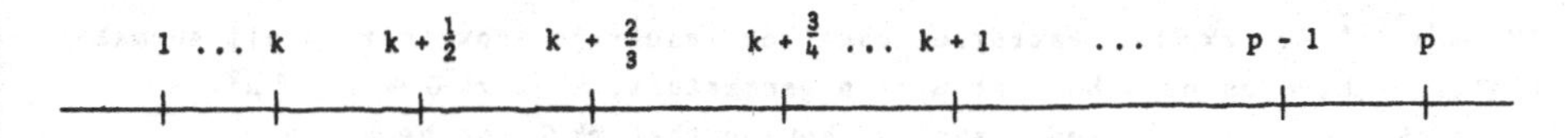

We then have: the equality $\mathfrak{M}_\alpha = \mathfrak{M}_\beta$ holds if, and only if, α and β lie in the same special interval.

The smallest class $\mathfrak{M}_{3/2}$ ($\mathfrak{M}_2$ for $p = 2$) satisfies (2) for all $\alpha > 1$, so satisfies $d(H) - 1 \leq d(K) - 1$, so this is just the class of monotone groups. For $p = 2$, this is the only class, but for odd p we have infinitely many, and if we add to them the class of all p-groups (corresponding to $\alpha > p$) we have $\omega(p-2) + 3$ classes. These form a hierarchy, as $\mathfrak{M}_\alpha \subseteq \mathfrak{M}_\beta$ for $\alpha < \beta$, and we may classify groups by the classes $\mathfrak{M}_\alpha$ to which they belong. We have

$$\operatorname{Inf}\{\alpha \mid G \in \mathfrak{M}_\alpha\} = \operatorname{Max}\left\{\frac{d(K) - 1}{d(H) - 1} \mid H \subseteq K \subseteq G,\ |K:H| = p,\ d(K) > 1\right\}$$

and this number is always either an integer, or of the form $k - \frac{1}{\ell}$.

Consider a group $G \in \mathfrak{M}_2$, and a 2-generator subgroup K of G. Then (2) shows that $\operatorname{rk} K = 2$, and so the same argument that was applied for monotone groups applies also here, and again we obtain that, for $p > 3$, groups in $\mathfrak{M}_2$, of exponent p^3 or higher, are modular. Thus, they are actually monotone, and only for exponents p and p^2 we get $\mathfrak{M}_2$ groups that are not monotone. For $p = 3$, however, our results apply only to monotone groups.

The next class, $\mathfrak{M}_{5/2}$, contains the groups of rank 3, and these groups satisfy no non-trivial laws. Thus $\alpha = 2$ is a sort of critical value. Another interesting value is $\alpha = p$. By a remark made at the beginning of this section, regarding generators of a maximal subgroup of a 2-generator regular group, the regular groups lie in $\mathfrak{M}_p$. Actually this is true also for power-closed groups, and even for wider classes.

A word about proofs. If $\mathfrak{C}$ is any class of p-groups, closed under taking subgroups and homomorphic images, we denote by bd $\mathfrak{C}$, the *boundary* of $\mathfrak{C}$, the set of groups which do not lie in $\mathfrak{C}$, but all of whose proper sections belong to $\mathfrak{C}$. Two classes are equal if and only if there boundaries are equal. Now the boundaries of the classes $\mathfrak{M}_\alpha$ are not difficult to determine. Indeed they consist of at most four groups. In this way the equality of various classes $\mathfrak{M}_\alpha$ is proved.

We close this section with a result about infinite groups, which seems appropriate here. Lubotzky [L1] has considered finitely generated pro-p-groups, G, in which Schreier's inequality is always an equality: if H has finite index in G, then $d(H) = 1 + |G:H|(d(G) - 1)$. He has shown that such groups are free, and asked whether a similar result holds for ordinary groups. This was answered by R. Strebel: if G is a residually finite finitely generated group in which Schreier's equality holds, then G is free. On the other hand, this is not true in the category of pro-finite groups.

5. *FURTHER RESULTS*

It has been observed by both Patterson [P] and Laffey [Lf] that a p-group G ($p > 2$) always contains a subgroup H such that $\operatorname{cl} H \leq 2$ and $d(H) = \operatorname{rk} G$. Laffey's result is more precise, and shows that H also satisfies

$|H| \leq p^{2r-2}$ $(r = \mathrm{rk}\,G)$. Patterson used her result to show that if all normal abelian subgroups of G have at most n generators, then $\mathrm{rk}\,G \leq n + \frac{1}{4}n^2$. Olshanskii has constructed examples showing that rk G can be as much as $\frac{1}{8}n^2 - 1$ [O]. Not much seems to be known about the relationships between the number of generators of different maximal abelian subgroups (or between other invariants of such subgroups) but it is shown in [JK] that if all normal abelian subgroups satisfy $d(H) \leq n$, with $n \leq 4$, then all abelian subgroups satisfy $d(H) \leq n$. This is not true for larger values of n, e.g. for $n = 6$. (Some further results on maximal abelian subgroups have just been obtained by T.J. Laffey.)

We end by describing another class of groups that can be characterized by conditions on generators. A group G is *order-closed* if, in all sections of G, the order of a product is no more than the orders of the factors [M4]. It suffices to consider products of elements of order p, so the definition can be put as follows: If $N \lhd H \subseteq G$, then the set of elements a of H satisfying $a^p \in N$ is a subgroup. Now term a group s-*order-closed*, if for any $N \subseteq G$, the set of elements a satisfying $a^p \in N$ is a subgroup (i.e. we remove the assumption $N \lhd H$).

The following are equivalent:

(a) G *is* s-*order-closed.*

(b) *If* $H = \langle a_1, \ldots, a_d \rangle \subseteq G$, *then* $\Pi_1(H) = \langle a_1^p, \ldots, a_d^p \rangle$.

(c) G *is order-closed, and if* $H \subseteq G$, *then* $d(\Pi_1(H)) \leq d(H)$.

Proof. (a) $\Rightarrow$ (b) $\Rightarrow$ (c) is immediate. Assume (c). It suffices to prove, that if $H \subseteq G$ and $a^p, b^p \in H$, then $(ab)^p \in H$. We may assume $G = \langle a,b \rangle$. Let $Z \lhd G$, $|Z| = p$. By induction, $(ab)^p \in HZ$. We may take $Z \subseteq \Pi_1(G)$, and assume that $H = \langle a^p, b^p \rangle$ and $Z \cap H = 1$. By assumption, $d(\Pi_1(G)) \leq 2$, so, by a previous result, $\Pi_1(G)$ is metacyclic, and so is $ZH = Z \times H$. Then H is cyclic, say $H = \langle a^p \rangle$. Then b^p is a power of a^p, and so commutes with both a and b, so $b^p \in Z(G)$. If $b^p \neq 1$, we can take $Z \subseteq \langle b^p \rangle$, so $Z \subseteq H$, while if $b^p = 1$, then a^p commutes with b, by properties of order-closed groups, so now $a^p \in Z(G)$ and we take $Z \subseteq \langle a^p \rangle$.

We have seen in the course of the proof, that if $a,b \in G$, then $\langle a^p, b^p \rangle$ is metacyclic. But order-closed groups are power-closed, so a^p and b^p can be any elements of $\Pi_1(G)$, and thus all 2-generator subgroups of $\Pi_1(G)$ are metacyclic, and $\Pi_1(G)$ is modular.

Footnote. This article is an upgraded version of the author's talk at Groups-St Andrews 1985, the talk having dealt mostly with the contents of Section 3 of the paper.

REFERENCES

[Ar] D.E. Arganbright, The power-commutator structure of finite p-groups, *Pacific J. Math.* 29 (1969), 11-17.

[Bl] N. Blackburn, On prime-power groups in which the derived group has two generators, *Proc. Cambridge Philos. Soc.* 53 (1957), 19-27.

[B2] N. Blackburn, On prime power groups with two generators, *Proc. Cambridge Philos. Soc.* 54 (1958), 327-337.

[B3] N. Blackburn, On a special class of p-groups, *Acta Math.* 100 (1958), 45-92.

[B4] N. Blackburn, Generalizations of certain elementary theorems on p-groups, *Proc. London Math. Soc.* 11 (1961), 1-22.

[Ha] P. Hall, A note on SI-groups, *J. London Math. Soc.* 39 (1964), 338-344.

[Ho1] C. Hobby, A characteristic subgroup of a p-group, *Pacific J. Math.* 10 (1960), 853-858.

[Ho2] C. Hobby, Generalizations of a theorem of N. Blackburn on p-groups, *Illinois J. Math.* 5 (1961), 225-227.

[Ho3] C. Hobby, Abelian subgroups of p-groups, *Pacific J. Math.* 12 (1962), 1343-1345.

[HM] J.F. Humphreys & J.J. McCutcheon, A bound for the derived length of certain polycyclic groups, *J. London Math. Soc.* (2) 3 (1971), 463-468.

[Hu] B. Huppert, *Endliche Gruppen 1*, Springer Verlag, Berlin (1967).

[JK] D. Jonah & M. Konvisser, Abelian subgroups of p-groups, a projective approach, *J. Algebra* 34 (1975), 309-330.

[Jo] M.R. Jones, Some inequalities for the multiplicator of a finite group II, *Proc. Amer. Math. Soc.* 45 (1974), 167-172.

[K1] B.W. King, Normal subgroups of groups of prime-power order, Proc. 2nd Internat. Conf. Theory of Groups, Lecture Notes in Mathematics 372, Springer-Verlag (1974), 401-408.

[K2] B.W. King, Normal structure of p-groups, *Bull. Austral. Math. Soc.* 10 (1974), 317-318.

[Lf] T.J. Laffey, The minimum number of generators of a finite p-group, *Bull. London Math. Soc.* 5 (1973), 288-290.

[Lz] M. Lazard, Groupes analytiques p-adiques, Inst. Hautes Etudes Sci. Publ. Math. 26 (1965), 389-603.

[LMK] C.R. Leedham-Green & S. McKay, On p-groups of maximal class I, II, III, *Quart. J. Math.* (2) 27 (1976), 297-311; 29 (1978), 175-186, 281-299.

[LGM] C.R. Leedham-Green & A. Mann, Space groups and groups of prime-power order VII, to appear.

[L1] A. Lubotzky, Combinatorial group theory for pro-p-groups, *J. Pure Appl. Algebra* 25 (1982), 311-325.

[L2] A. Lubotzky, Dimension function for discrete groups, these Proceedings.

[LM] A. Lubotzky & A. Mann, Powerful p-groups I, II, *J. Algebra*, to appear.

[M] A.J.S. Mann, On the orders of groups of exponent four, *J. London Math. Soc.* (2) 26 (1982), 64-76.

[M1] A. Mann, Groups with dense normal subgroups, *Israel J. Math.* 6 (1968), 13-25.

[M2] A. Mann, Regular p-groups I, II, *Israel J. Math.* 10 (1971), 471-477; 14 (1973), 294-303.

[M3] A. Mann, Regular p-groups III, *J. Algebra* 70 (1981), 89-101.

[M4] A. Mann, The power structure of p-groups I, *J. Algebra* 42 (1976), 121-135.

[M5] A. Mann, Regular p-groups and groups of maximal class, *J. Algebra* 42 (1976), 136-141.

[M6] A. Mann, Generators of 2-groups, *Israel J. Math.* 10 (1971), 158-159.

[M7] A. Mann, The number of generators of p-groups, in preparation.

[MC] J.J. McCutcheon, A class of p-groups, *J. London Math. Soc.* (2) 5 (1972), 79-84.

[NN] B.H. Neumann & H. Neumann, Embedding theorems for groups, *J. London Math. Soc.* 34 (1959), 465-479.

[O] A. Yu. Olshanskii, The number of generators and orders of abelian subgroups of finite p-groups, *Math. Notes* 23 (1978), 183-185 (original Russian: *Mat. Zametki* 23 (1978), 337-341).

[P] A.R. Patterson, The minimal number of generators for p-subgroups of GL(n,p), *J. Algebra* 32 (1974), 132-140.

[S] M. Suzuki, *Structure of a group and the structure of its lattice of subgroups*, Springer, Berlin (1956).

[Wi] J. Wiegold, The Schur multiplier: an elementary approach, *Groups-St Andrews 1981*, Cambridge University Press (1982).

[Ws] J. Wisliceny, Zur darstellung von pro-p-gruppen und lieschen algebren durch erzeugenden und relationen, *Math. Nachr.* 102 (1981), 57-78.

Note. I since learned that A.I. Kastrikin is preparing a book on the subject.

ON THE MATRIX GROUPS ASSOCIATED TO THE ISOMETRIES OF THE HYPERBOLIC PLANE

Ernesto Martínez
Universidad Complutense, 28040-Madrid, Spain

1. *PRELIMINARIES*

We note by $\mathbb{H}$ the halfplane $\{x+iy \in \mathbb{C},\ y > 0\}$ and by $\overline{\mathbb{H}}$ the closed halfplane $\{x+iy \in \mathbb{C},\ y \geq 0\} \cup \{\infty\}$. Let $\mathfrak{G}$ be the group of all transformations of one of the two forms

$$w = \frac{az+b}{cz+d} \quad ab - bc = 1, \quad a,b,c,d \quad \text{real} \tag{1.1}$$

$$w = \frac{a\bar{z}+b}{c\bar{z}+d} \quad ab - bc = -1, \quad a,b,c,d \quad \text{real} \tag{1.2}$$

$\mathfrak{G}$ is the group of isometries of $\mathbb{H}$ [1]. Isometries are of the following types [3]:

One of the form (1.1) is:

Hyperbolic, if $|a+d| > 2$, *it has two distinct fixed points in* $\overline{\mathbb{H}} - \mathbb{H}$.

Parabolic, if $|a+d| = 2$, *it has one fixed point in* $\overline{\mathbb{H}} - \mathbb{H}$

Elliptic, if $|a+d| < 2$, *it has one fixed point in* $\mathbb{H}$.

One of the form (1.2) is:

A glide-reflection, if $a+d \neq 0$, *it has two distinct fixed points in* $\overline{\mathbb{H}} - \mathbb{H}$.

A reflection, if $a+d = 0$, *it has a hyperbolic line of fixed points in* $\overline{\mathbb{H}}$ *called the axis.*

We also call on axis the hyperbolic line which joins the two fixed points of a hyperbolic isometry or a glide-reflection, and we denote it by γ. There are two opposite matrices

$$\begin{pmatrix} a & b \\ c & d \end{pmatrix} \quad \text{and} \quad \begin{pmatrix} -a & -b \\ -c & -d \end{pmatrix}$$

associated to an isometry w, with determinant 1 or -1, according to w is of the form (1.1) or (1.2). Here, we work with these matrices rather than the isometries.

In [2] Doyle and James classified the groups of matrices with determinant 1. In the present work we classify the groups of matrices with determinant ±1. Let Γ be one of these groups. We denote by Γ^+ the subgroup consisting of the elements with determinant +1, Γ^+ being a subgroup of index two in Γ.

By convention, we say that a matrix is hyperbolic, parabolic,..., and that it has fixed points or axis, whenever this happens for the corresponding isometry.

It is known that a hyperbolic matrix V is conjugate to

$$\begin{pmatrix} \lambda & 0 \\ 0 & 1/\lambda \end{pmatrix} \quad \lambda \neq -1,0,1 \quad \text{(with fixed points 0 and } \infty\text{)}$$

or

$$\begin{pmatrix} a & b \\ c & a \end{pmatrix} \quad \text{for some } a,b,c,\ |a| > 1$$

whose fixed points are symmetric with respect to the axis $x = 0$.

Each elliptic matrix of trace zero is conjugate to a matrix

$$\begin{pmatrix} 0 & k \\ -1/k & 0 \end{pmatrix} \quad \text{for some } k \neq 0.$$

Each reflection is conjugate to

$$\begin{pmatrix} 1 & 0 \\ 0 & -1 \end{pmatrix} \quad \text{with axis } x = 0 \quad \text{or} \quad \begin{pmatrix} 0 & k \\ 1/k & 0 \end{pmatrix} \quad \text{for some } k \neq 0$$

whose axis is perpendicular to $x = 0$.

Each glide-reflection is conjugate to

$$\begin{pmatrix} \lambda & 0 \\ 0 & -1/\lambda \end{pmatrix} \quad \text{with } \lambda \neq -1,0,1 \quad \text{(the fixed points are 0 and } \infty\text{)}$$

or

$$\begin{pmatrix} a & b \\ c & a \end{pmatrix} \quad \text{for some } a,b,c,\ a \neq 0$$

whose fixed points are symmetric with respect to the line $x = 0$. We make use of these matrices later.

For reference, we write the classification of [2] in the next theorem.

Theorem 1. *Every group of matrices with real entries and determinant* 1, *is in one of the four following classes:*

(i) *Groups generated by hyperbolic matrices, not all of which have the same pair of fixed points;*

(ii) *Groups which consist at most of two types of matrices, hyperbolics all of which have identical fixed points and elliptics with trace zero which interchange the fixed points of the hyperbolics;*

(iii) *Groups which consist of parabolics, all of which have the same fixed point;*

(iv) *Groups which consist of elliptics, all of which have the same fixed point.*

In each group the matrix $-I$ is permitted. The second class can be split into three subclasses:

(iia) Groups which only have hyperbolic elements;

(iib) Groups which have both hyperbolics and elliptics;

(iic) Groups $\{\pm I, \pm E\}$, where E is elliptic with trace zero.

2. *CLASSIFICATION*

First, we give several lemmas.

Lemma 1 ([1]). *Let* C *and* C' *be two reflections. The product* CC' *is hyperbolic, parabolic or elliptic according to whether* γ_C *and* $\gamma_{C'}$ *are disjoint, parallel or meet, respectively. In the first case, the axis of the hyperbolic element is the common perpendicular to* γ_C *and* $\gamma_{C'}$.

Lemma 2 ([1]). *Let* C *be a reflection and* V *an element of determinant* 1. *The product* CV *is a reflection if and only if* γ_C *is perpendicular to* γ_V *if* V *is hyperbolic, or the fixed point of* V *belongs to* γ_C *if* V *is elliptic or parabolic. Otherwise,* CV *is a glide-reflection.*

Lemma 3. *Let* C *be a reflection and* V *hyperbolic. Let us suppose that the product* CV *is a glide-reflection* R, *then* R *and* V *have the same fixed points if and only if* γ_C *and* γ_V *(and hence* γ_R*) coincide.*

Proof. If $\gamma_C = \gamma_V$ by conjugation we may suppose that C and V are the matrices

$$\begin{pmatrix} 1 & 0 \\ 0 & -1 \end{pmatrix} \text{ and } \begin{pmatrix} \lambda & 0 \\ 0 & 1/\lambda \end{pmatrix}.$$

Then, the product CV is

$$\begin{pmatrix} \lambda & 0 \\ 0 & -1/\lambda \end{pmatrix}$$

and the axis γ_R coincides with γ_V.

If $\gamma_R = \gamma_V$, as $R = CV$ then $C = RV^{-1}$ hence the fixed points of V and R are also fixed points of C and the three axes coincide.

Lemma 4. *Let* R *be a glide-reflection and* C *a reflection such that* γ_R *is perpendicular to* γ_C. *Then* RC *is an elliptic element with trace zero.*

Proof. By conjugation, we may suppose that R and C are the matrices

$$\begin{pmatrix} \lambda & 0 \\ 0 & -1/\lambda \end{pmatrix} \text{ and } \begin{pmatrix} 1 & 0 \\ 0 & -1 \end{pmatrix}.$$

The product RC is

$$\begin{pmatrix} 0 & \lambda k \\ -1/\lambda k & 0 \end{pmatrix}$$

which is an elliptic element with trace zero.

Theorem 4 of [2] proves that if a group contains a matrix of infinite order, then the group can be generated by matrices of infinite order, together with at most one matrix M such that $M^2 = -I$. Thus, as Γ^+ is a subgroup of index two in Γ, if the group Γ contains a matrix of infinite order then, Γ can be generated by matrices of infinite order together with at most two matrices of finite order.

We are now ready to enunciate the classification theorem.

Theorem 2. *Let Γ be a group of matrices with real entries and determinant 1 or -1. Then Γ is in one of the following classes:*

(I) *Groups generated by hyperbolics, not all of which have the same fixed points and one reflection or one glide-reflection. These groups cannot have only reflections.*

(II) *Groups generated by hyperbolics, all of which have the same fixed points, and one reflection* C *whose asix coincides with the axis of the hyperbolics. In these groups there exist glide-reflections with the same axis and there are no other reflections different from* C.

(III) *Groups generated by hyperbolics, all of which have the same fixed points and one reflection* C *whose axis is perpendicular to the axis of the hyperbolics. In these groups there are no glide-reflections.*

(IV) *Groups generated by hyperbolics, all of which have the same fixed points and one glide-reflection* R *with the same fixed points as the hyperbolics and* $\mathrm{tr}^2(R) \neq \mathrm{tr}^2(V) - 4$ *for all hyperbolic elements* V. *In these groups there are no reflections.*

(V) *Groups generated by hyperbolics, all of which have the same fixed points, one reflection* C *whose axis is perpendicular to the axis of the hyperbolics and one glide-reflection* R *whose axis coincides with the axis of the hyperbolics.*

(VI) *Groups of the type* $\langle E,C \mid E^2 = -I,\ C^2 = I \rangle$ *where* E *is elliptic and* C *is a reflection whose axis contains the fixed point of* E.

(VII) *Groups generated by parabolics, all of which have the same fixed point and one reflection whose axis contains the fixed point of the parabolics.*

(VIII) *Groups generated by elliptics, all of which have the same fixed point and one reflection whose axis contains the fixed point of the elliptics.*

Proof. (I) Let us suppose Γ^+ lies in the first class in Theorem 1. We suppose there exists a reflection C in Γ. By conjugation C is the matrix

$$\begin{pmatrix} 1 & 0 \\ 0 & -1 \end{pmatrix}.$$

We shall see that there are glide-reflections in Γ.

Let V be a hyperbolic element of Γ. If γ_V and γ_C are not perpendicular then CV is a glide-reflection by Lemma 2. Now, let us suppose γ_V and γ_C are perpendicular for every hyperbolic V of Γ, so that V is a matrix of the form

$$\begin{pmatrix} a & b \\ c & a \end{pmatrix}.$$

Let V' be another hyperbolic element,

$$\begin{pmatrix} a' & b' \\ c' & d' \end{pmatrix}.$$

The product VV' is

$$\begin{pmatrix} aa' + bc' & * \\ * & cb' + aa' \end{pmatrix}.$$

We can suppose that VV' is hyperbolic, as if it were not, there always would be a number n such that VV'^n is hyperbolic and then, we should take V'^n instead of V'. Now, $\gamma_{VV'}$ must be perpendicular to $\gamma_{C'}$ otherwise CVV' would be a glide-reflection. Then, $aa' + bc' = cb' + aa'$ and hence $bc' = b'c$. The fixed points of V and V' are respectively

$$\alpha = \frac{(a^2 - 1)^{\frac{1}{2}}}{c}, \quad \beta = -\frac{(a^2 - 1)^{\frac{1}{2}}}{c}$$

and similarly α' and β'. By the condition expressed by the determinant we have $\alpha\beta = -b/c$ and $\alpha'\beta' = -b'/c'$ and hence $\alpha\beta = \alpha'\beta'$. As $|\alpha| = |\beta|$ and $|\alpha'| = |\beta'|$ we deduce that V and V' have the same fixed points.

Since Γ^+ is generated by hyperbolic elements not all of which have the same pair of fixed points, Γ always contains glide-reflections.

(II) Now Γ^+ lies in the class (iia) of Theorem 1. By conjugation we can suppose that C is

$$\begin{pmatrix} 1 & 0 \\ 0 & -1 \end{pmatrix}$$

and the hyperbolics elements V are

$$\begin{pmatrix} \lambda & 0 \\ 0 & 1/\lambda \end{pmatrix}.$$

Then, for every V the product CV is

$$\begin{pmatrix} \lambda & 0 \\ 0 & -1/\lambda \end{pmatrix}$$

As we see CV is a glide-reflection whose axis coincides with γ_C and γ_V. If there were another reflection C' it must be $\gamma_V \neq \gamma_C$. So that C'V would be a glide-reflection R with $\gamma_R \neq \gamma_V$, by Lemma 3, if γ_C and γ_V are not perpendicular. But then R^2 would be a hyperbolic element with different fixed points than V. If γ_C and γ_V are perpendicular CVC' would be an elliptic element with trace zero by Lemma 4. It contradicts the fact that Γ^+ lies in the class (iia).

(III) Γ^+ belongs to the class (iia). By Lemma 2 the product CV for every hyperbolic V is a reflection. In order to see that there is not a glide-reflection R in Γ we can repeat the argument of II.

(IV) Γ^+ belongs to the class (iia). If there were a hyperbolic V with $tr^2(V) = tr^2(R) + 4$, then by a trivial computation we could see that VR would be a reflection with the same axis and we should be in the class (II).

(V) Γ^+ lies in the class (iib). By Lemma 4, the elliptic generator is obtained as the product of C and R.

In the remaining classes there are no glide-reflections.

(VI) Γ^+ belongs to the class (iic). The result is trivial.

(VII) Γ^+ lies in the class (iii). The result is evident by Lemma 2.

(VIII) Γ^+ lies in the class (iv). The result is again evident by Lemma 2.

This work belongs to a part of the author's doctoral thesis.

Acknowledgement. This work was partially supported by "Comisión Asesora de Investigación Científica y Técnica".

REFERENCES

[1] A.F. Beardon, *The geometry of discrete groups*, Graduate Texts in Mathematics 91, Springer-Verlag, New York (1983).
[2] C. Doyle & D. James, Discreteness criteria and high order generators for subgroups of SL(2,ℝ), *Illinois J. Math.* 25 (1981), 191-200.
[3] H.C. Wilkie, On non-Euclidean crystallographic groups, *Math. Z.* 91 (1966), 87-102.

A CHARACTERISTIC SUBGROUP OF $\underline{N}$-STABLE GROUPS

Dolores Perez Ramos
University of Valencia, Valencia, Spain

INTRODUCTION. NOTATION

In this note, all groups will be finite. $\underline{N}$ will denote the class of nilpotent groups. The concept of semisimple groups is taken from Gorenstein-Walter's paper ([7]). Our notation is standard and is taken mainly from ([6]). In particular, J(G) is the Thompson subgroup of the group G, that is, the subgroup of G generated by all abelian subgroups of maximum order.

Glauberman, ([6]), basing his work on some results and concepts of Thompson, obtained the following theorem:

Let G *be a group with* $O_p(G) \neq 1$ *which is* p*-constrained and* p*-stable,* p *odd. If* P *is an* S_p*-subgroup of* G*, then*

$$G = O_{p'}(G)N_G(Z(J(P))).$$

In particular, if $O_{p'}(G) = 1$, then $Z(J(P)) \lhd G$.

Using this subgroup one obtains the Thompson factorization of the group

$$G = N_G(J(P))C_G(Z(J(P)) = N_G(J(P))C_G(Z(P))$$

whenever G is a p-constrained and p-stable group with $O_{p'}(G) = 1$, p odd.

Later, Mann, ([8]), proved the following theorem:

Let G *be an* $\underline{N}$*-stable and* $\underline{N}$*-constrained group, with* $|F(G)|$ *odd. Let* S *be an* $\underline{N}$*-injector of* G*. Then* $Z(J(S)) \lhd G$.

Under the same assumptions as in the above theorem he proved:

$$G = N_G(J(S))C_G(Z(J(S))) = N_G(J(S))C_G(Z(S)).$$

Some related results were obtained by Arad in ([1]) and by Ezquerro in ([3]) by replacing the classes of p-groups and nilpotent groups by the class of π-groups (π a set of primes), and by a saturated Fitting formation, respectively.

Throughout these results we note that stability, constraint and existence of an unique conjugacy class of injectors are constant assumptions to obtain a factorization of Thompson type of a group.

Let $\tilde{\underline{N}}$ be the class of quasinilpotent groups, i.e., $\tilde{\underline{N}} = (G|G = F(G)L(G) = F^*(G))$. Blessenohl and Laue, ([2]), proved that all groups have an unique conjugacy class of $\tilde{\underline{N}}$-injectors. Moreover, it is well known that $C_G(F^*(G)) \leq F^*(G)$ for every group G, ([9]).

Definition. *Let* $\underline{F}$ *be a Fitting class. A group* G *is said to be* $\underline{F}$*-stable if whenever* A *is a* $\underline{F}$*-subgroup of* G *and* B *is a* $\underline{F}$*-subgroup of* $N_G(A)$ *such that* $[A,B,B] = 1$ *then* $B \leq (N_G(A) \bmod C_G(A))_{\underline{F}}$.

We obtain the following results, which generalize Mann's theorem. Full details including Proofs will appear in a forthcoming issue of the Israel J. Math.

Theorem A. *Let* G *be an* $\underline{N}$*-stable group, where* $1 \neq F(G)$ *is not a 2-group. Let* K *be an* $\tilde{\underline{N}}$*-injector of* G. *Then* $1 \neq O_{2'}(Z(J(K))) \triangleleft G$.

Corollary 1. *Under the same assumptions as in Theorem A*

$$G = N_G(J(K))C_G(O_{2'}(Z(J(K)))) = N_G(J(K))C_G(O_{2'}(Z(K))).$$

Corollary 2. *Let* G *be an* $\underline{N}$*-stable group, with* $1 \neq |F(G)|$ *odd. Let* K *be an* $\tilde{\underline{N}}$*-injector of* G. *Then* $1 \neq Z(J(K)) \triangleleft G$.

Corollary 3. *Under the same assumptions as in Corollary 2*

$$G = N_G(J(K))C_G(Z(J(K))) = N_G(J(K))C_G(Z(K)).$$

The proofs of these results depend on the following lemmas.

Lemma 1. *Let* A *be a semisimple subgroup of a group* G. *Let* B *be a subgroup of* $N_G(A)$ *such that* $[A,B;n] = 1$, $n \geq 1$. *Then* $[A,B] = 1$, i.e. $B \leq C_G(A)$.

Lemma 2. *If* G *is an* $\underline{N}$*-stable group then* G *is an* $\tilde{\underline{N}}$*-stable group.*

Lemma 3. *Let* G *be an* $\underline{N}$*-stable group. Let* K *be an* $\tilde{\underline{N}}$*-injector of* G. *Then* $Z(J(K)) \leq F(G)$.

Remark. (a) The converse of Lemma 2 is not true in general. It is enough to take $G = SA(2,5) \cong [C_5 \times C_5]\, SL(2,5)$, which is $\tilde{\underline{N}}$-stable but is not $\underline{N}$-stable.

(b) In general, it is not possible to obtain Theorem A and its corollaries under the weaker assumption of $\tilde{\underline{N}}$-stability. The group in Remark (a) is an example of this.

Recently, Förster, ([4]), obtained the following theorem:

(1.a) *For every group* G *and every*

$$T/Z(L(G)) \in Syl_2(L(G)/Z(L(G))),$$

$$\emptyset \neq Inj_{\tilde{\underline{N}}}(N_G(T)) \subseteq Inj_{\underline{N}}(G).$$

Using this result we obtain the following generalization of Mann's theorem for not necessarily $\underline{N}$-constrained groups ([9]):

Corollary 4. *Let* G *be an* $\underline{N}$*-stable group where* $1 \neq F(G)$ *is not a 2-group. Let* N *be an* $\underline{N}$*-injector as in* (1.α). *Then*

$$G = L(G)N_G(J(N))C_G(O_{2'}(Z(J(N)))) = L(G)N_G(J(N))C_G(O_{2'}(Z(N))).$$

Acknowledgements. This note is part of the author's thesis, which is being written at the University of Valencia (Spain) under the direction of Professor Pérez Monasor. The author wishes to express her gratitude to her advisor Professor Pérez Monasor and to Professor Iranzo Aznar for their devoted guidance and encouragement. The author was supported by a scholarship from the "Ministerio de Educación y Ciencia".

REFERENCES

1. Z. Arad, A characteristic subgroup of π-stable groups, *Canad. J. Math.* 26 (1974), 1509-1514.
2. D. Blessenohl & H. Laue, Fittingklassen endlicher Gruppen, in denen gewisse Haupfaktoren einfach sind, *J. Algebra* 56 (1979), 516-532.
3. L.M. Ezquerro, F-*estabilidad, Constricción y Factorización de grupos finitos*, Tesis Doctoral, Valencia (1983).
4. P. Förster, preprint.
5. G. Glauberman, A characteristic subgroup of a p-stable group, *Canad. J. Math.* 20 (1968), 1101-1135.
6. D. Gorenstein, *Finite Groups*, Harper Row, New York (1968).
7. D. Gorenstein & J. Walter, The π-layer of a finite group, *Illinois J. Math.* 15 (1971), 555-564.
8. A. Mann, Injectors and normal subgroups of finite groups, *Israel J. Math.* 9 (1971), 554-558.
9. F. Pérez Monasor, Grupos finitos separados respecto de una Formación de Fitting, *Rev. Acad. Cienc. Zaragoza* (2), 28 (1973).

THE ISOMORPHISM PROBLEM FOR INTEGRAL GROUP RINGS OF FINITE NILPOTENT GROUPS

K.W. Roggenkamp
University of Stuttgart, D-7000 Stuttgart-80, West Germany
L.L. Scott
University of Virginia, Charlottesville, Virginia 22903, U.S.A.

1. *STATEMENT OF THE RESULTS*

Ever since Graham Higman's notable thesis [4] the "isomorphism problem for integral group rings" has withstood many attacks:

Given two finite groups G *and* H, *is it true that* $\mathbb{Z}G \simeq \mathbb{Z}H$ *implies* $G \simeq H$? Higman gave a very strong positive answer, in case G is abelian:

(1.1) *Every finite subgroup* U *in the normalized units* $V(\mathbb{Z}G)$ - i.e. *units of augmentation one - is already a subgroup of* G.

Obviously, one can not expect such a strong statement in general, since for any subgroup H of G and a unit $u \in \mathbb{Z}G$, uHu^{-1} is a finite subgroup of $V(\mathbb{Z}G)$. Therefore the most one could ask in general is:

(1.2) *Is every finite subgroup* U *of* $V(\mathbb{Z}G)$ *conjugate in* $V(\mathbb{Z}G)$ *to a subgroup of* G.

But already the dihedral group of order 8, D_8, has in $V(\mathbb{Z}D_8)$ two conjugacy classes of D_8's. It was Berman and Rossa [1] who in 1966 speculated about (1.2), in case G is a finite p-group and $\mathbb{Z}$ is replaced by $\hat{\mathbb{Z}}_p$, the ring of p-adic integers, and $|U| = |G|$. There are similar considerations in the last chapter of Whitcomb's thesis [6], which he notes as inspired by his advisor, John Thompson.

Our main result in its basic form is a positive answer to this question of Berman and Rossa:

Theorem 1.3. *Let* G *be a finite* p*-group and* H *any subgroup of* $V(\hat{\mathbb{Z}}_p G)$ *with* $|H| = |G|$, *then* H *is conjugate to* G *by an inner automorphism of* $\hat{\mathbb{Z}}_p G$.

We should note that (1.3) answers the isomorphism problem in a very strong form. Actually Berman and Rossa ask if "Sylow's second theorem" holds for $V(\hat{\mathbb{Z}}_p G)$. One could interpret this as (1.3) or ask for a stronger answer dealing with maximal p-subgroups. In this direction we have made some progress for $p = 2$, and indeed (tentatively) believe we have obtained the following result.

Theorem 1.4. *Let* G *be a 2-group and* U *a finite subgroup of* $V(\hat{\mathbb{Z}}_2 G)$. *Then* U *is conjugate in* $V(\hat{\mathbb{Z}}_2 G)$ *to a subgroup of* G.

Richard Brauer once observed that groups don't know any sophisticated number theory. At the same time, they often can distinguish 2 from the odd primes, so we think that (1.4) does not allow us to make any conjecture about odd primes. At this stage it is worthwhile to recall a conjecture of Zassenahus [7] 1976, which is not as strong as (1.2) but still gives important information about how finite subgroups U of $V(\mathbb{Z}G)$ with $|U| = |G|$ are embedded in $V(\mathbb{Z}G)$:

If $U < V(\mathbb{Z}G)$ *is finite with* $|U| = |G|$, *then there exists an invertible element* $a \in \mathbb{Q}G$ *with* $aUa^{-1} = G$; i.e. U *is rationally conjugate to* G.

We verify this for nilpotent groups:

Theorem 1.5. *If* G *is finite nilpotent and* $\mathbb{Z}G \simeq \mathbb{Z}H$ *for some finite group* H, *then* $G \simeq H$. *Moreover, the Zassenhaus conjecture is true for* $\mathbb{Z}G$.

In order to prove the Zassenhaus conjecture for nilpotent groups, it is necessary to generalize (1.3) to the following situation - this generalization is by no means trivial:

(1.6) S *is a local or semilocal Dedekind domain of characteristic zero with a* unique *maximal ideal containing* pS.

We point out that the general version of (1.3) becomes false if there are at least two primes above p. We also prove (1.5) for $\mathbb{Z}$ replaced by an integral domain S of characteristic zero, in which no prime divisor of $|G|$ is invertible, in the following more general form:

(1.7) *Let* G *be nilpotent and* H *a finite group, such that the category of* SG-*modules is* S-*linearly equivalent to the category of* SH-*modules (as occurs if a full matrix-algebra over* SG *is isomorphic to a full matrix-algebra over* SH*). Then* $G \simeq H$.

The most general but least explicit result which we have obtained is:

Theorem 1.8. *Let* G *be a finite group with an abelian normal subgroup* N *and* G/N *nilpotent. Then* $\mathbb{Z}G \simeq \mathbb{Z}H$ *implies* $G \simeq N$.

For a survey on results concerning the isomorphism theorem, we refer to Sandling's paper [5].

The result (1.3) in its general form (1.6) is so strong that it allows us to compute the semilocal Piccard-groups $\mathrm{Pic}_S(SG)$, G nilpotent and S semilocal Dedekind with no prime divisor of $|G|$ invertible. We recall that in such a semilocal situation we have

$$\mathrm{Pic}_S(SG) \simeq \mathrm{Out}_S(SG),$$

the outer automorphism group. If $\mathrm{nOut}_S(SG)$ is the image in $\mathrm{Out}_S(SG)$ of the augmentation preserving S-automorphisms, then

$$\mathrm{Out}_S(SG) \simeq \mathrm{Hom}_{group}(G,S^{\times}) \cdot \mathrm{nOut}_S(SG).$$

In general neither factor is obviously normal: however, for nilpotent groups the result below, though, shows that the left hand factor is normal. Following Fröhlich [3] $\mathrm{Outcent}(SG) \simeq \mathrm{Picent}(SG)$ is the image of the central automorphisms in $\mathrm{Out}_S(SG)$.

If now G is nilpotent, (1.5) and (1.7) tell us

$$\mathrm{nOut}_S(SG) = \mathrm{Outcent}(SG) \cdot \mathrm{Out}(G), \tag{1.9}$$

where Out(G) is the group of outer automorphisms.

Now for nilpotent groups the natural map $\mathrm{Out}(G) \to \mathrm{Out}_S(SG)$ is injective as follows from [2] using [3]. The intersection

$$\mathrm{Out}_c(G) = \mathrm{Outcent}(SG) \cap \mathrm{Out}(G)$$

is induced from homomorphisms of G stabilizing the conjugacy classes of G. Hence to describe $\mathrm{Out}_S(SG)$ for G nilpotent it is - in view of the above reductions - enough to describe $\mathrm{Outcent}(SG) = \mathrm{Picent}(SG)$. So let $G = P \times N$ where P is a Sylow p-subgroup of G. For the sake of simplicity, we formulate the results only for

$$\mathbb{Z}_\pi = \mathbb{Q} \cap \left(\bigcap_{i=1}^{n} \hat{\mathbb{Z}}_{p_i}\right),$$

where $\{p_i\}_{1 \le i \le n}$ is a finite set of primes containing all prime divisors of $|G|$. (The interested reader can easily formulate the results in the above more general situation.)

We have

$$\mathrm{Picent}(\mathbb{Z}_\pi G) = \prod_{i=1}^{n} \mathrm{Picent}(\hat{\mathbb{Z}}_{p_i} G).$$

For $p_i \nmid |G|$, $\mathrm{Picent}(\hat{\mathbb{Z}}_{p_i} G) = 0$; so it suffices to compute $\mathrm{Picent}(\hat{\mathbb{Z}}_p G)$ for $p \mid |G|$. Consider for a moment the group ring

$$\hat{\mathbb{Z}}_p G = \hat{\mathbb{Z}}_p P \otimes_{\hat{\mathbb{Z}}_p} \hat{\mathbb{Z}}_p N,$$

and note that

$$\hat{\mathbb{Z}}_p N \simeq \prod_{j=1}^{t} (R_j)_{n_j},$$

where R_j, $1 \le j \le t$, are unramified extensions of $\hat{\mathbb{Z}}_p$. Thus

$$\mathrm{Picent}(\hat{\mathbb{Z}}_p G) \simeq \prod_{j=1}^{t} \mathrm{Picent}((R_j P)_{n_j})$$

$$\simeq \prod_{j=1}^{t} \mathrm{Picent}(R_j P).$$

Now because of (1.3) in the generality (1.6) we have $\mathrm{Picent}(R_j P) \simeq \mathrm{Out}_c(P)$. Thus

$$\mathrm{Picent}(\hat{\mathbb{Z}}_p G) \simeq \overset{t}{\times} \mathrm{Out}_c(P). \tag{1.10}$$

Putting things together, we get a complete description of $\mathrm{Out}_{\mathbb{Z}_\pi}(\mathbb{Z}_\pi G)$ for G nilpotent.

2. *SOME REMARKS TO THE PROOFS*

When we started out to work on the isomorphism problem we wanted to find counterexamples to the Zassenhaus conjecture. They were indeed very difficult to find, much more difficult than we had anticipated. Slowly we began to believe that some positive results were possible. Though the Zassenhaus conjecture was at first a tempting target for counterexamples, it was later a guide in the formulation of (1.3). We should note, however, that (1.2) is much better suited for induction than the Zassenhaus conjecture.

A. *Philosophy and some positive evidence*

A.1 *Questions about isomorphisms tend to reduce to questions about automorphisms*

In this section we assume that G is a finite p-group and we consider - for the sake of simplicity - the p-adic integral group ring. By induction on $|G|$ we want to prove (1.3). We have $\hat{\mathbb{Z}}_p G = \hat{\mathbb{Z}}_p H$ and we first want to show $G \simeq H$. Since the class sums of G and H are the same, we have $Z(G) = Z(H)$ for the centers of G and H. Thus $\hat{\mathbb{Z}}_p G = \hat{\mathbb{Z}}_p H$ implies $\hat{\mathbb{Z}}_p \overline{G} = \hat{\mathbb{Z}}_p \overline{H}$, where $\overline{G} = G/Z(G)$ and $\overline{H} = H/Z(H)$. Thanks to (1.3) applied to $\hat{\mathbb{Z}}_p \overline{G}$, there is a unit $\overline{u} \in \hat{\mathbb{Z}}_p \overline{G}$ with $\overline{u}\,\overline{H}\,\overline{u}^{-1} = \overline{G}$. This unit can be lifted to a unit u in $\hat{\mathbb{Z}}_p G$. If we replace H by $u H u^{-1}$, then in the "small group ring" s(G,Z(G)) we have

$$\begin{array}{ccccccccc} 0 & \longrightarrow & Z(G) & \longrightarrow & \hat{\mathbb{Z}}_p G/I(Z(G))\cdot I(G) & \longrightarrow & \hat{\mathbb{Z}}_p \overline{G} & \longrightarrow & 0 \\ & & \| & & \| & & \| & & \\ 0 & \longrightarrow & Z(H) & \longrightarrow & \hat{\mathbb{Z}}_p H/I(Z(H))\cdot I(H) & \longrightarrow & \hat{\mathbb{Z}}_p \overline{H} & \longrightarrow & 0 \end{array}$$

and thus $G = H$; in particular $G \simeq H$. Here I(X) denotes the augmentation ideal of X.

Thus, (1.3) *for* G/Z(G) *implies a positive answer to the isomorphism problem.*

This observation is essentially the same as that one credited by Whitcomb [6] to Thompson.

We should point out at this stage that one can use the discussion above on Piccard groups of nilpotent groups to construct finite non-isomorphic groups G and G' with abelian normal subgroups N and N' and G/N nilpotent such that the semilocal small group rings s(G,N) and S(G',N') are isomorphic.

Let us return to the proof of (1.3). We have $\hat{\mathbb{Z}}_p G = \hat{\mathbb{Z}}_p H$ and $\rho : G \simeq H$ an isomorphism by the above arguments. Now ρ induces an automorphism of $\hat{\mathbb{Z}}_p G$, also denoted by ρ, and in order to prove (1.3) we shall show:

(2.1) ρ *is the composition of a group-isomorphism of* G *and conjugation with a unit in* $\hat{\mathbb{Z}}_p G$.

A.2 *Philosophy: Why should the isomorphism problem be true?*

If (2.1) is true, then the group ring $\hat{\mathbb{Z}}_p G$ has very few automorphisms which are not inner. A glimpse of evidence is provided by the

Observation. $H^1(G, \hat{\mathbb{Z}}_p G) = 0$, *where* $\hat{\mathbb{Z}}_p G$ *is acted upon by* G *via conjugation.*

Consequently, all S-derivations from the algebra $\hat{\mathbb{Z}}_p G$ to itself are inner. Recalling some ideas from Lie-theory, the above property of derivations suggests a recipe to show that a given automorphism α of $\hat{\mathbb{Z}}_p G$ is inner:

(2.2) (1) *Take the logarithm of* α, *if possible.*
(2) *Show that* $\log \alpha$ *is a derivation.*
(3) *By the above property about derivations*

$$\log \alpha = \operatorname{ad} a, \ \operatorname{ad} a : x \to ax - xa$$

for some $a \in \hat{\mathbb{Z}}_p G$.
(4) *Form* $\exp a$, *if possible.*
(5) *Show* α *is conjugation by* $\exp a$.

Though it is not possible to follow the recipe through, it has been a guide for us.

Again we want to use induction on $|G|$. So pick a subgroup $C = \langle c : c^p = 1 \rangle$ in the center of G. As in (A.1) one can modify the automorphism α by group-automorphisms and inner automorphisms such that:

(2.3) α *is trivial on* $\hat{\mathbb{Z}}_p G/C$; *even* α *is trivial on* $\hat{\mathbb{Z}}_p G/I(C)I(G)$.

(We should pause at this moment to point out that for more general rings of coefficients, one can only arrange α to be trivial on $\hat{\mathbb{Z}}_p G/C$.) Because of (2.3)

$$\alpha(x) = x + (c-1)\tilde{\phi}(x)$$

for some $\hat{\mathbb{Z}}_p$-linear map $\tilde{\phi}$ which satisfies - α being an automorphism -

$$\tilde{\phi}(xy) = x\tilde{\phi}(y) + \tilde{\phi}(x)y + (c-1)\tilde{\phi}(x)\tilde{\phi}(y). \tag{2.4}$$

At this stage we have the good news that $\log \alpha$ exists: but the bad news is that c-1 is a zero divisor, which is an obstruction to (2.2) (2). The remedy to this is to pass to the order

$$\Lambda = \mathbb{Z}_p G/\underline{C}\,\mathbb{Z}_p G,$$

where $\underline{C} = \sum_{c \in C} c$. Now Λ is the pullback

$$\begin{array}{ccccccccc} 0 & \longrightarrow & I(C)\hat{\mathbb{Z}}_p G & \longrightarrow & \hat{\mathbb{Z}}_p G & \longrightarrow & \hat{\mathbb{Z}}_p G/C & \longrightarrow & 0 \\ & & \| & & \downarrow & & \downarrow & & \\ 0 & \longrightarrow & I(C)\hat{\mathbb{Z}}_p G & \longrightarrow & \Lambda & \longrightarrow & \mathbb{F}_p G/C & \longrightarrow & 0. \end{array}$$

Moreover, in Λ, $(c-1) = \pi$ is not a zero divisor any more, in fact, Λ is an R-order, where $R = \hat{\mathbb{Z}}_p[\xi]$, ξ a primitive p-th root of unity, and c acts on Λ as multiplication with ξ; note $\pi = \xi - 1$. α induces an automorphism on Λ, also denoted by α with:

(2.5) α *is the identity modulo* $\pi \cdot \operatorname{rad} \Lambda$ (as follows from (2.2)).

Since Λ is the above pullback, and since the center of $\hat{\mathbb{Z}}_p G/C$ maps onto the center of $\mathbb{F}_p G/C$, we have:

(2.6) *In order to show that* α *on* $\hat{\mathbb{Z}}_p G$ *is of the form described in* (2.1) *it is enough to show the corresponding result for* α *on* Λ.

By passing to Λ, $c-1$ is not any more a zero divisor; however, we have lost the property that derivations on Λ are inner. What is inherited from $\hat{\mathbb{Z}}_p G$ to Λ is

$$\pi H^1(\Lambda,\Lambda) = 0. \tag{2.7}$$

The conditions (2.6) and (2.7) almost give the desired result - *at first glance*. In fact we have:

Theorem 2.8. *Let* Λ *be a local* $R = \hat{\mathbb{Z}}_p[\xi]$*-order satisfying* $\pi H^1(\Lambda,\Lambda) = 0$. *Suppose* α *is an automorphism of* Λ *satisfying*

$$\alpha = 1 + \pi\phi$$

for some R-*linear map* ϕ *with*

$$\phi^n(\Lambda) \subset \mathrm{rad}^2\Lambda \text{ for some } n \in \mathbb{N}.$$

Then α *is inner.*

We point out that this result is in the spirit of Fröhlich [3]. The proof of (2.8) is done by following through the steps in (2.2); and everything works except steps (4) and (5); in fact in the above situation $\log\alpha = \mathrm{ad}\,a$, with $a \in \mathrm{rad}\,\Lambda$; but there is no reason why $\exp a$ should exist. We circumvent this problem by approximating α by conjugation with units - in the first step we use conjugation with $1 + a$. Then the modified $\alpha = 1 + \pi\phi$ has $\phi^n \subset \mathrm{rad}^3\Lambda$ etc.

Let us look at where we stand now in the proof of (1.3). We have Λ satisfying the hypothesis of (2.8), and we have

$$\alpha = 1 + \pi\phi$$

with $\phi(\Lambda) \subset \mathrm{rad}\,\Lambda$.

It appears that it should not be difficult to modify α in order to have the new ϕ satisfy $\phi^n(\Lambda) \subset \mathrm{rad}^2\Lambda$ for some n. However, this is indeed the *most difficult part of the paper.*

Previously we had an argument for a class two case by using Lie filtrations on the group and on $\mathbb{F}_p G$ in connection with transfer to find strong enough conditions on α to force it into some family of inner automorphisms. We found out that this class 2 group argument can be used in general, with the center of G replaced by a large subgroup on which α is the identity, namely on

$$C_G(Z(G,C))$$

where $Z(G,C)$ is the inverse image of $Z(G/C)$ in G.

To make α trivial on $C_G(Z(G,C))$ is a *very hard* induction. Assume α is the identity of an $\Omega \subset C_G(Z(G,C))$. Pick $t \in C_G(Z(G,C))$, centralizing G/Ω and of order p modulo Ω. Try to modify α to make the new α trivial on t as well. The main theme here is the interrelation - on sections - between multiplicative and additive 1-cohomology:

$$\mu(g) = \alpha(g)g^{-1}$$

is a multiplicative 1-cocycle and if we write

$$\mu = 1 + \pi\gamma,$$

then γ is an additive 1-cocycle modulo $\pi\,\mathrm{rad}^2\Lambda$. Because of our induction, μ and hence also γ take values in Λ^Ω, the Ω-fixed points of Λ. We now decompose Λ^Ω under the t-action into

$$\Lambda^\Omega = \Lambda_o^\Omega \oplus \Lambda_f^\Omega \oplus \bigoplus_{i=1}^{p-1} \Lambda_{\xi,i}^\Omega,$$

where Λ_o^Ω are the t-fixed points, t acts freely on Λ_f^Ω and t acts as ξ^i on $\Lambda_{\xi,i}$. Accordingly, the cocycle γ decomposes. In a triple see-saw mechanism, we very slowly make $\alpha(t)$ look more and more like the identity on Λ^Ω, starting with $\Lambda_{\xi,i}$.

In this process we do not only have to modify α by inner automorphisms but also by group automorphisms which arise from transfer. *This is notable,* for the following reason: If α is induced from a group automorphism $\sigma : G \to G$, and if α is the identity module $I(C)I(G)$ as we have assumed in (2.3), then $\sigma = \mathrm{id}_G$. So one might expect that α is inner, once $\alpha \equiv \mathrm{id} \bmod (\pi\,\mathrm{rad}\,\Lambda)$. But we have examples that this is definitely not the case; hence the above transfer homomorphisms do definitely occur.

Once this induction is carried through, we have:

(2.9) *α is the identity on $\Omega = C_G(Z(G,C))$ (in Λ) and α is the identity modulo $\pi\,\mathrm{rad}\,\Lambda$.*

In order to apply (2.8) we want to reach the following goal.

(2.1) Goal. *The automorphism α may be composed with inner automorphisms of Λ (note that group automorphisms do not occur any more) so that the new α is still the identity modulo $\pi\,\mathrm{rad}\,\Lambda$, and in addition, the new ϕ satisfies*

$$\phi^3(\Lambda) \subset \mathrm{rad}^2\Lambda.$$

Once this goal is reached, α has the form described in (2.1). In order to establish the goal, we recall that α is trivial on Ω and hence on the Frattini subgroup Fr(G). Using Tate's Mackey formula on transfer, one shows that the transfer map

$$H^1(L,F) \longrightarrow H^1(E,F) \tag{2.11}$$

is zero, where E is an elementary abelian p-group, L is a proper subgroup of E and F is an $\mathbb{F}_p$-vectorspace. Using (2.11) one can show:

$$\phi(\Lambda) \subset J(Z(G,C)) + \mathrm{rad}^2\Lambda, \tag{2.12}$$

where $J(Z(G,C))$ is the image of $I(Z(G,C))$ in Λ. Now put

$$V = \frac{J(Z(G,C)) + \mathrm{rad}^2\Lambda}{J(Z(Z(G,C))) + \mathrm{rad}^2\Lambda} \cong \frac{Z(G,C)}{Z(Z(G,C))}.$$

Then V is an $\mathbb{F}_p$-vectorspace, and to get

$$\phi^3(\Lambda) \subset \mathrm{rad}^2\Lambda,$$

it is enough

(2.13) *to modify* ϕ *such that* $\bar{\phi}$ *becomes zero on* V, *where* $\bar{\phi}: V \to V$ *is induced from* ϕ.

$V = Z(G,C)/Z(Z(G,C))$ comes in a natural way with a *non-degenerate bilinear form*

$$(\ ,\ ) : V \times V \longrightarrow \mathbb{F}_p = \Lambda/\mathrm{rad}\,\Lambda \tag{2.14}$$

with

$$(v,w) = \frac{1}{\pi}[x-1,\ y-1] \text{ modulo rad } \Lambda.$$

Here $x,y \in Z(G,C)$ and v and w represent x and y in V and [,] is the ring commutator.

Now, the function $\bar{\phi}$ is not any $\mathbb{F}_p$-linear map, but it arises from α and thus satisfies

$$\phi(xy) = \phi(x)y + \alpha(x)\phi(y).$$

This has as a consequence

$$(\bar{\phi}(u),v) + (u,\bar{\phi}(v)) = 0, \quad u,v \in V. \tag{2.15}$$

If $p = 2$ then we have in addition

$$(u,\bar{\phi}(u)) = 0, \quad u \in V. \tag{2.16}$$

But then it follows from classical Lie algebra (symplectic Lie algebras in char $\neq 2$ and orthogonal Lie algebras in char = 2) that:

2.17 $\bar{\phi}$ *is a linear combination of linear transformations of the form*

$$u \longrightarrow (u,v)w + (u,w)v, \quad u \in V$$

where $v,w \in V$.

The final argument is now as follows: Let $b = \sum_i r_i(y_i-1)(z_i-1)$ $y_i, z_i \in Z(G,C)$. Let β be the inner automorphism which is conjugation by $1+b$. Let $\bar{\phi}'$ be the new function $\bar{\phi}$ associated with the automorphism $\alpha' = \alpha\beta$. Then

$$\bar{\phi}'(u) = \bar{\phi}(u) + \sum \bar{r}_i\ ((v_i,u)w_i + (w_i,u)v_i)$$

where v_i represents y_i-1 and w_i represents x_i-1 in V.

Thanks to (2.17) we can thus make $\bar{\phi}$ equal to zero on V. This concludes the outline of the proof.

REFERENCES

1. S.D. Berman & A.R. Rossa, Integral group rings of finite and periodic groups (Russian), *Alg. and Math. Logic and Studies in Algebra* (Russian), 44-53, Izad. Kiev. Univ., Kiev (1966), as reported in MR 35-541.
2. D.B. Coleman, On the modular group ring of a p-group, *Proc. Amer. Math. Soc.* 5 (1964), 511-514.
3. A. Fröhlich, The Picard group of noncommutative rings, in particular of orders, *Trans. Amer. Math. Soc.* 180 (1973), 1-95.
4. G. Higman, *Units in group rings*, D. Phil. thesis, Oxford Univ., (1940).
5. R. Sandling, The isomorphism problem for group rings, a survey, *Proceedings of the 1984 Oberwolfach conference on Orders and their Applications*, Lecture Notes in Mathematics 1148, Springer-Verlag (1985), 256-289.
6. A. Whitcomb, *The group ring problem*, Ph.D. thesis, Univ. of Chicago (1968).
7. H. Zassenhaus, On the torsion units of finite group rings, *Studies in Math.* (in honour of A. Almeida Costa), Instituto de Alta Cultura, Lisboa (1974), 119-126.

EMBEDDING THE ROOT GROUP GEOMETRY OF $^2F_4(q)$

John Sarli
California State University at San Bernardino, San Bernardino, CA 92407, U.S.A.

In the standard interpretation of the 2-local geometry of a group of Lie type defined over a field of even characteristic it is well-known that the geometry of the finite Ree groups of type F_4 is that of a generalized octagon with parameters (q,q^2), q the order of the field. Outside of the theory of buildings itself this is perhaps most easily seen within an irreducible F_q-module of minimal dimension where each maximal 2-local P is identified with the subspace centralized by $O_2(P)$. It is also well-known that this geometry can be realized as a point-line incidence structure inside the group by taking the centers of the non-abelian root groups as the points and the groups generated by the pairs of such centers in a distinguished orbital as the lines. (This latter approach, of Cooperstein, was the subject of a considerable body of work on the long root groups of exceptional Lie type, [4].) An internal refinement (without reference to a representation space) has been given in [8], explaining the roles of the various root groups (there are three root lengths to consider in a non-reduced system; cf. Tits [11]) and resulting in a configuration built on the octagon which resembles a metasymplectic space. The purpose here is to give a direct description of the fundamental elements of this geometry inside a 26-dimensional F_q-space for these groups. Section 1 gives some terse background and suggests a basis for the module which minimizes calculations. Section 2 is an overview of the embedding of certain key configurations. We omit the calculations many of which are cumbersome but all of which follow from the commutator relations, the theorem of Borel-Tits, and the results of [1]. Some remarks on the global properties of the embedding are given in Section 3.

1. Throughout, $k = F_q$ and $G = {}^2F_4(q)$. Thus $q = 2^{2n+1}$ for some $n \geq 0$ and G is simple for $n \geq 1$. We use the notation of Ree [6] but refer the reader also to Carter [2]; cf. Tits [11] for a more general viewpoint encompassing non-perfect fields. Thus, denoting the standard generators of G by $\alpha_{\pm i}(t)$, $i = 1,\ldots,12$, let $V_j = \langle \alpha_j(t) \mid t \in K\rangle$ and $U_i = \langle V_j \mid i \leq j \leq 12\rangle$. V_j is elementary abelian of order q unless $j = \pm1,\pm4,\pm5,\pm6$ in which cases it is isomorphic to a 2-Sylow of Sz(q) and $V_j' = Z(V_j) = V_i$ for $i = \pm2,\pm8,\pm12,\pm11$ respectively. We have $U_i \triangleleft U_1 = U$ for each i. Also,

$$\langle V_j, V_{-j}\rangle \cong \begin{cases} SL_2(q), & j = 3,7,9,10; \\ Sz(q), & j = 1,4,5,6. \end{cases}$$

Set $B = UH$ with H the 'Cartan' torus. Then $P_1 = \langle B, V_{-1}\rangle$ and $P_2 = \langle B, V_{-3}\rangle$ represent the two classes of maximal parabolics. The term "root group" refers to any conjugate of any V_j. The standard root groups are pictured below; those of the same length are conjugate to one another under the Weyl group $W \cong D_{16}$. The class of long root groups is denoted by Γ, of those of intermediate length by Ω, and of the short groups by Θ.

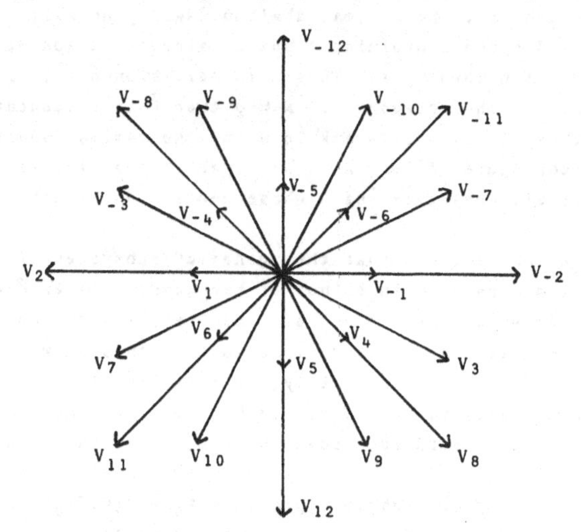

The 52-dimensional adjoint module for the Chevalley group of type F_4 in even characteristic can be written as the direct sum of the isomorphic irreducible modules $V(\lambda_1)$ and $V(\lambda_4)$; these summands remain irreducible under the action of G, cf. [12]. A convenient basis for a module V isomorphic to these summands is obtained from the 24 short roots of a root system of type F_4 which are usually described in terms of an orthonormal set in real 4-space: $\pm e_i (1 \leq i \leq 4)$ and $\frac{1}{2}(\pm e_1 \pm e_2 \pm e_3 \pm e_4)$. The F_q-space spanned by these roots together with the two co-roots h_3 and h_4 corresponding to the short fundamental roots $a_3 = e_4$ and $a_4 = \frac{1}{2}(e_1 - e_2 - e_3 - e_4)$ is G-invariant (note, however, that the space spanned by the long roots is not).

If $v \in V$ let $C_G(v)$ denote its centralizer in G. If v_r is a basis vector corresponding to the short root r it can be shown that $C_G(v_r) = C_G(V_j)$ for some $j = \pm 1, \ldots, \pm 12$. Thus we order the basis as follows: set $C_G(x_k) = C_G(V_j)$ with $k + j = 13$ for $j > 0$, $k + j = 14$ for $j < 0$. Set $x_{13} = h_3$ and $x_{14} = h_4$. With this ordering the matrices representing the unipotent generators $\alpha_i(t)$ $(\alpha_{-i}(t))$ are upper (lower) triangular, and those which are involutions are symmetric about the anti-diagonal. We have $C_G(x_1) = \langle U, V_{-1}\rangle < P_1$, $C_G(x_3) = \langle U, V_{-3}\rangle < P_2$, $C_G(x_8) = \langle V_1, V_{-1}\rangle\, O_2(C_G(x_1))$. Also, $C_G(x_{13}) \cong Sz(q) \times Sz(q)$ and $C_G(x_{14}) \cong Sp(4,q)\backslash Z_2$. The basis is as follows:

$x_1 = e_1$ $\quad x_5 = \frac{1}{2}(e_1 + e_2 - e_3 - e_4)$ $\quad x_9 = \frac{1}{2}(e_1 - e_2 - e_3 + e_4)$

$x_2 = \frac{1}{2}(e_1 + e_2 + e_3 + e_4)$ $\quad x_6 = \frac{1}{2}(e_1 - e_2 + e_3 + e_4)$ $\quad x_{10} = \frac{1}{2}(e_1 - e_2 - e_3 - e_4)$

$x_3 = \frac{1}{2}(e_1 + e_2 + e_3 - e_4)$ $\quad x_7 = \frac{1}{2}(e_1 - e_2 + e_3 - e_4)$ $\quad x_{11} = e_3$

$x_4 = \frac{1}{2}(e_1 + e_2 - e_3 + e_4)$ $\quad x_8 = e_2$ $\quad x_{12} = e_4$

2. Let K be a 2-group of G generated by root groups and R(K) the set of subgroups of K generated by root groups. Let $\widetilde{K} = Z(O_2(N_G(K)))$. The property that $K_0 = \widetilde{K}_0$ for all $K_0 \in R(K)$ limits K to either a member of Γ, or ⟨L,M⟩ where L,M ∈ Γ and K is the union of the members of Γ it contains. In this case $\hat{K} = C_V(O_2(N_G(K)))$ is a (projective) point or line of the octagon, resp. We refer to K (or $\hat{K}$) as a *singular* element. If K ∈ Ω then $\widetilde{K}$ is elementary abelian of order q^3, the maximal abelian 2-group normalized by a conjugate of P_2; $\hat{K}$ is a 3-space containing a unique singular 2-space, and if $L < \widetilde{K}$ is another member of Ω then $\widetilde{L} = \widetilde{K}$. Thus Ω is partitioned into classes represented by the lines of the octagon. If K ∈ Θ then $\widetilde{K}$ is elementary abelian of order q^5, thus of maximal 2-rank in G and the maximal abelian 2-group normalized by a conjugate of P_1. $\hat{K}$ is a 5-space containing all singular lines on $\widehat{Z(K)}$; all conjugates of K whose centers equal Z(K) are contained in $\langle K,\widetilde{K}\rangle$.

These observations suggest that the "generic" subspaces of V may be those associated with 2-groups of G in either of two ways: (1) as $\overline{K} = C_V(C_G(K))$ for some K; (2) as a subspace all of whose proper subspaces are afforded as in (1). The following constructions justify this view. The identification in (1) is reasonable, certainly, if K is singular. In general, one obvious difficulty is that $\overline{K}$ may be "larger" than K, especially if K is non-abelian, making it hard to recover K from V. We thus generalize the notion of singularity.

We say that K is a *focal* subgroup provided $K_0 = Z(C_G(K_0))$ for every $K_0 \in R(K)$; equivalently, $K_0 = Z(C_G(\overline{K}_0))$. The spaces identified with these groups turn out to be precisely those which satisfy (2) as well. Further:

(2.1) *The focal subgroups of* G *are precisely the elements of* $\bigcup_K R(\widetilde{K})$ *where* K ∈ Ω.

This follows from the classification of G-orbits on Ω × Ω which has been carried out in [7]. We identify all focal groups with their images in V and refer to them projectively. Thus members of Ω, like those of Γ, are points and Ω is partitioned into focal planes the collection of which provides a "stellation" of the octagon. This suggests that Γ ∪ Ω is the set of vertices of a larger complex. This can be realized by defining the *intrinsic* geometry of G to consist of the subspaces of V all of whose 1-spaces are focal.

(2.2) *The flag space of the intrinsic geometry of* G *is a rank* 3 *chamber complex in the sense of Tits. As an embedded incidence structure there are two orbits on points, three orbits on lines, and two orbits on planes.*

It is shown in [8] that this intrinsic geometry arises from the point set Γ ∪ Ω by considering two points adjacent provided the group they generate is the union of such root groups, thus defining the line they determine. Hence the intrinsic geometry can be derived from the "Cooperstein functor".

Having determined the generic points, any remaining orbits of generic lines must be lines realizable as $\overline{K}$ for some K. It is not difficult to show that any such K must be a subgroup of $\langle L,\widetilde{L}\rangle$ where L ∈ Θ, and from there that the only remaining possibility is that $K \in \Theta$ as well. (Again the determination proceeds from the classification of orbits on $\Omega \times \Omega$.) If $K = V_5$ then $\overline{K} = \langle x_1, x_8 \rangle$. The orbit of $\overline{K}$ by P_1 consists of the q^4 lines represented as $\langle x_1, (sv + tu)x_1 + sx_2 + tx_3 + ux_4 + vx_5 + x_8 \rangle$, s,t,u,v fixed but arbitrary, from which one can calculate that the plane spanned by $\overline{K}$ and $g.\overline{K}$ for $g \in P_1$ contains another line of the orbit iff $sv + tu = 0$, in which case q of the lines on $\overline{Z(K)}$ in this plane are from the orbit. The remaining line is afforded by the center of $\langle K, K^g \rangle$, and is a focal line. In this case $\langle K, K^g \rangle$ is the disjoint union of its center and the conjugates of K it contains minus their centers. Thus an adjacency criterion of Cooperstein type applies to members of Θ, and it is only for such pairs that $\langle K, K^g \rangle$ affords a plane. We remark that the $q^2 + 1$ focal lines on $\overline{Z(K)}$ together with the q^4 lines from conjugates of K can be used to construct a Luneburg plane acted upon by a Levi complement of P_1 (the construction can also be done inside the intrinsic geometry). This "dualizes" the stellation of the octagon.

Finally, as for lines, remaining generic planes are all afforded by subgroups of $\langle K, \widetilde{K} \rangle$, $K \in \Theta$.

(2.3) *All generic planes are either afforded by compatible pairs* (K, K^g) *with* $K \in \Theta$, *or generated by a pair of focal lines on a singular point. Any such pair of focal lines generates a generic plane.*

3. Suppose V is expressed as the direct sum of spaces V^+, M, and V^- with M the "zero weight space" $\langle h_3, h_4 \rangle$ and V^+, V^- defined in the obvious manner. Then the ordering chosen for the basis of V makes the following observation almost immediate:

(3.1) *Let* V_j *and* V_k *be standard abelian root groups,* $k \neq -j$, *and denote by* $V_{[j,k]}$ *the 2-group generated by all standard root groups which lie between* V_j *and* V_k *in the root diagram and inclusive of them. If* K *is an abelian root group contained in* $V_{[j,k]}$ *then* $\overline{K}$ *lies in the space spanned by the images of all the standard root groups in* $V_{[j,k]}$.

The B-N structure of G and the irreducibility of V also yield:

(3.2) *If* K *is any root group not contained in a Weyl conjugate of* U *then* $\overline{K}$ *is not contained in the space spanned by* V^+ *and* V^-.

Now suppose K,L are both in Γ or both in Ω, with $[K,L] \neq 1$, and draw the line m on $\overline{K}, \overline{L}$ in the projective space of V. If X is any other abelian root group with $\overline{X}$ on m then $X^{\#} \subset C_G(C(K,L)) - (C_G(K) \cup C_G(L))$, where $C(K,L) = C_G(K) \cap C_G(L)$, the point stabilizer of m in G. This observation together with (3.1) is enough to eliminate $\overline{X}$ from m provided $\langle K, L \rangle$ is a 2-group. Otherwise (3.2) can be used, the only somewhat tricky case being if $C(K,L) = 1$ so the point stabilizer of m is trivial (there is a 1-parameter family of such orbits on $\Omega \times \Omega$). In any case there can be no such X. Combining with the commuting case which is subsumed by the intrinsic geometry we obtain:

(3.3) *Any line of the projective space of* V *which meets the intrinsic geometry in* 3 *points lies entirely within it. Thus any line not in the geometry is either exterior to it, a tangent, or a secant.*

In light of the remark by Tits in [9] regarding the likelihood of the octagon consisting of the totally isotropic elements relative to some sort of "polarity" it is reasonable to surmise that the intrinsic geometry is perhaps the isotropic "envelope" of the octagon relative to that map. Hopefully the above constructions will lead to an explicit description[(*)].

The abundance of counting arguments in the proofs of these propositions indicates that some caution is necessary in attempting to generalize to infinite fields. In all likelihood, however, the theorems which are not inherently finite in nature should go through, particularly if k is perfect.

REFERENCES

1. M. Aschbacher & G. Seitz, Involutions in Chevalley groups over fields of even order, *Nagoya Math. J.* 63 (1976), 1-92.
2. R.W. Carter, *Simple Groups of Lie Type*, John Wiley and Sons, London (1972).
3. A.M. Cohen, An axiom system for metasymplectic spaces, *Geom. Dedicata* 12 (1982), 417-433.
4. B.N. Cooperstein, The geometry of root subgroups in exceptional groups, I; *Geom. Dedicata* 8 (1979), 317-381.
5. D. Parrot, A characterization of the Ree groups ${}^2F_4(q)$, *J. Algebra* 27 (1973), 341-357.
6. R. Ree, A family of simple groups associated with the simple Lie algebra of type F_4, *Amer. J. Math.* 83 (1961), 401-420.
7. J. Sarli, Ph.D. Thesis, Univ. of California at Santa Cruz (1984).
8. J. Sarli, The geometry of root subgroups in ${}^2F_4(q)$, submitted.
9. J. Tits, Les groupes simples de Suzuki et de Ree, *Seminaire Bourbaki* 13 (1960), Exposé 210.
10. J. Tits, A local approach to buildings, in *The Geometric Vein: the Coxeter Festschrift*, Springer-Verlag, New York, Heidelberg, Berlin (1981).
11. J. Tits, Moufang octagons and the Ree groups of type ${}^2F_4$, in *Geometry and Number Theory*, Johns Hopkins Univ. Press, Baltimore, London (1983).
12. W.J. Wong, Irreducible modular representations of finite Chevalley groups, *J. Algebra* 20 (1972), 355-367.

[(*)] *Note added in proof.* The author has recently obtained such a description by embedding the geometry in a metasymplectic space and defining an appropriate polarity.

ON GENERALIZED FROBENIUS COMPLEMENTS

Carlo M. Scoppola
University of Trento, 38050 Povo (Trento), Italy.

In [LP] the following definition was given:

Definition. *Let* K *be a field,* G *a finite group. Let* M *be a* KG*-module. Let* N(G,M) *be the (normal) subgroup generated by the elements of* G *that fix a nontrivial vector in* M. *Then* G/N(G,M) *is called a* ***generalized Frobenius complement (in short: GFC) for*** G ***(with respect to*** M***)***.

Remark. Via a well known theorem by Wielandt, A. Espuelas has shown in [E] that in the more general situation of a finite group G acting on a group Q such that all the elements outside a proper normal subgroup N of G act fixed-point-freely on Q, G/N is actually isomorphic to a factor group of some GFC for G. Two natural problems arise:

Problem 1. *Given a class of groups, study the properties of a GFC for a group* G *in the class.*

Problem 2. *Given a finite group* H, *find a group* G *such that* H *is isomorphic to a generalized Frobenius complement for* G.

The aim of this note is to state some answers to Problems 1 and 2, when G is a finite p-group. This restriction is a natural starting point, since in [LP] it is shown that the Sylow p-subgroups of a GFC for an arbitrary finite group G are factor groups of suitable GFCs for the Sylow p-subgroups of G. As in [LP], and without loss of generality, it is assumed throughout that $K = \mathbb{C}$.

Connections between Problems 1 and 2 and other topics in group theory, such as the generalized Hughes problem, or the study of the 'secretive' p-groups defined in [KNN], are explained in [M].

For Problem 1, it has been shown in [LP] and [M] that regular p-groups (in the sense of P. Hall) have cyclic GFC. On the other hand, it is remarked in [M] that the (irregular) p-group G appearing in [H, III.10.15] has an elementary abelian GFC of order p^2 (this can be easily seen inducing to G a faithful linear module of Z(G)); G is abelian-by-cyclic, and of maximal class, since its order is p^{p+1} and its nilpotency class is p. The above is relevant in view of the following results in [S]:

Theorem A. *Every GFC for a* p*-group of odd order and maximal class is abelian.*

Theorem B. *Every GFC for an abelian-by-cyclic* p*-group of odd order is abelian.*

The proof of Theorem A rests on Blackburn's theory of the p-groups of maximal class (see e.g. [H, III.14]), and on the results in [LP], [M]. Some of the techniques used in the proof of Theorem B give as a by-product the following:

Proposition C [S]. *Let* p *be any prime, let* G, H *be* p-*groups, and let* H *be isomorphic to a GFC for* G. *If* H *has more than one subgroup of order* p, *then the nilpotency class of* H *is strictly smaller than the nilpotency class of* G.

The known answers to Problem 2 can be summarized as follows. In [W], [M] it is shown that some abelian groups of large rank can occur as GFC for p-groups. Generalizing those techniques, one can show the following:

Theorem D. *Let* F *be the free group on two generators, and let* p *be an odd prime. Let* $F_i^{p^k}$ *be the subgroup of* F *generated by the* p^k-*th powers of elements of the* i-*th term of the lower central series of* F. *Then* $G = F/F_{p^2+1} F^{p}_{p+1} F^{p^2}_{2} F^{p^3}$ *has a nonabelian GFC.*

In the proof of Theorem D some results in commutator calculus, as those illustrated in [HB, VIII], can be used, once the problem is reduced to a purely group-theoretical one by the following:

Lemma E [S]. *Let* G *be a* p-*group,* $N \lhd G$. *Then there exists a* $\mathbb{C}G$-*module* M *such that* $N(G,M) < N$ *if and only if there exists a cyclic section* H/K *of* G *such that every element in* $G-N$ *has a power in* $H-K$.

It should be remarked that Lemma E is implicitly contained in some of the literature quoted above, and in a private communication by G. Busetto. The proof of Lemma E is elementary: it only uses Mackey's induction theorem and Frobenius reciprocity.

REFERENCES

[E] A. Espuelas, The complement of a Frobenius-Wielandt group, *Proc. London Math. Soc.* (3) 48 (1984), 564-576.

[H] B. Huppert, *Endliche Gruppen* I, Springer-Verlag, Berlin (1967).

[HB] B. Huppert & N. Blackburn, *Finite Groups* II, Springer-Verlag, Berlin (1982).

[KNN] L. G Kovacs, J. Neubuser & B. H. Neumann, On finite groups with 'hidden' primes, *J. Austral. Math. Soc.* 12 (1971), 287-300.

[LP] B. Lou & D. S. Passman, Generalized Frobenius Complements, *Proc. Amer. Math. Soc.* 17 (1966), 1166-1172.

[M] T. Meixner, Verallgemeinerte Hughes-Untergruppen endlicher Gruppen, *Arch. Math.* 36 (1981), 104-112.

[S] C. M. Scoppola, *Generalized Frobenius Complements*, thesis.

[W] G. E. Wall, Secretive prime-power groups of large rank, *Bull. Austral. Math. Soc.* 12 (1975), 363-369.

SUBGROUPS OF FINITE INDEX IN SOLUBLE GROUPS : I

Dan Segal
All Souls College, Oxford OX1 4AL, England

1. *EXPLANATION*

Let $a_n(G)$ denote the number of subgroups of index n in the group G. If G is finitely generated then $a_n(G)$ is finite for all n, so for each such group we have a number-theoretic function $n \to a_n(G)$, and it would be interesting to know something about these functions.

Perhaps the first question concerns the average rate of growth. Let

$$s_G(k) = \sum_{n \leq k} a_n(G) \quad \text{and define} \quad \alpha(G) = \limsup_{k \to \infty} \frac{\log s_G(k)}{\log k} .$$

It is easy to see that $\alpha(G)$ is finite if and only if there exist positive constants C and α such that

$$s_G(k) \leq Ck^{\alpha} \quad \text{for all } k, \tag{1}$$

and that in that case $\alpha(G)$ is the infimum of all such α.

Question. *For which finitely generated groups* G *is* $\alpha(G)$ *finite?*

In considering this question, it is reasonable to restrict attention to residually finite groups. It is easy to find examples where $\alpha(G) = \infty$, e.g. free groups, and examples where $\alpha(G) < \infty$, e.g. polycyclic groups. To give a precise general answer to the question, however, does not seem at all easy. As a first step, I have tried to settle the problem for soluble groups; but even here I have had to introduce an extra hypothesis.

Theorem. *Let* G *be a finitely generated soluble group which is almost residually nilpotent. Then* $\alpha(G)$ *is finite if and only if* G *has finite rank.*

This can be made to look more conclusive if we think of it as a characterization of finite rank:

Corollary 1. *Let* G *be a finitely generated, residually finite soluble group. Then* G *has finite rank if and only if* $\alpha(G)$ *is finite and* G *is almost residually nilpotent.*

(I say that G has rank $\leq$ d if every finitely generated subgroup of G can be generated by d elements, and G is almost $\underline{P}$ if G has a normal subgroup of finite index with the property $\underline{P}$.)

2. *AN EXAMPLE*

In this section we consider groups which need not be finitely generated.

Proposition 1. *If* G *is a soluble group of finite rank then* $\alpha(G)$ *is finite.*
Proof. Let A be an abelian normal subgroup of G. I will show that

$$\alpha(G) \leq \alpha(G/A) + 2d + 2 \tag{2}$$

where d is the rank of G; the result then follows by induction on derived length.

Suppose $|G:L| = n$ and put $H = AL$, $B = A \cap L$. Then $|G/A : H/A|.|A:B| = n$, and $B \triangleleft H$; conversely, given H and B with these properties, the number of possible choices for L is either zero or $\mathrm{Der}|(H/A, A/B)|$. Therefore

$$a_n(G) \leq \sum_{r|n} a_{n/r}(G/A)\, a_r(A) r^d .$$

Now $a_r(A) \leq a_r(\mathbb{Z}^d) \leq r^d$ (easy exercise!), so $a_n(G) \leq n^{2d} s_{G/A}(n)$. An elementary calculation then yields (2).

The converse, however, is false. Let q be an odd prime and consider the ring

$$R = \mathbb{Z}[q^{-1}, q^{2^{-n}} \mid n \in \mathbb{N}].$$

Let

$$X = \langle \{x_i \mid i \in \mathbb{Z}\} \mid x_i^2 = x_{i-1} \quad \text{for all } i\rangle$$

so X is the additive group of dyadic rationals written multiplicatively. Make R into an X-module by mapping $\mathbb{Z}X$ onto R via $x_i \to q^{2^{-i}}$ for all i.

Proposition 2. *Let* G *be any extension of the* X*-module* R *by the group* X. *Then* G *has infinite rank,* G *is residually finite, and every finite quotient of* G *is metacyclic. Hence* $\alpha(G) \leq 2$.

If G can be embedded as a normal subgroup in some finitely generated, residually finite soluble group Γ so that Γ/G has finite rank, then, by the argument of Proposition 1, we would have $\alpha(\Gamma) < \infty$ while Γ has infinite rank. I doubt the existence of such a group Γ, but have been unable to disprove it.

To prove Proposition 2, suppose S is a finite quotient ring of R/pR where p is a prime distinct from q. Let $\overline{X}$ denote the image of X in the unit group of S. Since $\overline{X}$ is finite and 2-divisible, it has odd order, and $\overline{X}$ is therefore cyclic and generated by the image of x_0. But this image is the identity element of S; since $\overline{X}$ additively generates S it follows that $S \cong \mathbb{Z}/p\mathbb{Z}$.

We conclude that *every* finite quotient ring of R is additively a cyclic group. As every normal subgroup of finite index in G meets R in an ideal of finite index, and every finite quotient of $G/R \cong X$ is cyclic, it follows that every finite quotient of G is metacyclic. It is easy to see that this implies $\alpha(G) \leq 2$.

Now let $\mathfrak{P}$ be the set of primes p such that q has odd order modulo p; it is easy to see that $\mathfrak{P}$ is infinite (using quadratic reciprocity). For $p \in \mathfrak{P}$, let $\overline{q}$ be the image of q in $\mathbb{F}_p$. The cyclic group $\langle \overline{q} \rangle$ contains a unique 2^nth root of $\overline{q}$ for each $n \in \mathbb{N}$, giving a unique homomorphism of R onto $\mathbb{F}_p$. Let K_p be the kernel of this homomorphism. For each n, every non-zero ideal of the ring $R_n = \mathbb{Z}[q^{-1}, q^{2-n}]$ has finite index, so for each n we have $R_n \cap K_p = 0$. Hence $\bigcap_{p\in\mathcal{P}} K_p = 0$. Since G/K_p is residually finite for each $p \in \mathcal{P}$ it follows that G is residually finite.

3. *LEMMAS*

The first lemma collects together some results due to Mal'cev, Robinson and Wehrfritz. Following Wehrfritz I write $\mathfrak{S}_t$ for the class of soluble groups of finite rank having no infinite periodic normal subgroup.

Lemma 1. *Let G be a soluble group of finite rank.*

(i) $G \in \mathfrak{S}_t$ *if and only if* G *is almost torsion-free.*

(ii) *If* $G \in \mathfrak{S}_t$ *then* G *is almost nilpotent-by-(free abelian).*

(iii) *If* G *is a minimax group, then* $G \in \mathfrak{S}_t$ *if and only if* G *is residually finite.*

(iv) *If* G *is a minimax group and* $G \in \mathfrak{S}_t$ *then* G *is almost residually nilpotent.*

(v) *If* G *is finitely generated then* G *is a minimax group.*

By now, these facts are all well known. The proofs can be found in [M], [R1], [R2], [R3] and [W]. Parts (v) and (iv) show that Corollary 1 follows from the main theorem.

I shall use the following notation: for a group G, $\gamma_i(G)$ denotes the ith term of the lower central series of G, $G' = \gamma_2(G)$, and spec(G) is the set of those primes p such that G has a section isomorphic to C_{p^∞}. If G acts on a group A then [A,G] denotes the group generated by all $a^{-1}a^g$ $(a \in A, g \in G)$, and for $n \geq 1$

$$[A,_n G] = [[A,_{n-1} G],G] \quad (\text{where } [A,_0 G] = A).$$

Lemma 2. *Let* N *be a group such that* N/N' *has finite rank, and let* A *be an abelian normal subgroup of* N *with* $N' \geq A \geq \gamma_c(N)$ *for some* $c \geq 2$. *Put* $A_i = [A,_i N]$ *and let* T_i/A_i *be the torsion subgroup of* A/A_i, *for each* i. *Then*

(i) $N/\gamma_i(N)$ *and* N/A_i *have finite rank, for each* i.

(ii) $\mathrm{spec}(A/A_i) \subseteq \mathrm{spec}(N/N')$, *for each* i.

(iii) *If* N/N' *is a minimax group and the groups* A/T_i *have bounded rank, then there exist* $m \in \mathbb{N}$ *and a finite set of primes* π *such that* T_m/A_i *is a* π*-group for all* $i \geq m$.

Proof. For each i, $\gamma_i(N)/\gamma_{i+1}(N)$ is a homomorphic image of $\gamma_{i-1}(N)/\gamma_i(N) \otimes N/N'$, and A_i/A_{i+1} is a homomorphic image of $A_{i-1}/A_i \otimes N/N'$. These observations reduce the lemma to an exercise on abelian groups, which I leave to the reader.

Lemma 3. *Let* G *be a group,* A *an* $\mathbb{F}_p G$*-module and* $N \triangleleft G$. *Write* $\mathfrak{n}$ *for the augmentation ideal of* $\mathbb{F}_p N$ *and put* $H = C_G(A/[A,N]) \cap C_G(\mathbb{F}_p N/\mathfrak{n}^2)$. *Then*

$$[[A,_j N],H] \leq [A,_{j+1} N] \text{ for all } j \in \mathbb{N};$$

and if $N/N'N^p$ *is finite and* $A/[A,N]$ *is finite then* $|G:H|$ *is finite.*

Proof. If $\mathfrak{h}$ is the autmentation ideal of H, we have

$$\mathfrak{n}\mathfrak{h} \subseteq \mathfrak{h}\mathfrak{n} + H\mathfrak{n}^2 \text{ and } A\mathfrak{h} \subseteq A\mathfrak{n}.$$

It follows by induction on j that $A\mathfrak{n}^j\mathfrak{h} \subseteq A\mathfrak{n}^{j+1}$, which is the first claim. The second claim is clear, since $\mathfrak{n}/\mathfrak{n}^2 \cong N/N'N^p$.

Lemma 4. *Let* Γ *be an* $\mathfrak{S}_t$ *group and* p *a prime. Then there is a positive integer* $\gamma = \gamma_p(\Gamma)$ *with the following property: if* Δ *is any subgroup of* Γ *and* R *is a central extension of an elementary abelian* p*-group* Z *by* Δ*, then* R *has a normal subgroup* S *such that* $S \cap Z = 1$ *and* $|R:ZS| \le \gamma$.

Proof. By Lemma 1, Γ has normal subgroups $\Gamma_1 \le \Gamma_2$ with Γ_1 torsion-free and nilpotent of class c, say, Γ_2/Γ_1 free abelian, and Γ/Γ_2 finite of order s, say. Let n_1 be an upper bound for the orders of sections of Γ_2 of exponent dividing p^{3c+3}, and let n_2 be an upper bound for the orders of automorphism groups of such sections. I claim that $\gamma = sn_1n_2$ has the required property. To see this, suppose Z and R are as described and identify R/Z with $\Delta \le \Gamma$. Put $R_1/Z = \Gamma_1 \cap \Delta$ and $R_2/Z = \Gamma_2 \cap \Delta$. Then R_1 is nilpotent of class at most $c+1$, so $R_1^{p^{c+1}}$ consists of pth powers in R_1 [S2, page 113]. Therefore $R_1^{p^{c+1}} \cap Z = 1$. Put $K = Z R_1^{p^{c+1}}$. Then $|R_1:K| \le n_1$, and putting $C = C_{R_2}(R_1/K)$ we have $|R_2:C| \le n_2$.

Since $C' \le R_1$ we have $\gamma_3(C) \le K$ and $\gamma_4(C) \le R_1^{p^{c+1}}$. So every element of $C^{p^{3c+3}}$ is a p^{c+1}th power in C modulo $R_1^{p^{c+1}}$, whence $C^{p^{3c+3}} \cap R_1 \le R_1^{p^{c+1}}$. Putting $S = C^{p^{3c+3}}$ we thus have $S \cap Z = 1$ and

$$|R:ZS| = |R:R_2|.|R_2:C|.|C:ZC^{p^{3c+3}}| \le sn_2n_1$$

as required.

The final lemma in this section is the key to the main theorem:

Lemma 5. *Let* G *be a group with normal subgroups* $A \le N$ *such that* G/A *is an* $\mathfrak{S}_t$ *group and* A *is an elementary abelian* p*-group, for some prime* p*. Suppose* d *is a positive integer such that*

$$p^d \le |[A,_iN] : [A,_{i+1}N]| < \infty \text{ for all } i \ge 0.$$

Then $\alpha(G) \ge d/\mu$*, where* μ *is the sum of the ranks of the factors in the derived series of* G/A.

Proof. Put $A_i = [A,_iN]$ for each i. By Lemma 3, G has a normal subgroup H of finite index such that $[A_i,H] \le A_{i+1}$ for each i. For each $j \in \mathbb{N}$ put $H_j = AH^{p^{j-1}}$. Then H_j centralizes A/A_j, hence by Lemma 4 there exists $S_j \triangleleft H_j$ such that

$$A \cap S_j = A_j \text{ and } |H_j:AS_j| \le \gamma = \gamma_p(G/A).$$

Now $AS_j/S_j \cong A/A_j$ is elementary abelian of order at least p^{jd}, so AS_j has at least p^{jd-1} subgroups of index p. Since $|H:H_j| \le p^{\mu(j-1)}$, it follows that G has at least p^{jd-1} subgroups of index at most

$$|G:H|p^{\mu(j-1)}\gamma p = p^{\mu j}.\gamma|G:H|p^{(1-\mu)}.$$

Given $\alpha > \alpha(g)$, there exists C such that $s_G(n) \le Cn^\alpha$ for all n. Then

$$p^{jd-1} \le C.p^{\mu j\alpha}.\gamma^\alpha|G:H|^\alpha p^{(1-\mu)\alpha},$$

whence

$$p^{(d-\mu\alpha)j} \leq C\gamma^{\alpha}|G:H|^{\alpha}p^{(1-\mu)\alpha+1} \text{ for all } j \in \mathbb{N}.$$

Therefore $d-\mu\alpha \leq 0$ and the result follows.

4. *A SUFFICIENT CONDITION FOR FINITE RANK*

In one direction, the main theorem follows from Proposition 1. In the other direction, it is a special case of a more general result, which can also be applied in other contexts.

Definition. *A* <u>generalized rank function</u> *or* <u>g.r.f.</u> *is a function*

$$\rho : \{\text{groups}\} \to \mathbb{N} \cup \{\infty\}$$

with the following properties:
(A) $\rho(G/N) \leq \rho(G)$ *if* $N \triangleleft G$;
(B) *if* H *is a subgroup of finite index in* G *then* $\rho(H) = \infty \leftrightarrow \rho(G) = \infty$;
(C) *for each finitely generated* $\mathfrak{S}_t$ *group* Γ *there is an unbounded function* $\mu_\Gamma : \mathbb{N} \to \mathbb{R}_+$ *such that following holds:*
if H *is a finitely generated extension of an elementary abelian* p-*group* A *by* Γ, $N \triangleleft H$, *and* $d \in \mathbb{N}$ *satisfies*

$$p^d \leq |[A,{}_iN] : [A,{}_{i+1}N]| < \infty \textit{ for all } i \geq 0,$$

then $\rho(H) \geq \mu_\Gamma(d)$.

Proposition 3. *Let* G *be a finitely generated, almost residually nilpotent soluble group. If* ρ *is a g.r.f. and* $\rho(G)$ *is finite then* G *has finite rank.*

It is easy to see that $\rho = \alpha$ has properties (A) and (B); that it satisfies (C) is the content of Lemma 5: we take $\mu_\Gamma(d) = d/\mu$ where μ is the sum of the ranks of the factors in the derived series of Γ. Thus Proposition 3 completes the proof of the Theorem.

To establish Proposition 3, we first deal with some special cases. Henceforth, ρ will denote a g.r.f.

Lemma 6. *Let* G *be a finitely generated, almost nilpotent-by-abelian group. If* $\rho(G)$ *is finite then* G *has finite rank.*

Proof. Suppose G has infinite rank, and let F be a nilpotent normal subgroup of G with G/F almost abelian. By Lemma 2, F/F' has infinite rank, and then Hall's lemma [H, Lemma 5.2] shows that $F/F'F^p$ is infinite for some prime p. Let

$$\bar{\ } : G \to G/F'F^p$$

denote the natural homomorphism. Then $\bar{G}$ is almost residually nilpotent [S1, Theorem A]. We may choose a normal subgroup H of finite index in G such that $H' \leq F \leq H$ and $\bar{H}$ is residually nilpotent (this follows from the proof of [S1, Theorem A]).

Put $A = \bar{F}$, and for $d \in \mathbb{N}$ put $N_d = A\,\bar{H}^{p^{d-1}}$. Now fix d, and write $A_j = [A,{}_jN_d]$. Then for each j we have $\bigcap_{i=0}^{\infty} [A_j,{}_i\bar{H}] = 1$; and $A/[A_j,{}_i\bar{H}]$ is finite for all i, since $[A_j,_i\bar{H}] \geq [A,_{j+i} N_d] \geq \gamma_{i+j+2}(N_d)$ and $N_d/\gamma_{i+j+2}(N_d)$ is a finitely generated nilpotent group. Therefore

$$A_j > [A_j,\bar{H}] > \dots > [A_j,_d\bar{H}] \geq [A_j,N_d] = A_{j+1}.$$

This implies that $p^d \leq |A_j : A_{j+1}| < \infty$. Hence by properties (C) and (A) we have $\rho(H) \geq \rho(\bar{H}) \geq \mu_{H/F}(d)$. Since $\mu_{H/F}$ is unbounded and d was arbitrary, $\rho(H)$ must be infinite. Thus by property (B) we have $\rho(G) = \infty$ as required.

(Essentially the same proof gives the result, more generally, for finitely generated nilpotent-by-polycyclic groups; in place of [S1] one has to quote the main result of D. Segal, *J. London Math. Soc.* 11 (1975), 445-452.)

Lemma 7. *Let* G *be a finitely generated extension of an infinite elementary abelian* p*-group* A *by an* $\mathfrak{S}_t$ *group. Suppose that* G/A *has a nilpotent normal subgroup* F/A *with* G/F *almost abelian and* $\bigcap_{i=0}^{\infty} [A,_iF] = 1$. *Then* $\rho(G) = \infty$.

Proof. Fix $d \in \mathbb{N}$, put $N_d = AF^{p^{d-1}}$, and for each j write $A_j = [A,_jN_d]$. Then $\bigcap_{i=0}^{\infty} [A_j,_iF] = 1$, $[A_j,_iF] \geq A_{i+j}$ for all i, and $[A_j,_dF] \geq A_{j+1}$.

Suppose first that A/A_n is infinite for some n. Then G/A_n is almost nilpotent-by-abelian and has infinite rank, so by Lemma 6

$$\rho(G) \geq \rho(G/A_n) = \infty.$$

In the alternative case, A_j/A_{j+1} is finite and $A_j > [A_j,F] > \dots > [A_j,_dF] \geq A_{j+1}$ for each j; as in the proof of Lemma 6 we infer that $\rho(G) \geq \mu_{G/A}(d)$. Thus again $\rho(G)$ must be infinite.

Lemma 8. *Let* E *be an abelian group, let* p *be a prime, and suppose* $(E_i)_{i\in\mathbb{N}}$ *is a descending chain of subgroups of* E *of finite* p*-power index, such that the rank of* E/E_i *is bounded for all* i. *For each* n, *put*

$$L_n = \varprojlim_{i \geq n} E_n/E_i.$$

Then there exist k *and* s *such that* $L_n \cong \mathbb{Z}_p^s$ *for all* $n \geq k$.

Proof. Put $r_i = \sup_{j\geq i} \operatorname{rank}(E_i/E_j)$ and let $s = \inf_{i\geq 1} r_i$. The sequence (r_i) is decreasing, so there exists $k \in \mathbb{N}$ such that $r_i = s$ for all $i \geq k$. Given $n \geq k$, put $F = F_0 = E_n$, and define a sequence (F_i) recursively by

$$F_{i+1} = pF_i + E_{j(i)}$$

where j(i) is sufficiently large to ensure that $F_i/E_{j(i)}$ has rank s. In this way we obtain a sequence $(F_i)_{i\in\mathbb{N}}$ cofinal with the sequence $(E_j)_{j>n}$: and for each i we have $F/F_i \cong (\mathbb{Z}/p^i\mathbb{Z})^s$. The result follows.

Proof of Proposition 3. We may suppose that G is residually nilpotent and that $\rho(G)$ is finite. If G is abelian the result is clear; if not, let B be a maximal abelian normal subgroup containing the last non-trivial term of the derived series of G. Then G/B is again residually nilpotent, by a well-known lemma of A. Learner, and arguing by induction on derived length we may suppose that G/B has finite rank. By Lemma 1, G/B is a minimax group, $G/B \in \mathfrak{S}_t$, and G/B has a nilpotent normal subgroup N/B with G/N almost abelian.

Lemma 6 shows that $G/\gamma_i(N)$ has finite rank for each i; also G/N' is residually finite by Hall's theorem [H]. Put $A = B \cap N'$. Then G/A is minimax and in $\mathfrak{S}_t$ (by Lemma 1), and $N' \geqslant A \geqslant \gamma_c(N)$ for some $c \geqslant 2$. For each i, write

$$A_i = [A, {}_iN], \quad T_i/A_i = \text{torsion group of } A/A_i.$$

Note that A/A_i has finite rank for each i, since $A_i \geqslant \gamma_{c+i}(N)$.

Now spec(N/N') is finite, since G/N' is minimax, and spec$(A/A_i) \subseteq$ spec(N/N') for all i, by Lemma 2. Choose a prime $q \notin$ spec(N/N'). Then for each i we have

$$|A : A_iA^q| \geqslant |A : T_iA^q| \geqslant q^{r_i}$$

where r_i is the rank of A/T_i. Suppose that $r_i \to \infty$ as $i \to \infty$. Putting $K = \bigcap_{i=0}^{\infty} A_iA^q$ we then have A/K infinite, and Lemma 7 gives $\rho(G) \geqslant \rho(G/K) = \infty$.

It follows that r_i is bounded as $i \to \infty$; so by Lemma 2 there exist $m \in \mathbb{N}$ and a finite set of primes π such that T_m/A_i is a π-group for all $i \geqslant m$. For each $i \geqslant m$, let D_i/A_i be the divisible part of T_m/A_i: thus D_i/A_i is the unique minimal subgroup of finite index in T_m/A_i.

Now G is finitely generated and residually nilpotent, so G, and consequently also N, is residually finite-nilpotent. Since $\gamma_{i+2}(N) \geqslant A_i$ for each i, it follows easily that $\bigcap_{i=m}^{\infty} D_i = 1$. Suppose that for some $p \in \pi$, the p-rank of T_m/D_i tends to ∞ with i. Then putting $K = \bigcap_{i=m}^{\infty} D_iT_m^p$, we can apply Lemma 7, as above, to infer that $\rho(G) \geqslant \rho(G/K) = \infty$ (here T_m/K plays the role of A). It follows that the p-rank of T_m/D_i is bounded as $i \to \infty$, for each $p \in \pi$.

Thus we are in a position to apply Lemma 8. Since π is finite, this shows that for a suitable k, and suitable integers s(p),

$$L(p) = \varprojlim_{i \geqslant k} E_k(p)/E_i(p) \cong \mathbb{Z}_p^{s(p)}$$

for each $p \in \pi$, where $E_i(p)/D_i$ denotes the p'-component of T_m/D_i. The action of G by conjugation on the factors $E_k(p)/E_i(p)$ lifts to an action on L(p). By the Lie-Kolchin-Mal'cev theorem, G has a normal subgroup U, with G/U almost abelian, such that $[L(p), {}_sU] = 0$, where $s = \max_{p\in\pi} s(p)$.

Now let $E_\infty(p) = \bigcap_{i=k}^{\infty} E_i(p)$. Then $E_k(p)/E_\infty(p)$ is isomorphic to a G-submodule of L(p); therefore

$$[D_k, {}_sU] \leqslant \bigcap_{p\in\pi} [E_k(p), {}_sU] \leqslant \bigcap_{p\in\pi} E_\infty(p)$$

$$\leqslant \bigcap_{i=k}^{\infty} D_i = 1.$$

Thus

$$\gamma_{c+k+s}(N \cap U) \leqslant [A_k, {}_sU] \leqslant [D_k, {}_sU] = 1$$

and so G is almost nilpotent-by-abelian. The result follows by Lemma 6.

REFERENCES

[H] P. Hall, On the finiteness of certain soluble groups, *Proc. London Math. Soc.* (3) 9 (1959), 595-622.

[M] A.I. Mal'cev, On certain classes of infinite soluble groups, *A.M.S. Translations* (2) 21 (1956), 1-21.

[R1] D.J.S. Robinson, On soluble minimax groups, *Math. Zeit.* 101 (1967), 13-40.

[R2] D.J.S. Robinson, Residual properties of some classes of infinite soluble groups, *Proc. London Math. Soc.* (3) 18 (1968), 495-520.

[R3] D.J.S. Robinson, On the cohomology of soluble groups of finite rank, *J. Pure Appl. Algebra* 6 (1975), 155-164.

[S1] D. Segal, A residual property of finitely generated abelian-by-nilpotent groups, *J. Algebra* 32 (1974), 389-399.

[S2] D. Segal, *Polycyclic Groups*, Cambridge University Press (1983).

[W] B.A.F. Wehrfritz, Groups of automorphisms of soluble groups, *Proc. London Math. Soc.* (3) 20 (1970), 101-122.

SUBGROUPS OF FINITE INDEX IN SOLUBLE GROUPS: II

Dan Segal
All Souls College, Oxford OX1 4AL, England

1. *INTRODUCTION*

If G is a group and n is a positive integer, G^n denotes the subgroup of G generated by all nth powers in G. Even if G is finitely generated, the index $|G:G^n|$ can be infinite (Novikov and Adian). If G is soluble, and either finitely generated or of finite rank, then $|G:G^n|$ is of course finite for all n, and one may ask how fast this index can grow with n.

Let us define, for each group G,

$$\beta(G) = \limsup_{n \to \infty} \frac{\log|G:G^n|}{\log n} .$$

Then $|G:G^n|$ has 'polynomial growth' if and only if $\beta(G)$ is finite, and in that case $\beta(G)$ is the infimum of all β such that for some C,

$$|G:G^n| \leq Cn^\beta \text{ for all } n \in \mathbb{N} .$$

Question. *For which soluble groups* G *is* $\beta(G)$ *finite?*
As in the companion paper, henceforth referred to as [I], I can answer this question subject to an extra hypothesis:

Theorem 1. *Let* G *be a finitely generated, almost residually nilpotent soluble group. Then* $\beta(G)$ *is finite if and only if* G *has finite rank.*
Just as in [I], there is a more satisfying corollary:

Corollary 1. *Let* G *be a finitely generated residually finite soluble group. Then* G *has finite rank if and only if* G *is almost residually nilpotent and* $\beta(G)$ *is finite.*
Another consequence which is perhaps slightly surprising is:

Corollary 2. *Let* G *be as in* Theorem 1. *Then the following are equivalent:*
(a) G *is the product of finitely many subgroups of finite rank,*
(b) G *is the product of finitely many cyclic subgroups,*
(c) G *is a minimax group.*
In one direction this is due to Kropholler [K, Proposition 1]; in the other, it follows from Theorem 1 and Lemma 2, below (together with Derek Robinson's result stated in [I, Lemma 1v]).

The proof of Theorem 1, a simple adaptation of methods from [I], is given in Section 2. The rest of the paper is devoted to:

Theorem 2. *Let* G *be an almost torsion-free soluble group of finite rank. Assume that infinitely many primes do not belong to* spec(G). *Then* $\beta(G)$ *is equal to the Hirsch length of* G.

Here, spec(G) denotes the set of primes p such that G has a section isomorphic to C_{p^∞}. The *Hirsch length* h(G) of G is the sum of the torsion-free ranks of the factors in the derived series of G (or choose your favourite definition). (I am grateful to Peter Kropholler for suggesting this problem to me; he proved Theorem 2 for the special case of finitely generated nilpotent groups.)

It is worth remarking that the function α, studied in [I], appears to admit of no such simple evaluation.

Corollary 3. *If* G *is a finitely generated soluble group of finite rank then* $h(G) = \beta(G)$.

This follows from Theorem 2 by results of Robinson and Wehrfritz (see [I, Lemma 1]). As Kropholler has observed, in connection with his result [K, Proposition 1], Corollary 3 implies that such a group G cannot be equal to the product of fewer than h(G) cyclic subgroups.

In Theorem 2, some condition on spec(G) is unavoidable: if we partition the set of all primes into two non-empty subsets π and π' and take

$$G = \mathbb{Z}[p^{-1} | p \in \pi] \oplus \mathbb{Z}[p^{-1} | p \in \pi'],$$

we get a residually finite group G with $h(G) = 2$ and $\beta(G) = 1$.

2. *THEOREM 1*

The first two lemmas are left as exercises for the reader:

Lemma 1. *If* H *is a subgroup of finite index in a group* G *then* $\beta(H) = \beta(G)$.

Lemma 2. *Let* G *be a group with a finite subnormal series in which* ℓ *of the factors are infinite and locally cyclic, and the remaining factors are finite. Let* m *be the product of the orders of these finite factors. Then*

$$|G:G^n| \leq n^\ell m \text{ for all } n.$$

Lemma 2 shows that if G is soluble of finite rank then $\beta(G)$ is finite. The reverse implication in Theorem 1 is a direct application of Proposition 3 of [I]: we take $\rho = \beta$, define $\mu_\Gamma(d) = d$ for all d and each group Γ, and only have to verify:

Lemma 3. *Let* G *be a finitely generated group with a normal elementary abelian* p*-subgroup* A *such that* G/A *is an* $\mathfrak{S}_t$ *group. Suppose that for some* $N \triangleleft G$ *and* $d \in \mathbb{N}$ *we have*

$$p^d \leq |[A,_i N]:[A,_{i+1} N]| < \infty \text{ for all } i \geq 0.$$

Then $\beta(G) \geq d$.

(Recall that an $\mathfrak{S}_t$ group is a soluble group of finite rank without infinite periodic normal subgroups.)

Proof. Let $A_j = [A,_j N]$ for each j. By [I, Lemma 3], G/A has a normal subgroup H/A of finite index such that $[A_i,H] \leq A_{i+1}$ for each i. Put $H_j = AH^{p^{j-1}}$.

Then A/A_j is central in H_j/A_j, so Lemma 4 of [I] provides us with a normal subgroup D_j of H_j such that

$$A \cap D_j = A_j \quad \text{and} \quad |H_j : AD_j| \leqslant \gamma$$

where γ depends only on G/A and on p.

If $m = |G:H|$ we have, for each $j \in \mathbb{N}$,

$$G^{m\gamma p^j} \leqslant (AD_j)^p \leqslant D_j,$$

$$|G:D_j| \geqslant |AD_j : D_j| = |A:A_j| \geqslant p^{dj}.$$

Therefore

$$\beta(G) \geqslant \limsup_{j \to \infty} \frac{jd \log p}{j \log p + \log m\gamma} = d.$$

3. *GROUPS OF FINITE RANK*

Theorem 2 will follow from a sharper result:

Theorem 3. *Let* G *be an* $\mathfrak{S}_t$ *group. Then for each sufficiently large prime* $p \notin \operatorname{spec}(G)$, *there is a subgroup* G_p *of finite index in* G *such that*

$$|G_p : G_p^{p^n}| = p^{nh(G)} \quad \textit{for all } n \in \mathbb{N}.$$

This will be proved in Section 5. Now let G be an $\mathfrak{S}_t$ group, and suppose that some sufficiently large prime p does not belong to $\operatorname{spec}(G)$ (what this means, exactly, will emerge in Section 5). Then Lemma 1 and Theorem 3 show that

$$\beta(G) = \beta(G_p) \geqslant h(G).$$

On the other hand, G has a finite subnormal series in which the factors are either finite or isomorphic to additive subgroups of $\mathbb{Q}$. The number of factors of the latter kind is $h(G)$, so Lemma 2 implies that $\beta(G) \leqslant h(G)$. Thus Theorem 2 follows.

4. *LEMMAS*

Lemma 4. *Let* N *be a nilpotent group of class* c *and let* p *be a prime with* $p > c$. *Then* N^{p^n} *consists of* p^nth *powers in* N, *for each* $n \in \mathbb{N}$.

Proof. N is locally a homomorphic image of a finitely generated torsion-free nilpotent group of class c (e.g. a free nilpotent group). So we may assume that N is finitely generated and torsion-free, and now argue by induction on $h(N)$. If $h(N) \leqslant 2$ the result is clear, so suppose $h(N) > 2$. Let $x,y \in N$. Then

$$x^{p^n} y^{p^n} = (xy)^{p^n} c_2^{r_2} \cdots c_{p^n-1}^{r_{p^n-1}} c_{p^n}$$

where $r_j = \binom{p^n}{j}$ and $c_j \in \gamma_j \langle x,y \rangle$ for each j (the Hall-Petrescu formula, see [H], page 317). Since $p > c, c_j = 1$ for $j \geqslant p$, and for $2 \leqslant j < p$ we have $p^n | r_j$. Therefore $x^{p^n} y^{p^n} \in H^{p^n}$ where $H = N'\langle xy \rangle$. Since $h(N) > 2$ we have $h(H) < h(N)$, so by inductive hypothesis H^{p^n} consists of p^nth powers in H.

Thus a product of p^nth powers in N is a p^nth power, and the result follows.

Lemma 5. *Let* N *be a torsion-free nilpotent group of class* c *and of finite rank. Let* $p \notin \mathrm{spec}(N)$ *be a prime with* $p > c$. *Then*

$$|N:N^{p^n}| = p^{nh(N)} \text{ for all } n.$$

Proof. If $c = 1$ this is clear, from the structure of torsion-free abelian groups. If $c > 1$, let Z be the centre of N. Then N/Z is torsion-free, so Lemma 4 implies that $N^{p^n} \cap Z = Z^{p^n}$ for each n. Since

$$|N:N^{p^n}| = |N:ZN^{p^n}|.|Z:N^{p^n} \cap Z|$$

the result now follows by induction on c.

Definition. *Let* G *be a group with an automorphism* x. *For* $a \in G$ *and* $r \in \mathbb{N}$,

$$\phi_r(a,x) = a.a^x....a^{x^{r-1}}.$$

Lemma 6. *Let* N *be a group,* x *an automorphism of* N, *and* p *a prime. Suppose* $N = N_0 \geq N_1 \geq N_2 \geq \ldots$ *is a descending chain of normal subgroups of* N *such that*

$$N_i^p[N_i,x] \leq N_{i+1} \text{ for all } i \geq 0.$$

Then

$$\phi_{p^n}(a,x) \in N_n \text{ for all } a \in N \text{ and } n \in \mathbb{N}.$$

Proof. Put $v_i = \phi_{p^i}(a,x)$ for each i. Then $v_o = a \in N_o$. Now suppose $i \geq 0$ and $v_i \in N_i$. Then $v_i^x \equiv v_i \pmod{N_{i+1}}$ and $v_i^p \equiv 1 \pmod{N_{i+1}}$, so

$$v_{i+1} = v_i \,.\, v_i^{x^{p^i}} \,.\, v_i^{x^{2p^i}} \,....\, v_i^{x^{(p-1)p^i}} \equiv v_i^p \equiv 1 \pmod{N_{i+1}}.$$

The result follows by induction.

5. PROOF OF THEOREM 3

We have an $\mathfrak{S}_t$ group G. Thus G has a torsion-free nilpotent normal subgroup N such that G/N is almost free abelian ([I, Lemma 1]; see the proof of [R, Theorem 2.11]). A result of Newell, [N, Theorem 1], then implies that G has a nilpotent subgroup C with $|G:NC|$ finite. I will call a prime number *sufficiently large* if it exceeds the nilpotency classes of both N and C.

Now let p be a sufficiently large prime with $p \notin \mathrm{spec}(G)$. Since G has finite rank, there is a normal subgroup G_1 of finite index in G such that $[K,G_1] \leq K^p$ for every $K \triangleleft G$; choose G_1 so that NG_1/N is free abelian, and put

$$H = G_p = (N \cap G_1).(C \cap G_1).$$

Write $M = N \cap G_1$, $X = C \cap G_1$ and for $i \geq 0$ let $M_i = M^{p^i}$. Put $r = h(M)$ and $s = h(H/M)$. By Lemma 5, $|M:M_n| = p^{rn}$ for all $n \in \mathbb{N}$. Since H/M is free abelian,

$$|H:MH^{p^n}| = p^{sn} \text{ for all } n \in \mathbb{N}.$$

Now $|H:H^{p^n}| = |H:MH^{p^n}|.|M:M\cap H^{p^n}|$ and $M\cap H^{p^n} \geq M_n$ for each n. So if we can show that

$$M\cap H^{p^n} \leq M_n \text{ for all } n, \tag{1}$$

it will follow that $|H:H^{p^n}| = p^{rn}.p^{sn} = p^{nh(G)}$, which is what we have to prove.

So it only remains to establish (1). By Lemma 4 we have $M_i^p \leq M_{i+1}$, and the choice of G_1 then ensures that $[M_i, X] \leq M_{i+1}$, for each i. Hence Lemma 6 shows that for $a \in M$ and $x \in X$,

$$(ax^{-1})^{p^n} = \phi_{p^n}(a,x)x^{-p^n} \equiv x^{-p^n} \pmod{M_n}$$

for each n. Now suppose $w \in H^{p^n}$. Then

$$w = (a_1x_1^{-1})^{p^n} \dots (s_tx_t^{-1})^{p^n}$$

for some $a_1,\dots,a_t \in M$ and $x_1,\dots,x_t \in X$. Thus

$$w \equiv x_1^{-p^n} \dots x_t^{-p^n} \pmod{M_n};$$

and Lemma 4 shows that $x_1^{-p^n} \dots x_t^{-p^n} = y^{p^n}$ for some $y \in X$. But $X/(X\cap M)$ is torsion free, so if $w \in M$ then $y \in M$, and so $w \in M_n$. Thus (1) is established.

REFERENCES

[I] D. Segal, Subgroups of finite index in soluble groups: I, These Proceedings.

[H] B. Huppert, *Endliche Gruppen I*, Springer-Verlag, Berlin - Heidelberg - New York (1967).

[K] P. Kropholler, On finitely generated soluble groups with no large wreath product sections, *Proc. London Math. Soc.* (3) 49 (1984), 155-169.

[R] D.J.S. Robinson, On soluble minimax groups, *Math. Zeit.* 101 (1967), 13-40.

[N] M.L. Newell, Nilpotent projectors in $\mathfrak{S}_1$-groups, *Proc. Royal Irish Acad.* (A) 75 (1975), 107-114.

SOME INTERCONNECTIONS BETWEEN GROUP THEORY AND LOGIC

Theodore Tollis
University of Illinois, Urbana, Illinois 61801, U.S.A.

In his classic paper [5] Mal'cev applied certain methods of mathematical logic in order to prove local theorems for various classes of groups. In [1] Cleave simplified certain ideas of Mal'cev and proved the local theorems for the same classes of groups. In this paper we describe a metamathematical method similar to that of Mal'cev and Cleave. Using this method we study properties of various classes of finite and infinite groups expressed by second order sentences which will be called strongly boolean-(universal-existential).

We say that a class of groups $\mathfrak{J}$ has the B-*property* if $G_1, G_2 \in \mathfrak{J}$ and $G_1 \leq G \leq G_2$ imply $G \in \mathfrak{J}$. (B-property is the abbreviation of Betweenness property, meaning that whenever a group G is "between" two groups which belong to a class then necessarily G is in the class. As usual $H \leq G$ means that H is a subgroup of G.) Then, an immediate consequence of Theorem 1 in Section 1 is that classes of groups defined by strongly boolean-(universal-existential) sentences have the B-property. Such sentences are closely related to crypto-universal sentences [1]. In particular, second order sentences of the form $(QP)(\forall v)\phi$ - where QP denotes a sequence of predicate quantifiers, $\forall v$ denotes a sequence of universal individual quantifiers and ϕ is quantifier free - are called *crypto-universal*. Properties defined from such sentences are hereditary. Now sentences of the form $(QP)\theta$ - where QP denotes a sequence of predicate quantifiers and θ is formed from crypto-universal sentences using sentential connectives except the rule $\Psi \rightarrow \Omega$ - will be called *strongly boolean-universal* if Q is $\forall$, and *strongly boolean-existential* if Q is $\exists$. Moreover, Section 1 deals with relational structures; therefore Theorem 1 will be proved for relational structures in general.

The connection between the finite simple groups and strongly boolean-(universal-existential) has yet to be discovered. This will appear in Section 2 and the results proved therein are rather technical. Two of them require the classification theorem. The other is independent of the classification but requires a recent result by V.D. Mazurov and A.N. Fomin [7].

Finally, in Sections 3 and 4 it will be proved that certain classes containing generalized solvable groups, generalized nilpotent groups, and orderable groups have the B-property. Moreover, we raise some related problems and we show that no answer can be derived using the metamathematical methods of Section 1.

1. *PROPERTIES OF RELATIONAL STRUCTURES*

Let α,β,γ be ordinals. A *relational structure* A is a sequence $\langle A,\{R_i\}_{i<\alpha}, \{f_i\}_{i<\beta}, \{c_i\}_{i<\gamma}\rangle$ where

(i) A, the domain of A, is a non-empty set;

(ii) for all $i<\alpha$, $R_i \subseteq A^m$ for some m (the relations of A);

(iii) for all $i<\beta$, $f_i : A^n \to A$ for some n (the operations of A);

(iv) for all $i<\gamma$, $c_i \in A$ (the constants of A).

Groups, rings, fields, vector spaces, partially-ordered sets are a few examples of relational structures.

A substructure of $\langle A,\{R_i\}_{i<\alpha}, \{f_i\}_{i<\beta}, \{c_i\}_{i<\gamma}\rangle$ is a relational structure $B = \langle B,\{S_i\}_{i<\alpha}, \{g_i\}_{i<\beta}, \{d_i\}_{i<\gamma}\rangle$ satisfying the following conditions:

(i) $B \subseteq A$;

(ii) for all $i<\alpha$, S_i is a restriction of R_i to B;

(iii) for all $i<\beta$, g_i is the restriction of the function f_i to B;

(iv) for all $i<\gamma$, $c_i = d_i$.

If B is a substructure of A, we write $B \leqslant A$; and $B < A$ if in addition $B \neq A$. If $B < A$ then B is called a *proper substructure* of A. Here by a *maximal substructure* we mean a proper substructure which is not contained in any larger proper substructure.

Two relational structures $A = \langle A,\{R_i\}_{i<\alpha}, \{f_i\}_{i<\beta}, \{c_i\}_{i<\gamma}\rangle$, $B = \langle B,\{S_i\}_{i<\alpha'}, \{g_i\}_{i<\beta'}, \{d_i\}_{i<\gamma'}\rangle$ are similar if $\alpha = \alpha'$, $\beta = \beta'$, $\gamma = \gamma'$ and

(i) for all $i<\alpha$, $R_i \subseteq A^n$, $S_i \subseteq B^n$ for some n;

(ii) for all $i<\beta$, f_i, g_i take the same number of arguments.

Similarity is an equivalence relation. A collection of similar structures together with all of their isomorphs is called a similarity class of structures.

For a member A of a similarity class $\mathfrak{K}$ the *first order predicate logic* (FOPL) $T_1(A)$ is associated which consists of:

(i) a sequence $(v_1,\ldots,v_n,\ldots)_{\eta<\omega}$ of individual variables;

(ii) the set of sentential connectives $\neg$, $\vee$, $\&$, $\to$, the universal quantifier $\forall$, the existential quantifier $\exists$, and the identity symbol $=$;

(iii) the nonlogical symbols: $\{P_i\}_{i<\alpha}$ the predicate letters, $\{F_i\}_{i<\beta}$ the function letters and $\{a_i\}_{i<\gamma}$ the constant symbols.

$T_2(A)$ denotes the *second order predicate logic* (SOPL) obtained from $T_1(A)$ by adjoining sequences $(P_1,\ldots,P_n,\ldots)_{\eta<\omega}$ of k-place predicate variables.

We assume familiarity with the notions of formula, free variable, etc.; see e.g. [9, Ch.1]. A formula containing no free variables is called a sentence. We think of all formulae as being in prenex form, and when we speak of quantifiers we shall have prenex quantifiers in mind. Also, satisfaction of a sentence of $T_1(A)$ or $T_2(A)$ in a relational structure A is to be understood in the usual sense; for satisfaction of $T_2(A)$ sentences see [4].

Consider now the following classes of formulae:

(A) universal formulae: $T_1(A)$ formulae of the form $(\forall v_1)\ldots(\forall v_n)\phi$, where ϕ is a quantifier free formula.

(B) crypto-universal formulae: $T_2(A)$ formulae of the form $Q\phi$, where Q is a succession of expressions of the form $(\forall P_i)$ or $(\exists P_i)$ or $(\forall v_i)$ and ϕ is a quantifier free formula.

(C) boolean-universal formulae: $T_2(A)$ formulae of the form $(\forall P_1)\ldots(\forall P_n)\psi$, where ψ is formed from formulae of type B using only sentential connectives.

(D) strongly boolean-(universal-existential) formulae: $T_2(A)$ formulae of the form $Q\theta$, where Q is a succession of expressions of the form $(\forall P_i)$ or $(\exists P_i)$ and θ is formed from formulae of type B using only sentential connectives *except* the rule $\Psi \to \Omega$, where Ψ, Ω are formulae of type B.

Crypto-universal and boolean-universal formulae were introduced by Cleave [1, pp.122-123]. These formulae are simplified forms of the original formulae "object-universal" and "quasi-universal" first introduced by Mal'cev [5].

We shall need the following well-known results:

Lemma 1 (Mal'cev). *If Φ is a crypto-universal sentence which is true in a relational structure A, then Φ is true in every substructure of A.*
Proof. See [5] or [1].

Lemma 2 (Mal'cev). *Crypto-universal sentences and boolean-universal sentences define local properties.*
(Clearly strongly boolean-universal sentences define local properties. Then it is easy to see that strongly boolean-existential sentences also define local properties.)
Proof. See [5] or [1].

The following result gives the importance of strongly boolean-(universal-existential) sentences.

Theorem 1. *Let A_1, A, A_2 be relational structures such that $A_1 < A < A_2$. If Φ is a strongly boolean-(universal-existential) sentence which is true in both A_1 and A_2, then Φ is also true in A.*
Proof. Suppose that Φ is a strongly boolean-universal sentence. Then by definition Φ has the form

$$\Phi = (\forall P_1)\ldots(\forall P_n)\theta$$

where θ is formed from crypto-universal sentences using only sentential connectives except the rule $\Psi \to \Omega$, where Ψ, Ω are crypto-universal sentences. We wish to show that given any substitution of the predicates $P_1,\ldots,P_n$ on A and given also that θ is true on A_1 and A_2 for the predicates $P_1,\ldots,P_n$ on A_1 and A_2 induced for those substituted on A, then θ is a true sentence in A.

Thus, if we regard $P_1, \ldots, P_n$ as predicates on A_2 and use the same symbols for their restrictions on A and on A_1, then the Theorem is reduced to proving that θ is a true sentence in A.

If θ were built up by means of the connectives $\neg$, $\vee$, $\&$, $\rightarrow$, from crypto-universal sentences then by an elementary fact of logic θ could be written in an equivalent form as

$$\theta = \vee\theta_\alpha, \quad \theta_\alpha = \&\theta_{\alpha\beta}$$

where the $\theta_{\alpha\beta}$ are crypto-universal sentences or the negations of such sentences. Now, the sentence $\Psi \rightarrow \Omega$ (Ψ, Ω crypto-universal sentences) is equivalent to $\neg\Psi\vee\Omega$ and since strongly boolean-universal sentences do not permit this rule, it is easy to see that θ can be written in one of the following forms:

$$\theta = \Psi_1 \vee \ldots \vee \Psi_\eta;$$

$$\theta = \neg\Omega_1 \vee \ldots \vee\neg\Omega_m;$$

$$\theta = \Psi_1 \& \ldots \& \Psi_\eta \& \neg\Omega_1 \& \ldots \& \neg\Omega_m \tag{1}$$

where $n, m \geq 1$, and the Ψ_i, Ω_j are crypto-universal sentences.

We shall prove only the third of these cases. The other cases can be proved using simpler arguments. Assume now that (1) is true in A_1 and in A_2 but false in A. Then either some Ψ_i is false in A or some Ω_j is true in A. Using Lemma 1, we obtain that either some Ψ_i is false in A_2 or some Ω_j is true in A_1. Hence, (1) is either false in A_2 or false in A_1; contrary to our hypothesis.

A similar argument yields the result in the case where Φ is a strongly boolean-existential sentence. Hence, the proof of the Theorem is complete.

The following terminology will be used in Sections 3 and 4. We give the definitions now in terms of similarity classes of relational structures.

Definitions. (a) *Let $\mathbf{R}$ be a similarity class of relational structures. We say that $\mathbf{R}$ has the* B-*property if for all $A_1, A_2 \in \mathbf{R}$ and $A_1 \leq A \leq A_2$ then $A \in \mathbf{R}$.*

(b) *We say that $\mathbf{R}$ has the* wB-*property* (wB-*property is the abbreviation of weak Betweenness property) if for all $A_1, A_2 \in \mathbf{R}$ and A_1 is not a maximal substructure of A_2, then there exists $A \in \mathbf{R}$ such that $A_1 < A < A_2$.*

An immediate consequence of Theorem 1 is that similarity classes of relational structures defined by means of strongly boolean-(universal-existential) sentences have the B-property. On the other hand, Theorem 1 is not true in the case where Φ is a boolean-universal sentence in general. A counterexample is the class of simple groups. Indeed, the class of simple groups can be distinguished within the class of groups by means of a boolean-universal sentence of the form $(\forall P)(\Psi(P) \rightarrow \Omega(P))$ where P is a 2-place predicate variable, Ψ, Ω are both sentences of the form $(\forall v_1)\ldots(\forall v_n)\theta$; θ is a quantifier-free formula containing the predicate variable P. Clearly, the class of simple groups does not have the B-property. Classes of groups which have the B-property exist, more details will be given in Sections 3 and 4.

2. *APPLICATIONS TO SIMPLE GROUPS*

In this section we shall use the classical notation for the finite simple groups of Lie type. Also, as a general reference for the finite simple groups we use [3].

Let $\Phi = \Phi(x_1,\ldots,x_n)$ be an SOPL formula with free object variables $x_1,\ldots,x_n$. If $G \cong \alpha H$, then $\Phi = \Phi(x_1,\ldots,x_n)$ is true in G if and only if $\Phi' = \Phi'(x_1\alpha,\ldots,x_n\alpha)$ is true in H, see e.g. [5, pp.123-124]. So, by the previous result, and instead of using a complicated notation, we shall use the same letter for Φ and Φ'; meaning of course that the object variables of Φ' are those substituted by α.

The following result is an easy consequence of Corollary 1 [12, p.388] and of Theorem 1.

Theorem 2. *Let Φ be a strongly boolean-(universal-existential) sentence. Assume that Φ is true in the alternating groups $A_n, n \geq 5$. Assume also that Φ is true in each of the following simple groups:*

(i) $L_2(2^p)$, p: *any prime;*

(ii) $L_2(3^p)$, p: *any odd prime;*

(iii) $L_2(p)$, p: *prime,* $p > 3$ *and* $p^2 + 1 \equiv 0 (\mathrm{mod}\ 5)$;

(iv) $Sz(2^p)$, p: *any odd prime;*

(v) $L_3(3)$.

Then Φ is a true sentence in the class of nonabelian finite simple groups. Moreover, Φ is a true sentence in the class of countable locally finite simple groups.

Proof. It is well-known that every finite group can be embedded in A_n for some n. By [12] the groups listed in (i), (ii), (iii), (iv) and (v) comprise the minimal simple groups, and since every finite simple group contains a minimal simple group the result follows from Theorem 1. Using Lemma 2 we obtain that Φ is true in the countable locally finite simple groups, and this completes the proof of the Theorem.

Let $m_1,\ldots,m_{26}$ be the minimal degrees of the alternating groups for which the 26 sporadic groups are embedded into the alternating groups. Under this notation we state the following:

Theorem 3. *Let Φ be a strongly boolean-(universal-existential) sentence. Suppose that Φ is a true sentence in the alternating groups $A_{m_1},\ldots,A_{m_{26}}$. Suppose also that is a true sentence in the simple groups A_{12} and $L_2(p)$ for* $p = 7,11,17,29$. *Then Φ is a true sentence in the sporadic groups.*

Proof. By Theorem 1, it is sufficient to show that each sporadic group contains a simple group in which Φ is a true sentence. Some knowledge of the subgroup structure of the sporadic groups is required. We shall use the survey article [11]. The table at the end of this section might be useful.

$L_2(11)$ is a maximal subgroup of M_{11}. M_{11} is a maximal subgroup of M_{12}. M_{22} is a maximal subgroup of M_{23}. $L_2(11)$ is isomorphic to a maximal subgroup of M_{22}. M_{23} is a maximal subgroup of M_{24}. Hence Φ is a true sentence in the Mathieu groups.

$L_2(11)$ is a maximal subgroup of J_1. $U_3(3)$ is a maximal subgroup of J_2. $U_3(3) \supset L_2(7)$ [12, p.409]. So $L_2(7)$ can be embedded in J_2. $L_2(17)$ is isomorphic to a maximal subgroup of J_3. Let v_2 be a non-central representative of the conjugacy class of J_4. Then $C_{J_4}(v_2)$ acts indecomposably on an elementary abelian group V of order 2^{11}. Also $N_{J_4}(V) = VK$, where $C_{J_4}(V) = V$ and $K \cong M_{24}$. Hence M_{24} can be embedded in J_4. This proves that Φ is a true sentence in the Janko groups.

M_{23} is a maximal subgroup of $\cdot 3$. Also M_{23} is isomorphic to a subgroup of $\cdot 2$. Now $\cdot 2$ is a subgroup of $\cdot 1$. Hence Φ is a true sentence in the Conway groups.

$M(22)$ contains an elementary abelian group E of order 2^{10} for which $C_{M(22)}(E) = E$ and $N_{M(22)}(E) = EK$ where $E \cong M_{22}$, so M_{22} can be embedded in $M(22)$. Now S_{12} is contained in $M(23)$ and so does A_{12}. Finally, $M(23)$ is isomorphic to a subgroup of $M(24)'$ and $\cdot 2$ is isomorphic to a subgroup of F_2. This proves that Φ is a true sentence in the Fischer's groups.

M_{11} is a maximal subgroup of HS and isomorphic to a maximal subgroup of Mc. Now J_2 is isomorphic to a subgroup of Sz. Also, $L_2(29)$ is isomorphic to a subgroup of Ru. Hence Φ is a true sentence in the groups HS, Mc, Sz and Ru.

Aut(Mc) can be embedded in Ly and so does Mc. $S_4 \times L_2(17)$ is isomorphic to a maximal subgroup of He and so $L_2(7)$ can be embedded in He. J_1 is isomorphic to a subgroup of O'N. Now in F_3 all elements of order 7 are conjugate. Their centralizers are isomorphic to $Z_2 \times L_2(7)$ and so $L_2(7)$ can be embedded in F_3. Also $\mathrm{Aut}(M_{12})$ is a maximal subgroup of F_5 and so F_5 contains a subgroup isomorphic to M_{12}. Finally, the centralizer of a certain element of order 5 in F_1 is isomorphic to $Z_5 \times F_5$ and so F_1 contains a subgroup isomorphic to F_5. This completes the proof of the Theorem.

We mention that the proofs of Theorems 2 and 3 require the Classification Theorem of finite simple groups. In conclusion, we shall give a result similar to that of Theorem 2 which does not depend on the classification. First we need some definitions.

Following V.D. Mazurov and A.N. Fomin [7] a nonabelian simple subgroup A of a finite group G is an extensive subgroup, if the following conditions are satisfied:

(i) if $A \leq B < G$ then $B \leq N_G(A)$;

(ii) if $1 \neq B \leq G$ and $A \leq N_G(B)$ then $A \leq B$;

(iii) if $M < G$ and M is of smallest index, then
  (a) $G = MA$,
  (b) $|A:X| > |A:M \cap A|$ for every proper subgroup X of A.

Definition. *A finite group* G *is called a minimal extensive group if* G *does not possess any extensive subgroups.*

Let $\mathfrak{M}$ be a class of nonabelian finite simple groups which contains the alternating groups A_n, $n \geq 5$. Let also $\mathfrak{M}$ contain all extensive subgroups of every group $G \in \mathfrak{M}$ for which every nonabelian simple section of every proper subgroup of smallest index of G belongs to $\mathfrak{M}$. The following theorem gives the connection between $\mathfrak{M}$ and strongly boolean-(universal-existential) sentences.

The Sporadic Simple groups

Notation	Order	Name of Discoverers
M_{11}	$2^4 \cdot 3^2 \cdot 5 \cdot 11$	Mathieu
M_{12}	$2^6 \cdot 3^3 \cdot 5 \cdot 11$	"
M_{22}	$2^7 \cdot 3^2 \cdot 5 \cdot 7 \cdot 11$	"
M_{23}	$2^7 \cdot 3^2 \cdot 5 \cdot 7 \cdot 11 \cdot 23$	"
M_{24}	$2^{10} \cdot 3^3 \cdot 5 \cdot 7 \cdot 11 \cdot 23$	"
J_1 = Ja	$2^3 \cdot 3 \cdot 5 \cdot 7 \cdot 11 \cdot 19$	Janko
J_2 = HJ	$2^7 \cdot 3^3 \cdot 5^2 \cdot 7$	"
J_3 = HJM	$2^7 \cdot 3^5 \cdot 5 \cdot 17 \cdot 19$	"
J_4	$2^{21} \cdot 3^3 \cdot 5 \cdot 7 \cdot 11^3 \cdot 23 \cdot 29 \cdot 31 \cdot 37 \cdot 43$	"
Ly = LyS	$2^8 \cdot 3^7 \cdot 5^6 \cdot 7 \cdot 11 \cdot 31 \cdot 37 \cdot 67$	Lyons
He = HHM	$2^{10} \cdot 3^3 \cdot 5^2 \cdot 7^3 \cdot 17$	Held
O'N = O'NS	$2^9 \cdot 3^4 \cdot 5 \cdot 7^3 \cdot 11 \cdot 19 \cdot 31$	O'Nan
F_3 = E	$2^{15} \cdot 3^{10} \cdot 5^3 \cdot 7^2 \cdot 13 \cdot 19 \cdot 31$	Thompson
F_5 = D	$2^{14} \cdot 3^6 \cdot 5^6 \cdot 7 \cdot 11 \cdot 19$	Harada & Notron
F_1 = M	$2^{46} \cdot 3^{20} \cdot 5^9 \cdot 7^6 \cdot 11^2 \cdot 13^3 \cdot 17 \cdot 19 \cdot 23 \cdot 29 \cdot 31 \cdot 41 \cdot 47 \cdot 59 \cdot 71$	Fischer & Griess
HS = HiS	$2^9 \cdot 3^2 \cdot 5^3 \cdot 7 \cdot 11$	Higman & Sims
Mc = McL	$2^7 \cdot 3^6 \cdot 5^3 \cdot 7 \cdot 11$	McLaughlin
Sz = Suz	$2^{13} \cdot 3^7 \cdot 5^2 \cdot 7 \cdot 11 \cdot 13$	Suzuki
Ru	$2^{14} \cdot 3^3 \cdot 5^3 \cdot 7 \cdot 13 \cdot 29$	Rudvalis
$\cdot 3 = Co_3$	$2^{10} \cdot 3^7 \cdot 5^3 \cdot 7 \cdot 11 \cdot 23$	Conway
$\cdot 2 = Co_2$	$2^{18} \cdot 3^6 \cdot 5^3 \cdot 7 \cdot 11 \cdot 23$	"
$\cdot 1 = Co_1$	$2^{21} \cdot 3^9 \cdot 5^4 \cdot 7^2 \cdot 11 \cdot 13 \cdot 23$	"
$M(22) = F_{22}$	$2^{17} \cdot 3^9 \cdot 5^2 \cdot 7 \cdot 11 \cdot 13$	Fischer
$M(23) = F_{23}$	$2^{18} \cdot 3^{13} \cdot 5^2 \cdot 7 \cdot 11 \cdot 13 \cdot 17 \cdot 23$	"
$M(24)' = F'_{24}$	$2^{21} \cdot 3^{16} \cdot 5^2 \cdot 7^3 \cdot 11 \cdot 13 \cdot 17 \cdot 23 \cdot 29$	"
F_2 = B	$2^{41} \cdot 3^{13} \cdot 5^6 \cdot 7^2 \cdot 11 \cdot 13 \cdot 17 \cdot 19 \cdot 23 \cdot 31 \cdot 47$	"

Theorem 4. *Let Φ be a strongly boolean-(universal-existential) sentence. Assume that Φ is true in the alternating groups $A_n, n \geq 5$. Assume also that Φ is true in the minimal extensive groups which belong to $\mathfrak{M}$. Then Φ is a true sentence in the class of nonabelian finite simple groups. Moreover, Φ is a true sentence in the class of countable locally finite simple groups.*

Proof. By Proposition 1 of [7] $\mathfrak{M}$ is the class of all nonabelian finite simple groups. Hence, it suffices to prove the result for the members of $\mathfrak{M}$. Let $G \in \mathfrak{M}$. Using Theorem 1 it is sufficient to show that G contains a simple subgroup in which Φ is a true sentence.

If G is a minimal extensive group, then there is nothing to prove. Suppose now that G possess an extensive subgroup H. If H does not possess any extensive subgroups, then H is a minimal extensive group and so Φ is true in H. Then Φ is true in G by Theorem 1. If H possesses extensive subgroups then continue this process until we find a minimal extensive group. Therefore using Theorem 1 we obtain that Φ is true in $\mathfrak{M}$. Now using Lemma 1 we obtain that Φ is true in the class of countable locally finite simple groups; this completes the proof of the theorem.

3. *GENERALIZED SOLVABLE AND NILPOTENT GROUPS*

First we recall some definitions. For further details and all unexplained terminology the reader is referred to [10]. Let G be an Ω-operator group. By an *Ω-series* in G we mean a set of subgroups S which is totally ordered by inclusion and which satisfies the following conditions:

(i) if $1 \neq x \in G$, the union of all members of S which do not contain x is a member V_x of S;

(ii) if $1 \neq x \in G$, the intersection of all members of S which contain x is a member Λ_x of S;

(iii) $V_x \triangleleft \Lambda_x$;

(iv) every member of S is of the form V_x or Λ_x for some $1 \neq x \in G$. The V_x and Λ_x are the terms of S and the Λ_x/V_x are the factors of S. We shall usually write $S = \{\Lambda_\sigma, V_\sigma | \sigma \in \Sigma\}$ where Σ is the order-type of S. If $\Omega = \text{Inn}G$, then an Ω-series is called a *normal series*, all terms of S being normal in G. A *serial* Ω-subgroup of G is an Ω-subgroup of G which occurs in an Ω-series of G.

If S and S* are Ω-series in an Ω-group G, then S* is said to be a *refinement* of S if every term of S is also a term of S*. An *Ω-composition series* in G is an Ω-series which has no proper refinements. When $\Omega = \emptyset$ then the composition series are precisely the series whose factors have no proper non-trivial serial subgroups; that is series with *absolutely simple* factors. An InnG-composition series of G is called a *chief* series.

A group is called an SN-group if it possesses a series with abelian factors and an $\overline{\text{SN}}$-group if all of its composition factors are abelian.

A group is called an SI-group if it possesses a normal series with abelian factors. An $\overline{\text{SI}}$-group is a group in which every principal factor is abelian.

If a group G has a central series $\{\Lambda_\sigma, V_\sigma | \sigma \in \Sigma\}$, that is $[\Lambda_\sigma, G] < V_\sigma$ for all $\sigma \in \Sigma$, then G is called a Z-group. A $\overline{Z}$-group is a group in which every principal factor is central.

Consider the classes of groups $\mathfrak{J}_1$, $\mathfrak{J}_2$, $\mathfrak{J}_3$ defined as follows:

(i) $G \in \mathfrak{J}_1$ if and only if G is an SN-group but not an SI-group;

(ii) $G \in \mathfrak{J}_2$ if and only if G is an SI-group but not a Z-group;

(iii) $G \in \mathfrak{J}_3$ if and only if G is an SN-group but not a Z-group.

By well-known results $\mathfrak{J}_1$, $\mathfrak{J}_2$ and $\mathfrak{J}_3$ are not the trivial classes. In particular, $\mathfrak{J}_1$ and $\mathfrak{J}_3$ contain non-cyclic simple groups and $\mathfrak{J}_2$ contains non-nilpotent solvable groups, see e.g. [10, Vol. 2, pp.110-116].

Theorem 5. *The classes* $\mathfrak{J}_1$, $\mathfrak{J}_2$ *and* $\mathfrak{J}_3$ *have the* B-*property*.
We shall prove the result in detail only for the class $\mathfrak{J}_1$. Two proofs will be given.
First proof. Let $G_1, G_2 \in \mathfrak{J}_1$ and let $G_1 < G < G_2$. Since the property SN is inherited by subgroups, G is an SN-group. The property of being an SI-group can be expressed by a crypto-universal sentence of the form $(\exists P)(\forall x)(\forall y)(\forall z)\phi$ where ϕ is quantifier free. Hence the property of not being an SI-group can be expressed by a strongly boolean-universal sentence of the form $(\forall P)(\neg\Psi)$ where Ψ is the crypto-universal sentence $(\forall x)(\forall y)(\forall z)\phi$. Since G_1 is not an SI-group, then by the proof of Theorem 1 we obtain that G is not an SI-group. Consequently $\mathfrak{J}_1$ has the B-property.

Second proof. Let $G_1, G_2 \in \mathfrak{J}_1$ and let $G_1 < G < G_2$. Suppose $G \notin \mathfrak{J}_1$. Then G is either not an SN-group or an SI-group. If G is not an SN-group then by Lemma 1 G_2 is not an SN-group, a contradiction. If G is an SI-group then by Lemma 1 G_1 is an SI-group, a contradiction. Hence $G \in \mathfrak{J}_1$.

A similar argument yields the result for the classes $\mathfrak{J}_2$ and $\mathfrak{J}_3$.

Definition. *A group* G *is a* $\complement\overline{SN}$-*group (resp.* $\complement\overline{SI}$, $\complement\overline{Z}$) *if* G *is an* SN-*group (resp.* SI, Z) *but not an* $\overline{SN}$-*group (resp.* $\overline{SI}$, $\overline{Z}$).

Proposition 1. *Let* G_1, G, G_2 *be groups such that* $G_1 < G < G_2$. *Let* $G_2 \in \complement\overline{SN}$ *(resp.* $\complement\overline{SI}$, $\complement\overline{Z}$). *Suppose also that for the restriction of the series of* G_2 *on* G_1 *for which* $G_2 \in \complement\overline{SN}$ *(resp.* $\complement\overline{SI}$, $\complement\overline{Z}$), $G_1 \in \complement\overline{SN}$ *(resp.* $\complement\overline{SI}$, $\complement\overline{Z}$). *Then* G *is a* $\complement\overline{SN}$-*group (resp.* $\complement\overline{SI}$, $\complement\overline{Z}$).
Proof. We shall give a detailed proof only for the $\complement\overline{SN}$ case. Clearly G is an SN-group. Now the property of being an $\overline{SN}$-group can be expressed by a boolean-universal sentence of the form $(\forall P)(\Psi \to \Omega)$ where P is a 2-place predicate variable and Ψ, Ω are crypto-universal sentences [1, pp.127-128]. Hence the property of not being an $\overline{SN}$-group can be expressed by a strongly boolean-existential sentence of the form $(\exists P)(\neg(\Psi \to \Omega))$. The result now follows from Theorem 1. A similar argument yields the result in the other two cases.

Clearly Proposition 1 says that the classes of $\complement\overline{SN}$, $\complement\overline{SI}$, $\complement\overline{Z}$-groups have the B-property under certain conditions. We raise now the following problems.

Problem 1. *Do the classes of* $\mathfrak{C}\overline{SN}$, $\mathfrak{C}\overline{SI}$, $\mathfrak{C}\overline{Z}$*-groups have the* B*-property in general?*

More generally, one might ask:

Problem 2. *Do the classes of* $\mathfrak{C}\overline{SN}$, $\mathfrak{C}\overline{SI}$, $\mathfrak{C}\overline{Z}$*-groups have the* wB*-property?*

One might expect a negative answer to Problem 1. Indeed, the known examples for which the properties $\overline{SN}$, $\overline{SI}$ and $\overline{Z}$ are not inherited by subgroups, can provide a negative answer to Problem 1. (See Note at end.)

In conclusion, we would like to raise some further related problems. At the end of Section 1 we proved that similarity classes of relational structures defined by

boolean-universal sentences which are not equivalent to strongly boolean-universal sentences (*)

do not necessarily have the B-property. The properties $\overline{SN}$, $\overline{SI}$ and $\overline{Z}$ can be expressed by sentences which satisfy (*), see [5, pp.332-334] or [1, pp.127-128]. On the other hand, the property of being an $\tilde{N}$-group, i.e. a group in which every subgroup is serial, can be expressed by a sentence which satisfies (*). In particular, the property $\tilde{N}$ can be expressed by a sentence of the form $(\forall P)(\Psi \rightarrow \Omega)$ where P is a 2-place predicate variable and Ψ, Ω are crypto-universal sentences [5, p.334]. Clearly the property $\tilde{N}$ is inherited by subgroups and so the class of $\tilde{N}$-groups trivially has the B-property. The following problems are of particular interest.

Problem 3. *Find conditions for which the classes of* $\overline{SN}$, $\overline{SI}$, $\overline{Z}$*-groups have the* B*-property.* (The answer in general is in the negative.)

Problem 4. *Do the classes of* $\overline{SN}$, $\overline{SI}$, $\overline{Z}$*-groups have the* wB*-property?*
The result of P. Hall [13, p.577], that the property $\overline{SI}$ is inherited by normal subgroups of finite index, might be useful.

4. *ORDERABLE GROUPS*

A group G is called an O-group if there exists a total order on the set G. An R*-group is a group in which the condition $a^{g_1}\dots a^{g_n} = b^{g_1}\dots b^{g_n}$ implies $a = b$ for all a, b, $g_1,\dots,g_n$ in G. Here a^g denotes the conjugate $g^{-1}ag$ of a.

Clearly an O-group is always an R*-group. The opposite implication is not correct and there exist R*-groups which are not O-groups, see e.g. [8, pp.89-93]. Also by a well-known result an O-group is an SN-group. An SN-group need not be an O-group; torsion abelian groups are solvable groups but not O-groups.

Consider the classes of groups $\mathfrak{J}_4$ and $\mathfrak{J}_5$ defined as follows:

(i) $G \in \mathfrak{J}_4$ if and only if G is an R*-group but not an O-group;

(ii) $G \in \mathfrak{J}_5$ if and only if G is an SN-group but not an O-group.

The following result is closely related to Theorem 5.

Theorem 6. *The classes* $\mathfrak{J}_4$ *and* $\mathfrak{J}_5$ *have the* B*-property.*

Proof. Clearly the properties R* and O are inherited by subgroups. The same argument we used in the proof of Theorem 5 yields the result.

A partial order on G is called *isolated* if for all a in G and all $n > 0$, $a^n > e$ implies $a > e$; and *strongly isolated* if $a^{g_1} \ldots a^{g_n} > e$ implies $a > e$, for all $a, g_1, \ldots, g_n$ in G [2, p.15]. Here e denotes the identity element of G.

Definition. *An I**-group is a group in which every partial order can be extended to a strongly isolated partial order.*
Clearly a total order is a strongly isolated order. I**-groups are closely related to I*-groups, i.e. groups in which every partial order can be extended to an isolated partial order; see e.g. [8, p.73].

An O*-group is a group in which every partial order can be extended to a total order. Mal'cev proved [5, pp.334-335] that the property of being an O*-group can be expressed by a boolean-universal sentence of the form $(\forall P)(\Psi \to \Omega)$ where P is a 2-place predicate variable and Ψ, Ω are crypto-universal sentences. The same is true for I**-groups. This result is neither contained in [5] nor in [1]. The proof now follows.

Let G be a group. Let P denote the 2-place predicate on G defined by

$$P(x,y) = T \longleftrightarrow x < y.$$

Let $\Phi(P)$ be the FOPL formula

$$P(x,x) \& (P(x,y) \& P(y,z) \to P(x,z)) \& (P(x,y) \& P(y,x) \to x = y) \& P(x,y) \to P(gxg', gyg'),$$

meaning that P is a partial order preserved by the group operation. Let $\Psi(P)$ be the conjunction of $\Phi(P)$ and the formula

$$P(e, a^{g_1} \ldots a^{g_n}) \to P(e,a).$$

Now G is an I**-group if and only if G satisfies

$$(\forall P)(\Phi(P) \to (\exists Q)[\Psi(Q) \& (\forall x)(\forall y)(P(x,y) \to Q(x,y)]).$$

It is easy to see that this sentence is boolean-universal.

Definition. *A group* G *is a* $\complement$O*-*group (resp.* $\complement$I**) *if* G *is an* O-*group but not an* O*-*group (resp.* I**).

The following result states that the classes of $\complement$O*, $\complement$I** groups have the B-property under certain conditions. The proof is similar to the proof of Proposition 1 and will be omitted.

Proposition 2. *Let* G_1, G, G_2 *be groups such that* $G_1 < G < G_2$. *Let* $G_2 \in \complement O^*$ *(resp.* $\complement I^{**}$*). Suppose also that for the restriction of the partial order of* G_2 *on* G_1 *for which* $G_2 \in \complement O^*$ *(resp.* $\complement I^{**}$*),* $G_1 \in \complement O^*$ *(resp.* $\complement I^{**}$*). Then* G *is a* $\complement$O*-*group (resp.* $\complement$I***).*

It is well-known that the property O* is not inherited by subgroups. Moreover, it can be derived from Theorem 3.4.7 [8, p.76] that the property I** is not inherited by subgroups. The following problems are similar to Problems 1 and 2 of Section 3.

Problem 5. *Do the classes of* $\mathfrak{C}$O*, $\mathfrak{C}$I***-groups have the* B*-property?*

More generally, one might ask:

Problem 6. *Do the classes of* $\mathfrak{C}$O*, $\mathfrak{C}$I***-groups have the* wB*-property?*

In conclusion, we would like to raise some further related problems.

Problem 7. *Do the classes of* O*, I***-groups have the* B*-property?*

Problem 8. *Do the classes of* O*, I***-groups have the* wB*-property?*

The answer to Problem 7 is negative in general. This can be derived from a result due to Kopytov, see the book by A.I. Kokorin and V.M. Kopytov, Fully Ordered Groups; Proposition 5, p.84.

Acknowledgement. I would like to thank Professor Derek J.S. Robinson for his encouragement while preparing this paper. This work was supported in part by the National Science Foundation.

REFERENCES

1. J.P. Cleave, Local properties of systems, *J. London Math. Soc.* (1) 44 (1969), 121-130; Addendum, *J. London Math. Soc.* (2) 1 (1969), 384.
2. L. Fuchs, *Partially ordered algebraic systems*, Pergamon Press (1963).
3. D. Gorenstein, *Finite Simple Groups*, Plenum (1982).
4. L. Henkin, Some interconnections between modern algebra and mathematical logic, *Trans. Amer. Math. Soc.* 79 (1953), 410-427.
5. A.I. Mal'cev, Model correspondences, *Izves. Acad. Nauk. USSR Ser. Mat.* 23 (1959), 313-336.
6. A.I. Mal'cev, *Algebraic systems*, Springer-Verlag (1973).
7. V.D. Mazurov & A.N. Fomin, Finite simple non-Abelian groups, *Mat. Zametki* 34 (1983), 821-824.
8. R.B. Murra & A. Rhemtulla, *Orderable Groups*, Marcel Dekker (1977).
9. A. Robinson, *Introduction to Model Theory and to Metamethematics of Algebra*, North-Holland (1963).
10. D.J.S. Robinson, *Finiteness conditions and generalized soluble groups*, 2 Vols., Springer-Verlag (1972).
11. S.A. Syskin, Abstract properties of the simple sporadic groups, *Uspekhi Mat. Nauk* 35:5 (1980), 181-212, Translated *Russian Math. Surveys* 35:5 (1980), 209-246.
12. J.G. Thompson, Nonsolvable finite groups all of whose local subgroups are solvable, *Bull. Amer. Math. Soc.* 74 (1968), 383-437.
13. J.S. Wilson, On Normal Subgroups of $\overline{\text{SI}}$-groups, *Arch. Math.* 25 (1974), 574-577.

Note added in proof. The answer to Problem 1 is negative in general. We give a sketch of the proof for the class of $\mathfrak{C}\overline{\text{SN}}$-groups. Let G be an $\overline{\text{SN}}$-group with $\mathfrak{C}\overline{\text{SN}}$-subgroups. If H is an $\mathfrak{C}\overline{\text{SN}}$-subgroup of G, then the group G×H is an $\mathfrak{C}\overline{\text{SN}}$-group, since direct products of SN-groups are SN-groups and {e}×H is a normal subgroup of G×H. Hence, we have $H \leqslant G \hookrightarrow G\times H$, where H and G×H are $\mathfrak{C}\overline{\text{SN}}$-groups but G is an $\overline{\text{SN}}$-group. A similar argument yields the result for the classes of $\mathfrak{C}\overline{\text{SI}}$, $\mathfrak{C}\overline{\text{Z}}$-groups. The same construction can be used to derive a negative answer to Problem 5.

More details will appear in a forthcoming paper.

GROUPS COVERED BY ABELIAN SUBGROUPS

M.J. Tomkinson
University of Glasgow, Glasgow G12 8QW, Scotland

Using a theorem of B.H. Neumann [4] on groups which are covered by finitely many cosets, R. Baer (see [5]) showed that a group G has centre of finite index if and only if G is covered by finitely many abelian subgroups. As is usual in this area we use the term 'is covered by' to mean 'is the set-theoretic union of'.

As part of an investigation of a certain graph associated with a group G, P. Erdös asked whether a group G covered by κ abelian subgroups (κ an infinite cardinal) must have centre of bounded index. This question was answered positively by Faber, Laver and McKenzie [2] who showed that

$$|G/Z(G)| \leq 2^{2^{2^\kappa}}.$$

They also showed that the direct product of 2^κ finite groups can be covered by κ abelian subgroups so that the best bound we could hope for is 2^κ.

Although we have not been able to completely close the gap between these two values we will reduce the bound to 2^{2^κ} giving a partial answer to Problem 2 on p.944 of [2] and our results indicate that 2^κ is the correct bound in certain situations.

Our basic result is a consequence of the following Partition Theorem of Erdös, Hajnal and Rado [1]. (See, for example, [6, Theorem 2.2.5].)

Theorem 1. *Let κ be an infinite cardinal and S a set with $|S| > 2^\kappa$ and suppose that the family $[S]^2$ of 2-element subsets of S is expressed as a union of κ subfamilies, $[S]^2 = \cup_{\alpha<\kappa} \Delta_\alpha$. Then there is an $\alpha < \kappa$ and a subset T of S with $|T| = 3$ and $[T]^2 \subseteq \Delta_\alpha$.*

In the original version of this theorem the subset T is constructed with cardinality κ^+ but we shall only require $|T| = 3$. It should be noted that even with $|T| = 3$ the size of S can not be reduced in the hypotheses of this theorem, see [6, Corollary 2.5.2].

Theorem 2. *Let G be covered by κ cosets, $G = \cup_{\alpha<\kappa} H_\alpha x_\alpha$. If $H_\alpha x_\alpha \not\subseteq \cup_{\beta\neq\alpha} H_\beta x_\beta$, then $|G:H_\alpha| \leq 2^\kappa$.*

Proof. Multiplying on the right by x_α^{-1} and writing g_β for $x_\beta x_\alpha^{-1}$, we have

$$G = H_\alpha \cup (\cup_{\beta\neq\alpha} H_\beta g_\beta)$$

and $H_\alpha \not\subseteq \cup_{\beta\neq\alpha} H_\beta g_\beta$.

There is an element $h \in H_\alpha - (\cup_{\beta \neq \alpha} H_\beta g_\beta)$ and

$$G = H_\alpha \cup (\cup_{\beta \neq \alpha} H_\beta g_\beta h^{-1}).$$

Suppose $|G:H_\alpha| > 2^\kappa$; then there is a set S of $(2^\kappa)^+$ distinct coset representatives t_i of H_α in G. The set S can be well-ordered and, if $i < j$, then $t_i t_j^{-1} \notin H_\alpha$ and so there is a least $\beta = \beta(i,j)$ such that $t_i t_j^{-1} \in H_\beta g_\beta h^{-1}$. We define the subsets Δ_β $(\beta \neq \alpha)$ of $[S]^2$ by

$$\Delta_\beta = \{\{t_i,t_j\} : \beta(i,j) = \beta\}.$$

Using theorem 1, there is a 3-element set $T = \{t_i,t_j,t_k\}$ and a $\beta \neq \alpha$ such that

$$t_i t_j^{-1},\ t_i t_k^{-1},\ t_j t_k^{-1} \in H_\beta g_\beta h^{-1}.$$

But then

$$hg_\beta^{-1} = (t_j t_k^{-1} h g_\beta^{-1})(t_i t_k^{-1} h g_\beta^{-1})^{-1}(t_i t_j^{-1} h g_\beta^{-1}) \in H_\beta$$

contrary to $h \notin H_\beta g_\beta$.

Corollary 3. *Let* G *be covered by* κ *abelian subgroups,* $G = \cup_{\alpha<\kappa} A_\alpha$. *If, for each* α, $A_\alpha \not\subseteq \cup_{\beta \neq \alpha} A_\beta$, *then* $|G/Z(G)| \leq 2^\kappa$.

Proof. By Theorem 2, $|G:A_\alpha| \leq 2^\kappa$, for each $\alpha<\kappa$, and so

$$|G:\cap_{\alpha<\kappa} A_\alpha| \leq (2^\kappa)^\kappa = 2^\kappa.$$

The result now follows since $\cap_{\alpha<\kappa} A_\alpha \leq Z(G)$.

When G is covered by κ abelian subgroups it is sometimes possible to arrange for a covering satisfying the conditions of Corollary 3. However this does not seem to be always possible and a slightly more complicated argument is needed to obtain the improved bound on $|G/Z(G)|$.

Theorem 4. *Let* G *be covered by* κ *abelian subgroups,* $G = \cup_{\alpha<\kappa} A_\alpha$. *Then* $|G/Z(G)| \leq 2^{2^\kappa}$.

Proof. For each $x \in G$, let $\sigma(x) = \{\alpha<\kappa : x \in A_\alpha\}$ and let $\Sigma = \{\sigma \subseteq \{\alpha:\alpha<\kappa\} : \sigma = \sigma(x)$ for some $x \in G\}$; then $|\Sigma| \leq 2^\kappa$.

For each $\sigma \in \Sigma$, define $H_\sigma = \langle A_\alpha : \alpha \in \sigma \rangle$ and $A_\sigma = \cap_{\alpha\in\sigma} A_\alpha$. It is then clear that $G = \cup_{\sigma\in\Sigma} A_\sigma$ and that $H_\sigma \leq C_G(A_\sigma)$ so that $Z(G) \leq \cap_{\sigma\in\Sigma} H_\sigma$. Also $G = H_\sigma \cup (\cup_{\alpha\notin\sigma} A_\alpha)$ and, if $\sigma(x) = \sigma$, then $x \in H_\sigma - \cup_{\alpha\notin\sigma} A_\alpha$. By Theorem 2, we have $|G:H_\sigma| \leq 2^\kappa$ and hence

$$|G:\cap_{\sigma\in\Sigma} H_\sigma| \leq (2^\kappa)^{2^\kappa} = 2^{2^\kappa}.$$

Much of the work in [2] relied on the fact that groups covered by κ abelian subgroups have bounded conjugacy classes. More precisely, they proved:

Lemma 5 (Faber, Laver and McKenzie [2, Lemma 3]). *If* G *is covered by* κ *abelian subgroups then* $|G:C_G(x)| \leq \kappa$, *for each* $x \in G$.

It is perhaps surprising that this fact has not been required in the proof of Theorem 4 and it seems quite possible that any improvement in the bound would need this result. Although we have not used it considering $|G/Z(G)|$ we can make use of it to obtain a bound on $|G'|$.

Theorem 6. *Let* G *be covered by* κ *abelian subgroups,* $G = \cup_{\alpha<\kappa} A_\alpha$. *Then* $|G'| \leq 2^\kappa$.

Proof. It follows from Lemma 5 that any set of 2^κ elements is contained in a normal subgroup of cardinality 2^κ. Since $|G:H_\sigma| \leq 2^\kappa$, there is therefore a normal subgroup N_σ of G such that $G = H_\sigma N_\sigma$ and $|N_\sigma| \leq 2^\kappa$.

If $\sigma(x) = \sigma$, then

$$[x,G] = [x,H_\sigma N_\sigma] = [x,N_\sigma] \leq N_\sigma.$$

Thus $G' \leq \cup_{\sigma\in\Sigma} N_\sigma$ and so $|G'| \leq 2^\kappa 2^\kappa = 2^\kappa$.

This result is of course of much less significance than bounds on $|G/Z(G)|$ as there are groups with finite derived subgroups G' equal to Z(G) and $|G|$ arbitrarily large. Some of the relationships between $|G'|$ and $|G/Z(G)|$ are discussed in [3].

REFERENCES

1. P. Erdös, A. Hajnal & R. Rado, Partition relations for cardinal numbers, *Acta Math. Acad. Sci. Hung.* 16 (1965), 93-196.
2. V. Faber, R. Laver & R. McKenzie, Coverings of groups by abelian subgroups, *Canad. J. Math.* 30 (1978), 933-945.
3. V. Faber & M.J. Tomkinson, On theorems of B.H. Neumann concerning FC-groups, II, *Rocky Mountain J. Math.* 13 (1983), 495-506.
4. B.H. Neumann, Groups covered by permutable subsets, *J. London Math. Soc.* 29 (1954), 236-248.
5. B.H. Neumann, Groups covered by finitely many cosets, *Publ. Math. Debrecen* 3 (1954), 227-242.
6. N.H. Williams, *Combinatorial set theory,* Studies in Logic, Vol. 91, North Holland (1977).

EMBEDDINGS OF INFINITE PERMUTATION GROUPS

J.K. Truss
University of Leeds, Leeds LS2 9JT, England

1. *INTRODUCTION*

Throughout this paper I shall be considering permutation groups G acting on a countably infinite set Ω. By an *embedding* of one such permutation group (G_1,Ω_1) into another, (G_2,Ω_2), will be understood a bijection from Ω_1 onto Ω_2 which induces a group monomorphism of G_1 into G_2. This seems the appropriate notion of embedding when considering permutation groups, and is not equivalent to asserting the existence only of a group monomorphism from G_1 into G_2. For example, it was shown in [8] that there is a group monomorphism from $S_{\mathbb{N}}$, the full symmetric group on $\mathbb{N} = \{0,1,2,...\}$ into Aut Γ, the automorphism group of Rado's universal graph. However it is clear that there can be no embedding in the above sense, since Aut Γ has no non-identity elements of finite support. In a similar way one can trivially define a group monomorphism of $S_{\mathbb{N}}$ into $H(\mathbb{Q})$, the group of homeomorphisms of the rationals $\mathbb{Q}$ to themselves, by allowing $S_{\mathbb{N}}$ to act on the clopen (i.e. closed-and-open) sets $\{(n+\pi, n+1+\pi) : n \in \mathbb{Z}\}$ in some natural way, but again, as $H(\mathbb{Q})$ has no non-identity elements of finite support, there can be no (permutation) embedding of $S_{\mathbb{N}}$ into $H(\mathbb{Q})$.

This latter example, of $H(\mathbb{Q})$, together with its evidently high degree of transitivity led P.M. Neumann to propose, in the Kourovka Notebook [4], that if (G,Ω) is countable and highly transitive (i.e. n-fold transitive for every n), then it can be embedded in $H(\mathbb{Q})$ if and only if all non-identity members of G have infinite support. (Actually he originally made this proposal with $H(\mathbb{Q}\times\mathbb{Q})$ in place of $H(\mathbb{Q})$; however $\mathbb{Q}$ and $\mathbb{Q}\times\mathbb{Q}$ are homeomorphic.) In [5] Mekler gave an example of a 3-generator Abelian permutation group all of whose non-identity elements have infinite support and which cannot be embedded in $H(\mathbb{Q})$, and deduced a negative answer to Neumann's question. He did however show that a slight (?) strengthening of the infinite support condition is necessary and sufficient for a countable (G,Ω) to be embeddable in $H(\mathbb{Q})$. We work with a reformulation of his condition which we call "Mekler's Criterion".

In addition to considering the infinite support condition proposed by Neumann, and Mekler's Criterion, we also wish to discuss two strengthenings of the latter which can be similarly handled, one of which we call the "strong Mekler Criterion", and the other the notion of a "sharp" permutation group, due to Hickin. The four conditions on (G,Ω) to be considered are therefore the following.

NC, *Neumann's Criterion:* every non-identity member of G has infinite support.

MC, *Mekler's Criterion:* if $\sigma_1,\dots,\sigma_n$ are finitely many members of G, then $\cap$ {support $\sigma_i : 1 \leq i \leq n$} is empty or infinite.
SMC, *Strong Mekler Criterion:* if $\sigma_1,\dots,\sigma_n$ are finitely many non-identity members of G, then $\cap$ {support $\sigma_i : 1 \leq i \leq n$} is infinite.
SH, *Sharp:* if σ is a non-identity member of G, then Ω-support σ is finite.

(Mekler called the condition he gave in [5] the "mimicking property" which was as follows: if $\sigma_1,\dots,\sigma_n$ are finitely many members of G, and $x \in \Omega$, then $\{y : (\forall i,j)\ (\sigma_i y = \sigma_j y \rightarrow \sigma_i x = \sigma_j x)\}$ is infinite. The reformulation as MC is due to Neumann.)

It is evident that SH $\Rightarrow$ SMC $\Rightarrow$ MC $\Rightarrow$ NC for any group G.

We shall give in §2 a fairly straightforward proof of Mekler's Theorem. The difference between his proof and ours is as follows. In each case the desired embedding is obtained by means of a sequence of approximations. In his case the full structure of $\mathbb{Q}$ is recovered, including the order, metric, and convergence structure. The idea of our proof is to aim only for the "skeleton" of the structure of $\mathbb{Q}$, which is taken to consist of a countable Boolean algebra $\mathbb{B}$ of clopen sets which is a base for its topology. The rest of the structure is then recovered by means of a uniqueness theorem of Sierpinski and Neumann. This approach, apart from being technically simpler, allows us to pinpoint the precise place in the proof where MC is used.

In [8] a study was made of various properties of Aut Γ, the automorphism group of Rado's universal graph. The method used to prove Mekler's Theorem also allows us to deduce quite easily that Aut Γ can be embedded in $H(\mathbb{Q})$. We have been unable to determine whether or not the same is true of AAut Γ, the group of "almost automorphisms" of Γ, though we can show that AAut Γ fulfils SMC, and hence MC. (We cannot immediately deduce the embeddability of AAut Γ in $H(\mathbb{Q})$ from Mekler's Theorem because AAut Γ is uncountable.) Indeed this is our main reason for wishing to consider SMC.

An interesting biproduct of Mekler's work is some information on extensions of permutation groups to highly transitive permutation groups. Specifically, he shows that if (G,Ω) is countable and fulfils NC then (G,Ω) can be embedded in a highly transitive permutation group on Ω also fulfilling NC. The same result for MC easily follows from the main theorem, and we prove in §3, using a method similar to Mekler's, that it also holds for SMC and SH.

Are there "naturally occurring" permutation groups (G_1,Ω), (G_2,Ω) fulfilling the same rôle as regards NC, SMC and SH as $H(\mathbb{Q})$ does for MC? At present we can only show that they exist assuming the continuum hypothesis (or Martin's Axiom will do), and so they can hardly be said to be "naturally occurring". The key idea here is to use the techniques mentioned in the previous paragraph to prove the relevant joint embedding property. This also raises the question as to whether $H(\mathbb{Q})$ is the only permutation group which will serve in Mekler's Theorem. At any rate the proof we supply may be taken as evidence that $H(\mathbb{Q})$ is at least the most *natural* such group.

A final word on cycle types. One of the main results of [6] gives a classification of the cycle types occurring in $H(\mathbb{Q})$. A similar analysis was carried out in [8] for Aut Γ, and it turns out that the possible cycle types of members of Aut Γ forms a proper subset of those of H($\mathbb{Q}$). One may regard the problem of classifying cycle types as that of asking which cyclic permutation groups can be embedded in H($\mathbb{Q}$) or Aut Γ. Thus the cycle types occurring in H($\mathbb{Q}$) may be derived from Mekler's Theorem. In a similar way we are able to classify the cycle types occurring in groups fulfilling NC, SMC and SH. We shall see that there is a natural progression from NC to MC, to SMC, to SH, and also from SMC to Aut Γ. This immediately enables us to deduce some *non*-embeddability results. In addition we have some information on the cycle types of AAut Γ, sufficient for us to show that its cycle types lie strictly between those of Aut Γ and an arbitrary group fulfilling SMC. Thus it will not serve as a "universal" permutation group for countable groups fulfilling SMC.

Further information on AAut Γ, including a proof that it is not simple, will be given in a forthcoming paper.

I would like to thank Peter Neumann for introducing me to the topics discussed in this paper, and for acquainting me with the papers [5] and [6].

2. *UNIQUENESS OF THE RATIONAL TOPOLOGY, AND PROOF OF MEKLER'S THEOREM*

The proof we give of Mekler's Theorem is based on a result of Sierpinski [7] as modified by Neumann in [6]. In a similar spirit to Mekler's proof, Neumann proves the uniqueness of the usual topology on $\mathbb{Q}$, subject to certain conditions, by explicitly reconstructing the metric. To give the flavour of the approach here advocated, we shall reprove this result by another method. The essential point about the usual topology on $\mathbb{Q}$ is that it has a countable base $\mathbb{B}$ of clopen sets such that $\mathbb{B}$ is a Boolean algebra (under the usual operations of $\cup$, $\cap$, and complementation in $\mathbb{Q}$) all of whose non-empty members are infinite, and such that for any distinct rationals x and y there is $b \in \mathbb{B}$ with $x \in b$ & $y \notin b$. For example we may take $\mathbb{B}$ to be the Boolean algebra generated by the set of intervals of the form $(\pi + a, \pi + b) \cap \mathbb{Q}$ for $a,b \in \mathbb{Q}$. We show that the structure just described is unique up to isomorphism.

Theorem 2.1. *Suppose that* $\mathbb{B}_1$ *and* $\mathbb{B}_2$ *are countable Boolean algebras satisfying the following.*

(i) $\mathbb{B}_i$ *is a Boolean algebra of subsets of the countable set* X_i.

(ii) *All members of* $\mathbb{B}_i$ *are empty or infinite.*

(iii) *If* x, y *are distinct members of* X_i *then there is* $b \in \mathbb{B}_i$ *such that* $x \in b$ & $y \notin b$.

Then there is a bijection θ *from* X_1 *onto* X_2 *which induces an isomorphism from* $\mathbb{B}_1$ *onto* $\mathbb{B}_2$.

Proof. Here and later we shall adopt the terminology of forcing; partly because that is how we think of it, and partly because this will facilitate the application of Martin's Axiom.

A *condition* is a pair (p,q) such that p is a 1-1 map from a finite subset of X_1 into X_2, q is an isomorphism from a finite subalgebra of $\mathbb{B}_1$

onto a finite subalgebra of $\mathbb{B}_2$, and

$$x \in \operatorname{dom} p \;\&\; b \in \operatorname{dom} q \;\&\; x \in b \rightarrow px \in qb.$$

The idea is that p approximates the bijection θ, and q ensures that p (so far) preserves the Boolean algebra structure. P is the set of all conditions, partially ordered by co-ordinatewise extension.

We show that given any condition (p,q) and $x_i \in X_i$, $b_i \in \mathbb{B}_i$ $(i = 1,2)$ there is an extension (p',q') of (p,q) such that $x_1 \in \operatorname{dom} p'$, $x_2 \in \operatorname{range} p'$, $b_1 \in \operatorname{dom} q'$, $b_2 \in \operatorname{range} q'$. Since we may successively extend, it suffices to do each of these separately. We just consider $i = 1$, the case in which $i = 2$ being handled similarly.

If $x_1 \in \operatorname{dom} p$, no extension is necessary here. Otherwise observe that dom q, being finite, is completely atomic, so that x lies in a unique atom b of dom q. Then qb is an atom of range q and since all non-empty members of $\mathbb{B}_2$ are infinite, qb is infinite, and we may let y be a member of qb not in range p. The desired extension of (p,q) is then $(p \cup \{(x,y)\}, q)$.

If $b_1 \in \operatorname{dom} q$, again no extension is necessary. Otherwise let dom q' be $\mathbb{C}$, the subalgebra of $\mathbb{B}_1$ generated by $\operatorname{dom} q \cup \{b_1\}$. Let $\{c_1, c_2, \ldots, c_n\}$ be the set of atoms of dom q. Then $\mathbb{C}$ is the set of unions of members of the set

$$\{c_1 \cap b_1,\ c_2 \cap b_1,\ \ldots,\ c_n \cap b_1,\ c_1 - b_1,\ c_2 - b_1,\ \ldots,\ c_n - b_1\}.$$

If we specify $q'(c_i \cap b_1)$ and $q'(c_i - b_1)$ for each i, as disjoint sets with union $q(c_i)$, then q' immediately extends to the whole of $\mathbb{C}$. We choose an extension p' of p, as in the previous paragraph, if necessary, so that whenever $c_i \cap b_1$ and $c_i - b_1$ are both non-empty, so are $c_i \cap b_1 \cap \operatorname{dom} p'$ and $(c_i - b_1) \cap \operatorname{dom} p'$.

If $c_i \subseteq b_1$ or $c_i \cap b_1 = \emptyset$, then $c_i \cap b_1$ and $c_i - b_1$ already lie in dom q, and no new definition is required. Otherwise $c_i \cap b_1$ and $c_i - b_1$ are both infinite, since they are non-empty members of $\mathbb{B}_1$. We have to find a subset d of $q(c_i)$ such that d and $q(c_i) - d$ are both infinite and lie in $\mathbb{B}_2$, and such that p' maps $c_i \cap b_1 \cap \operatorname{dom} p'$ into d and $(c_i - b_1) \cap \operatorname{dom} p'$ into $q(c_i) - d$. We shall then be able to let $q'(c_i \cap b_1) = d$ and $q'(c_i - b_1) = q(c_i) - d$.

Let $\{x_1, \ldots, x_k\}$ and $\{y_1, \ldots, y_\ell\}$ be the images of $c_i \cap b \cap \operatorname{dom} p'$ and $(c_i - b) \cap \operatorname{dom} p'$ respectively, under p'. By the above proviso, $k, \ell \geq 1$. Since p' is 1-1, $x_i \neq y_j$, all i, j. By the hypothesis on $\mathbb{B}_2$ there is $d_{ij} \in \mathbb{B}_2$ containing x_i but not y_j. Then

$$d = \left(\bigcup_{i=1}^{k} \bigcap_{j=1}^{\ell} d_{ij} \right) \cap q(c_i)$$

is a member of $\mathbb{B}_2$ contained in $q(c_i)$ containing all x_i's and no y_j's, as required.

A standard back-and-forth argument now completes the proof. Namely we form an increasing sequence $(p_0,q_0) < (p_1,q_1) < (p_2,q_2) < \ldots$ of conditions so that for every $x_1 \in X_1$, $x_2 \in X_2$, $b_1 \in \mathbb{B}_1$, $b_2 \in \mathbb{B}_2$ there is n such that $x_1 \in \operatorname{dom} p_n$, $x_2 \in \operatorname{range} p_n$, $b_1 \in \operatorname{dom} q_n$, $b_2 \in \operatorname{range} q_n$. This is possible in view of the countability conditions on X_i and $\mathbb{B}_i$. The union $\theta = \bigcup_{n \in \mathbb{N}} p_n$ is then a bijection from X_1 onto X_2, and it induces an isomorphism $\phi = \bigcup_{n \in \mathbb{N}} q_n$ from $\mathbb{B}_1$ onto $\mathbb{B}_2$. For if $b \in \mathbb{B}_1$, it follows from the definition of "condition" that $\theta(b) \subseteq \phi(b)$, and since $\theta(X_1 - b) \subseteq \phi(X_1 - b) = X_2 - \phi(b)$ and $\theta(b) \cap \theta(X_1 - b) = \emptyset$, $\theta(b) = \phi(b)$.

Corollary 2.2 (Sierpinski, Neumann [6]). *Let (X,τ) be a second countable, 0-dimensional, T_1 topological space with no isolated points and with X countable. Then (X,τ) is homeomorphic with $\mathbb{Q}$.*

Proof. We let $\mathbb{B}_1$ be the Boolean algebra generated by a countable base of clopen sets for X. Saying that X has no isolated points is equivalent to saying that all non-empty members of $\mathbb{B}_1$ are infinite. Letting $\mathbb{B}_2$ be the corresponding Boolean algebra for the topology on $\mathbb{Q}$, derived from a countable base for the clopen sets, it follows from the theorem that there is an isomorphism between $\mathbb{B}_1$ and $\mathbb{B}_2$ induced by a bijection from X onto $\mathbb{Q}$. This bijection is the desired homeomorphism.

Theorem 2.3 (Mekler [5]). *The countable permutation group (G,Ω) can be embedded in $H(\mathbb{Q})$ if and only if it fulfils MC.*

Proof. If $G \leqslant H(\mathbb{Q})$ then G certainly fulfils MC, since the support of any homeomorphism is an open set in $\mathbb{Q}$, and any open set is empty or infinite.

Conversely we suppose that (G,Ω) fulfils MC and construct an embedding of (G,Ω) into $H(\mathbb{Q})$. Let the distinct pairs of members of Ω be enumerated as $\{(x_k,y_k) : k \in \mathbb{N}\}$. We shall show that there is a sequence $(\mathbb{B}_k : k \in \mathbb{N})$ of finite or countably infinite Boolean algebras of subsets of Ω such that

(I) $X \in \mathbb{B}_k \;\&\; \sigma \in G \rightarrow \sigma X \in \mathbb{B}_k$.

(II) $\mathbb{B}_0 \subseteq \mathbb{B}_1 \subseteq \mathbb{B}_2 \subseteq \ldots$.

(III) there is $X \in \mathbb{B}_{k+1}$ such that $x_k \in X \;\&\; y_k \notin X$.

(IV) if $b \in \mathbb{B}_k$ and $\sigma_1,\ldots,\sigma_n \in G$ then $b \cap \bigcap_{i=1}^{n}$ support σ_i is empty or infinite (in which case we say that b *fulfils MC*).

Suppose this has been established, and let $\mathbb{B} = \bigcup_{k \in \mathbb{N}} \mathbb{B}_k$. Since the vacuous case $n = 0$ will be included in (IV), it follows that all non-empty members of $\mathbb{B}$ are infinite, and $\mathbb{B}$ fulfils (i), (ii) and (iii) of Theorem 2.1. Hence the topology having the members of $\mathbb{B}$ as a base for its open sets is homeomorphic to that on $\mathbb{Q}$ and we obtain a corresponding bijection from Ω onto $\mathbb{Q}$. Since every member of G preserves $\mathbb{B}$, this bijection induces an embedding of G into $H(\mathbb{Q})$.

To arrange for (I)-(IV) to hold we define $\mathbb{B}_k$ inductively. We let $\mathbb{B}_0 = \{\emptyset,\Omega\}$. Then (I) holds trivially for $\mathbb{B}_0$, and (IV) is just MC, which we are assuming.

Now suppose that $\mathbb{B}_k$ has been defined, satisfying (I)-(IV). $\mathbb{B}_{k+1}$ will be the Boolean algebra of subsets of Ω generated by $\mathbb{B}_k$ and $\{\sigma X : \sigma \in G\}$, where X is a single subset of Ω containing x_k but not y_k, yet to be defined.

That (I) and (III) hold is immediate. The main problem is to choose X in such a way that all members of $\mathbb{B}_{k+1}$ fulfil MC.

Now a general non-empty member of $\mathbb{B}_{k+1}$ may be written in disjunctive normal form; i.e. it will be a finite disjunction (union) of conjuncts (intersections) of generators or their complements. To show that such an element fulfils MC it is sufficient to show that each of the conjuncts does. Such a conjunct will be of the form

$$c = b \cap \bigcap_{i=1}^{\ell} \sigma_i X \cap \bigcap_{j=\ell+1}^{m} (\Omega - \sigma_j X)$$

for some $b \in \mathbb{B}_k$ and (distinct) $\sigma_1, \ldots, \sigma_\ell, \sigma_{\ell+1}, \ldots, \sigma_m \in G$. Our task is therefore to choose X in such a way that all non-empty sets of this form fulfil MC.

The idea is to build up the sets X and $\Omega - X$ in stages, finitely many of each at a time. We shall ensure that if the set c ever becomes non-empty, then it keeps on growing. Thus we must make sure that we keep on coming back to it. In addition we must ensure that every member of Ω is assigned either to X or to $\Omega - X$ at some stage. We therefore consider two enumerations. Firstly, let all quadruples of the form

$$(b, (\sigma_1, \ldots, \sigma_\ell), (\sigma_{\ell+1}, \ldots, \sigma_m), (\sigma'_1, \ldots, \sigma'_r))$$

where $b \in \mathbb{B}_k$ and $\sigma_i, \sigma'_i \in G$ be listed in sequence so that each occurs infinitely often. We take $\sigma_1, \ldots, \sigma_m$ to be distinct, and also $\sigma'_1, \ldots, \sigma'_r$, but there may be an overlap between the two lists. Also, as indicated above, the cases where $\ell = 0$, $m - \ell = 0$ or $r = 0$ are not excluded. Since $\mathbb{B}_k$ and G are countable this is possible. Secondly we let $\{z_n : n \in \mathbb{N}\}$ be an enumeration of Ω.

We choose finite sets $p_0 \subseteq p_1 \subseteq p_2 \subseteq \ldots$ and $q_0 \subseteq q_1 \subseteq q_2 \subseteq \ldots$ by induction, so that $p_n \cap q_n = \emptyset$, all n. (p_n, q_n) are "conditions" approximating $(X, \Omega\text{-}X)$ in a rather similar manner to Theorem 2.1. In this case the back-and-forth aspect of the construction is being spelt out in rather more detail. We shall ensure that $z_n \in p_{2n+1} \cup q_{2n+1}$, so that $X = \bigcup_{n \in \mathbb{N}} p_n$ and $Y = \bigcup_{n \in \mathbb{N}} q_n$ are disjoint sets whose union is Ω.

Let $p_0 = \{x_k\}$ and $q_0 = \{y_k\}$. Assume p_{2n} and q_{2n} have been chosen. If $z_n \in p_{2n} \cup q_{2n}$, let $p_{2n+1} = p_{2n}$ and $q_{2n+1} = q_{2n}$. If $z_n \notin p_{2n} \cup q_{2n}$, let $p_{2n+1} = p_{2n} \cup \{z_n\}$ and $q_{2n+1} = q_{2n}$.

Assume p_{2n+1} and q_{2n+1} have been chosen. Let

$$(b, (\sigma_1, \ldots, \sigma_\ell), (\sigma_{\ell+1}, \ldots, \sigma_m), (\sigma'_1, \ldots, \sigma'_r))$$

be the nth quadruple in the listing above, and let c be the finite set

$$b \cap \bigcap_{i=1}^{\ell} \sigma_i\, p_{2n+1} \cap \bigcap_{j=\ell+1}^{m} \sigma_j\, q_{2n+1} \cap \bigcap_{i=1}^{r} \text{support } \sigma'_i.$$

If $c = \emptyset$, let $p_{2n+2} = p_{2n+1}$ and $q_{2n+2} = q_{2n+1}$. If $c \neq \emptyset$, let $x \in c$. Then for $1 \leq i \leq \ell$, $\sigma_i^{-1} x \in p_{2n+1}$, and for $\ell + 1 \leq j \leq m$, $\sigma_j^{-1} x \in q_{2n+1}$. Hence $\sigma_i^{-1} x \neq \sigma_j^{-1} x$ and $\sigma_j \sigma_i^{-1} x \neq x$. It follows that $x \in d$, where

$$d = b \cap \bigcap \{\text{support}(\sigma_j\sigma_i^{-1}) : 1 \leq i \leq \ell,\ \ell + 1 \leq j \leq m\} \cap \bigcap_{i=1}^{r} \text{support } \sigma_i'.$$

Now by (IV) for $\mathbb{B}_k$, MC holds for b, so d is infinite. Therefore there is some y in d but not in

$$c \cup \bigcup_{i=1}^{\ell} \sigma_i q_{2n+1} \cup \bigcup_{j=\ell+1}^{m} \sigma_j p_{2n+1}.$$

We let $p_{2n+2} = p_{2n+1} \cup \{\sigma_i^{-1}y : 1 \leq i \leq \ell\}$ and $q_{2n+2} = q_{2n+1} \cup \{\sigma_j^{-1}y : \ell+1 \leq i \leq m\}$. Then p_{2n+2} and q_{2n+2} are finite sets containing p_{2n+1} and q_{2n+1} respectively. We show they are disjoint. That $p_{2n+1} \cap q_{2n+1} = \emptyset$ is given by the induction hypothesis. From $y \notin \sigma_i q_{2n+1}$ we deduce that $\sigma_i^{-1}y \notin q_{2n+1}$ and from $y \notin \sigma_j p_{2n+1}$ we deduce that $\sigma_j^{-1}y \notin p_{2n+1}$. Finally, $\sigma_i^{-1}y \neq \sigma_j^{-1}y$ holds since $y \in \text{support}(\sigma_j\sigma_i^{-1})$.

The effect of this construction is as follows. If

$$b \cap \bigcap_{i=1}^{\ell} \sigma_i X \cap \bigcap_{j=\ell+1}^{m} \sigma_j Y \cap \bigcap_{i=1}^{r} \text{support } \sigma_i'$$

is non-empty, then an element of it must appear at some stage of the construction of X and Y. From that point on, a new element must be added to it each time $(b, (\sigma_1,\ldots,\sigma_\ell), (\sigma_{\ell+1},\ldots,\sigma_m), (\sigma_1',\ldots,\sigma_r'))$ is enumerated in the sequence of quadruples, so it must be infinite.

This concludes the definition of p_n and q_n for all n, and hence of X and Y. Clearly $x_k \in X$ and $y_k \notin X$, and as remarked in the previous paragraph,

$$b \cap \bigcap_{i=1}^{\ell} \sigma_i X \cap \bigcap_{j=\ell+1}^{m} \sigma_j Y \cap \bigcap_{i=1}^{r} \text{support } \sigma_i'$$

is either empty or infinite. Hence, MC holds for each set of the form

$$b \cap \bigcap_{i=1}^{\ell} \sigma_i X \cap \bigcap_{j=\ell+1}^{m} \sigma_j Y = b \cap \bigcap_{i=1}^{\ell} \sigma_i X \cap \bigcap_{j=\ell+1}^{m} (\Omega - \sigma_j X),$$

and hence for every member of $\mathbb{B}_{k+1}$. This concludes the proof.

Let us remark that it would be possible to perform the construction above in a single induction, producing sets X_k such that $\mathbb{B}_{k+1}$ = the algebra generated by $\mathbb{B}_k$ and $\{\sigma X_k : \sigma \in G\}$ all in one go. The technical details however become considerably more complicated, and we prefer to use the double induction adopted above.

3. *OTHER UNIVERSAL PERMUTATION GROUPS*

Let us refer to a permutation group (G,Ω) as *universal* for a class $\mathfrak{G}$ of permutation groups if the countable permutation groups which can be embedded in (G,Ω) are precisely those lying in $\mathfrak{G}$. Thus Mekler's Theorem states that $H(\mathbb{Q})$ is universal for the class of countable groups fulfilling MC. It is not clear whether $H(\mathbb{Q})$ is the only group with this property, but at least the proof of Theorem 2.3 shows it to be a very natural such group. We now turn to the other properties NC, SMC and SH and show that they also have universal groups. Unfortunately our proof requires the continuum hypothesis or Martin's Axiom. It would be greatly preferable to have "naturally occurring" representations for such groups, similar to $H(\mathbb{Q})$, which would also avoid the use of extra set-theoretical hypotheses. The key property is "joint embedding" for the three classes, which we prove by a technique similar to that of [5 Lemma 2.3].

Theorem 3.1. *Let* (G_1,Ω_1) *and* (G_2,Ω_2) *be countable permutation groups, both fulfilling* SMC. *Then there is a bijection* θ *from* Ω_1 *onto* Ω_2 *such that the group generated by* $G_1 \cup \theta^{-1} G_2\theta$ *also fulfils* SMC.

Remark. $\theta^{-1}G_2\theta$ is a permutation group on Ω_1 isomorphic (qua permutation group) to G_2. Thus this result essentially says that two groups fulfilling SMC can be jointly embedded in a third.

Proof. In fact we show that θ can be chosen so that the group thus generated "is" the free product $G_1 * G_2$, i.e. the canonical homomorphism from $G_1 * G_2$ onto the group generated by G_1 and $\theta^{-1}G_2\theta$ is bijective. Recall that the free product $G_1 * G_2$ consists of the set of reduced words, i.e. finite strings of non-identity members of $G_1 \cup G_2$ with entries alternating from G_1 and G_2. The empty word is denoted by Λ (and this is the identity of $G_1 * G_2$).

If p is a 1-1 map from a subset of Ω_1 into Ω_2 and w is a reduced word in $G_1 * G_2$ we may determine a 1-1 map from a subset of Ω_1 into Ω_2, denoted by w^p, by $w^p = \sigma_1^p\sigma_2^p \dots \sigma_n^p$, where $w = \sigma_1\sigma_2 \dots \sigma_n$ and $\sigma_i^p = \sigma_i$ or $p^{-1}\sigma_i p$ according as $\sigma_i \in G_1$ or G_2. If θ is a bijection from Ω_1 onto Ω_2 then w^θ is a permutation of Ω_1. The idea is that if p is an approximation to θ, w^p gives as much information as possible about w^θ.

Let P be the set of 1-1 maps from a finite subset of Ω_1 into Ω_2. Let $p \in P$ and let T be a finite tree of reduced words of $G_1 * G_2$. This means that if $\sigma w \in T$ then $w \in T$. Observe that it follows from this that $\Lambda \in T$. We show that there is an extension q of p in P and $x \in \mathrm{dom}\, q - \mathrm{dom}\, p$ such that for every non-identity $w \in T$, $w^q x$ is defined and $w^q x \neq x$.

Let $<$ be a linear ordering on T extending the tree ordering, i.e. such that if $w = \sigma v$ then $v < w$. We define x_w, y_w for $w \in T$ by induction on the position of w in this ordering. In doing this we take account of which of G_1 and G_2 the first entry of w lies in.

If w is the first element in the ordering, then $w = \Lambda$. Let Σ_i be the set of all (non-identity) members of G_i occurring in a word in T and let $\Sigma_i^* = \{\tau^{-1}\sigma : \sigma, \tau \in \Sigma_i, \sigma \neq \tau\} \cup \Sigma_i$. Choices of x_Λ and y_Λ are then made in $\bigcap \{\mathrm{support}\, \sigma : \sigma \in \Sigma_1^*\} - (\mathrm{dom}\, p \cup \bigcup\{\sigma^{-1}\mathrm{dom}\, p : \sigma \in \Sigma_1\})$ and $\bigcap \{\mathrm{support}\, \tau : \tau \in \Sigma_2^*\} - (\mathrm{range}\, p \cup \bigcup\{\tau^{-1}\mathrm{range}\, p : \tau \in \Sigma_2\})$ respectively. These sets are non-empty (infinite actually) by SMC.

Now suppose that x_v, y_v have been defined for all $v < w$ in the ordering.

Case 1. $w = \sigma v$, $\sigma \in G_1$.

We let $x_w = \sigma x_v$ and we let y_w be some member of

$$\bigcap \{\mathrm{support}\, \tau : \tau \in \Sigma_2^*\} - [\mathrm{range}\, p \cup \bigcup\{\tau^{-1}\mathrm{range}\, p : \tau \in \Sigma_1\}$$

$$\cup \{y_u : u < w\} \cup \{\tau^{-1}y_u : \tau \in \Sigma_2,\ u < w\} \cup \{\tau y_u : \tau \in \Sigma_2^*,\ u < w\}].$$

This set is non-empty by SMC.

Case 2. $w = \tau v$, $\tau \in G_2$.

Similarly we let x_w be some member of

$$\bigcap \{\text{support } \sigma : \sigma \in \Sigma_1^*\} - [\text{dom } p \cup \bigcup \{\sigma^{-1} \text{dom } p : \sigma \in \Sigma_1\}$$

$$\cup \{x_u : u < w\} \cup \{\sigma^{-1} x_u : \sigma \in \Sigma_1,\ u < w\} \cup \{\sigma x_u : \sigma \in \Sigma_1^*,\ u < w\}]$$

and $y_w = \tau y_v$.

Finally $q = p \cup \{(x_w, y_w) : w \in T\}$.

We check firstly that $x_w \notin \text{dom } p$, $y_w \notin \text{range } p$, and $w_1 \neq w_2 \to x_{w_1} \neq x_{w_2}$ & $y_{w_1} \neq y_{w_2}$. This will show that q is a 1-1 function.

If $w = \Lambda$, $x_w \notin \text{dom } p$ and $y_w \notin \text{range } p$ are immediate from the definition. If Case 1 applies, $y_w \notin \text{range } p$ is also immediate. Also $x_w = \sigma x_v$, and since the entries of w are alternating from G_1 and G_2, $v = \Lambda$ or Case 2 will be applicable to the definition of x_v. In each of these instances $x_v \notin \sigma^{-1} \text{dom } p$, so $x_w = \sigma x_v \notin \text{dom } p$. A similar proof works if Case 2 applies to the definition of x_w, y_w.

Now suppose that $x_{w_1} = x_{w_2}$ where $w_1 < w_2$. Examining the relevant clause of Case 2 we see that Case 1 must apply to the definition of x_{w_2}. Letting $w_2 = \sigma_2 v_2$ with $\sigma_2 \in \Sigma_1$, $x_{w_2} = \sigma_2 w_{v_2}$. Therefore $x_{v_2} = \sigma_2^{-1} x_{w_2} = \sigma_2^{-1} x_{w_1}$ so $v_2 \leqslant w_1$ (since otherwise this would contradict the choice of x_{v_2} according to Case 2). Since $x_{v_2} \in \text{support } \sigma_2$, $v_2 \neq w_1$, and it follows that $v_2 < w_1$. Again Case 1 must apply to the definition of x_{w_1}, so $w_1 = \sigma_1 v_1$ where $\sigma_1 \in \Sigma_1$, and $x_{w_1} = \sigma_1 x_{v_1}$. Thus $\sigma_1 x_{v_1} = \sigma_2 x_{v_2}$ and $\sigma_2^{-1} \sigma_1 x_{v_1} = x_{v_2}$. If $v_1 = v_2$, since $x_{v_1} \in \bigcap \{\text{support } \sigma : \sigma \in \Sigma_1^*\}$ we must have $\sigma_2^{-1} \sigma_1 \notin \Sigma_1^*$, i.e. $\sigma_1 = \sigma_2$. This gives $\sigma_1 v_1 = \sigma_2 v_2$ and $w_1 = w_2$, a contradiction. Otherwise suppose for example that $v_1 < v_2$. Thus Case 2 must apply to the definition of x_{v_2} and this is again a contradiction, since $\sigma_2^{-1} \sigma_1 \neq 1$ and $\sigma_2^{-1} \sigma_1$ thus lies in Σ_1^*.

Similarly $w_1 \neq w_2$ implies that $y_{w_1} \neq y_{w_2}$.

Next we show that for each $w \in T$, $w^q x_\Lambda = x_w$, by induction on the length of w. If $w = \Lambda$ this is immediate.

Suppose $w = \sigma v$, $\sigma \in \Sigma_1$. Then $w^q x_\Lambda = (\sigma v)^q x_\Lambda = \sigma v^q x_\Lambda = \sigma x_v = x_{\sigma v} = x_w$. Suppose $w = \tau v$, $\tau \in \Sigma_2$. Then $w^q x_\Lambda = (\tau v)^q x_\Lambda = q^{-1} \tau q v^q x_\Lambda = q^{-1} \tau q x_v = q^{-1} \tau y_v = q^{-1} y_{\tau v} = q^{-1} y_w = x_w$.

Now since all x_w's are distinct, as shown above, $x_w \neq x_\Lambda$ for each non-identity w in T. Hence $w^q x_\Lambda \neq x_\Lambda$, so q and $x = x_\Lambda$ are therefore as desired.

Let us enumerate all finite trees of reduced words so that each occurs infinitely often, and also all members of Ω_1, Ω_2. We form a sequence (p_n) of members of P. At the 2n-th stage we ensure that the nth members of Ω_1, Ω_2 are in $\text{dom } p_{2n}$, $\text{range } p_{2n}$ respectively. At the (2n+1)th stage we ensure that an extra element is adjoined to $\bigcap \{\text{support } w^{p_{2n+1}} : w \neq 1,\ w \in T_n\}$ where T_n is the nth tree in the listing. The union $\theta = \bigcup_{n \in \mathbb{N}} p_n$ is thus a bijection from Ω_1 onto Ω_2, and for each T, $\bigcap \{\text{support } w^\theta : w \neq 1,\ w \in T\}$ is infinite.

Clearly any finite subset of $G_1 * G_2$ can be extended to a finite tree of reduced words, so SMC is verified for the action of $G_1 * G_2$ given by $w \to w^{\theta}$. This action is faithful and is generated by $G_1 \cup \theta^{-1}G_2\theta$.

Theorem 3.2 (CH). *There is a permutation group* (G,Ω) *such that a countable permutation group can be embedded in* (G,Ω) *if and only if it fulfils* SMC.

Proof. Let Ω be a fixed set of cardinality $\aleph_0$. There are $2^{\aleph_0}$ permutations of Ω, and hence $\leq (2^{\aleph_0})^{\aleph_0} = 2^{\aleph_0}$ countable permutation groups on Ω. Assuming the continuum hypothesis we may therefore enumerate all countable permutation groups on Ω fulfilling SMC as $(G_\alpha : \alpha < \omega_1)$. Define H_α by transfinite induction by

$H_0 = \{1\}$

$H_{\alpha+1}$ = the group of permutations of Ω generated by H_α and $\theta^{-1}G_\alpha\theta$ where $\theta : \Omega \to \Omega$ is some bijection for which this group fulfils SMC.

$H_\lambda = \bigcup_{\alpha<\lambda} H_\alpha$ (λ a limit ordinal).

It is clear that each H_α for $\alpha \leq \omega_1$ fulfils SMC and that each H_α for $\alpha < \omega_1$ is countable. We let $G = H_{\omega_1}$. Then any countable permutation group fulfilling SMC is permutation isomorphic to some G_α, so can be embedded in G. Conversely, since G itself fulfils SMC, so does any subgroup.

We now analyse the proof of Theorem 3.1 with a little more care, in the manner of [5], so as to be able to derive the conclusion of Theorem 3.2 from Martin's Axiom (a weakening of the continuum hypothesis compatible with $2^{\aleph_0} = \aleph_2, \aleph_3, \ldots$; see [3]).

Let (G_1,Ω_1) and (G_2,Ω_2) be permutation groups fulfilling SMC, of cardinality $\leq \kappa$. Let P, as before, be the set of 1-1 maps from a finite subset of Ω_1 into Ω_2. For each $x \in \Omega_1$, $y \in \Omega_2$ let

$$D_x = \{p \in P : x \in \mathrm{dom}\ p\} \text{ and } D'_y = \{p \in P : y \in \mathrm{range}\ p\}.$$

Then clearly D_x and D'_y are dense open subsets of P. We also let $\mathfrak{T}$ be the set of finite trees of reduced words in $G_1 * G_2$, and for $T \in \mathfrak{T}$ and $n \in \mathbb{N}$, let

$$D_{T,n} = \{p \in P : |\{x : (\forall w \in T)(w \neq \Lambda \to w^p x \neq x)\}| \geq n\}.$$

The proof of 3.1 shows that each $D_{T,n}$ is also dense open.

Martin's Axiom (MA) asserts that for any c.c.c. partial ordering P, and for any family $\mathfrak{D}$ of less than $2^{\aleph_0}$ dense open subsets of P there is a filter $\mathcal{F}$ on P which meets every member of $\mathfrak{D}$. In this case the partial ordering is actually countable, not merely c.c.c., so we may if we wish appeal to a weaker hypothesis to achieve our conclusion. We recall that a subset of the real line is said to be *meagre* (or of *first category*) if it is a countable union of nowhere dense sets. It follows easily from MA that any union of less than $2^{\aleph_0}$ meagre sets is meagre, and hence, by the Baire category theorem, is not equal to $\mathbb{R}$. For an account of present knowledge of the strength of the statement "$\mathbb{R}$ is not the union of κ meagre sets" see [2].

Theorem 3.3. *Let* (G_1,Ω_1) *and* (G_2,Ω_2) *be permutation groups of cardinality* $\leq \kappa$, *both fulfilling* SMC. *Assume that* $\mathbb{R}$ *is not the union of* κ *meagre sets. Then there is a bijection* θ *from* Ω_1 *onto* Ω_2 *such that the group generated by* $G_1 \cup \theta^{-1}G_2\theta$ *also fulfils* SMC.

Theorem 3.4. *Assume that* $\mathbb{R}$ *is not the union of less than* $2^{\aleph_0}$ *meagre sets. Then there is a universal permutation group for countable groups fulfilling* SMC.

Proof. Letting $2^{\aleph_0} = \aleph_\alpha$, we may again enumerate the countable permutation groups on Ω fulfilling SMC as $(G_\beta : \beta < \aleph_\alpha)$. H_β are defined as before, using Theorem 3.3 in place of Theorem 3.1.

With the full Martin's Axiom we can actually do rather better.

Theorem 3.5 (MA). *There is a universal permutation group for groups of cardinality less than* $2^{\aleph_0}$ *fulfilling* SMC.

Proof. The additional point here is that $\aleph_\alpha < 2^{\aleph_0} \rightarrow 2^{\aleph_\alpha} = 2^{\aleph_0}$ is a well-known consequence of Martin's Axiom, so that there are $\leq (2^{\aleph_0})^{< 2^{\aleph_0}} = 2^{\aleph_0}$ permutation groups on Ω of cardinality less than $2^{\aleph_0}$ fulfilling SMC, and they can all be incorporated into the sequence (G_β).

We now briefly consider NC, the condition that every non-identity member of (G,Ω) has infinite support. We remark, without going into all the details, that Theorems 3.1 - 5 carry through in this case too. Theorem 3.1 here is to all intents and purposes the same as [5 Lemma 2.3] (except that there G_2 is actually an infinite cyclic group). It is less involved in that one only needs to consider one word at a time, so the "tree" is already linearly ordered. In a similar way, these results may be proved for SH too.

We conclude this section by pursuing a little the extensions of permutation groups to highly transitive groups. This is the substance of Lemma 2.3 and Theorem 2.4 of [5], where it is shown that any countable permutation group fulfilling NC can be extended to a highly transitive permutation group fulfilling NC. We derive the same result for MC, SMC and SH.

Theorem 3.6. *Any countable permutation group fulfilling* MC *can be extended to a (countable) highly transitive permutation group fulfilling* MC.

Proof. We recall that (G,Ω) is "highly transitive" if it is n-fold transitive for all n. The result follows easily from Theorem 2.3, since we may assume that $G \leq H(\mathbb{Q})$. Now $H(\mathbb{Q})$ is itself obviously highly transitive, so it has a countable highly transitive subgroup. Namely for each pair $\underline{x}$, $\underline{y}$ of finite sequences of distinct members of $\mathbb{Q}$ of equal length, let $\sigma(\underline{x},\underline{y}) \in H(\mathbb{Q})$ take $\underline{x}$ to $\underline{y}$, and let H be the subgroup of $H(\mathbb{Q})$ generated by all such $\sigma(\underline{x},\underline{y})$. The subgroup of $H(\mathbb{Q})$ generated by G and H is then as desired.

Theorem 3.7. *Any countable permutation group fulfilling* SMC *can be extended to a (countable) highly transitive permutation group fulfilling* SMC. *Similarly for* SH.

Proof. If we are assuming CH or MA then the same proof as in 3.6 will do (using 3.2 or 3.4). The object is to prove the result directly, without these extra assumptions, and so we use Theorem 3.1. Let (G_1,Ω_1) fulfil SMC and let (G_2,Ω_2) be some countable highly transitive permutation group fulfilling SMC. Then by Theorem 3.1, if G_1 is countable (G_1,Ω_1) and (G_2,Ω_2) can be jointly embedded in a countable permutation group G on Ω_1 fulfilling SMC. Since $G > G_2$ it is highly transitive.

It therefore suffices to show that *some* highly transitive countable permutation group fulfils SMC. Though there are presumably many such groups, we shall in fact show in §5 that AAut Γ, the almost automorphism group of Rado's universal graph, fulfils SMC. Since it is highly transitive (see [1]), an appropriate countable highly transitive subgroup of AAut Γ will serve.

The proof for SH is similar. We remark that Neumann has shown that the free group of countable rank has a sharp, highly transitive permutation representation on a countable set. Theorems 3.1 and 3.7 for the case of sharp groups were proved independently by S. Adeleke. His proof is given in the additional reference [9].

We remark that, as in Mekler's paper [5], "countable" may be replaced by "of cardinality less than $2^{\aleph_0}$" in Theorems 3.6 and 3.7 under the assumption of Martin's Axiom, or just "$\mathbb{R}$ is not the union of less than $2^{\aleph_0}$ meagre sets".

Let us further remark that it is easy to formulate another condition on the supports of members of G, superficially only slightly stronger than SH, for which joint embedding fails and there is no universal group. We refer to G semi-regular, which we may express by requiring that every non-identity member of G has Ω as its support. Here if we let G_1 and G_2 be the cyclic groups generated by permutations having just one infinite cycle, and infinitely many cycles of length 2, respectively, then it is clear that G_1, G_2 are semi-regular, and G_1 is maximal semi-regular (i.e. regular) so that G_1 and G_2 cannot be jointly embedded in a semi-regular group.

4. *CYCLE TYPES*

The progression from groups fulfilling NC to those fulfilling MC, SMC or SH is neatly illustrated by the possible cycle types of their members. Indeed the classification of the cycle types occurring in groups of a certain class corresponds to determining its one-generator members, so comprises an important special case. It was this special case which was treated by Neumann for MC in [6]. Moreover it is often easiest to establish non-embeddability results by looking at cycle types, as will be done in §5 for Aut Γ and AAut Γ.

By a cycle type we understand a sequence $(k_\infty,k_1,k_2,\ldots)$ where for each $n = \infty,1,2,\ldots, 0 \leq k_n \leq \infty$. For a permutation σ of Ω, the cycle type of σ is given by k_n = the number of cycles of σ of length n.

Theorem 4.1. k *is the cycle type of a member of a group fulfilling* NC *if and only if one of the following holds.*

(i) $k_\infty \neq 0$.

(ii) $\{n : k_n \neq 0\}$ *is infinite.*

(iii) $\{n : k_n \neq 0\}$ *is finite, and if* $L = \mathrm{LCM}\{n : k_n = \infty\}$ *then* $k_n \neq 0 \to n$ *divides* L.

Theorem 4.2. k *is the cycle type of a member of a group fulfilling* MC *if and only if one of the following holds.*

(i) $k_\infty \neq 0$.

(ii) $\{n : k_n \neq 0\}$ *is infinite.*

(iii) *Whenever* $k_n \neq 0$ *for some finite* n, *there is* m *such that* $k_m = \infty$ *and* n *divides* m.

Theorem 4.3. k *is the cycle type of a member of a group fulfilling* SMC *if and only if one of the following holds.*

(i) $k_\infty \neq 0$.

(ii) $\{n : k_n \neq 0\}$ *is infinite.*

(iii) $\{n : k_n \neq 0\}$ *is finite, and if* $L = \mathrm{LCM}\{n : k_n \neq 0\}$ *then* $k_L = \infty$.

Theorem 4.4. k *is the cycle type of a member of a group fulfilling* SH *if and only if one of the following holds.*

(i) k_n *is finite for each* $n < \infty$.

(ii) *For some finite* n, $k_n = \infty$, $\Sigma\{k_m : m \neq n\}$ *is finite, and* $k_m = 0$ *for all* m *not dividing* n *(including* ∞*).*

To give examples: the cycle type k for which $k_{12} = k_{18} = \infty$, $k_{36} = 1$, $k_n = 0$ otherwise (quoted in [6]) fulfils the condition (iii) of 4.1 but not that of 4.2. The cycle type k for which $k_{12} = k_{18} = \infty$, $k_n = 0$ otherwise fulfils the condition (iii) of 4.2 but not that of 4.3. The cycle type for which $k_\infty = 1$, $k_1 = \infty$, $k_n = 0$ otherwise fulfils (i) of 4.3 but neither (i) nor (ii) of 4.4.

Proof of Theorem 4.1. If σ has an infinite cycle, then for each $n \neq 0$, σ^n moves all points in that infinite cycle. Hence $\bigcap_{n \neq 0}$ support σ^n is infinite, and the group $\langle \sigma \rangle$ generated by σ fulfils SMC, hence also MC and NC. If σ has arbitrarily large finite cycles, then $\cap\{\text{support } \sigma^n : 0 \neq |n| < N\}$ contains all cycles of length greater than N, so again $\langle \sigma \rangle$ fulfils SMC, MC and NC.

Now suppose condition (iii) holds, and let $\sigma^r \neq 1$. Then for some n, $k_n \neq 0$ and n does not divide r. Now by (iii), n divides L, so L does not divide r. Hence for some m with $k_m = \infty$, m does not divide r. Therefore support σ^r contains all cycles of length m and so is infinite.

Conversely, suppose that σ lies in a group G fulfilling NC, and that (i), (ii) are false. Thus $\{n : k_n \neq 0\}$ is finite and L is defined. We see that σ^L is the product of finitely many finite cycles, i.e. it has finite support. By NC, $\sigma^L = 1$. Therefore if $k_n \neq 0$, n divides L.

Proof of Theorem 4.2. We may deduce this from [6], since by Theorem 2.3 $\langle \sigma \rangle$ fulfils MC if and only if σ has the cycle type of a member of $H(\mathbb{Q})$. However we can also derive the result directly.

As shown above, if (i) or (ii) holds, $\langle\sigma\rangle$ fulfils MC. Suppose (iii) holds, and let $A \subseteq \mathbb{Z}$ be finite and such that $\bigcap_{r\in A}$ support σ^r is non-empty. Let $x \in$ support σ^r, all $r \in A$. Let n be the length of the cycle of σ containing x. Thus $k_n \neq 0$ and so by (iii) there is m such that $k_m = \infty$ and n divides m. If $r \in A$, $\sigma^r x \neq x$, so n does not divide r. Hence m does not divide any member of A and every element of a cycle of length m lies in $\bigcap_{r\in A}$ support σ^r. This set is therefore infinite.

Conversely, suppose $\sigma \in G$ where G fulfils MC, and suppose that (i) and (ii) are false. Then, as before, $L = \mathrm{LCM}\{n : k_n \neq 0\}$ exists. Let $k_n \neq 0$ and let x lie in a cycle of length n. Then $x \in X = \cap\{\text{support } \sigma^r : 1 \leqslant r \leqslant L \ \&\ n$ does not divide $r\}$ so by MC, X is infinite. But $y \in X$ implies that y is in a cycle of length m where n divides m. Since there are only finitely many non-zero values of k_m, $k_m = \infty$ for some multiple m of n.

Proof of Theorem 4.3. Suppose that (iii) holds, and let $A \subseteq \mathbb{Z}$ be finite and such that $\sigma^n \neq 1$ for $n \in A$. Then L does not divide any member of A, so all members of cycles of length L lie in $\bigcap_{n\in A}$ support σ^n. This set is therefore infinite, verifying SMC for $\langle\sigma\rangle$.

Conversely, assume that σ lies in a group G fulfilling SMC, and suppose that (i) and (ii) are false. Thus $L = \mathrm{LCM}\{n : k_n \neq 0\}$ is defined. If $\sigma = 1$ the result is clear. Otherwise $L > 1$ and $\sigma^r \neq 1$ for all r with $1 \leqslant r \leqslant L-1$. By SMC, $\cap\{\text{support } \sigma^r : 1 \leqslant r \leqslant L-1\}$ is infinite. But any member of this set lies in a cycle of length L, so $k_L = \infty$.

The proof of Theorem 4.4 follows along similar lines and is omitted.

Theorem 4.5. (i) *There is a cyclic permutation group* $\langle\sigma\rangle$ *fulfilling* NC *which cannot be embedded in* $H(\mathbb{Q})$.
(ii) *There are cyclic permutation groups* $\langle\sigma_1\rangle$ *and* $\langle\sigma_2\rangle$ *such that* $\langle\sigma_1\rangle$ *fulfils* MC *but not* SMC *and* $\langle\sigma_2\rangle$ *fulfils* SMC *but not* SH.
Proof. (i) This improves Mekler's result [5] where he showed that there is a 3-generator Abelian permutation group fulfilling NC which cannot be embedded in $H(\mathbb{Q})$. By Theorem 2.3, $\langle\sigma\rangle$ can be embedded in $H(\mathbb{Q})$ if and only if it fulfils MC. By Theorem 4.2 this holds if and only if one of (i), (ii), (iii) of 4.2 holds. As remarked above is suffices to consider σ having cycle type k such that $k_{12} = k_{18} = \infty$, $k_{36} = 1$, $k_n = 0$ otherwise.
(ii) We take σ_1 having cycle type k given by $k_{12} = k_{18} = \infty$, $k_n = 0$ otherwise, and σ_2 having cycle type k given by $k_\infty = 1$, $k_1 = \infty$, $k_n = 0$ otherwise.

5. *THE UNIVERSAL GRAPH*

A study of the automorphism group Aut Γ of Rado's universal graph Γ and of the related universal coloured graphs Γ_C was carried out in [8]. In §4 of [8] a classification of the possible cycle types of members of Aut Γ_C was given. One can verify that every cycle type occurring in Aut Γ_C is one of those allowed by Theorem 4.2, but that there is a cycle type occurring in $H(\mathbb{Q})$ and not in Aut Γ_C, for example that for which $k_\infty = k_1 = 1$, $k_n = 0$ otherwise. It follows that $H(\mathbb{Q})$ cannot be embedded in Aut Γ_C, but the possibility that Aut Γ_C might be embeddable in $H(\mathbb{Q})$ is left open. Similar remarks apply to AAut Γ_C, the group of almost automorphisms of Γ_C (these are permutations of Γ_C which preserve the colour of all but a finite set of edges). A positive answer is further suggested once one has shown that Aut Γ_C and AAut Γ_C both fulfil MC, though one cannot immediately deduce from Theorem 2.3 that they can be embedded in $H(\mathbb{Q})$, since these groups are uncountable. However a direct proof on the same lines of that of the theorem does establish embeddability for Aut Γ_C. For AAut Γ_C the position is less clear. As in the case of Aut Γ_C we can establish just by looking at cycle types that $H(\mathbb{Q})$ cannot be embedded in AAut Γ_C. We still do not know however whether AAut Γ_C can be embedded in $H(\mathbb{Q})$. The best we can do is to show that it fulfils SMC. Finally this raises the question of whether it is a universal group for countable groups fulfilling SMC, but again examination of cycle types shows otherwise.

For details on Γ, Γ_C, Aut Γ_C and AAut Γ_C we refer the reader to [1] and [8].

Theorem 5.1. *For each* C *with* $2 \leq |C| \leq \aleph_0$, Aut Γ_C *can be embedded in* $H(\mathbb{Q})$.
Proof. If $|C_1| \leq |C_2|$ it is easy to embed Aut Γ_{C_2} into Aut Γ_{C_1}, so we may suppose $|C| = 2$, i.e. $\Gamma_C = \Gamma$, the universal graph. We let the Boolean algebra $\mathbb{B}$ be generated by $\{\Gamma_x : x \in \Gamma\}$, where Γ_x is the set of members of Γ joined to x. In the spirit of our proof of Theorem 2.3 we check that $\mathbb{B}$ is a countable Boolean algebra of empty or infinite subsets of Γ closed under the action of Aut Γ, such that for any distinct $x,y \in \Gamma$ there is $b \in \mathbb{B}$ with $x \in b$ & $y \notin b$. It will follow that the topology with $\mathbb{B}$ as base is homeomorphic to that on $\mathbb{Q}$, and so Aut Γ can be embedded in $H(\mathbb{Q})$.

Firstly to see that all non-empty members of $\mathbb{B}$ are infinite it suffices to look at a member of $\mathbb{B}$ of the form

$$X = \Gamma_{x_1} \cap \ldots \cap \Gamma_{x_m} \cap (\Gamma - \Gamma_{y_1}) \cap \ldots \cap (\Gamma - \Gamma_{y_n}).$$

If $X \neq \emptyset$ it follows that $x_i \neq y_j$, each i, j. Let A be an arbitrary finite subset of Γ. By universality there is $z \notin A$ joined to each x_i and to no y_j. Thus $z \in X$ and so $X - A \neq \emptyset$. As A was an arbitrary finite subset of Γ, X must be infinite.

Next, for any $x,y \in \Gamma$ and $\sigma \in$ Aut Γ, $y \in \sigma\Gamma_x \leftrightarrow \sigma^{-1}y \in \Gamma_x \leftrightarrow x$, $\sigma^{-1}y$ joined in $\Gamma \leftrightarrow \sigma x$, y joined in $\Gamma \leftrightarrow y \in \Gamma_{\sigma x}$, so that $\sigma\Gamma_x = \Gamma_{\sigma x} \in \mathbb{B}$. Hence $\mathbb{B}$ is closed under the action of Aut Γ.

Finally, if $x \neq y$, it follows from universality that there is z joined to x but not to y, so that $x \in \Gamma_z$ & $y \notin \Gamma_z$.

Theorem 5.2. AAut Γ *contains no element of cycle type* k *where* $k_\infty = 1$, $k_1 = 2$, *and* $k_n = 0$ *otherwise. Hence* $H(\mathbb{Q})$ *cannot be embedded in* AAut Γ_C *or* Aut Γ_C, *and there is a cyclic permutation group fulfilling* SH *not embeddable in* AAut Γ_C *or* Aut Γ_C.
Proof. Suppose $\sigma \in$ AAut Γ has the given cycle type. Let x lie in the infinite cycle, and y_1, y_2 be the two fixed points. Since σ is an almost automorphism there is a finite N such that if a or b is outside $\{\sigma^i x : |i| < N\} \cup \{y_1, y_2\}$ then σ preserves the edge or non-edge $\{a,b\}$. By universality there

are n_1, n_2, n_3 such that $|n_1|$, $|n_2|$, $|n_3| > N$ and

$\sigma^{n_1}x$ is joined to y_1 and y_2

$\sigma^{n_2}x$ is joined to y_1 but not y_2

$\sigma^{n_3}x$ is joined to neither y_1 nor y_2.

Two of n_1, n_2 and n_3 must have the same sign. Suppose n_1 and n_2 are both positive, and $n_1 < n_2$ (other cases being handled similarly). Then since $n_1 > N$, σ preserves the edge $\{\sigma^{n_1}x, y_2\}$. Applying σ $n_2 - n_1$ times we find that $\{\sigma^{n_2}x, y_2\}$ is an edge, contrary to the choice of $\sigma^{n_2}x$.

(Note however that AAut Γ does contain a permutation whose cycle type satisfies $k_\infty = k_1 = 1$, $k_n = 0$ otherwise.)

$H(\mathbb{Q})$ cannot be embedded in AAut Γ since it does contain an element of cycle type k satisfying $k_\infty = 1$, $k_1 = 2$, $k_n = 0$ otherwise. For the same reason we see from Theorem 4.4 that not every group fulfilling SH can be embedded in AAut Γ. The same results apply to AAut Γ_C and Aut Γ_C, since they *can* be embedded in AAut Γ.

Theorem 5.3. AAut Γ_C *fulfils* SMC.

Proof. To fix our notation we recall that the defining property of the countable universal C-coloured graph $\Gamma_C = (\Gamma_C, F_C)$, where F_C maps unordered pairs of members of Γ_C into C, is that for any finite partial map α from Γ_C into C there should exist $x \notin \text{dom}\,\alpha$ such that $(\forall y \in \text{dom}\,\alpha)(\alpha(y) = F_C\{x,y\})$. Since $|C| \geq 2$ we may fix distinct c_1, c_2 in C.

Let $\sigma_1, \ldots, \sigma_n$ be non-identity members of AAut Γ_C. We show by induction on n such that $\bigcap_{i=1}^{n}$ support σ_i is infinite.

For $n = 1$, suppose on the contrary that A = support σ_1 is finite. As $\sigma_1 \neq 1$, $A \neq \emptyset$. Let $x \in A$ and let B be a finite subset of Γ_C such that any endpoint of an edge whose colour is not preserved by σ_1 lies in B. By universality there is $y \notin A \cup B$ such that $F_C\{x,y\} = c_1$, $F_C\{z,y\} = c_2$ all $z \in A \cup B - \{x\}$. As $y \notin B$, $F_C\{\sigma_1 x, \sigma_1 y\} = c_1$. As $y \notin A$, $\sigma_1 y = y$, so $F_C\{\sigma_1 x, y\} = c_1$. But $\sigma_1 x \in A$ and $\sigma_1 x \neq x$, contrary to $F_C\{\sigma_1 x, y\} = c_2$.

Now assume the result for n. Hence $\bigcap_{i=1}^{n}$ support σ_i is non-empty. Let $x \in \bigcap_{i=1}^{n}$ support σ_i. Since support σ_{n+1} is infinite we may let $y \in$ support σ_{n+1}, $y \neq \sigma_i x$ $(1 \leq i \leq n)$ and $y \neq \sigma_{n+1}^{-1}x$. This time we take B to be finite and such that any endpoint of an edge not preserved by some σ_i lies in B. Let A be any (arbitrarily large) finite set containing x, y, $\sigma_{n+1}y$, and $\{\sigma_i x : 1 \leq i \leq n\}$. By universality there is $z \in \Gamma$ such that $F_C\{z,x\} = F_C\{z,y\} = c_1$ and $F_C\{z,t\} = c_2$ for $t \in A \cup B - \{x,y\}$. As $x \in$ support σ_i and $y \neq \sigma_i x$, $F_C\{z, \sigma_i x\} = c_2$. Hence $z \neq \sigma_i z$ $(1 \leq i \leq n)$ as before. Also $\sigma_{n+1}y \neq x, y$, so $F_C\{z, \sigma_{n+1}y\} = c_2$ and $z \neq \sigma_{n+1}z$. Thus $z \in \bigcap_{i=1}^{n+1}$ support σ_i. Therefore $\bigcap_{i=1}^{n+1}$ support $\sigma_i - A$ is non-empty. Since A was taken to be an arbitrarily large finite set, $\bigcap_{i=1}^{n+1}$ support σ_i is infinite.

Remark. The fact that H(ℚ) cannot be embedded in AAut Γ_C (and hence not in Aut Γ_C either) follows from this result too, since any group embeddable in one fulfilling SMC must also fulfil SMC itself.

Finally we deduce from Theorem 5.3 a further curiosity concerning AAut Γ_C. Whether it is actually any help in proving the embeddability of AAut Γ_C in H(ℚ) is doubtful however.

Theorem 5.4. *Let* $\sigma_1,\ldots,\sigma_n$ *be finitely many non-identity members of* AAut Γ_C. *Then* $X = \bigcap_{i=1}^{n}$ support σ_i *is itself a universal* C-*coloured graph.*

Proof. By Theorem 5.3, X is infinite. Let α be a finite partial map from X into C. Since X is infinite there is $x \in X - (\mathrm{dom}\,\alpha \cup \bigcup_{i=1}^{n} \sigma_i^{-1}\,\mathrm{dom}\,\alpha)$. Let $B \subseteq \Gamma_C$ be a finite set such that any endpoint of an edge whose colour is not preserved by some σ_i lies in B. As $x \in$ support σ_i, $x \neq \sigma_i x$, and $x, \sigma_i x \notin$ dom α for each i. By universality of Γ_C there is therefore $z \notin B$ such that $F_C\{x,z\} = c_1$, $F_C\{\sigma_i x,z\} = c_2$, and $F_C\{y,z\} = \alpha(y)$, all $y \in$ dom α. Since $z \notin B$, $F_C\{\sigma_i x,\sigma_i z\} = c_1 \neq F_C\{\sigma_i x,z\}$. Hence $z \neq \sigma_i z$ each i, and $z \in X$. Therefore z is the required witness in X corresponding to the map α, and the universality of X is established.

REFERENCES

1. P.J. Cameron, *Random structures, universal structures, and permutation groups*, handwritten notes.
2. D.H. Fremlin, Cichoń's diagram, *Séminaire Initiation à l'Analyse*, 23e année (1983/84) 5, Université Pierre et Marie Curie, Paris.
3. D.A. Martin & R.M. Solovay, Internal Cohen extensions, *Ann. Math. Logic* 2 (1970), 143-178.
4. V.D. Mazurov, Yu.I. Merzlyakov & V.A. Churkin (editors), *The Kourovka Notebook*, Novosibirsk, 5th edition (1976); 8th edition (1982).
5. A. Mekler, Groups embeddable in the autohomeomorphisms of ℚ, *J. London Math. Soc.* (2) 33 (1986), to appear.
6. P.M. Neumann, Automorphisms of the rational world, *J. London Math. Soc.* (2) 32 (1985), 439-448.
7. W. Sierpiński, Sur un propriété topologique des dénombrables denses en soi, *Fund. Math.* 1 (1920), 11-16.
8. J.K. Truss, The group of the countable universal graph, *Math. Proc. Cambridge Philos. Soc.* 98 (1985), 213-245.

ADDITIONAL REFERENCE

9. S.A. Adeleke, Embeddings of infinite permutation groups in sharp, highly transitive, and homogeneous groups, *Proc. Edinburgh Math. Soc.*, to appear.

MAXIMAL SUBGROUPS OF SPORADIC GROUPS

R.A. Wilson
University of Cambridge, Cambridge CB2 1SB, England

In this paper I shall try to summarise the current situation regarding the problem of finding the maximal subgroups of the sporadic simple groups, and to give some idea of the techniques that have been used to attack this problem. The fundamental lemma underlying the method is:

Lemma 1. *If* G *is a simple group, and* M *is a maximal subgroup of* G, *and* K *is a minimal normal subgroup of* M, *then*

(i) K *is a characteristically simple group* (i.e. *a direct product of isomorphic simple groups*),

(ii) $M = N_G(K)$.

Proof. Elementary.

The general strategy is therefore to classify the characteristically simple subgroups into conjugacy classes, and find their normalizers. We then have a list which contains all the maximal subgroups, and it is usually a straightforward matter to eliminate the non-maximal subgroups from this list.

There is a fundamental dichotomy between the cases when K is an elementary Abelian p-group (in this case M is called a p-*local* subgroup), and the cases when K is non-Abelian. The most difficult problems are concerned with proving uniqueness rather than existence, and with non-local rather than local subgroups.

Before I go into details of the methods perhaps I should give a survey of the results that have been obtained to date. The actual lists of maximal subgroups are of course much too long to include here, but can all be found in the ATLAS [3]. The complete maximal subgroup problem has been solved for 19 of the 26 sporadic groups (and their automorphism groups [21]):

Group	Date	Reference(s)
M_{11}		R.J. List (unpublished)
M_{12}		R.J. List (unpublished)
M_{22}		R.J. List (unpublished)
M_{23}		R.J. List (unpublished)
J_1	1966	Z. Janko [10]
HS	1970	S.S. Magliveras [12]
McL	1970	L. Finkelstein [6]
Co_3	1970	L. Finkelstein [6]
M_{24}	1972	Chang Choi [2], R.T. Curtis [4]
J_2	1973	L. Finkelstein and A. Rudvalis [7]
J_3	1974	L. Finkelstein and A. Rudvalis [8]

HN	1976	S.P. Norton (unpublished, see also [13])
He	1980	G. Butler [1]
Suz	1981	R.A. Wilson [16], S. Yoshiara [29]
Ru	1981	R.A. Wilson [15], S. Yoshiara [30]
Co_2	1982	R.A. Wilson [17]
Co_1	1982	R.A. Wilson [18], * R.T. Curtis [5]
O'N	1984	R.A. Wilson [22], S. Yoshiara [31], * O'Nan [14]
Ly	1984	R.A. Wilson [23]

In this table, * denotes a reference to the local analysis, which has often been completed long before the full list of maximal subgroups has been proved.

For the remaining groups some partial results are known. For example, we know all the maximal subgroups of Fi_{22} and Th with the possible exception of some very small ones [19, 25]. Furthermore, most of the maximal p-local subgroups of all the sporadic groups are known, the only exceptions being the 2-local subgroups of Fi_{24}, B and M. The 2-local subgroups of Fi_{22} and Fi_{23} were classified by D. Flaass [9], while those of J_4 were classified by W. Lempken [11]. The 3-local subgroups of Fi_{23}, Fi_{24}, B and M were classified by the present author [24, 26, 27].

Now I want to discuss some of the techniques by which these results have been obtained, and I will begin by describing the non-local arguments. The first task is to use Lagrange's theorem and the classification of finite simple groups to produce a list of those finite simple groups whose order divides $|G|$. We can then eliminate many of these very quickly by showing that there is no character restriction, or that they contain some local subgroup which is not in G, or for some other reason. We then have a list of candidates for non-local subgroups, which must be considered case by case. The most important arguments are of three types:

1. *Structure constant arguments.*
2. *Constructions from smaller subgroups.*
3. *Vector stabilizers.*

In the remainder of the paper I will describe these individually, then briefly discuss the local analysis, and finally consider extensions to automorphism groups of simple groups.

1. *STRUCTURE CONSTANT ARGUMENTS*

The basic result used here is a standard one from ordinary character theory, namely:

Lemma 2.

$$\sum \frac{1}{|C(x,y)|} = \frac{|G|}{|C(x)||C(y)||C(z)|} \sum \frac{\chi(x)\cdot\chi(y)\cdot\chi(z)}{\chi(1)}$$

where the left-hand sum is taken over all conjugacy classes of ordered triples (x,y,z) *with* $xyz = 1$ *and* x, y, z *in the conjugacy classes* X, Y, Z *respectively, and the right-hand sum is taken over all irreducible characters* χ *of* G.

The value of this expression (the *symmetrized structure constant*) is denoted by $\xi_G(X,Y,Z)$, and is calculated from the character table. It can be used to give information about groups generated by elements x, y, z in this way, and about their centralizers. For example, if a subgroup H can be generated by elements x, y, z as above, and $\xi_G(X,Y,Z) < 1$, then it is clear that the centralizer of H is non-trivial. In groups of moderate size it will be obvious that $C_G(H)$ is soluble, and hence we can apply the following trivial lemma to show that $N_G(H)$ is contained in a local subgroup of G:

Lemma 3. *If* $H < G$ *and* $C_G(H)$ *contains a non-trivial elementary Abelian characteristic subgroup* K *(in particular, if* $C_G(H)$ *is a non-trivial soluble group), then* $N_G(H)$ *is contained in the local subgroup* $N_G(K)$ *of* G.

The standard example is when X, Y and Z are classes of elements of orders 2, 3 and 5, for then x, y and z generate A_5, since A_5 has a presentation $\langle x,y,z \mid x^2 = y^3 = z^5 = xyz = 1 \rangle$. Hence the structure constant is equal to the sum over the conjugacy classes of A_5 of the reciprocal of the centralizer order. (Actually, this is not quite true, as A_5 has an outer automorphism: we have to double the contribution from any A_5 whose normalizer does not contain an element realizing this automorphism.)

In general however it is more complicated than this, as the relations $x^a = y^b = z^c = xyz = 1$ do not suffice to define a finite group, and there may be many different isomorphism classes of groups contributing to a single structure constant.

2. *CONSTRUCTIONS FROM SMALLER SUBGROUPS*

In this second method, if we want to find all subgroups of G isomorphic to H, then we first find some way to generate H by two subgroups K_1 and K_2 intersecting in L, say. Assuming that we have already classified all subgroups isomorphic to K_1 and K_2 then we can work out how many conjugacy classes of the situation $K_1 > L < K_2$ there are in G. Hopefully, given a good choice of K_1, K_2 and L, there will not be too many of these, so that we can identify them all, and then we will *inter alia* have completed the classification of subgroups isomorphic to H.

Obviously this is a very general method, and the art lies in choosing the groups K_1, K_2 and L. If H is large, then in general it is a good idea to choose these to be local subgroups. If H is smaller then it seems to be easier to use non-local subgroups. However, there is a problem here in that if H is too small, then there are not enough non-local subgroups. This is the main reason why subgroups like A_6, $L_2(7)$, $L_3(3)$, ... are the hardest to classify completely.

If there is no choice of K_1, K_2, and L for which the number of conjugacy classes of the situation $K_1 > L < K_2$ is small enough for all cases to be identified by hand, it may still be possible to do it by computer. In this case, we start with any subgroup K_1 which can easily be found computationally, or which has already been found by this method, and take L to be a small subgroup of K_1, preferably of order 2 or 3. Then we take K_2 to be a (preferably small) subgroup of $N_H(L)$. We can find $N_G(L)$ fairly easily computationally, and can then run through all the possibilities for K_2.

For example, more or less the only way to classify subgroups isomorphic to $L_2(7)$ computationally is to construct it as follows:

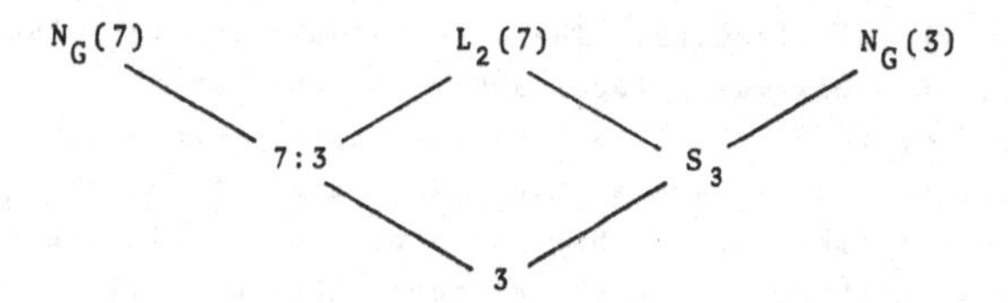

3. *THE VECTOR STABILIZER METHOD*

This method depends on a very trivial result from modular representation theory - so trivial in fact that I'm almost embarrassed to write it down. However, I think it is worth while to do so, since until I had stated it in such generality that it became transparently obvious, no-one would actually believe that it could possibly work.

Lemma 4. *Let* H *be a finite group and* V *be any finite-dimensional* $\mathbb{C}$H*-module, and let* $\mathcal{L}$ *be any lattice of submodules of* V. *Let* $^-$: $V \to \overline{V}$ *be any reduction of* V *modulo* p. *Then* $^-$ *induces a map* $\mathcal{L}$ *into* $\overline{\mathcal{L}}$*, which is a lattice of submodules of* V.

Proof. This is an immediate consequence of the definition of reduction modulo p.

In our applications, V is usually a $\mathbb{Q}$H-module, $p = 2$, and $\mathcal{L}$ consists of a single 1-dimensional submodule. The idea is to take V to be an ordinary irreducible representation of the simple group G, and use the character table to investigate possible restrictions to subgroups isomorphic to H. If we can show that any such subgroup H must fix a non-zero vector in V, then by Lemma 4, H fixes a non-zero vector in any reduction $\overline{V}$ of V modulo 2. We can assume that $C_G(H) = 1$, since it is easy, using Lemma 3, to classify simple groups with non-soluble centralizers, and so $N_G(H)$ is a subgroup of Aut(H). Now any automorphism of order 2 either fixes the fixed vector of H, or interchanges it with another fixed vector, in which case it fixes their sum. In either case, the resulting group H.2 fixes a non-zero vector. So if, as is usually the case, Out(H) is a 2-group, then by induction $N_G(H)$ fixes a non-zero vector in $\overline{V}$.

So the problem has now been reduced to the geometrical one of finding the orbits of G on vectors in $\overline{V}$. The only drawback is that for this to be feasible we need to have a small number of orbits, which requires $|G| \ll 2^{\dim(V)}$. Unfortunately, very few groups have such small representations, and really the method is only useful for the groups Co_1, Co_2, Co_3, and to some extent McL and Ru (see [17,18]).

4. *LOCAL ANALYSIS*

The general principle here is straightforward: one determines the conjugacy classes of elementary Abelian subgroups, and their centralizers, by induction on their order, simply by working inside the centralizers of the elements of prime order. There are various things that may happen. For example, one may obtain the same group more than once, in which case its normalizer is not transitive on its non-trivial elements. Or one may have only one group with a given centralizer, in which case its normalizer is transitive on its non-trivial elements. These techniques are in principle sufficient to determine the maximal p-local subgroups completely.

The only problem is that if the Sylow p-subgroup has order 3^{20}, say, as in the Monster, then there are so many cases that it becomes quite impracticable to proceed in this way. There is a plethora of little tricks which enable one to deal with several cases at once, though it is difficult to describe them systematically. The most obvious one is that it is only necessary to consider elementary Abelian groups whose normalizer acts irreducibly. Another is that we need only consider those which are equal to the centre of their centralizer. (But one must be careful that these two steps do not introduce a circular argument, as the first puts the normalizer inside that of a smaller group, and the second puts it inside that of a larger group!)

Perhaps the most important thing, though, is to have an efficient notation for working in the various p-element centralizers, or the Sylow p-subgroups. For example, one could use a power-commutator presentation of the Sylow p-subgroup, which lends itself readily to computer calculations. Another notation, which I have found very productive, is most useful in the case when the O_p-subgroup of the centralizer $C = C(x)$ of a p-element x is an extraspecial group. Then $C/\langle x\rangle$ is an extension of a vector space by a group of matrices acting on it. Furthermore, the representation is symplectic, so the matrices can be written as matrices of half the size over quaternions, where it is understood that the vectors are always reduced modulo a quaternionic prime divisor of p. Thus for example we can write an arbitrary element of $3^{12}.2Suz$ as the product of a vector of length 6 and a 6×6 quaternionic matrix, which is extremely useful in classifying the 3-local subgroups of the Monster.

5. *AUTOMORPHISM GROUPS*

Suppose we wish to find the maximal subgroups of an automorphism group of a simple group. We might hope to be able to read them off directly from the maximal subgroups of the simple group. However, almost any conceivable conjecture here is false, and the only sensible way to proceed is to find the maximal subgroups of all automorphism groups at the same time as those of the simple group. We need to modify Lemma 1 slightly:

Lemma 5. *If* $S \leq G \leq Aut(S)$, S *is a simple group, and* M *is a maximal subgroup of* G, *then*

(i) $M \cap S \neq 1$,

(ii) *if* K *is a minimal characteristic subgroup of* $M \cap S$ *then* $M = N_G(K)$.

Since the proof of (i) is case by case, this uses the classification of finite simple groups, or alternatively it can be thought of as a result about the "known" simple groups. In any case the main burden of the problem is again finding the conjugacy classes of characteristically simple subgroups of S, so the maximal subgroups of G can usually be read off from the proof of completeness of the list of maximal subgroups of S, though not from the list itself (see [21] for more details).

6. *OPEN PROBLEMS*

The open problems are of two different kinds:

1. *The 2-local subgroups of* Fi_{24}, B *and* M.
2. *The non-local subgroups of* Fi_{22}, Fi_{23}, Fi_{24}, Th, B, M *and* J_4.

It seems possible that methods similar to those used for the 3-local subgroups of the Monster and its large subgroups can be used to attack the 2-local problems, though they are certainly much harder.

In some of the non-local cases it may be feasible to complete the enumerations by computer, depending on the sizes of the representations. The easiest cases would seem to be Fi_{22} (dimension 27 over $\mathbb{F}_4$, or 77 over $\mathbb{F}_3$) and J_4 (dimension 112 over $\mathbb{F}_2$), followed by Th (dimension 248) and Fi_{23} (dimension 253 over $\mathbb{F}_3$). The other cases seem to be out of reach at present, even Fi_{24} (dimension 783).

REFERENCES

1. G. Butler, The maximal subgroups of the sporadic simple group of Held, *J. Algebra* 69 (1981), 67-81.
2. Chang Choi, On subgroups of M_{24}, *Trans. Amer. Math. Soc.* 67 (1972), 1-27; 29-47.
3. J.H. Conway, R.T. Curtis, S.P. Norton, R.A. Parker & R.A. Wilson, *An ATLAS of Finite Groups*, Oxford Univ. Press (1985).
4. R.T. Curtis, The maximal subgroups of M_{24}, *Math. Proc. Cambridge Philos. Soc.* 81 (1977), 185-192.
5. R.T. Curtis, On subgroups of ·O. II. Local structure, *J. Algebra* 63 (1980), 413-434.
6. L. Finkelstein, The maximal subgroups of Conway's group C_3 and McLaughlin's group, *J. Algebra* 25 (1973), 58-89.
7. L. Finkelstein & A. Rudvalis, Maximal subgroups of the Hall-Janko-Wales group, *J. Algebra* 24 (1973), 486-493.
8. L. Finkelstein & A. Rudvalis, The maximal subgroups of Janko's simple group of order 50,232,960, *J. Algebra* 30 (1974), 122-143.
9. D. Flaass, The 2-local subgroups of Fischer's groups F_{22} and F_{23}, *Mat. Zametki* 35 (1984), 333-342.
10. Z. Janko, A new finite simple group with Abelian Sylow 2-subgroups, and its characterization, *J. Algebra* 3 (1966), 147-186.
11. W. Lempken, The 2-local and some non-local subgroups of J_4, personal communication.
12. S.S. Magliveras, *The maximal subgroups of the Higman-Sims groups*, Ph.D. thesis, Birmingham (1970).
13. S.P. Norton & R.A. Wilson, Maximal subgroups of the Harada-Norton group, *J. Algebra*, to appear.
14. M.E. O'Nan, Some evidence for the existence of a new simple group, *Proc. London Math. Soc.* (3) 32 (1976), 421-479.
15. R.A. Wilson, The geometry and maximal subgroups of the simple groups of A. Rudvalis and J. Tits, *Proc. London Math. Soc.* (3) 48 (1984), 533-563.
16. R.A. Wilson, The complex Leech lattice and maximal subgroups of the Suzuki group, *J. Algebra* 84 (1983), 151-188.
17. R.A. Wilson, The maximal subgroups of Conway's group ·2, *J. Algebra* 84 (1983), 107-114.
18. R.A. Wilson, The maximal subgroups of Conway's group Co_1, *J. Algebra* 85 (1983), 144-165.
19. R.A. Wilson, On the maximal subgroups of the Fischer group Fi_{22}, *Math. Proc. Cambridge Philos. Soc.* 95 (1984), 197-222.
20. R.A. Wilson, The subgroup structure of the Lyons group, *Math. Proc. Cambridge Philos. Soc.* 95 (1984), 403-409.
21. R.A. Wilson, On maximal subgroups of automorphism groups of simple groups, *J. London Math. Soc.* (2) 32 (1985), 460-466.
22. R.A. Wilson, The maximal subgroups of the O'Nan group, *J. Algebra* 97 (1985), 467-473.

23. R.A. Wilson, The maximal subgroups of the Lyons group, *Math. Proc. Cambridge Philos. Soc.* 97 (1985), 433-436.
24. R.A. Wilson, Some subgroups of the Baby Monster, *Invent. Math.*, submitted.
25. R.A. Wilson, Some subgroups of the Thompson group, preprint (1985).
26. R.A. Wilson, The 3-local subgroups of the Monster, in preparation.
27. R.A. Wilson, The 3-local subgroups of the Fischer groups, in preparation.
28. A. Woldar, Some subgroups of the Lyons group, personal communication.
29. S. Yoshiara, The complex Leech lattice and sporadic Suzuki group, in *Topics in Finite Group Theory*, Kyoto (1982), 26-46.
30. S. Yoshiara, The maximal subgroups of the sporadic simple group of Rudvalis, Tokyo (1984), preprint.
31. S. Yoshiara, The maximal subgroups of the sporadic simple group of O'Nan, *J. Fac. Sci. Univ. Tokyo* 32 (1985), 105-141.

Note added in proof. The maximal subgroups of J_4 and Fi_{22} have now been completely enumerated (see [33] and [34]). The following references have come to my attention since this paper was written:

32. A.A. Ivanov, S.V. Tsaranov & S.V. Shpektorov, The maximal subgroups of the O'Nan-Sims sporadic simple group and its automorphism group, *Doklady Mat. Nauk.*, to appear.
33. P.B. Kleidman & R.A. Wilson, The maximal subgroups of J_4, *Proc. London Math. Soc.*, submitted.
34. P.B. Kleidman & R.A. Wilson, The maximal subgroups of Fi_{22} and $Fi_{22}:2$, in preparation.
35. W. Lempken, Die Untergruppenstruktur der endlichen, einfachen Gruppe J_4, Thesis, Mainz (1985).
36. R.J. List, On the maximal subgroups of the Mathieu groups. I. M_{24}, *Atti Accad. Naz. Lincei Rend. Cl. Sci. Fis. Mat. Natur.* (8) 62 (1977), 432-438.
37. A. Woldar, On subgroups of the Lyons group, *Comm. Algebra*, to appear.